The Design of High-Efficiency Turbomachinery and Gas Turbines

second edition, with a new preface

David Gordon Wilson and Theodosios Korakianitis

The MIT Press
Cambridge, Massachusetts
London, England

Library of Congress Cataloging-in-Publication Data

Wilson, David Gordon, 1928-
   The design of high-efficiency turbomachinery and gas turbines / David Gordon Wilson and Theodosios Korakianitis. -- Second edition, with a new preface.
         pages cm
   Includes bibliographical references and index
      ISBN 978-0-262-53668-5 (pbk. : alk. paper)
1.   Turbomachines--Design and construction.
2.   Gas-turbines--Design and construction. I. Karakianitis, Theodosios. II. Title.
   TJ267.W47 2014
   621.43'3--dc23

      2014007225

DEDICATIONS

Dave Wilson dedicates his part of this book to his wonderful creative spouse Ellen (who should be writing her own book) and to their daughter Susan Speck Wilson, who at a young age is showing a desire to be an engineer.

Theodosios Korakianitis dedicates his work on this book to Christine, his loving and patient child-development-expert budding-artist spouse; and to their gifted children Nataliè, Demètrios, and Karina.

Contents

List of figures

List of illustrations

List of tables

Preface

The first edition of this text was published in 1984 by the MIT Press. It was received with kindness, and forty-six instructors in charge of turbomachinery and turbine courses reported that they had adopted the book at their universities. I spent a few weeks basking in the glow of what I happily thought was a long job well done. While I was working on a sabbatical at GE Aircraft Engines, someone told me that he had ordered over sixty copies for the company's design engineers. I felt that this was equivalent to winning a Pulitzer prize, and my modesty was severely challenged.

However, this wide use brought in a great deal of feedback, soon indicating that a second edition was needed to incorporate improvements in treatment, update examples, and correct some errors. The second edition, for which Theodosios Korakianitis joined as coauthor, was published in 1998 by Prentice Hall. We had several aims for a second edition. We wanted to reduce errors to a minimum, of course. We also introduced new flow correlations into chapter 3 (cycle thermodynamics), and extended the cycles to include the principal aircraft-engine types. We wanted to rationalize the treatment of preliminary and detailed design by taking energy transfer in turbomachines (chapter 5) further so that a great deal of preliminary design (arriving at the overall size and shape of a machine) could be done with the material in that chapter. More detailed design of axial-flow turbines and compressors and of radial-flow turbomachines in chapters 7, 8, and 9 became, we hope, more useful and more consistent. We also took three-dimensional design (chapter 6) to a more practical level, and extended heat-exchanger design (chapter 10) considerably. Every chapter was updated in various ways. Many chapters had what we have termed "illustrations" to distinguish them from "figures." The latter are part of and are referred to in the text. "Illustrations" are photographs or cutaway drawings of machines or components, generally accompanied by a commentary on their design aspects.

Many people helped us. Some we have acknowledged in the text. We have been allowed to use a wide variety of graphs, diagrams, and photos for the figures. Our students then at Washington University in St. Louis and at MIT contributed materially, sometimes unwittingly, as tryouts for our methods. Andrew R. Mech of the Rose-Hulman Institute of Technology with his students J. Lawrence Elkin and William Mathies went through the first edition and through a draft of the second edition in great detail and dedication, giving us many useful recommendations and feedback. Aristide F. Massardo, on the faculty of the University of Genoa, also reviewed the second edition and offered valuable suggestions.

Before I turn to the current printing of this book, I would also like to take this opportunity to acknowledge longer-term debts. I became fascinated by turbomachinery when I was a student, but my enthusiasm was given depth and breadth through the generosity of Ian Goodlet, chief engineer of gas turbines at the Brush company in Loughborough, Leicestershire, UK, where I was working on my postgraduate apprenticeship and had my first professional job. Brush awarded me a research fellowship at Nottingham University, where there were at that time no faculty in the gas-turbine field. Ian put a lot of effort into helping me find a good thesis topic. I took one suggested by A. G. Smith at the (UK) National Gas Turbine Establishment on the chordwise variation of heat-transfer coefficient on gas-turbine blades. The fellowship work on that project at Nottingham University became the most exhilarating experience of my professional life up to that point.

Now, in 2014, we are reprinting this book. The publication date of the second edition, 1998, is relevant to the present situation for the following reason. University instructors in power and propulsion, and particularly in high-efficiency design, know that when their courses in these topics are not required of all students, the enrollment numbers fairly closely rise and fall with the price of oil. The delay from problems of the authors' making resulted in the publication occurring at a time when the price of oil had fallen. Class enrollments in power and propulsion fell throughout the country and probably throughout the world. Sales of the rather beautiful second edition were therefore disappointing. Prentice Hall produced the book for two or three years, but then notified me around 2001 that it was taking the book off regular publishing and was putting it on "print on demand" (POD) status.

This turned out to be a considerable drawback for the book. The second edition was already considerably more expensive than the first, and the cost of the POD (paperback) version was further increased substantially. The quality also decreased considerably. The photographs were so dark in most cases as to be indecipherable. After a short time of use large numbers of pages were apt to fall out. Bunches of pages appeared twice in early printings. Prentice Hall was not making money on the book, and in 2011 was gracious enough to return the rights to me.

Obviously, the book should have been allowed a respectful death at that point. However, it had faithful followers. We were repeatedly asked to find fifty or so copies for new classes. The strong design nature of the book gave students and others the ability to arrive at hardware to a greater degree, they stated, than could be given by alternative texts. They asked us to try to make the book available in some form.

We considered trying to produce the text as an electronic book with a great deal of help, but were delighted when Clay Morgan, acquisitions editor at the MIT Press, agreed to take the book back under the wing of the press. This book is highly similar to the published second edition (not the POD version), but has a large number of small corrections and improvements incorporated.

We hope that instructors of turbomachinery and gas-turbine design courses enjoy the problems in this book. A solutions manual is available to instructors who request it via the MIT Press website, http://mitpress.mit.edu, or by telephone at 617-253-3620. We often used to challenge students in our classes by giving them more open problems than are present in the book. We would ask them individually to choose values that would normally be a designer's responsibility, for instance, the number of blades in a centrifugal compressor or the pressure ratio of a gas turbine. Such open questions are more challenging for students but are very educational.

The instructor has to spend a little more time grading the submitted responses, but learns immediately how well the students understand the material. There is usually some guidance in the text, but if there does not seem to be enough, wild guesses by the students provide wonderful educational opportunities.

We hope that instructors, students, companies, and individual engineers find our book useful. Please let us know of any attributes that are not of a high standard.

David Gordon Wilson
Email address: dgwilson@mit.edu

Theodosios Korakianitis
Email address: talexander@slu.edu

Note to readers

Much of chapter 2 may be skipped by readers who are familiar with thermodynamics of the "Keenan school." However, even they should check to ensure that they are comfortable with our approach. Author DGW remembers arriving as a supposedly bright post-doc at MIT in 1955 only to feel confused in every course involving thermodynamics and fluid mechanics: the approach, the understanding of previous knowledge, and even the terminology were all quite different from those to which he was accustomed.

We have tried here to use the most commonly used terminology. After strong recommendations to drop the inclusion of g_c in equations derived from Newton's law, we have decided to retain it. It may seem to academics that everyone works in SI units; however, major US aircraft and aircraft-engine companies, at least, still use US units, and with them g_c. We could not find any evidence of consistency in the treatment of subscripts; therefore, we devised what we hope is a consistent system. All subscripts are of one or, at most, two symbols, e.g., "in" for "inlet" and "ex" for "exit". If a term needs to carry two or more subscripts, e.g., the stagnation temperature at outlet, $T_{0,ex}$, the subscripts are separated by a comma. We have tried to reduce the confusion that exists in many thermodynamic texts (and in the minds of many more students) in equations that include both the ratio of specific heats, γ, and the specific heat at constant pressure, C_p. Most people treat γ as constant with temperature and C_p as varying with temperature. In fact, one is a function of the other. We have therefore derived the one-dimensional-flow equations in terms just of C_p, or perhaps more strictly of C_p/R, and we have included a tabulation of this quantity against temperature in appendix A. If we had had just a little more courage we would have devised a single symbol for C_p/R, so that not only the usage but also the appearance of equations written this way would have been simpler. However, as a concession to those who have been schooled in other ways, we have included the "pure-gamma" equations (i.e., not having a mix of γ and C_p/R) alongside the final C_p/R equation in all significant derivations.

The difference between "stagnation" (static plus velocity terms) and "total" conditions (static plus velocity plus elevation terms) has also been emphasized, with subscript 0 for stagnation and T for total. We hope that this will resolve some inconsistencies in earlier versions of the text.

We unceremoniously threw overboard the somewhat archaic treatment of one-dimensional compressible-flow functions using the virtual Mach number $M*$ followed in the first edition in favor of a simpler and more-powerful approach.

We have started chapter 2 with the first law of thermodynamics for a closed system, even though the book is entirely about open systems. We go rapidly from this to the SFEE, the steady-flow energy equation, the basis for open-system turbomachinery calculations. We have also derived Gibbs' equation from a simple statistical treatment of the second law. We have taken these steps because, again, of DGW's past confusions: even if he could occasionally remember the SFEE and Gibbs' equations, he could not remember if either or both were applicable to all substances and processes, or just to "simple" substances and reversible processes, and so on. By going through the derivations and, it is hoped, remembering the specifications or assumptions underlying them, readers should be able to apply them with greater assurance.

Confident that, after this thorough grounding, readers will know when and how to use the various equations, we have grouped them in appendix B.

Nomenclature

It is impossible to avoid having more than one use for a symbol. We have tried to reduce confusion by:

1. giving the location of each symbol's first use in each definition, so that the reader can find a full explanation;

2. limiting compound symbols to no more than two primary symbols, the second of which will be a subscript;

3. limiting subscripts to no more than two symbols for each meaning; and

4. separating multiple subscripts by commas.

Latin symbols

a	velocity of sound (eqn. 2.33)
a	a constant (eqn. 6.7)
A	area (eqn. 2.44)
A_f	face area (eqn. 10.38)
A_L	altitude (eqn. 3.33)
b	breadth (eqn. 1.1; fig. 4.5)
b	a constant (eqn. 6.7)
b_c	blade backbone length (fig. 7.9)
b_r	turbofan bypass ratio (eqn. 3.40)
b_x	blade-row axial chord (eqn. 7.2)
B	boundary-layer blockage (fig. 4.9)
c	blade chord (eqn. 7.6)
C	absolute flow velocity (eqn. 2.2)
C_1	compressor temperature ratio (eqn. 3.1)
C_f	skin-friction coefficient (eqn. 10.11)
C_L	lift coefficient (eqn. 7.4)

C_p	specific heat at constant pressure (eqn. 2.30)
C_{pr}	pressure-rise coefficient (eqn. 4.1)
C_r	conductance ratio (eqn. 10.29)
C_{ra}	heat-exchanger capacity-rate ratio (eqn. 10.19)
C_{ro}	periodic-flow heat-exchanger switching-rate ratio (eqn. 10.30)
C_v	specific heat at constant volume (eqn. 2.30)
C_w	specific heat capacity of water (section 2.4)
d	diameter (eqn. 5.10)
d_h	hydraulic diameter (sec. 4.5, eqn. 5.10)
d_x	stator-rotor axial gap (fig. 7.21)
D_{eq}	equivalent diffusion ratio (eqns. 8.3–8.6)
e	radius of curvature of blade convex surface at throat (eqn. 7.8)
E	energy, general (eqn. 2.1)
$E_1, \ldots$	expander temperature ratios (eqn. 3.2)
f	power-law index for axial-velocity variation (eqn. 9.2)
F	force (eqn. 7.1), thrust (eqn. 3.35)
g	gravitational acceleration (eqn. 2.2)
g_c	unit-matching constant in Newton's law (eqn. 2.2)
h	enthalpy per unit mass (eqn. 2.3)
$\hat{h}$	enthalpy per kg mole (eqn. 12.5)
h_b	height, blade height (eqn. 7.2)
h_t	heat-transfer coefficient (eqn. 10.8)
H	head in gravitational field (eqn. 2.12)

H'	$\Delta H_{ci}/\Delta H_T$ (eqn. 9.27)		(eqns. 2.67, 2.69)
ΔH_{ci}	head drop for cavitation inception (eqn. 9.21)	r_e	turbine (expander) pressure ratio (eqns. 2.68, 2.70)
H_L	head-loss ratio (eqn. 9.25)	r_f	fan pressure ratio (fig. 3.46)
H_{bl}	boundary-layer shape factor (eqn. 8.10)	r_j	jet pressure ratio (fig. 3.46)
		r_r	ram pressure ratio (fig. 3.46)
i	incidence (angle) of flow (eqn. 7.7)	$\Re$	universal gas constant (eqn. 2.48)
j	Colburn modulus (section 10.4)	R	$\equiv \Re/M_w$, specific gas constant (eqn. 2.29)
k	thermal conductivity (eqn. 10.2)		
l	length (eqn. 6.2)	R_e	Reynolds number (fig. 4.12)
L	length along a wall (fig. 4.5)	R_h	reheat factor (eqn. 2.78)
L_{HF}	lower heating value of the fuel (eqn. 3.42)	R_n	reaction (eqn. 5.6)
		$R_{sr,rr}$	stator-rotor pitch ratio (eqn. 7.19)
m	mass (eqn. 2.2)	s	entropy per unit mass (eqn. 2.23)
M	Mach number (eqn. 2.39)	S	entropy (eqn. 2.15)
M_w	molal mass (molecular weight) (eqn. 2.48)	S_p	spacing or pitch (eqn. 7.1)
		S_t	Stanton number (eqn. 10.14)
n	polytropic index (eqn. 2.60)	t	time (eqn. 2.2)
n	index (eqn. 6.7)	t_F	thrust specific fuel consumption (eqn. 3.38)
n'	$(n-1)/n$ (eqn. 2.74)		
$\hat{n}$	number of moles (eqn. 12.5)	t_{mx}	airfoil maximum thickness (eqn. 8.6)
n_c	a constant (eqn. 6.7)		
$\hat{n}$	number of moles (eqn. 12.5)	t_p	pressure thickness distribution (eqn. 7.18)
N_d	axial length of diffuser (fig. 4.9)		
N_s	specific speed, nondimensional (eqn. 5.10)	t_s	suction thickness distribution (eqn. 7.18)
N_{sv}	suction specific speed (eqn. 9.20)	t_w	wall thickness (eqn. 10.30)
N_{tu}	number of heat-transfer units (eqn. 10.18)	T	temperature (eqn. 2.24)
		T'	temperature ratio (eqn. 3.1)
N_u	Nusselt number (eqn. 10.10)	T_q	torque (fig. 5.1)
o	opening or throat width (eqn. 7.8)	u	internal thermal energy per unit mass (eqn. 2.2)
p	pressure (eqn. 2.2)		
p_e	perimeter of channel (eqn. 10.20)	u	peripheral speed (eqn. 5.1)
p_{or}	porosity (eqn. 10.38)	u_{fn}	$u_1^2/(C_p T_{0,ne})$ (eqn. 7.25)
P_r	Prandtl number (eqn. 10.12)	U	overall heat-transfer coefficient (eqn. 10.16)
q	heat per unit mass transferred (eqn. 2.1)		
		v	volume per unit mass (eqn. 2.2)
q_d	dynamic head or pressure (eqn. 4.9)	V	extensive volume (eqn. 10.25)
Q	heat transferred (eqn. 2.2)	V_a	aircraft velocity (eqn. 3.35)
$\dot{Q}$	rate of heat transfer (eqn. 2.2)	w	work transfer per unit mass (eqn. 2.1)
r	radius (eqn. 6.2)		
r'	radius relative to mean radius, nondimensional (eqn. 6.8)	W	fluid velocity relative to boundaries (eqn. 4.8)
r_c	compressor pressure ratio	$\dot{W}$	power or work transfer rate (eqn. 2.2)

$\dot{W}'$	specific power (eqn. 1.6)	σ_c	Thoma coefficient (section 9.12)
x	distance or length (fig. 2.2)	σ_s	surface tension (text ch. 9)
X	mole fraction (eqn. 12.4)	σ_w	slip for radial velocity
y	normal distance (eqn. 10.1)		diagrams (eqns. 5.22, 9.1)
Y	diffuser width (fig. 4.5)	τ	shear stress (eqn. 10.1)
z	height above datum in	ϕ	flow coefficient (eqn. 5.6)
	gravitational field (fig. 2.2)	χ	loss coefficient (eqn. 7.11)
z_c	compressibility factor (eqn. 2.83)	ψ	loading or work coefficient
Z	number of blades or passages		(eqn. 5.5)
	(eqn. 5.22, 9.14)	ψ^x	head coefficient (eqn. 5.16)
		ω_{bm}	blade cooling-flow parameter
			(eqn. 3.24)

Greek symbols

α	angle of flow with axial or radial direction (fig. 5.3)	ω	rotational speed (eqn. 6.1)
		Ω	number of alternative ways a
β	angle of blade with axial or radial direction (fig. 7.3)		collection of particles can be arranged (eqn. 2.14)
β_v	void fraction (eqn. 9.25)		
γ	ratio of specific heats (eqn. 2.34)		

Subscripts

Γ	circulation around a blade row (eqn. 8.6)	0	stagnation conditions (static term plus velocity term, eqn. 2.6)
δ	deviation of flow (fig. 7.3)	$1, \ldots$	thermodynamic points in a cycle
δ^*	displacement thickness of boundary layer (eqn. 8.13)		or process
		ac	actual (eqns. 2.64, 4.11)
ϵ	deflection of flow (fig. 7.3)	ad	adiabatic (eqn. 5.5)
ϵ_{hx}	heat-exchanger effectiveness (eqns. 3.2, 3.13)	an	annulus (eqns. 2.33, 7.13)
		be	best-efficiency point (eqn. 9.4)
η	efficiency (eqn. 2.51)	b	burner, combustion, heat
θ	angle; semidivergence angle of diffuser (fig. 4.5)		addition (eqn. 3.2)
		bl	boundary layer
θ	camber angle of blade (eqn. 7.7)	bm	blading material
Θ	momentum thickness of boundary layer (eqn. 8.8)	c	compressor or compression (eqn. 2.60)
ϑ	temperature difference driving heat transfer (eqn. 10.2)	cd	cold (eqn. 3.39)
		ci	cavitation inception (eqn. 9.21)
κ	a constant (eqn. 6.7)	CE	simple (no-heat-exchanger)
λ	blade setting or stagger angle (fig. 7.3)		cycle (eqn. 3.6a)
		CX	heat-exchanger cycle (eqn. 3.2)
Λ	hub-shroud diameter ratio (eqn. 5.10)	df	diffuser (eqn. 4.9)
		dp	design point (eqn. 9.23)
μ	absolute viscosity (eqn. 10.1)	e	expansion (eqn. 2.66)
ξ	a constant (eqn. 6.7)	en	environmetal (eqn. 2.3)
ρ	density (eqn. 2.42)	ex	exit or outlet (eqn. 4.1, 7.1)
σ	solidity (eqn. 8.4)	F	fuel (eqn. 3.37)

f	fan	ro	rotation (eqn. 10.30)
fc	face (eqn. 10.32)	s	isentropic (eqn. 2.52)
ff	free-face area (fig. 10.9)	S	shaft (sect. 3.10)
fl	flow	se	stage (eqn. 5.7)
fs	free stream (eqn. 8.13)	sh	outer (shroud) annulus diameter (eqn. 5.10)
$g1, \ldots$	group 1, 2, or 3 losses (eqn. 7.13)		
gt	gas turbine (eqn. 1.1)	sl	surge line (eq. 9.23)
h	heat transfer (eqn. 10.17)	sp	single phase (eqn. 9.25)
hb	inner (hub) annulus diameter (eqn. 5.10)	sq	thickness parameter (eqn. 7.18)
		sr	stator
hd	hydraulic diameter (eqn. 10.25)	st	static conditions (eqn. 2.2)
ht	hot (eqn. 10.32)	sy	secondary (eqn. 7.13)
hx	heat-exchanger	t	throat (eqn. 7.8)
ic	intercooler	tb	turbulent (eqn. 10.26)
id	induced (eqn. 2.64)	te	trailing edge (eqn. 8.11)
ie	ideal (eqn. 2.64)	th	thermal (eqn. 3.2)
in	into control volume (eqns. 2.1–2.4)	tl	theoretical (eqn. 4.9)
		tp	two phase (eqn. 9.25)
ip	incompressible (table 2.5)	ts	stagnation-to-static (efficiency) (eqn. 3.20–3.23)
it	isothermal (eqn. 2.56)		
le	leading edge (eqn. 8.12)	tt	stagnation-to-stagnation (efficiency) (table 2.2)
lk	leakage (fig. 3.11)		
lm	laminar (eqn. 10.27)	T	total conditions (static term plus velocity term plus elevation term, eqn. 2.6)
m	mean (fig. 5.15)		
mt	matrix (eqn. 10.30)		
mi	minimum (eqn. 10.16)	u	peripheral or tangential direction (eqn. 5.1)
mx	maximum (eqn. 8.14)		
ne	nozzle exit (eqn. 7.25)	w	wall (eqn. 10.1)
nf	nonflow (eqn. 10.32)	x	axial or x direction
ni	nozzle inlet (fig. P2.5)	xs	cross-sectional (eqn. 10.21)
N	nozzle (eqn. 3.34)	z	vertical direction (eqn. 5.5)
op	optimum (eqn. 7.5)		
ov	overall efficiency (eqn. 3.36)		

pu	pump
p	polytropic (eqn. 2.60)
pd	products (eqn. 12.5)
pe	piston engine (eqn. 1.5)
pl	profile (eqn. 7.11)
pr	propulsive (eqn. 3.36)
pu	pump (eqn. 5.21)
r	radial direction (eqn. 6.3)
rc	reactants (eqn. 12.5)
rl	relative (eqn. 9.4)
rr	rotor (eqn. (5.6)

Symbols and operators

d	derivative operator
∂	partial derivative operator
δ	small difference operator
Δ	finite difference operator
Σ	summation operator
$\overline{}$	mean value
$'$	nondimensional quantity (T, r, ...)
$\otimes$	hypothetical specified condition (eqn. 2.52)

A brief history of turbomachinery

Somewhat mischievously, experienced designers may ask this question of people who proudly describe how they have arrived at successful designs by going through successive iterations of refinement: "Why didn't you design your final machine right away, saving all the waste motion of the preliminary phases?"

One can similarly ask: "Why didn't the Romans make water turbines?" and the answers can be equally instructive. In general, there are two types of responses to these questions. In one, the answer may be that the best course was revealed by trial and error, or by structured experiments, or by analysis. These steps may take hours or centuries, but when the knowledge is acquired, it can be applied to the original design problem without other essential influences. Thus, the Romans had the technology to cast and machine and mount a high-efficiency water turbine. They also needed water power, and they made water mills, but they lacked an understanding of hydrodynamics and mechanics, an understanding humankind acquired over many centuries.

The second type of response is more complex and would answer what seems to be a sillier question: "Why didn't the Romans make gas-turbine engines?" The first part of the response is obvious: they had no need for them. But if we repeat the question and apply it to the engineers and entrepreneurs of the nineteenth century, when there was a need for engines of increasing power and efficiency, when there were repeated attempts to design and make gas-turbine engines, and when appropriate gas and liquid fuels were available, we find that a sufficient knowledge of fluid mechanics was still lacking, particularly of how to accomplish flow deceleration (diffusion) efficiently. But also needed was a large set of developments in other fields, such as metallurgy, high-speed gearing and bearings, and combustion. In part developments in related fields may allow developments to take place in turbomachinery and gas-turbine engines, and in part these external developments can be, and have been, stimulated by the need for better high-temperature turbine materials, for instance. Technological history is therefore not only a study of development but of the background and incentives for development.

It is not certain, of course, that anyone who is ignorant of history is condemned to repeat it. A reading of this summary is not an essential part of a program of instruction.

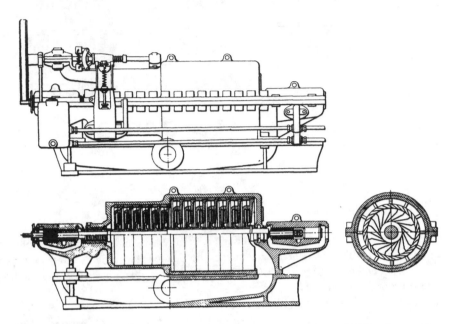

Illustration. Parsons' "Jumbo" radial-inflow steam turbine, 1889. From Parsons (1936)

But we should learn from mistakes. For instance, in the late 1980s there was much enthusiasm for radial-inflow gas-turbine expanders. When questioned by the authors, no member of the teams involved in the design and development of the engines using these expanders responded in the affirmative when questioned about knowledge of the disastrous erosion, in about one hour, of Parsons' first radial-inflow steam turbine (see illustration above). Droplets of water in the wet steam would not reach the level of radially-inward acceleration (tens of thousands times that of gravity) required for them to go through the expander blading. The designers questioned used flow and rotor speeds that would require well over a half-million "g"s of inward acceleration of every small particle, such as a grain of sand, for it to pass through the rotor. In at least three programs such particles produced erosion of the ceramic rotors to the extent that the radial-inflow turbines were essentially impracticable. Designers' skills may enable them to avoid these problems without a knowledge of history. But design skills are sharpened by experience, and vicarious experience (history) is still valid experience.

The material for this section has been culled from several sources (listed at the end of the section), including historical review articles in the *Chartered Mechanical Engineer* and *Power*, and from one on radial-inflow turbines by John D. Stanitz; but the major source has been an excellent book: *The Origins of the Turbojet Revolution* by Edward W. Constant II (Johns Hopkins University Press, 1980). Obviously many interesting details have had to be omitted in this short treatment.

Early turbomachinery

The Romans introduced paddle-type water wheels (pure "impulse" wheels) in around 70 B.C. for grinding grain. It seems likely that the Romans were the true initiators, because Chinese writings set the first use of water wheels there at several decades later.

The Greek geometrician and inventor Hero (or Heron) of Alexandria reported, and possibly invented, the first steam-powered engine, the aeolipile, a pure-reaction machine, around the time of Christ. A simple closed spherical vessel mounted on bearings carrying steam from a cauldron or boiler with one or more pipes discharging tangentially at the vessel's periphery is driven around by the reaction of the steam jets.

We know of little else on steam or gas turbines until Taqi-al-Din, writing in 1551 about Islamic engineering, described an impulse steam turbine that drove a spit (Ludlow and Bahrani, 1978). He may have inspired Giovanni Branca who proposed similar steam turbines in 1629. Branca was formerly considered to be the originator of the impulse steam turbine.

The movement toward modern turbomachinery really started in the eighteenth century. In 1705 Denys Papin published full descriptions of the centrifugal blowers and pumps he had developed (Smith, 1994). Four volumes on *Architecture hydraulique* were published in France in the period from 1737 to 1753 by Bernard Forest de Belidor, describing waterwheels with curved blades. These were also called "tub wheels" and were precursors of what are now called radial-inflow or Francis turbines.

The great Swiss mathematician Leonhard Euler (1707-1783), then at the Berlin Academy of Sciences, analyzed Hero's turbine and carried out experiments with his son Albert in the period around 1750. He published his application of Newton's law to turbomachinery, now universally known as "Euler's equation", in 1754, and thereby immediately permitted a more scientific approach to design than the previous cut-and-try methods.

The study of turbomachinery by models was introduced by the British experimenter John Smeaton (1724-1792) in 1752 and subsequently. He showed that the efficiency of "overshot" waterwheels could exceed 60 percent, versus the usual maximum of 30 percent for the more-usual "undershot" type. He also defined power as equivalent to the rate of lifting of a weight, a concept that is still fundamental in thermodynamics.

Stream-tube analysis and the study of ideal waterwheels were introduced in 1767 by Jean Charles Borda.

Up to this point there were no "turbines". The word was coined in 1822 by another Frenchman, Claude Burdin, from the Latin for "that which spins, as a spinning top: turbo, turbinis".

In the United States, Elwood Morris produced a turbine of the Fourneyron (radial-outflow) type in 1843 (see illustration on page 4), followed by Uriah A. Boyden's similar turbines in 1844 and 1846. Boyden added a vaneless radial diffuser, the first known use of this device (see illustration on page 5). This added three points to the efficiency, bringing it up to 88 percent, and his turbines were produced in sizes giving more than 190 hp.

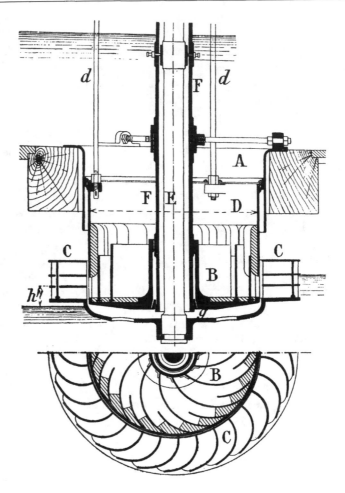

Illustration. Fourneyron radial-outflow hydraulic turbine, 1827. From Buchetti (1892)

In 1851 James B. Francis began a series of turbine tests with Boyden and A. H. Swaim in the Lowell, Massachusetts, canals, first published in 1855 as the Lowell Hydraulic Experiments. The high-efficiency radial-inflow turbines which resulted by 1875 are now almost universally known as Francis turbines (see illustration on page 6).

In Britain James Thomson (1822-1892) started work on an efficient radial-inflow turbine with a spiral inlet casing and adjustable inlet guide vanes in 1846-47, and patented it in 1850 as the Vortex turbine (see illustration on page 7). This continued in production for eighty years. In 1883 Osborne Reynolds established the foundations of flow similarity for laminar-turbulent transition in channels, and in 1892 Rayleigh introduced dimensional analysis.

Illustration. Boyden radial-outflow turbine with a vaneless radial diffuser, 1846. From Stanitz (1966)

Blowers and pumps

After Papin described his centrifugal blowers and pumps in 1705, there is little information on further developments until the nineteenth century. Crude centrifugal pumps were used in the United States and probably elsewhere by 1818, and improved versions were introduced in the New York dockyards in 1830. Eli Blake used a single-sided impeller (rather than paddle-wheel-type blading) in 1830, and in 1839 W. D. Andrews added a volute.

The first mention of a vaned diffuser was in a patent by Osborne Reynolds in 1875 for a so-called "Turbine pump", which was the combination of a centrifugal impeller with a vaned diffuser. In that year Reynolds (the son of a Belfast minister, made head of department at Owens College, later Manchester University, in 1868 at the age of twenty-six) also built a multistage axial-flow steam turbine, which ran at 12,000 rpm. In 1885 he described something of great relevance to future steam turbines, the convergent-divergent nozzle.

From 1887 his centrifugal pumps were made by Mather & Platt, a still-flourishing British manufacturer. In the next decade turbine pumps began being produced by Sulzer, Rateau, and Byron Jackson, and in the 1900s by Parsons (see illustration on page 8), de Laval, Allis Chalmers, and Worthington.

Compressors

The major fluid-mechanics problem in turbomachinery is the design of compressors. A turbine, with its flow usually going from a high to a low pressure, always works, and with reasonable care it works at high efficiency. A compressor, particularly the axial-flow

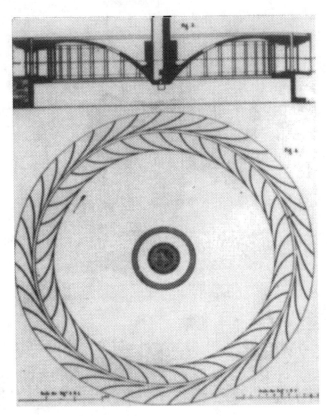

Illustration. Francis center-vent turbine. From Stanitz (1966)

type, may not produce a pressure rise at all. Until almost the beginning of the present century, compressor isentropic efficiencies were generally less than 50 percent.

We do not know the efficiencies of Papin's centrifugal blowers (1705). There appears to have been little development between that time and 1884, when Charles Parsons patented an axial-flow compressor. Three years later he designed and sold a three-stage centrifugal compressor for ship ventilation. He returned to experiments on axial-flow compressors in 1897, and made an eighty-one-stage machine in 1899 which attained 70 percent efficiency. The number of stages is probably an all-time record. By 1907 his company had made or had on order forty-one axial-flow compressors (see illustration on page 8), but they were plagued by poor aerodynamics, and he ceased production in 1908. Parsons used far too high a spacing/chord ratio for the rotor-blade settings, and all blade rows would likely be stalled over much of the operating range (see illustration on page 9). Parsons returned to making radial-flow compressors (see illustration on page 10).

The other major pioneer working on compressors at that time was Auguste Rateau, who published a major paper on turboblowers in 1892. However, a turbocompressor he designed to give a pressure ratio of 1.5 at 12,000 rpm and tested in 1902 gave an isentropic

Illustration. James Thomson's "Vortex" turbine, 1847-50. From Stanitz (1966)

efficiency of only 56 percent. Subsequently Rateau designed and built compressors of increasing pressure ratio and mass flow, and gradually increasing efficiency. In 1905 he demanded a more rigorous definition of efficiency. (See section 2.6.)

The continued story of compressor development is largely involved with the long struggle to produce a working gas-turbine engine. Many attempts failed simply because of poor compressor efficiency.

Steam turbines

Hero's turbine of A.D. 62 was a toy or experiment having no power output, and Giovanni Branca's proposal for an impulse steam turbine in 1629 was not built, so far as we know. Interest in turbine possibilities must have continued, because in 1784 James

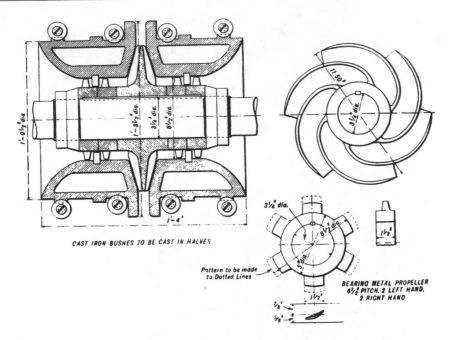

Illustration. Parsons' axial-radial double-flow pump. From Parsons (1936)

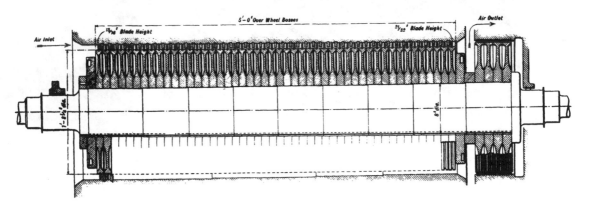

Illustration. Forty-eight-stage Parsons axial-flow compressor for a gold mine, 1904. From Parsons (1936)

Watt wrote to Matthew Boulton that he was glad he did not have to consider a turbine for his locomotives because of the high shaft speeds. The first useful steam turbine we know of was the spinning-arm reaction (Hero's-type) turbine of William Avery (U.S.) who in 1831 and following years made about fifty as drivers for circular saws and similar duties.

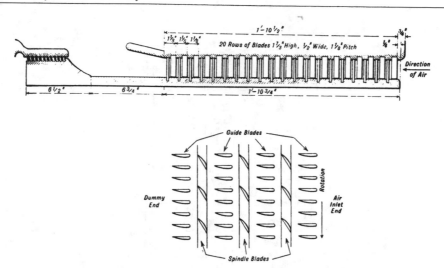

Illustration. Blade settings used in Parsons' axial-flow compressors, showing potential for stalled flow. From Parsons (1936)

Following the engineering tradition of Scottish clerics (Robert Stirling, for instance), Rev. Robert Wilson of Greenock patented in 1848 some forms of steam turbine, including a radial-inflow turbine. We have mentioned the multistage axial turbine constructed and run by Osborne Reynolds in 1875. We have no evidence that either of these efforts had any influence on other designers. In fact, the much earlier and still astonishingly advanced theory of ideal thermodynamic reversible cycles published by Sadi Carnot in 1824 also had no measurable influence on design in his century: in steam turbines, theory tended to follow practice.

The first steam turbine that had a major impact on the engineering world was that of Charles Parsons (1854-1931), who made a multistage axial-flow reaction turbine giving 10 hp at 18,000 rpm in 1884, his first year as a junior partner at Clarke, Chapman & Co. in Gateshead in northern England. He used brass blades on a steel wheel and an advanced self-centering bearing, presumably to avoid whirling-speed problems (also see the section on marine turbines). This was to fill a new market: a shipboard electrical generating set, and Clarke, Chapman made over 300 of Parsons' sets in the next five years. But at the end of that period he quarreled with Clarke, Chapman, dissolved the partnership, leaving the ownership of the axial-turbine patents with them, and formed C.A. Parsons & Co. He had in mind what he thought was a better idea, a multistage radial-inflow turbine. He made a thirteen-stage version called the "Jumbo" (illustration on page 2), which ran beautifully on first startup, but the power dropped almost to zero during the first hour. On taking it apart, he found that the blading had almost disappeared, because "foreign matter and water, imprisoned in each wheel case (was) thrown alternately inward by the steam and outward by centrifugal force" (from Richardson, quoted by Stanitz).

Parsons then turned to radial-outflow turbines but decided that the future lay with axial-flow machines, and in 1893 he bought back his patents from Clarke, Chapman

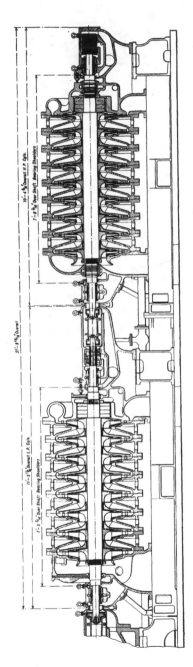

Illustration. Parsons' sixteen-stage radial-flow tandem intercooled compressor (partial cross-section). From Parsons (1936)

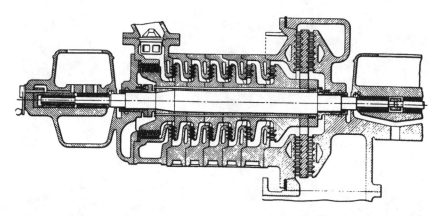

Illustration. Parsons' 100-kW multistage radial-outflow turbine, 1891. From Parsons (1936)

for 1,500 pounds. He had what seems to be an essential combination for a successful innovator: not only superb design skill and judgment but business acumen and a flair for showmanship. Having failed to persuade the Admiralty of the advantages of turbine power, he designed not only the propulsion machinery but the hull (after many model experiments) of a small vessel, which he named the *Turbinia*, powered by a multistage radial-outflow turbine giving 2,000 hp at 2,000 rev/min. At the great naval review of 1897 celebrating the diamond jubilee of Queen Victoria, the *Turbinia* dashed around the fleet at the astonishing speed of 34-1/2 knots (see illustration on page 12). The Admiralty was indeed impressed, along with everyone else, and placed an order for a 30-knot turbine-driven destroyer, *HMS Viper*. By 1907 his marine turbines were powering the 38,000-ton Atlantic liner *Mauretania* with an output of 70,000 hp. Parsons assigned U.S. rights to Westinghouse, and in 1900 a contract was signed by the Hartford Electric Light Company for a 1.5 MW unit, the first steam turbine to be produced in the U.S. It produced close to 2 MW, and the steam rate was 12.5 lbm/(kW · hr) versus the 20 lbm/(kW · hr) of the best contemporary reciprocating steam engines. The turbine was extremely temperamental, sometimes requiring blading replacement two or three times a week, and it was two years before *MaryAnn*, as the turbine came to be called, ran for a whole day. Nevertheless, *MaryAnn* was the progenitor of every steam-station turbine in the U.S., and brought about the end of the reciprocating steam engine. In 1912 Parsons' company supplied a 25-MW station for Chicago, and in 1925 his turbines reached 50 MW in size, with a plant thermal efficiency of 30 percent.

Parsons was not of course the only developer of steam turbines. In 1878 Carl Gustav Patrik de Laval (1845-1913) had made a new cream separator that needed to be driven at 12,000 rev/min. He had tried a Hero-type reaction turbine (see illustration on page 12), which was unsatisfactory. He turned to an impulse wheel and a convergent-divergent (supersonic) nozzle, which ran the turbine up to 30,000 rev/min, with a peripheral speed of 360 m/s (1,200 ft/s; see illustration on page 12). Realizing that the high kinetic energy of the nozzle was not being used efficiently, he constructed in 1897 a velocity-compounded

Illustration. The Turbinia at 34.5 knots, 1897. From Garnett (1906)

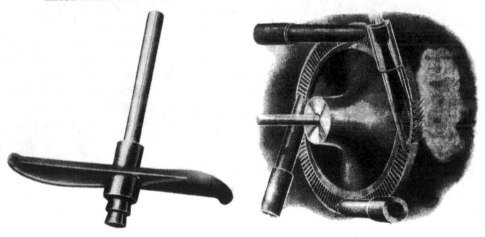

Illustration. De Laval "Hero-type" re-action turbine, 1893. From Parsons (1936)

Illustration. De Laval impulse steam turbine. From Garnett (1906)

impulse turbine (a two-row axial turbine with a row of turning-vane stators between them, all fed by a single set of high-velocity nozzles). In 1897 de Laval developed double-helical gearing, especially useful for turbine-powered ships, and used very high temperatures and pressures (3,400 psia, 23 MPa) for his turbines.

In France Auguste Rateau (1863-1930) experimented with a de Laval turbine in 1894, and developed the pressure-staged impulse turbine by 1900. As with Parsons and

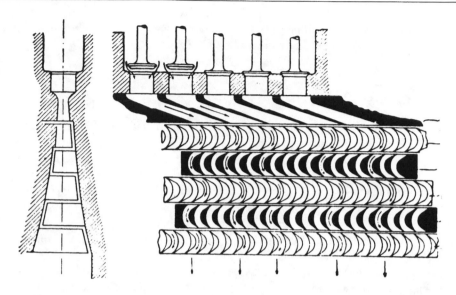

Illustration. Curtis velocity-compounded impulse turbine. From Stodola (1905)

de Laval, the company bearing his name is still very active. Zoelly started the manufacture of rather similar machines in Switzerland in 1904, and ten European manufacturers formed the "Zoelly Syndicate" to manufacture turbines of his design.

In the United States, Charles G. Curtis (1860-1953) patented in 1896 the velocity-staged turbine (illustration on page 13), similar to a two-stage de Laval turbine, but after further development he sold all his rights to General Electric in 1901 for $1.5 million. Meanwhile, Allis Chalmers and Westinghouse had taken licenses on the Parsons designs in 1895, followed by Brown-Boveri (a Swiss-German company founded in 1892 partly by a Briton) in 1899.

Some miscellaneous steam-turbine developments worth noting are these. Birger Ljungstrom (Sweden) introduced his counter-rotating radial-outflow turbine in 1912. DGW did part of his apprenticeship in Britain on Ljungstrom turbines, and remembers a 1913 turbine coming back for repair in 1949, illustrating the durability of that turbine configuration. W. L. R. Emmot recommended the mercury high-temperature cycle in 1914, and one was given a full-scale trial at Dutch Point in Hartford, Connecticut, in 1923. Several other sets followed, but they never achieved their thermodynamic promise.

Regenerative feed heating was utilized in 1920, and the reheat cycle in 1925. In the same year the Weymouth, Massachusetts, set used 1,200 psig (8.2 MPa) steam. Huntley station had a single-casing 75-MW set in 1929, and in the same year the State Line station installed a triple-cross-compound unit of 208 MW. It was 1958 before the units grew to 500 MW with a single steam generator.

A very influential educator and chronicler of turbine design was Aurel Stodola, the first edition of whose text on turbines was published in Switzerland around the turn of the century. In his second edition (1905) he added an appendix on gas turbines, and all subsequent editions were entitled *Steam and Gas Turbines*.

Marine turbines

Parsons' *Turbinia* turbine was supplied with saturated steam at 157 psig, 11.9 bar, and exhausted to a condenser at 1 psia, 52 mm Hg. It used 71 stages (each having a stationary row and a rotating row) of 50%-reaction blading (having the pressure dropping in each row, see chapter 5). Parsons used brass blades on steel disks or drums. Despite his entrepreneurial courage, he was a very conservative engineer. The peripheral speeds of the blades were around 40 m/s. (These and most other details in this section are from Nicholas (1995).) He used direct drive to the ship's propeller running at 130-450 rpm. This practice led to enormous machines: each low-pressure rotor of the 73,000-shp (54-MW) turbine for the *Mauretania* was almost 6 m in diameter and weighed over 300 t.

Parsons adopted gearing in 1910 partly because it was only then that he considered it reliable enough, and partly to combat competition from Curtis turbines, which had only three or four stages. Each Curtis stage consisted of a high-velocity nozzle followed by at least two rotor blade rows and at least one stationary blade rows that were simply "turning vanes" (there was no pressure drop across them). Rateau developed impulse turbines, having a nozzle-blade row and a single row of rotor turning blades. General Electric (US) bought patents and supplied turbines with an initial Curtis stage followed by impulse stages, far more compact than Parsons' reaction design. "At the Battle of Jutland (in World War I) every significant ship in both navies was turbine powered, but half the British ships were fitted with Brown-Curtis machinery" (Nicholas, 1995). (John Brown was the British licensee of GE.) Later, the higher-speed GE machines ran into severe vibration problems that gave the conservative, reliable, Parsons designs a temporary advantage.

Steam conditions at turbine entry up to the end of World War I were 200-250 psia, 13.6-17.0 bar, and efficiencies were low. The thermal efficiency of the *Mauretania* turbine engines was about 11.5%, and 200 stokers were needed to keep the boilers fired. Pressures increased with steam-generator development to 24 bar, 343 °C, by 1930.

GE's Wilfred Campbell (see the Campbell diagram in figure 7.22) solved the vibration problems plaguing impulse turbines, and in the 1930s GE became the foremost marine-turbine builder with compact designs running, with advanced gearsets, at up to 6000 rev/min. US companies then produced double-reduction gears that enabled a further increase in turbine speed coupled with the selection of optimum proportions (see "Specific speed" in chapter 5) for the turbines. "This range of standardized naval turbines produced by GE proved to be outstandingly successful during World War II and outperformed the propulsion equipment of any other navy" (Nicholas, 1995). Reaction turbines were eclipsed for marine use right through the 1960s, when the steam conditions reached 100 bar and 510 °C. The rapid increase in oil prices in the 1970s coupled with the development of low-speed marine Diesels of power levels over 35 MW and thermal efficiencies over 50% put an end to the marine steam turbine except for submarines and aircraft carriers powered by nuclear reactors. Naval vessels have since then been increasingly equipped with gas turbines, mostly converted from aircraft engines. Developments in marine turbines will always be linked to developments in the high-reduction-ratio gearing required to reduce the shaft speeds from those of steam and gas turbines to those required by the propellers. For example, this concern in the early 1980s resulted in a joint development program for the development of epicyclic gears between GE and the US Navy for use in US Navy combatants. In the early 1990s the US Navy sponsored the development of the WR-21, an intercooled-regenerative marine gas turbine jointly developed by Rolls Royce and Westinghouse (see chapter 3).

Gas-turbine engines

Just as the name "steam turbine" is used for both the complete plant and the turbine expander itself, so the name "gas turbine" is used for the gas-turbine engine and for the gas expander. We have discussed separately some of the developments in compressors and expanders; this section concerns the complete engine.

It has had a long and frustrating history. In comparison the steam engine, both piston and turbine, was relatively easy to design, build and run. Little work is required to force water into a boiler, little sophistication is required to boil the water, and when steam is formed at high pressure and led to either a piston engine or a turbine, it will produce more power than required by the "feed" pump. Internal-combustion piston engines also function fairly easily. Although relatively much more work is required to compress air than is needed to get water into a boiler, their maximum working-fluid temperatures are much higher than those for gas turbines. Therefore piston-expansion work is much greater than the piston-compression work, even if the compression and expansion processes themselves involve considerable flow losses. Gas-turbine engines could not use such high working-fluid temperatures at the start of expansion, and therefore, to produce net positive work, the compression and expansion losses had to be low. For many decades the compression losses in particular were just too high for positive work to be given at maximum working-fluid temperatures that turbines of the day could withstand. Consequently, many inventors produced machines that never even ran without an external power input.

The earliest patent for a gas-turbine engine was John Barber's of 1791, but nothing resulted from this. Franz Stolze (1836-1910) was influenced by an 1860 book by Ferdinand Redtenbacher on the theory and construction of turbines and first worked to reduce the losses in axial Jonval water turbines (Friedrich, 1991). Another Redtenbacher book, *Die calorische Maschine* (1853), about a double piston-and-cylinder machine apparently working on the Brayton-Joule cycle, led to Stolze designing a "fire turbine" that Friedrich (op. cit.) stated was exactly the principle and configuration of today's gas turbines. The patent office, however, stated that it was not new, and he did not receive patents until 1898. By that time Stolze had added a heat exchanger to heat the compressed air from the heat in the turbine exhaust, another advanced concept. The combustion chamber had air passing upward through a perforated grate through a bed of burning anthracite similar in layout to a fluidized bed, though it was probably not intended to operate in this

Illustration. Elling's centrifugal compressor stage with variable-setting diffuser vanes. From Johnson and Mowill (1968)

manner. The machine illustrated in the patent is indeed remarkably close to a modern configuration, having an axial-flow compressor and an axial turbine on a two-bearing rotor. The compressor has rather too few stages, and the turbine too many: the compressor blading diagram clearly shows passages that would be incapable of efficient diffusion. This otherwise very advanced gas (so-called "hot-air") turbine, when built and developed from 1900 to 1904, did not produce power, probably solely because of the lack of understanding of diffusion at that period.

It seems likely that Stolze's work influenced Aegidius Elling (1864-1949), whose achievements have been unaccountably overlooked in most historical accounts of turbomachinery development, and who avoided Stolze's problems by using a centrifugal, instead of an axial, compressor. He started working on gas-turbine designs in 1882 and filed his first patent in 1884. He constructed the first constant-pressure-cycle gas turbine to produce net output (11 hp), in the form of compressed air bled from the compressor, in 1903. It had a relatively advanced six-stage centrifugal compressor with variable-angle diffuser vanes (see illustration above) and water injection between stages; a combustion chamber; a small heat exchanger producing steam which was mixed with the combustion gases before the nozzles; and a centripetal turbine.

In 1904 Elling built a regenerative gas turbine: instead of the combustion gases raising steam in a heat exchanger, the turbine-outlet gases transferred heat to the compressor-delivery air. This turbine used a turbine-inlet gas temperature of 500 °C, versus the 400 °C of his first machine, and produced 44 hp. Thus Elling successfully designed and produced the two principal types of gas-turbine engine, one working on the "simple" or "non-regenerative" cycle, and one working on the "regenerative" or "heat-exchanger" cycle, well before any competitor. He went on to build a four-shaft machine with intercooling and reheat between three compressor-turbine groups on separate free-running shafts, and a "free" power turbine on a fourth shaft (1924). This was a highly advanced concept. Elling continued his pioneering developments to the end of his life, but unfortunately no commercial development, apparently, resulted. He illustrates the discouraging truth that technical brilliance is seldom enough to change the engineering world; the engineer must simultaneously be a brilliant showman and businessman.

The third successful gas turbine was produced in France by Charles Lemale, who was granted a patent for a constant-pressure (Brayton or Joule) cycle in 1901. He formed, with Rene Armengaud, the Societe Anonyme des Turbomoteurs in Paris in 1903. In 1905-06 the company commissioned Rateau to design a 25-stage centrifugal compressor in three casings on one shaft running at 4,000 rev/min and absorbing 245 kW (328 hp), giving a pressure ratio of three to one. This was made by Brown Boveri, and it achieved an isentropic efficiency of 65 to 70 percent. This would normally not be high enough to allow a gas-turbine engine to produce net power, but an astonishingly high combustion temperature of 1,800 °C was reported to be attained in a carborundum-lined combustor, and the two-stage Curtis turbine was water cooled. The steam raised in the cooling circuit was led to nozzles and passed through the same turbine wheel, a concept that is receiving renewed attention at the present time. This ambitious engine produced positive power, albeit at only 3.5 percent thermal efficiency. When Armengaud died in 1909, the experiments were stopped. Brown Boveri continued in the compressor business. The Societe des Turbomoteurs turned to making kerosene-combustion-heated compressed-air-powered torpedoes.

Another early engine to achieve partial success was proposed by Hans Holzwarth in 1906-08 and constructed in 1908-13 first by Koerting and later by Brown Boveri. This was an explosion or constant-volume cycle, in which the potentially high temperatures of combustion that are obtained in a periodic-firing system, as in an engine cylinder, can compensate for poor compression and expansion efficiencies. Holzwarth's 1,000-hp design produced only 200 hp, and both its size and efficiency were less favorable than then-available reciprocating engines. He worked with Brown Boveri and M. F. Thyssen, where he was chief engineer, gas turbines, from 1912 to 1927. Holzwarth continued to persevere with his engine, producing eight prototypes between 1908 and 1938, and the ninth, a 5-MW single-shaft unit with a turbine-inlet temperature of 930 °C, giving over 28 percent thermal efficiency, was made by Brown Boveri in 1938. It was installed in a steelworks to run on blast-furnace gas, and was destroyed in the war. One of these was converted to the constant-pressure cycle and led to the first successful industrial gas-turbine engines (see the discussion of the contributions of Stodola and Noack below). The association of gas turbines with Diesel engines in a modified form of the constant-volume cycle has continued to this day, ranging from simple turbocharging in which the turbomachinery is a small engine appendage (see below), to cases where the Diesel engine, in the form perhaps of a free-piston "gasifier", is simply a replacement for the more-conventional constant-pressure combustor.

A potentially significant, but, alas, fruitless, effort began in the United States when Sanford Moss, influenced by Professor Fred Hesse's thermodynamic classes at the University of California, devised a constant-pressure (Brayton or Joule) cycle in 1895. He wrote his master's thesis in 1900 on gas-turbine-engine design, with a proposal for a locomotive, and did his doctorate on a similar topic at Cornell. He used a steam-driven air compressor, but his engine did not produce positive work. Moss returned to his company, General Electric, in 1903, and carried out experiments on compressor design at Schenectady, New York, before moving to the Lynn, Massachusetts, works in 1904. He consulted with Professor Elihu Thomson and Richard H. Rice, with whom he made the important contribution of rationalizing diffusers into subsonic and supersonic types. General Electric ended work on his gas-turbine engines in 1907 when it seemed that they could not exceed a thermal efficiency of about 3.5 percent, well below the capability of competing engines. However, General Electric made and sold turbocompressors to Moss's designs from then until 1925, when the company sold the compressor business. Moss became a research engineer on steam turbines but worked also on aircraft-engine superchargers until the 1940s. Another gas-turbine-engine enthusiast, Glenn B. Warren, who wrote his bachelor's thesis on combined gas-steam cycles at the University of Wisconsin in 1919, joined General Electric in 1920 and subsequently made a design study of a 10,000 kW engine. Management decided that the company did not have sufficient knowledge in compressor design or heat transfer, however, and Warren was directed to work as well on steam-turbine problems. He made major advances in several areas.

There were many other abortive attempts to produce working gas-turbine engines in the first two decades of this century including Adolph Vogt (1904-05), Barbezat and Karavodine (making a constant-pressure engine which barely produced net power in 1908), and Hugo Junkers (who made free-piston engines with gas-turbine expanders, unsuccessfully, in 1914). Alfred Buchi studied under Stodola and later worked on the Vogt gas turbine built at Carels Freres in Ghent, Belgium, in 1904-5. After the failure of the gas turbine, Buchi turned to improving the company's Diesel engines, and devised a system of using the exhaust gases, expanding through a turbine wheel, to drive a supercharging compressor, thus devising the turbo-supercharger. The first commercial application to Diesel engines was for the propulsion plants of three ships, completed in 1925. Because they were relatively large units, they had axial turbines following centrifugal compressors, an arrangement now virtually universal on large Diesels. Meanwhile, Rudolf Birmann wrote his thesis in Zurich in 1922, presumably also under Stodola, on the subject of a radial-inflow turbine used for a gas-turbine engine. Birmann came to the U.S.A. and built and demonstrated an automotive turbocharger using a radial-inflow turbine in 1929. The turbines he designed at the DeLaval Steam Turbine Company (and an offshoot called the Turbo Engineering Corp.) were often more mixed-flow than radial-inflow (that is, the flow had a significant axial component at rotor entry); an early design achieved a peak efficiency of 85 percent at a pressure ratio of 4:1.

The various efforts leading to the modern successful gas-turbine engine can be said to have started in the few years from 1927 to 1936 by different people, some of whom were unaware of earlier or contemporary work. The first of these people, however, and one who has had a major impact was certainly fully cognizant of past attempts: Aurel Stodola, professor at Zurich Polytechnic. He was involved in testing a Holzwarth engine at the Thyssen steel works and noted the high heat transfer in the water jacket around the turbine. Noack of Brown Boveri recommended that the waste heat be used in a steam generator, and during 1933-36 the so-called Velox boiler was developed. This had an axial compressor "supercharging" a boiler in which gas or liquid fuel was burned, with the hot gases subsequently being expanded through a gas turbine that drove the compressor. For shop tests of the turbomachinery, a high-intensity combustor was substituted for the boiler in 1936, and net power was produced at a reasonable thermal efficiency. Thus the first successful industrial gas-turbine engine was arrived at by chance. Adolf Meyer designed true gas-turbine engines which Brown Boveri sold from 1939, and one was fitted in a locomotive in 1942, but the major gas-turbine sales were of Velox boilers for Houdry catalytic-cracking plants from 1936 on. The twenty-stage axial compressor gave the high isentropic efficiency of 86 percent, and was driven by a five-stage axial turbine.

Much of the other work on shaft-power gas-turbine engines also started in Switzerland. In 1936 Sulzer studied three alternative types of gas turbines and went on to produce axial-flow engines working on the constant-pressure open cycle. Jacob Ackeret and Curt Keller at Escher Wyss designed, in 1939, the closed-cycle gas turbine using, at that time, air. Escher Wyss continues to be the leading proponent of closed-cycle gas turbines and produces them now with principally helium as the working fluid for coal and nuclear heat input.

Christian Lorensen in Berlin began experiments on axial-flow turbines and hollow air-cooled blades, which were made by Brown Boveri in 1929. His work led directly to German turbojet cooled-blade designs.

In Sweden, A.J.R. Lysholm, chief engineer of the Ljungstrom Steam Turbine Company, began investigating gas-turbine engines and designed some units. One of his designs was made by Bofors in 1933-35, but success was prevented by severe surging in the centrifugal compressors. Lysholm turned to the Roots straight-lobe-type blower, which cannot surge but has no internal compression and is therefore inefficient for anything above a very small pressure ratio. Lysholm then invented the helical-screw compressor with internal compression, versions of which are now widely used in the niche between piston and centrifugal compressors. He did no further work that we know of on gas turbines. However, he sold his first U.S. license to the Elliott Company, which received a Navy contract to develop a one-thousand-hp gas turbine engine using a two-stage Lysholm screw compressor with an axial turbine. The project was never completed because of turbine-rotor problems. The U.S. Navy also rejected the compressor because of the extremely high noise levels.

A brilliant Hungarian engineer, George Jendrassik (1898-1954), was yet another who started work on gas turbines in the early 1930s.[1] He graduated in 1922 in Budapest and joined Ganz & Company in that city, making Diesel engines for rail traction. He brought about and patented substantial improvements, and Ganz-Jendrassik engines were licensed world-wide. His increasing interest in the gas turbine led him to leave Ganz and

---

[1] We are indebted to Prof.-Dr. Georg Gyarmathy of ETH, Zurich, for information on Stolze and Jendrassik. "The Jendrassik Combustion Turbine" appeared in *Engineering* (London) on February 17, 1939; and he was memorialized in *Engineering*, February 26, 1954, and the *Oil Engine and Gas Turbine* (London) in the March 1954 issue.

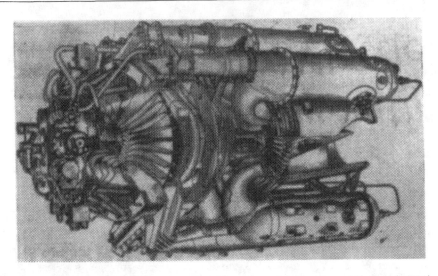

Illustration. Whittle W2/700 turbojet engine, 1943. From Wailes (1981)

to form his own company in 1936 and to start testing his first engine, producing 74 kW, in 1937. This was demonstrated in 1939. The journal *Engineering* (op. cit.) announced that "a new form of combustion turbine has been developed", apparently unaware of the others who had departed from Holzwarth's constant-volume system to develop constant-pressure engines. Jendrassik's test data show outstanding performance for an initial program: with a turbine-inlet temperature of only 748 K (887 °F), a leaking steel-plate heat exchanger, and a poor combustor, a thermal efficiency of 0.21 was obtained. The polytropic efficiency of his ten-stage axial compressor (pressure ratio 2.2) was 0.865. He went on to design and build a 750-kW turboprop engine of advanced design in all respects except the combustor, which greatly limited output on test; and a 225-kW recuperative free-power turbine for rail traction (adopted by Ganz). He held patents on axial-flow compressors (1937), the exhaust-heated cycle (1938, developed by Mordell in the 1950s), and rotary-regenerative heat exchangers (1943). He is credited with inventing the free power turbine. He left Hungary during World War II and eventually came to Britain in 1948, working on pressure exchangers for Power Jets from 1949 until his death in 1954. As with Elling, Jendrassik is often not given the credit that is his due.

The turbojet

In the same decade that saw the shaft-power gas-turbine engine reach successful operation, the turbojet was independently developed by four people: Whittle in Britain, and von Ohain, Wagner, and Schelp in Germany.

Frank Whittle, a Royal Air Force cadet, was first with his invention (1929) and patent (1930), although he was not first at successful running or in the first flight. His patent was for an axial-plus-centrifugal compressor and a two-stage turbine and included the possibility of using silica-ceramic blades. His proposals were rejected by all the major British aero-engine manufacturers (in fact, no turbojet development in Britain or Germany originated with an engine builder) and he was not given government support, except for an education at Cambridge, until 1936. He formed Power Jets Ltd. in 1935, but his patent lapsed because he could not afford the five pounds sterling required to renew it. His design included free-vortex-twist turbine blades, and his incorporation of blade-angle variations that allowed for radial-flow equilibrium gave British engines a higher efficiency than their German counterparts.

Whittle had two principal problems: to make a combustor with about ten times the previous maximum combustion intensity for liquid-fuel combustion; and to overcome the mechanical failures that plagued his turbines. A major contribution to solving the combustion problem was made by Ian Lubbock of Shell in 1940. A year later the Henry Wiggin Company produced its first Nimonic series of nickel-chromium-cobalt turbine-blade materials, effectively solving this problem. Meanwhile, the Whittle engine (see illustration on page 20) was taken to the United States and production started at the General Electric plant at Lynn, Massachusetts; it had its first U.S. run in 1942 and its first flight later that year in a Bell XP-59. This was only a year later than the first British flight. The Whittle engine was developed into the Rolls Royce Welland in Britain and the J 33 in the United States.

Whittle's counterpart in Germany was Hans von Ohain. Although he started later than Whittle (he received his Ph.D. in physics from Gottingen in 1936 and joined Ernst Heinkel that year) he achieved his first engine run and first flight sooner. This was partly due to better backing (Heinkel wanted to build the world's fastest plane); partly due to excellent assistance (Heinkel hired Max Hahn to help von Ohain); and partly due to wise tactical decisions. Von Ohain recognized the value of showmanship. He knew that, to get continued support from Heinkel, he had to demonstrate, quickly, something that showed promise. The configuration he chose was a centrifugal compressor plus a radial-inflow turbine made mostly of riveted mild-steel sheets (see illustration on page 22), and he avoided the liquid-fuel-combustion problem by using hydrogen for his first run in 1937 (see illustration on page 23). With greater support subsequently forthcoming, he was able to solve the combustion problem ahead of Whittle, and the first flight of a jet aircraft, the Heinkel 178, was on August 27, 1939, with an engine weighing 361 kg (795 lbm), giving 4,890 N (1,100 lbf) thrust at 13,000 rev/min.

In 1935 Herbert Wagner, professor of aerodynamics in Danzig and Berlin, was on leave at Junkers and designed a turboprop engine. In 1937 he and Max Adolf-Muller designed an engine with a five-stage axial-flow compressor, a single annular combustion chamber, and a two-stage turbine. The compressor had 50%-reaction blading and a pressure ratio of three to one (too great for only five subsonic stages) and accordingly had a poor operating range. The engine never ran under its own power. It was later included in the program of the German Air Ministry (RLM).

Helmut Schelp came to the U.S. for part of his education, and took a masters degree at Stevens Institute of Technology. He returned to Germany in 1936, joined the RLM and studied the problems of high-speed aircraft. In 1937 he decided that a jet engine incorporating an axial-flow compressor would be optimum. Unlike the other pioneers,

Illustration. Von Ohain's back-to-back compressor-turbine rotor, with rotating shroud removed. From U.S.A.F. Historical Publication (1986)

he was aware of preceding work. With his superior, Hans A. Mauch, he visited all the major German engine manufacturers in 1938 and persuaded Junkers and BMW to accept study contracts on reaction propulsion. (Nowadays governments have less difficulty in getting companies to accept funds.) A contract was also given to Messerschmidt for a turbojet fighter, which became the highly successful Me 262. Anselm Franz, head of supercharger development at Junkers, was put in charge of the new engine project. In the same way as von Ohain before him, he decided on a design with the best chance of running even if some sacrifice in performance must be taken. The Junkers Jumo 004 resulted, with an eight-stage axial-flow compressor giving a pressure ratio of three to one and an isentropic efficiency of 78 percent, and with a single-stage turbine. Initially, the turbine blades were solid, but later they were made from a manganese-alloy sheet and were air cooled. There was no nickel and only 2 kg of chromium in the engine. The combustor was aluminized mild steel and lasted twenty-five hours. Six-thousand Jumo 004 engines were made by the end of the war.

A similar diversity was taking place in Britain. Griffith at the Royal Aircraft Establishment had been very influential in axial-compressor fluid mechanics, and he and Hayne Constant designed an eight-stage axial-flow compressor in 1937 which ran in 1938 but failed mechanically. It was rebuilt but was destroyed in an air raid in August 1940. C. A.

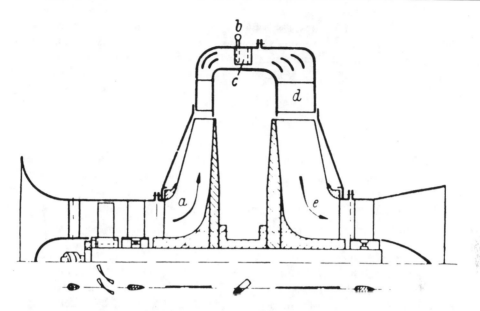

Illustration. Von Ohain's radial turbojet engine, He S-1, 1936-7. From U.S.A.F. Historical Publication (1986)

Parsons Ltd. made a similar test compressor in 1940. Armstrong Siddeley ran an engine in 1939, following Griffith's concept of having several "spools" (concentric shafts each with a turbine stage driving a few compressor stages), now an established procedure for large turbojets. In 1939 Griffith went from the RAE to Rolls Royce to speed up jet-engine work there. In January 1941 Frank Halford, designer of the highly advanced Napier Sabre piston engine, was asked to design a turbojet at de Havilland, and the DH Goblin was built by August 1941, ran in April 1942, and flew in a Gloster Meteor in March 1943. It was giving 13,450 N (3,000 lbf) thrust by 1945, with a very good fuel consumption. Bristol Engines Co., which rejected Whittle in 1930 because of its development problems with its double-sleeve-valved radial piston engines, started work on turbojets in 1944.

Since the end of the second world war the gas turbine has established itself as the power plant for civilian and military aircraft, including helicopters, other than for small general-aviation craft. Many companies entered the business and many have dropped out or consolidated. Three giant companies remain: General Electric (and GE Aircraft Engines) and Pratt & Whitney in the U.S., and Rolls Royce in Britain, whose largest fan jets are, at the end of this century, approaching 45 kN in thrust. There have been only three generations of engines since 1946, and it is now so expensive to develop a new engine that companies must form alliances with companies that recently were competitors.

In commercial power production one effect of the rising energy prices in the 1970s and the increased concern over emissions has been the near-cessation of steam-turbine manufacture except for "combined-cycle" plants with gas turbines, having outputs that can exceed 500 MW.

Gas turbines are used for most fossil-fueled naval vessels and some high-speed commercial vessels. Turbines have not yet made inroads in highway vehicles, although concerns over local and global pollution may bring major developments in that area also.

At the low-power end of the scale, a standby generator of 2 KW output is being successfully marketed in Japan.

The fallibility of experts' forecasts

Despite the successful flight of the Heinkel 138 in August 1939, a prestigious committee on gas turbines of the National Academy of Sciences, including Theodor von Karman, C. F. Kettering, Lionel S. Marks and R. A. Millikan, gave the following opinion on the feasibility of gas turbines for aircraft propulsion. (Technical Bulletin No. 2, published January 1941.)

> In its present state, and even considering the improvements possible when adopting the higher temperatures proposed for the immediate future, the gas turbine could hardly be considered a feasible application to airplanes mainly because of the difficulty in complying with the stringent weight requirements imposed by aeronautics.

> The present internal-combustion-engine equipment used in airplanes weighs about 1.1 pounds per horsepower, and to approach such a figure with a gas turbine seems beyond the realm of possibility with existing materials. The minimum weight for gas turbines even when taking advantage of higher temperatures appears to be approximately 13 to 15 pounds per horsepower.

(Kettering developed the automobile electric starter despite being assured by electrical engineers that it was impossible to produce it within a feasible weight limit.)

Young engineers who have ideas on how to change the future should consider the conclusions and composition of this committee. Creators of innovative ideas should not be discouraged by adverse opinions, no matter how prestigious their sources (unless these ideas rely on breaking the laws of thermodynamics). The future for gas turbines is promising and exciting, in full observation of thermodynamic laws.

References

A century of power progress—growth marks the years 1907-57. Power, June 1981.

A century of power progress—boosting powerplant performance. Power, July 1981.

Air Force Systems Command Historical Publication (1986). An encounter between the jet-engine inventors—Sir Frank Whittle and Dr. Hans von Ohain, 3-4 May 1978. Dayton, OH, 1986.

Agricola, Georgius (1556). De Re Metallica. Tr. by Herbert Clark Hoover and Lou Henry Hoover, Dover, NY, 1950.

Anderson, S.B. (1974). Development in air compression. I. Mech. E. (September): 66, 72.

Bamford, L.P., and S.T. Robinson (1945). Turbine engine activity at Ernst Heinkel AG Werk Hirth—Motoren. Combined Intelligence Objectives Sub-Committee, file no. XXIII-14, London, UK.

Bowden, A.T. (1964). Charles Parsons—purveyor of power. Chartered Mechanical Engineer (September): 433–438, London, UK.

Buchetti, J. (1892). Les moteurs hydrauliques. Published by the author, Paris, France.

Buchi, Alfred J. (1953). Exhaust turbocharging of internal-combustion engines. Monograph no. 1, Tr. of the Franklin Institute, Philadelphia, PA.

Constant, E.W., II (1980). The Origins of the Turbojet Revolution. Johns Hopkins University Press, Baltimore, MD.

Friedrich, R. (1991). Dokumente zur Erfindung der heutigen Gasturbine vor 118 Jahren. VGB-KRAFTWERKSTECHNIK Gmbh, Essen, Germany.

Garnett, W.H. Stuart (1906). Turbines. Bell, London, UK.

Holmes, R. (1967). The evolution of engineering—the rocket engine. Chartered Mechanical Engineer (October): 436–442, London, UK.

Johnson, Dag, and R.J. Mowill (1968). Aegidius Elling—A Norwegian Gas-Turbine Pioneer. Norwegian Technical Museum (March), Oslo, Norway.

Legat, E.S. (1950). The development of air and gas compressing plant. Institution of Mechanical Engineers (February): 468–471, London, UK.

Ludlow, C.G. and A.S. Bahrani (1978). Mechanical engineering during the early Islamic period. Chartered Mechanical Engineer (November): 79–84, London, UK.

Lysholm, A.J.R. (1964). An interview with Fredrik Ljungstrom. Chartered Mechanical Engineer (May): 261–263, London, UK.

Nicholas (1995). A brief history of the marine steam turbine. Transactions, The Institute of Marine Engineers, London, UK.

Parsons, R.H. (1936). The Development of the Parsons Steam Turbine. Constable, London, UK.

Redtenbacher, F. (1853). Die calorische Maschine. 2, vermehrte Auflage. Friedrich Basserman, Mannheim, Germany.

Schlaifer, Robert (1950). Development of Aircraft Engines. Harvard University Graduate School of Business Administration, Boston, MA.

Seippel, Claude (1983). Personal letter to David Gordon Wilson, with a translation of an article "Constant volume versus constant pressure combustion—a chapter in the history of the gas turbine". February 22, Zurich, Switzerland.

Smith, Alan (1994). "Engines moved by fire and water", Trans. The Newcomer Society, vol 66, pp 1–25, London, UK.

Smith, N. (1980). The origins of the water turbine. Scientific American, 242 (January): 138–148, New York, NY.

Stanitz, J.D. (1966). History of radial-inflow turbine development. Unpublished monograph.

Stodola, A. (1905). Steam Turbines (with an appendix on gas turbines and the future of heat engines). Second German edition translated by Louis C. Loewenstein, D. Van Nostrand, New York, NY.

Stodola, A. (1927). Steam and Gas Turbines. 6th German edition. Transl. by Louis C. Loewenstein. McGraw-Hill, New York, NY, 1927.

Wailes, Rex (1961). An interview with Sir Frank Whittle, Chartered Mechanical Engineer (June): 353–358, London, UK.

Weaver, Glenn (1969). The Hartford Electric Light Company, Hartford, CT.

Whittle, F. (1981). CME interview Sir Frank Whittle. Chartered Mechanical Engineer (July): 41–43, London, UK.

Chapter 1

Introduction

1.1 Aims

The goals of this text are: to make available to non-specialist engineers and to students a simple but fundamental approach to the design of turbomachinery, including the choice of configuration and the determination of close approximations to optimum dimensions and flow angles; to develop simple methods for finding the performance characteristics of turbomachinery under various operating conditions and to show how the designs may be modified to give favorable "off-design" operation; and to show how the interrelationships among material limitations, fluid-mechanic laws, thermodynamics, heat transfer, and such considerations as mechanical vibration set boundaries to design choices.

The particular combination of turbomachinery and other components (combustors and heat exchangers) that forms the power-producing gas turbine is considered in some detail.

This work has been undertaken because of the apparent lack of a modern text dealing with turbomachinery from the design, or problem-solving, standpoint. The problems arise when satisfactory answers cannot be obtained from handbooks or from manufacturers' offerings. If, for instance, the requirement is to compress a given flow of gas from given inlet conditions to a given discharge pressure, with power available at a given shaft speed, and given trade-offs for first cost, efficiency, size, and so forth, what type of compressor represents an optimum choice? Is it radial, mixed flow, axial flow, in a single stage, or in several stages? What are the outer and inner diameters, blade lengths, flow and blading angles, and the diffuser size? Or if the reader is handed a certain configuration of turbomachine, on paper or in hardware, s/he should be able to predict its off-design performance to at least an approximate extent.

It is one of the aims of this work to answer these and similar questions with sufficient accuracy for most purposes. The accuracy will be sufficient for many areas of industrial and experimental design, and for systems-engineering studies. In those applications where the value of fractions of a percentage point in efficiency or in specific power is very large (as it is for example in aircraft gas turbines and fans) it is necessary to treat the designs obtained by the methods given here as inputs to detailed analytical and experimental methods (not treated here) of flow analysis, stress analysis, and vibration prediction, in particular. Such advanced methods have been developed by or for government laboratories, for instance, the Lewis Laboratories of the National Aeronautics and

Space Administration in the United States and the National Gas-Turbine Establishment in Britain, as well as in private companies.

The savings to be realized by approaching such sophisticated and extremely expensive programs with a near-optimum design are very large. The technical literature and the unwritten histories of virtually all firms involved with turbomachinery manufacture are full of cases where many millions of dollars have been wasted because development work was started on designs that were very poor choices for one or more reasons.

In other cases needless expense was incurred and inappropriate machines produced simply because designers and engineers have become accustomed to thinking only of a certain range of types of turbomachinery and have neglected other types far better suited to their needs. A chief engineer lost his company a great deal of money and perhaps more important, time, because he refused to change his pre-disposition for very-high-reaction compressors even when faced with performance requirements that plainly called for lower-reaction machines with variable-setting-angle stator blades. Other examples could be the overlooking of partial-admission turbines by a generation of engineers brought up on full-admission machines and the automatic use of radial-flow machinery even when all the conditions cry out for the use of axial-flow units.

No amounts of refinement by advanced analytical methods or by experiment can overcome the severe disadvantage of an initially incorrect choice of configuration or type. A principal aim of this book is to give engineers the information and insight needed to avoid such errors.

1.2   Definitions

Turbomachine

A turbomachine produces a change in enthalpy[1] in a stream of fluid passing through it and transfers work through a rotating shaft: the interaction between the fluid and the machine is primarily fluid-dynamic lift.

A change in enthalpy can also be brought about by heat transfer. Although there is often considerable heat transfer in a turbomachine (for instance, in a high-temperature cooled gas turbine), its contribution to the overall enthalpy change through the engine is (virtually always) almost negligible.

It is not possible in the real world to have lift forces without at the same time having drag forces. Much of the skill in achieving good designs is in keeping the drag forces and associated energy losses small; but they are always far from negligible. By stating in the definition of a turbomachine that the interaction must be "primarily" lift, we exclude that class of machines in which the fluid and the rotor interact purely through viscous forces (so-called "drag" turbines and pump-compressors). This definition may not be universal: advocates of viscous-action machines may regard them as true turbomachines.

---

[1]Readers unfamiliar with enthalpy may think of it as related to the energy content of a substance. Enthalpy is defined thermodynamically in chapter 2.

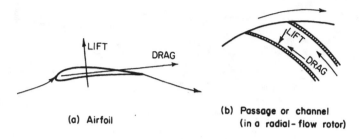

(a) Airfoil   (b) Passage or channel
                         (in a radial-flow rotor)

Figure 1.1. Lift and drag forces of fluids on solids

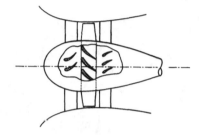

Figure 1.2. Axial-flow pump or compressor

Lift and drag

Lift and drag forces are normally defined with reference to airfoils (figure 1.1a), lift being normal to the direction of relative motion and drag along this direction. Airfoils very similar to airplane-wing sections can in fact be used in a particular class of turbomachines: axial-flow compressors and pumps (figure 1.2). However, most other classes of turbomachinery employ shaped channels rather than airfoils (figure 1.1b). Lift and drag forces must then be defined with reference to the normal and tangential directions of the channel walls.

Turbomachines can be divided into categories in several different ways. Shown in figure 1.3 are turbines on the left and pumps, fans, blowers, and/or compressors on the right. Pumps, fans, blowers, and compressors absorb shaft work from a "driver", and increase the enthalpy (and the pressure) of a flow of fluid.

The word "turbine" is used for all power-producing turbomachines, whether they work on liquid or gas, over a low or a high pressure ratio. There is no all-encompassing word for power-absorption machines. Pumps are pressure-increasing devices that operate on liquids. Fans work with gases, usually, but not always, giving so low an increase in pressure that the gas can be considered incompressible. In fact most fans merely consist of a single row of rotating blades which increase the velocity of the throughflow. The name "fan", however, is also given to the high-speed devices sometimes used on jet engines ("fan jets") to produce a combined jet of greatly increased size (and mass flow) and reduced velocity, which gives higher propulsion efficiency and lower noise than a small, high-velocity jet (see figure 13.17). Compressors work with gases and give

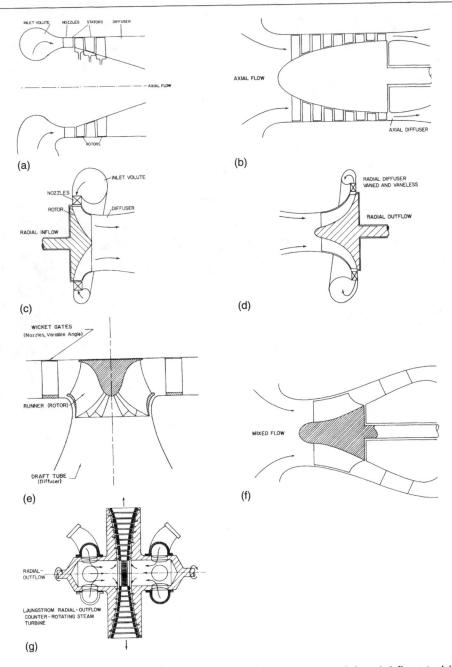

Figure 1.3. Illustrations of full-admission turbomachinery: (a) axial-flow turbine; (b) axial-flow compressor; (c) radial-inflow "centripetal" turbine; (d) radial-outflow "centrifugal" compressor; (e) radial-inflow hydraulic (Francis) turbine; (f) mixed-flow compressor; and (g) multi-stage radial-outflow counter-rotating Ljungstrom steam turbine

pressure ratios that result in a significant increase in density. A machine with a pressure ratio of 1.2 : 1 or above would be called a compressor. The word "blower" is, like "fan", used sometimes to mean a fan of negligible pressure ratio and sometimes (as in a blast-furnace blower) to mean a compressor of considerable pressure ratio.

"Full-admission" turbomachines in which the fluid flow through the blading is axisymmetric are shown in figure 1.3, while in figure 1.4 there are so-called "partial-admission" turbomachines. There is often a considerable advantage in designing a rotor that, if it "ran full", would be much too large for the flow available, and in ducting the flow to only a portion of the rotor blading. In gas-turbine engines the output power is often produced in a separate, or "free", power turbine, and occasionally it may be justifiable to use partial admission in order to produce a lower shaft speed.

A partial-admission turbine in the final drive of a ship-propulsion power plant, for instance, gives a lower shaft speed than would a full-admission design (because the same rotor-blade speed occurs at a larger radius). Another advantage is that the high-pressure fluid (usually steam for this application) may be ducted to several "nozzle boxes" covering different arcs of admission, and these may be connected, by valves, to the supply separately or in combination, giving a wide range of power outputs.

A special case of partial admission turbine is the Pelton wheel, used with water under high heads (pressures). The water leaves the nozzle as a jet with a so-called "free-surface" in air. In most turbomachinery the fluid stream is confined in ducts.

The Pelton jet can exist with a free surface because the entire pressure drop to atmospheric pressure takes place in the nozzle.

In partial-admission turbines all the pressure drop takes place in the nozzles, and the rotor converts only kinetic energy to shaft power. This type of arrangement, which can be used also in axisymmetric or full-admission machines, is called an "impulse" turbine. If, in a full-admission turbine, there is any pressure drop in the rotor, the fluid is accelerated in the rotor passages, and the acceleration results in a reaction force on the rotor blades. Even if the reaction force is small compared with the "impulse" force from turning (from the change of momentum of the fluid), a machine with flow that accelerates relative to the rotor is called a "reaction" turbine. Later in chapter 5 we shall introduce the concept of "degree of reaction", which can be used to quantify the extent to which a turbine, or a compressor, or fan, differs from "pure" impulse conditions.

The direction of flow forms another category that can be used to characterize turbomachines. The flow in the Pelton-wheel turbine is tangential to the rotor. In most cases the flow is either axial, that is, in an annulus whose axis coincides with that of the rotor and whose flow direction is approximately parallel to the axis, or radial. In radial-flow turbomachines the flow may travel radially inward or radially outward. Near the axis the flow in a radial-flow machine must turn to or from the axial direction. Radial-flow steam turbines usually have radially outward flow because of the rapidly increasing specific volume of steam during expansion. Radial-flow gas-turbine expanders use radially inward flow because the increase in volume flow is relatively small and can easily be accommodated in the flow path near the axis, and because there are advantages in locating the interaction of the nozzle exit flow in a region of high rotor-blade velocity.

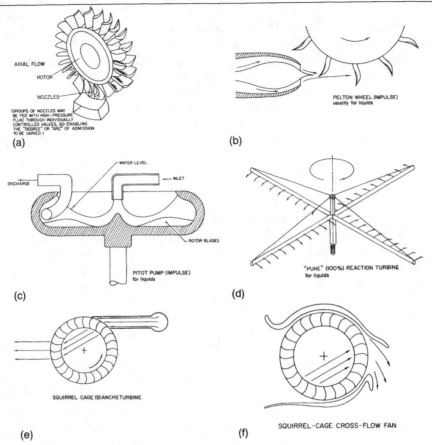

Figure 1.4. Illustrations of partial-admission turbomachinery: (a) axial-flow impulse turbine; (b) Pelton hydraulic (impulse) turbine; (c) Pitot pump (impulse) for liquids; (d) "pure" (100-percent) reaction turbine for liquids; (e) squirrel-cage (Bianchi) turbine; and (f) squirrel-cage crossflow fan

In some machines the flow at entrance to and/or exit from the rotor blades is close neither to the axial nor the radial direction. These are called "mixed-flow" turbomachines.

Another category concerns the number of separate rows or rings of blades. The "pure-reaction" machines have simply a rotor containing nozzles in the case of the rotating-pipe liquid distributor (frequently used over gravel filters in waste-water-treatment plants, or to water crops, as shown in figure 1.4d). Most turbomachines have a stator as well as a rotor. The stator may be as simple as the "snail-shell" volute of a centrifugal blower, or it may have the somewhat greater complexity of the volute, blade ring, casing, and ducting of the radial-flow turbine and compressor. The combination of rotor and stator (in either sequence) is called a "stage". Machines can therefore be categorized as being just rotors, or single stages, or having multiple stages.

The methods developed in this book apply to all these different machines, although we have the space to treat only a few particular classes with any degree of thoroughness.

Gas turbine

The term "gas turbine" is most often used as an abbreviation for a gas-turbine engine, which is a heat engine that accepts and rejects heat and produces work. The input heat is usually in the form of fuel that is burned (giving rise to the term "combustion turbine"), but may also come from another process via a heat exchanger. The rejected heat is usually in the form of hot engine-exhaust flow released to atmosphere, but may also be rejected to another process via a heat exchanger. The work may be given as output torque in a turning shaft or as the velocity and pressure energy in a jet, which would produce thrust on a moving airplane. Occasionally the output is in the form of compressed air from an oversized compressor. The term "gas turbine" can also be used more narrowly for just the turbine expander in a gas-turbine engine.

A gas-turbine engine (figure 1.5) consists of: a compressor, which continuously compresses gas from a low pressure to a higher pressure; a heater or heaters, in which the temperature of the compressed gas is raised; an expander, which continuously expands the hot gas to a lower pressure; and a cooler or cooling system, in which the temperature of the gas is reduced to that established for the compressor inlet. Figure 1.5 shows a closed-cycle engine, in which the same working fluid recirculates through the components without mixing with other working fluids from the surroundings. Two heaters and two coolers are shown in figure 1.5 because sometimes the first heater receives heat from the first cooler. (This combination of heater plus cooler is known as a heat exchanger.) It is defined more fully later in this section, but its design is treated in chapter 10.

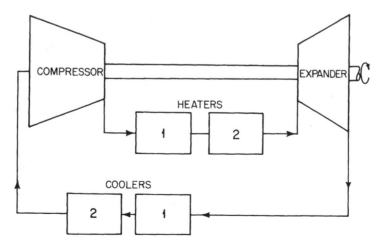

Figure 1.5. General gas-turbine engine

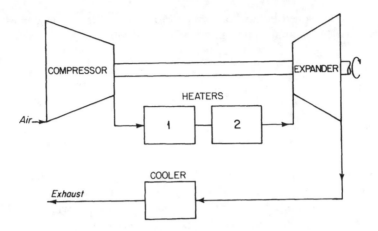

Figure 1.6. Open-cycle gas-turbine engine

An open-cycle gas turbine (figure 1.6) is one in which there is no engine cooling system; the atmosphere performs this function. The gas entering the compressor is atmospheric air, and the hot gas leaving the expander or the heat exchanger is discharged to the atmosphere. The earth's flora regenerate the oxygen from the carbon dioxide produced.

An internal-combustion gas turbine (figure 1.7) is one in which at least one of the coolers is dispensed with, and the gas is heated by combustion of fuel in the gas stream. For the fuel to burn, the gas must be either atmospheric air or oxygen. Internal-combustion gas turbines are invariably open cycle, working with air.

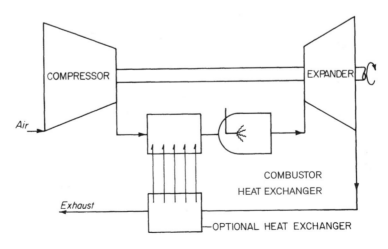

Figure 1.6. Principle of regenerative gas turbine

A regenerative gas turbine (illustrated in figure 1.7) supplies heat to the compressed gases in the first heater (after the compressor) from heat given by the hot expanded gases (in the first cooler). The heat exchanger so required is usually called a "regenerator" if it involves submerging a matrix alternately in the hot and then in the cold streams, and a "recuperator" if the heat is transferred through tube or duct walls (these definitions are accepted in Britain and are gaining ground in the United States, but they are not universal). A regenerative cycle may be open or closed.

Although only one shaft, one compressor and one expander are shown in figures 1.5-1.7, there may be several of each. In shaft-power engines, a commonly used arrangement is to split the turbine expansion. The high-pressure turbine produces just enough power to drive the compressor. The combination of compressor, combustor, and high-pressure turbine that drives the compressor is often called the "gasifer", and it often is identical to the components of a jet engine. The low-pressure turbine is called the "power" turbine because it supplies the output power. When the gas turbine drives an alternator, it is desirable to keep the power turbine integral with the high-pressure turbine and on the same shaft because it is easier to maintain constant speed and because this arrangement is lower in cost.

In engines of high pressure ratio it is necessary or desirable to divide the compressor into low-pressure and high-pressure units, separately driven on separate shafts by high-pressure and intermediate-pressure turbines, to give better starting and part-load characteristics (chapter 9). In aircraft engines these separate shafts are usually concentric, and the combination of a compressor, shaft and turbine is known as a "spool".

1.3   Comparison of gas turbines with other engines

The gas-turbine engine cycle has some resemblance to both other internal-combustion-engine cycles and to steam-turbine engines.

In a spark-ignition (Otto) or in a compression-ignition (Diesel) engine, air goes through a sequence of processes that is similar to that in an open-cycle internal-combustion gas turbine. Atmospheric air is compressed, heated, expanded, and discharged to the atmosphere for cooling and regeneration. There are two vital differences, however. First, in the Otto and Diesel engines the processes occur in the same place (the engine cylinder) at different times (the different strokes). In the gas-turbine engine the processes occur in different places (in the various components) at the same time (figure 1.8). The piston engine uses a "batch" process, while the gas-turbine engine works on a continuous process. The second difference is that much of the combustion in an engine cylinder takes place at constant (or nearly constant) volume. In a gas turbine combustion is carried out at constant (or nearly constant) pressure. (Constant-volume combustors for gas turbines are, however, possible. The Holzwarth gas turbine mentioned in the brief history had a constant-volume combustion.)

The result of the continuous nature of the processes in a gas turbine is that it can pass very much larger volume flows of gas than can a piston engine of the same engine volume. The turbine is very compact and has a high power-weight ratio. On the other hand, although it is easier to design the component efficiencies of the gas turbine to

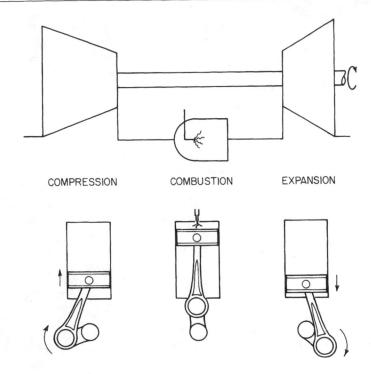

Figure 1.8. Comparison between gas-turbine and piston engines

be above the efficiencies of the individual processes of a piston engine, it is easier to use a higher peak cycle temperature in a piston engine because this peak temperature occurs momentarily in a cylinder rather than continuously in a combustor. In the past the thermodynamic benefit of higher peak temperatures in Diesel engines has outweighed the benefits of better component efficiencies in gas turbines. The use of turbine-blade cooling and higher-temperature materials is enabling the gas turbine to overtake the Diesel engine in maximum thermal efficiency in some cases.

The similarity of a gas turbine to a steam-turbine engine is perhaps more obvious. Both employ continuous processes taking place simultaneously in separate components. In some cases the components bear a strong resemblance. Both types of engines have turbine expanders which to the uninitiated eye may seem virtually identical. The steam-turbine engine has a pressurizer in the form of a feed pump paralleling the gas turbine's compressor. Feed pumps in large electrical-generating stations may absorb tens of thousands of kilowatts and look a little like multistage centrifugal compressors. And the steam generator or boiler of a steam-turbine engine can look rather similar to a gas heater of a closed-cycle gas-turbine engine.

The principal difference of course is that the steam-turbine engine works with a fluid that changes phase during the cycle, whereas the gas turbine, as its name implies (its name has nothing to do with the fuel it uses) works on fluids that remain in the gaseous phase.

The feed pump of a steam-turbine plant delivers pressurized liquid, which means that the work required is relatively small (less than two or three percent of the steam turbine output). A great deal of heat must then be added per unit mass of fluid in the steam-turbine engine to provide sensible and latent heat. In contrast, the compressor in a gas-turbine engine absorbs a large proportion of the power produced by the turbine expander. The gas-turbine combustor, being required to add a proportionally small amount of heat and working with high inlet-air density, shrinks to an almost negligible size in comparison with that of a steam generator in a steam-turbine cycle of similar net output.

The gas turbine as a Micawber engine

Mr. Micawber of Dickens' *David Copperfield* used to say (in very rough translation) "Income twenty shillings, expenditure twenty-one shillings; result, misery. Income twenty shillings, expenditure nineteen shillings: result, happiness". The gas-turbine engine, with its large compressor work input, has this kind of energy economics (figure 1.9).

Early gas turbines produced misery because their lossy compressors wanted to absorb more power than their lossy turbines would deliver. Successful, "happy", gas-turbine engines were made only when the compressor efficiencies had been improved to the point where the power required was less than the turbine power.

Such an engine, where the net power produced is the difference between two large quantities, is obviously very much affected by component efficiencies. When the compressor power almost equals the turbine power, a small change in the performance of any component can have a delightful or a disastrous effect on the net output of the machine. In early gas-turbine engines a 1 percent improvement in compressor efficiency could yield a 5 percent increase in engine thermal efficiency. And the turbine output could be increased either by improvement in its efficiency or by increasing the turbine-inlet temperature. Accordingly, there was an enormous effort made in many countries, in many engine companies, and in governmental and academic laboratories, to understand

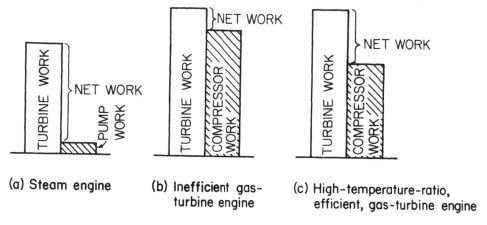

(a) Steam engine   (b) Inefficient gas-turbine engine   (c) High-temperature-ratio, efficient, gas-turbine engine

Figure 1.9. Energy accounting for different engines

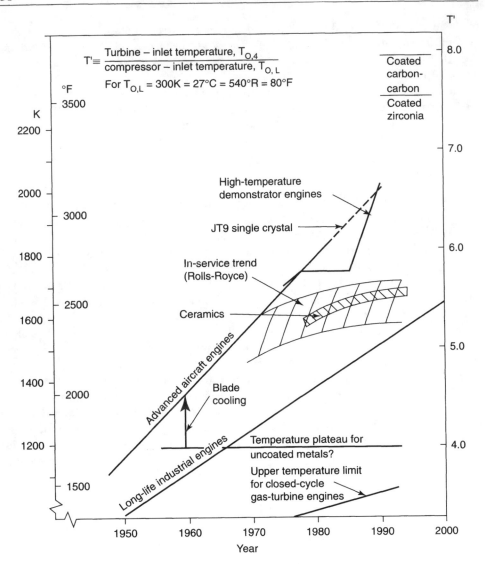

Figure 1.10. Increase of turbine-inlet temperature

the fluid mechanics of turbines and compressors, to produce higher-temperature materials, and to improve cooling of surfaces scrubbed by hot gases (figure 1.10). The gas turbine quickly became the most highly researched engine in history. The resulting design methods have become classics of the scientific method. In contrast, piston engines, which have received a much longer period of continuous, worldwide development, are still designed to a considerable extent by rule of thumb.

The "swallowing" capacity of turbomachinery

We mentioned earlier that the gas turbine can process much larger volume flows of gas than can piston engines of similar size because of the steady, rather than discontinuous, nature of the flow. In addition, the flow area used by a gas turbine is a larger proportion of the cross section than that of a piston engine.

As a simplification to compare the mass flows of air that could be used by gas turbines and piston engines, consider two engines that are contained in an envelope of square cross section of side b and of length $2b$, figure 1.11. The gas turbine draws in air through an annulus of outside diameter a little less than b, say $0.9b$, and inner diameter typically one-half of this, $0.45b$, with a Mach number, the ratio of the relative flow velocity to the speed of sound at the stream evaluated at "static" conditions a_{st} (see chapter 2 for an explanation of "static" and "stagnation" conditions), in the range of 0.3 to 0.5, say, 0.4. The volume flow handled by the gas turbine is, then:

$$\dot{V}_{gt} = \frac{\pi}{4}(0.9b)^2(1 - 0.5^2)0.4a_{st} = 0.191a_{st}b^2 \tag{1.1}$$

If the unit of static temperature T_{st} is in kelvin, K (SI name and symbol, respectively, for degrees Kelvin), then the speed of sound in air in m/s is given approximately by

$$a_{st} = 20\sqrt{T_{st}}$$

If we use a temperature of $T_{st} = 300\ K$, then $a_{st} = 346.4\ m/s$, and the corresponding inlet velocity is about 139 m/s. Therefore

$$\dot{V}_{gt} = \frac{\pi}{4}(0.9b)^2(1 - 0.5^2)139 = 63.5\,b^2\ m^3/s \tag{1.2}$$

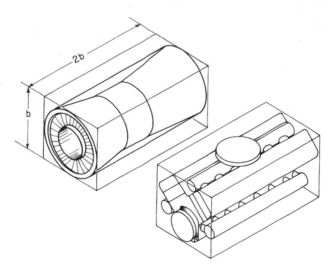

Figure 1.11. Gas-turbine and piston engine envelopes

In the same envelope the piston engine might be a V-8 arrangement. The cylinder bores, in a typical engine, are about $0.2b$ (and the stroke is similar).

The carburetor venturi throat has an area which, for standard automobiles, is about one seventy-fifth of the piston cross-sectional area. At maximum power the flow through the venturi throat is a fraction, say, seven-eighths of sonic velocity.

Then the volume flow of air handled by the piston engine is given by

$$\dot{V}_{pe} = 8\frac{\pi}{4}(0.2b)^2\frac{1}{75}\frac{7}{8}a_{pe} = 0.003a_{pe}b^2 \tag{1.3}$$

where a_{pe} is the sonic velocity. Then the ratio

$$\frac{\dot{V}_{gt}}{\dot{V}_{pe}} = \frac{0.191a_{gt}b^2}{0.003a_{pe}b^2} = 64\frac{a_{gt}}{a_{pe}} \tag{1.4}$$

Obviously the gas turbine engine can handle a vastly higher flow of air, and thus produces a far higher power output than can a piston engine of the same volume. We can refine this result by evaluating the acoustic velocities for the two engines. The speed of sound at the gas-turbine inlet will be higher than that at the piston-engine carburetor venturi because the temperature will not have been lowered to the same extent. From equation (2.57) in the next chapter,

$$\frac{a_{gt}}{a_{pe}} = \sqrt{\left(\frac{T_{gt}}{T_0}\right)\left(\frac{T_0}{T_{pe}}\right)} = \sqrt{\frac{1 + M_{pe}^2/\{2[(C_p/R) - 1]\}}{1 + M_{gt}^2/\{2[(C_p/R) - 1]\}}} \approx 1.06 \tag{1.5}$$

Then $\dot{V}_{gt}/\dot{V}_{pe} \approx 68$. The power output of an engine is a direct function of mass flow, rather than volume flow, but the densities of the intake air flows to the two engines are identical within the accuracy of this estimate.

This result is biased in favor of the gas-turbine engine in one respect. It has been arrived at by considering an engine without a heat exchanger. While most gas turbines run on the "simple cycle" (i.e., without a heat exchanger), to achieve Diesel-engine levels of efficiency a heat-exchanger cycle, perhaps with an intercooler, is required. Heat exchangers increase engine volume greatly, but the volume and weight of the resulting engines are still well below those of Diesel engines.

Relative cost of gas turbines and other engines

A comparison of the cost (in 1981 dollars) of Diesel and gas-turbine engines is shown in figure 1.12 by Woodhouse (1981). He demonstrates the correlation of both engines' specific cost with unit size and production rate. We believe that this chart, corrected for inflation, is still valid. The automobile industry shows repeatedly that the manufacturing cost of any hardware assembly that is produced at a rate of a million a year is about 25 percent greater than the material cost. This is a principal reason why future non-metallic gas-turbine engines should have a considerable cost advantage over Diesel engines. However, at the time of writing the cost per kilogram of ceramics and

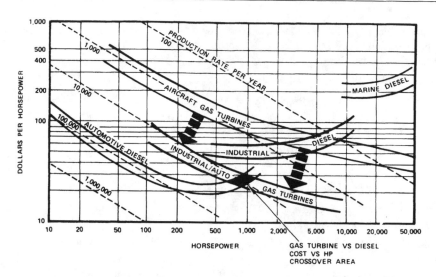

Figure 1.12. Specific cost of gas-turbine engines and competitors. From Woodhouse (1981)

carbon-graphite materials, two alternative non-metallics, is higher than that of the alloy metals they might replace.

The power output for a unit mass flow of air, termed the specific power $\dot{W}'$ and given nondimensionally as

$$\dot{W}' \equiv \frac{\dot{W}}{\dot{m}\,C_p\,T_{0,1}} \tag{1.6}$$

where $\dot{W}$ is the power output, $\dot{m}$ is the air mass-flow rate, C_p is the air specific heat capacity at constant pressure, and $T_{0,1}$ is the absolute temperature of the inlet air. Specific power can vary widely. In Brayton cycles it is generally in the range of 0.4 to 1.5 (see performance figures in chapter 3). A value of $\dot{W}' \approx 1.0$ is representative for current (1996) high-performance gas turbines. A typical value for Diesel engines is 3.0, and for spark-ignition engines, 2.5.

If we use representative specific-power values of 1.0 for the Brayton-cycle engines and 2.8 for the piston engines, we arrive at a ratio of 69/2.8 (or about 25), for the power output of a gas turbine relative to the power output of a piston engine of similar size. This ratio would apply reasonably well to the larger, axial-flow aircraft-engine-driven gas turbines but not to simple, small radial-flow gas turbines, whose volume tends to become dominated by ducts. Therefore, to ease manufacturing difficulties, they are purposely designed to be larger than the minimum possible size.

References

(This list includes some general references on turbomachinery and on comparative performance of engines, in addition to the specific references cited in the text of chapter 1).

Anderson, Richard H. (1979). Material properties and their relationship to critical jet-engine components. In Proc. Workshop in High-Temperature Materials for Advanced Military Engines. Institute for Defense Analyses, Arlington, VA.

Balje, O.E. (1981). Turbomachinery, Wiley, NY.

Bathie, William W. (1996). Fundamentals of Gas Turbines, 2nd ed., Wiley, NY.

Cohen, H., Rogers, G.F.C., and Saravanamuttoo, H.I.H. (1996). Gas-Turbine Theory, fourth edition, Longman Group Ltd, Essex, England.

Csanady, G.T. (1964). Theory of Turbomachines. McGraw-Hill, NY.

Dixon, S.L. (1975). Fluid Mechanics, Thermodynamics of Turbomachinery. Pergamon, NY.

Harman, Richard T.C. (1981). Gas-Turbine Engineering. Wiley, NY.

Hill, Philip G., and Carl R. Peterson (1992). Mechanics and Thermodynamics of Propulsion, 2nd ed. Addison-Wesley, Reading, MA.

Jennings, Burgess H., and Willard L. Rogers. (1969). Gas-Turbine Analysis and Practice. Dover, NY.

Karassik, Igor J., Krutzsch, William C., Fraser, Warren H. and Messina, Joseph P. (1986). Pump Handbook, 2nd ed. McGraw-Hill, NY.

Kerrebrock, Jack L. (1992). Aircraft Engines and Gas Turbines, 2nd ed. The MIT Press, Cambridge, MA.

Logan, Earl, Jr. (1995). Handbook of Turbomachinery. Marcel Dekker, NY.

Pfleiderer, C., and Petermann, H. (1964). Stromungsmaschinen, 4th ed. Springer-Verlag, Berlin.

Shepherd, D.G. (1969). Introduction to the Gas Turbine. Van Nostrand, NY.

Shepherd, D.G. (1956). Principles of Turbomachinery. Macmillan, NY.

Shepherd, Dennis G. (1972). Aerospace Propulsion. American Elsevier, NY.

Stepanoff, A.J. (1957). Centrifugal and Axial-Flow Pumps. Wiley, NY.

Vavra, M.H. (1960). Aero-Thermodynamics and Flow in Turbomachines. Wiley, NY.

Wilson, David Gordon (1978). Alternative automobile engines. Scientific American 239 (July), p. 39-49, NY.

Woodhouse, G.D. (1981). A new approach to vehicular-gas-turbine power-unit design. Paper 81-GT-152. ASME, NY.

Problems

This first chapter has dealt principally with definitions, and with some discussion of the characteristics and capabilities of turbomachinery and gas turbines. Answering questions on definitions does not engender a love for the subject matter. Accordingly, the following questions probe background knowledge. Few readers will have more than a few of the data asked for in the first question. Its purpose is to stimulate a survey of the engineering-society papers in the library, and perhaps the advertisements in some of the engineering magazines. Some of the other questions ask for opinions. Again, the purpose is to stimulate thought and perhaps some reading. There are not necessarily "correct" answers to these questions. August committees of eminent scientists and engineers have been frequently totally wrong when they have tried to forecast the future.

1. Complete as much of table P1.1 as possible, from your own knowledge or from library study. In general, every entry will be for a different machine, although in some cases there will be relationships between two figures. For instance, the highest-power steam turbine may not operate at the highest pressure used for steam turbines but may well have the largest low-pressure flow volume. Use numbered references to footnotes to identify your sources.

Table P1.1.

Give the maximum known value of	Steam turbine in generating station	Boiler-feed pump in generating station	Gas expander in gas-turbine engine	Compressor in gas-turbine engine	Axial compressor, any duty per single casing
Power (MW)					
Pressure (MPa)					
Temperature (deg C)					
Pressure ratio (max/min)					
Temperature ratio (max/min)					
Number of stages					
Low-pressure flow volume (m³/s)					
Mass flow (kg/s)					
Efficiency (percent, which?)					

Give the maximum known value of	Centrifugal compressor for any duty, per casing	Water pump for pumped-storage system	Pelton water turbine (high-head impulse)	Francis water turbine (medium-head inflow)	Kaplan water turbine (low-head axial-flow)
Power (MW)					
Pressure (MPa)					
Temperature (deg C)					
Pressure ratio (max/min)					
Temperature ratio (max/min)					
Number of stages					
Low-pressure flow volume (m³/s)					
Mass flow (kg/s)					
Efficiency (percent, which?)					

2. Gas-turbine engines have not reached the power-output levels of the largest steam turbines. Why?

3. Estimate the design power output of the smallest gas-turbine engine produced in the last decade. Why aren't smaller engines made?

4. Give your opinion of the two most promising new applications for gas-turbine engines in the next twenty years, and give reasons for your opinion.

5. Why is the maximum temperature of steam turbines so much lower than the current turbine-inlet temperature of gas-turbine engines?

6. What do you think are the two principal problems preventing gas-turbine engines from having a much wider application?

7. Do you think that gas-turbine engines will be used in outboard-motor boats by 2010? Why or why not?

8. For which of the applications in the list below do you think that the gas-turbine engine, as a prime mover, would be: (a) Suitable now, and, if so, how would it be used, or in what form? (b) Suitable after certain developments have been successfully completed, and if so, which? (c) Unsuitable, and if so, why?

As engines for	To exploit the heat output in
automobiles	geothermal energy
highway trucks	solar energy
city buses	nuclear energy
long-distance interurban buses	seawater thermal gradient with depth
motor cycles	
snowmobiles	
lawn mowers	

9. Discuss any present applications of, and future prospects for, vapor-cycle engines using fluids other than water.

Chapter 2

Review of thermodynamics

Turbomachinery and gas turbines may be fully analyzed, to the degree reached in this text, by rather simple thermodynamic relations. All are based on applications of the first and second laws of thermodynamics to so-called "flow" systems. From these, powerful and widely applicable tools such as the energy-enthalpy relations for various components and the one-dimensional compressible-flow functions can be quickly and easily derived.

The very simplicity of these tools can lead to their misuse. Serious errors can result when a relation derived for incompressible flow is used for a high-Mach-number gas stream, or when an isentropic-flow function is employed to relate upstream to downstream conditions in frictional flow. Rigor is vital. We shall derive the significant relations from first principles so that the conditions under which they may individually be used may be fully understood. This review will also serve to establish a language and a system of symbols that will be used in the remainder of the book.

2.1 The first law of thermodynamics

The law of conservation of energy is an axiom. Common experience and careful experiments have shown that it generally holds, but it cannot be formally proved. In words, it can be stated as: *The energy passing into a given mass of material in a given time is equal to the energy passing out of the material plus the energy stored within it.* The "mass of material" is any defined collection of molecules, whether solid, liquid, gas, plasma, or a mixture of more than one of these phases.

Thermodynamic textbooks often refer to this selection of molecules as a "system". There is no restriction on the form of the energy transfers (as work, heat, chemical, etc., energy transfers); they may follow some ideal patterns, or they may be far from ideal.

In symbols, the first law for a fixed mass of material A is written as follows (illustrated in figure 2.1):

$$\delta q = dE_A + \delta w \qquad (2.1)$$

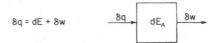

Figure 2.1. Energy flow and storage

where

$\delta q(= \delta q_{in})$ is the heat transferred in the process (positive for heat transfer into the material);

dE_A is the increase of energy level of the material inside the system A, termed the "internal" energy; and

$\delta w(= \delta w_{ex})$ is the work transferred out of the material to the surroundings (positive for work done by the material on the surroundings).

The change in internal energy, dE_A, includes all forms of energy storage, such as kinetic, gravitational, elastic, electric, magnetic, and thermal. In the following discussion we will give thermal-energy storage the special symbol du.

Of these quantities only E, the internal energy, is a property of the material. That is, its value depends only on the state of the material and not on how that state was reached. The energy transfers δq_{in} and δw_{ex} are not properties, and their values depend absolutely on the processes used to attain the final state. That is why different symbols are used for the differentials (δ and d). Later the apparently similar Gibbs' equation will be compared with this statement of the first law.

In many engineering examples there may be several identifiably different heat flows and work transfers, and the material may be composed of parts having different internal energies. The first law can be applied to the whole mass of material with the various quantities summed algebraically, or to individual components.

The first law derived for a flow process

The above statement of the first law is useful for the analysis of, for instance, positive-displacement machinery, because such devices (e.g., piston compressors) confine an identifiable mass of material within boundaries before undergoing various processes. It is less applicable to turbomachinery, in which material (working fluid) crosses control-volume boundaries.

We can easily adapt the first law to the material-crossing-boundary case, usually known as a "flow" system, by modelling a device or a space in which a process takes place in a box or "control volume". We can then look at the fixed boundaries around a defined mass of material (usually a fluid) as it moves into and out of the control volume. Mass and energy flow rates are considered at the inlets and outlets of the control volume, as well as energy flow rates through the surface of the control volume in some increment of time δt.

For simplicity we shall consider a control volume with a single inlet (station 1) and a single outlet flow (station 2), each using a duct of constant diameter (figure 2.2). Using

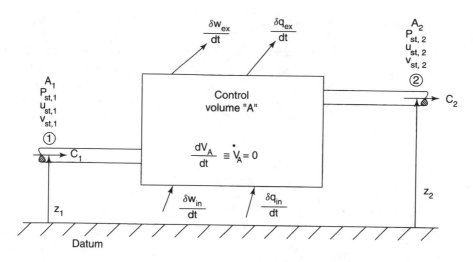

Figure 2.2. Energy flows for a steady stream of fluid passing through a control volume

only two ports is not a necessary assumption, and using it results in no loss of generality. Likewise, without loss of generality we shall consider only one of each heat transfer rate entering $(\delta q_{in}/dt)$ and exiting $(\delta q_{ex}/dt)$ the control volume, and one of each power transfer entering $(\delta w_{in}/dt)$ and exiting $(\delta w_{ex}/dt)$ the control volume.

For flow-process control volumes we have to take into account not only heat and work transfers through the surfaces of the control volume, but also the various categories of energy that the material takes with it as it passes through the inlets and outlets of the control volume. One of these categories of energy is kinetic (the effect of high flow velocity as material passes through the ports). Another category accounts for the displacement of mass through the port at pressure p. Another category accounts for potential-energy changes between ports due to changes of elevation z with respect to an arbitrary elevation level. There are of course other forms of energy, such as surface tension, electromagnetic forces, etc., which should be included in general formulations, but are seldom of significant influence in turbomachinery problems.

When applying energy balances to turbomachinery problems it is sometimes convenient to add to the thermal-energy term the kinetic-energy term; and at other times it is more convenient to add to the thermal-energy term the kinetic-energy and the potential-energy term, evaluated from an arbitrary elevation level. This leads to the definition of what are known as "static", "stagnation", and "total" properties, as explained further below.

We shall henceforth use the subscript *st* ("static") to denote property values that do not include the effects of kinetic-energy or potential-energy terms, the subscript 0 ("stagnation") to denote property values that include the effects of kinetic energy but do not include the effects of potential energy, and the subscript T ("total") to denote property values that include the effects of kinetic energy as well as potential energy. The following examples illustrate when it is convenient to use each one of these types of properties.

At any flow station with appreciable area-averaged flow velocity C perpendicular to the flow area A, incompressible liquids have (constant) density ρ and temperature T (corresponding to static conditions), the mass-flow rate is given by $\dot{m} = \rho A C$, and the fluid has different values of static, stagnation, and total pressure (p_{st}, p_0 and p_T). Perfect or semi-perfect gases (defined below) have different static and stagnation temperatures, pressures, and densities (T_{st}, T_0, p_{st}, p_0, ρ_{st} and ρ_0), and the mass-flow rate is given by $\dot{m} = \rho_{st} A C$.

The energy categories that usually need to be accounted for at the inlets and outlets of control volumes of turbomachinery problems are:

> internal thermal energy $\delta m\, u_{st}$
> kinetic energy $\delta m\, C^2/(2g_c)$
> potential energy $\delta m\, z\, g/g_c$
> flow energy (the work of the fluid pushing an incremental volume of material
>   into the control volume against the static pressure) $p_{st} A \delta x = \delta m\, v_{st} p_{st}$

where

> δm is the mass of material passing into or out of the control volume in time
> δt
> C is the velocity of the material
> g_c is a unit-matching constant in Newton's law ($g_c \equiv 1$ for SI and other
>   "consistent" systems of units[1])
> z is height above a datum
> v_{st} static specific volume (i.e., measured at the speed of the flowing materi-
>   als), where v_{st} is the reciprocal of static density ρ_{st}
> p_{st} is static pressure

Thus, if we exclude other energy forms such as chemical energy, electromagnetic forces, surface tension, etc., and we use the symbol u for the more restricted internal thermal energy, the change of energy δE associated with mass δm of fluid passing through an inlet or an outlet of the control volume (figure 2.2) is given by

$$\delta E = \delta m \left(u_{st} + p_{st} v_{st} + \frac{C^2}{2g_c} + \frac{gz}{g_c} \right) \tag{2.2}$$

Accounting for the heat and work flows, and the inlet and outlet flows of system A, shown in figure 2.2, with energy E_A and volume V_A in a region where the surrounding (atmospheric) environmental pressure is p_{en} at any time, the first law (energy balance)

---

[1] In Newton's law, force = (mass × acceleration)/g_c, $g_c \equiv 1$ for SI and other "consistent" systems of units. g_c is useful for unit systems having the same units for mass and weight (at sea level), e.g., pounds-mass and pounds-force (for which g_c=32.18 [lbm · ft]/[lbf · s^2]) and kilograms-mass and kilograms-force (for which g_c=9.807 [kgm · m]/[kgf · s^2]).

for the control volume becomes:

$$\frac{dE_A}{dt} + p_{en}\frac{dV_A}{dt} = \frac{\delta w_{in}}{dt} - \frac{\delta w_{ex}}{dt} + \frac{\delta q_{in}}{dt} - \frac{\delta q_{ex}}{dt}$$

$$+\frac{dm_1}{dt}\left(u_{st,1} + p_{st,1}v_{st,1} + \frac{C_1^2}{2g_c} + \frac{gz_1}{g_c}\right)$$

$$-\frac{dm_2}{dt}\left(u_{st,2} + p_{st,2}v_{st,2} + \frac{C_2^2}{2g_c} + \frac{gz_2}{g_c}\right) \qquad (2.3)$$

Equation 2.3 should be used whenever transients are of interest (such as transient flow through a compressor or turbine, hot-air balloon races, charging and discharging rigid or non-rigid vessels, etc.). For steady-flow processes $\dot{m} = dm/dt$, $\dot{Q} = \delta q/dt$, and $\dot{W} = \delta w/dt$. In practice turbomachinery components do not change volume with time ($\dot{V}_A = 0$). Also, the characteristic response time of the flow in turbomachinery is usually orders of magnitude shorter than the characteristic time of the overall transient problem (Korakianitis et al., 1994), so that most transients encountered can be defined as "slow". That is, most cases may be analyzed by examining a series of steady-state conditions, in which there is no storage of mass or energy inside the control volume ($dm_A/dt = 0$, or $\dot{m}_1 = \dot{m}_2$).

Therefore the energy flows into and out of the control volume in steady flow can be equated, and the steady-flow form of equation 2.3 becomes:

$$\dot{W}_{ex} + \dot{Q}_{ex} + \dot{m}_2\left(u_{st,2} + p_{st,2}v_{st,2} + \frac{C_2^2}{2g_c} + \frac{gz_2}{g_c}\right)$$

$$= \dot{W}_{in} + \dot{Q}_{in} + \dot{m}_1\left(u_{st,1} + p_{st,1}v_{st,1} + \frac{C_1^2}{2g_c} + \frac{gz_1}{g_c}\right) \qquad (2.4)$$

This is known as the "steady-flow energy equation", (SFEE). If there is only one inlet and one outlet flow port, as in figure 2.2, then the mass flow rate is $\dot{m} = \dot{m}_1 = \dot{m}_2$. The SFEE has very wide usefulness for the analysis of turbomachinery and associated steady-flow processes. Some examples of its use are given below. It is convenient to simplify equation 2.4 by defining the property enthalpy, mostly used for gas flows (compressors, gas turbines), and stagnation pressure, mostly used for liquid flows (hydraulic machinery).

Enthalpy

The static properties in $(u_{st} + p_{st}v_{st})$ are properties of the material at a given point. Properties, by definition, are functions only of the thermodynamic state of the material. This uniqueness means that new, unique, properties can be formed from any combination of other properties. It is obviously convenient to define $u_{st} + p_{st}v_{st}$ as one property. It is given the name "static enthalpy" and the symbol h_{st}:

$$h_{st} \equiv u_{st} + p_{st}v_{st} \qquad (2.5)$$

While pressure and volume can be defined in terms of measurable length and force quantities, the internal thermal energy, and therefore the enthalpy, can be measured only from a defined datum state. Different tables often use different datum states. When using two sets of data to cover widely differing states, one must take care that they have the same datum or base state.

The steady-flow energy equation incorporating this "new" property, enthalpy, is written

$$\frac{\dot{Q}_{in} + \dot{W}_{in} - \dot{Q}_{ex} - \dot{W}_{ex}}{\dot{m}} = \Delta_1^2 \left(h_{st} + \frac{C^2}{2g_c} + \frac{gz}{g_c} \right) \tag{2.6}$$

where the Δ operator implies that the difference of the quantity in parentheses should be taken from outlet station 2 to inlet station 1.

The enthalpy has the subscript st because, as mentioned previously, it is the value for so-called "static" or "stream" conditions. These are the conditions measured with instruments that are "static" or stationary with respect to the fluid. It is difficult to measure static properties (pressure and temperature) in high-speed flows. The stagnation properties are more easily measured by means of upstream-facing "Pitot" or "stagnation" probes, shielded to reduce heat transfer by radiation and conduction (figure 2.3). The steady-flow energy equation can be written for conditions (1) upstream and (2) within a stagnation probe, assuming that the velocity of flow within the probe is vanishingly small, that is $C_2 \to 0$:

$$\frac{\dot{Q}_{in} + \dot{W}_{in} - \dot{Q}_{ex} - \dot{W}_{ex}}{\dot{m}} = 0 = \left(h_{st,2} + 0 + \frac{gz_2}{g_c} \right) - \left(h_{st,1} + \frac{C_1^2}{2g_c} + \frac{gz_1}{g_c} \right) \tag{2.7}$$

If the streamline is horizontal $z_2 = z_1$. Let us give the stagnation enthalpy the suffix 0, h_0:

$$h_{st,2} = h_{0,2} \equiv h_{st,1} + \frac{C_1^2}{2g_c} \tag{2.8}$$

or in general, at any flow station

$$h_0 \equiv h_{st} + \frac{C^2}{2g_c} \tag{2.9}$$

We shall henceforth distinguish between stagnation and static conditions, using appropriate subscripts, except in those circumstances where all properties are of the same type. Where motion is slow, for instance, the static properties equal the stagnation properties. And in dealing with the relationships of a thermodynamic cycle, stagnation conditions should be used throughout.

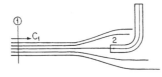

Figure 2.3. Isentropic flow into a Pitot tube

Using this definition, we can write the steady-flow energy equation as:

$$\frac{\dot{Q}_{in} + \dot{W}_{in} - \dot{Q}_{ex} - \dot{W}_{ex}}{\dot{m}} = \Delta_1^2(h_0) + \Delta_1^2\left(\frac{gz}{g_c}\right) \qquad (2.10)$$

In air and gas turbomachinery, changes of height have a negligible effect on the enthalpy change, and this general form of the steady-flow energy equation can be approximated by:

$$\frac{\dot{Q}_{in} + \dot{W}_{in} - \dot{Q}_{ex} - \dot{W}_{ex}}{\dot{m}} = \Delta_1^2(h_0) = h_{0,2} - h_{0,1} \qquad (2.11)$$

The steady-flow energy equation (SFEE) is usually simplified for most of the components of plants incorporating turbomachinery. This is because some are virtually thermally isolated, so that the heat transfer is negligible and the process undergone is adiabatic ($\dot{Q}_{in} = \dot{Q}_{ex} = \dot{Q} = 0$), for instance, compressors, pumps, and turbines. In other turbomachinery components there is considerable heat transfer but no work transfer ($\dot{W} = 0$), for instance, boilers, combustion chambers, and heat exchangers. In yet other turbomachinery components there is neither heat transfer nor work transfer, as in nozzles, throttles, diffusers, and ducts. Table 2.1 summarizes how the SFEE, equation 2.4, is used in components with gas, vapor, or liquid, and how the restricted SFEE, equation 2.11, is used in components with gas, vapor, or liquid in horizontal flow.

Use of the SFEE is especially convenient for steady-flow devices that have gases as their working fluids, because over a range of commonly encountered pressures, the enthalpy of gases at ranges well above their critical temperatures is a function of temperature only. Tabulated data, such as those in *Gas Tables* by Keenan and Kaye (1983),

Table 2.1. Applicability of the steady-flow energy equation

	$\dot{W}$	$\dot{Q}$	$\Delta_1^2 h_0$
Nozzles	0	0	0
Valves	0	0	0
Throttles	0	0	0
Ducts, pipes, diffusers	0	0	0
Boilers	0		$\dot{Q}_{in}/\dot{m}$
Heat exchangers	0		$\dot{Q}/\dot{m}$
Combustion chambers	0		$\dot{Q}_{in}/\dot{m}$
Compressors, pumps, etc.		0(?*)	$\dot{W}_{in}/\dot{m}$
Turbines		0(?*)	$\dot{W}_{ex}/\dot{m}$

* The queries are entered because in special cases there is substantial heat transfer in compressors and turbines. Compressors may incorporate intercoolers. Turbine blades, disks and duct walls may incorporate cooling circuits. Reheat combustors can be used between turbine stages or groups of stages.

give enthalpies for air and other common gases as functions only of temperature for "low pressures". Comparison with the only high-pressure data available when the tables were originally published (1948) confirms that the tabulated values are fully valid for pressures up to 1.4 MPa (200 psia) and that "the characteristics of the isentropic as given by table 1 [of the Gas Tables] are suitable for precise work even to pressures in the hundreds of pounds per square inch" (quoted from the appendix of the 1948 edition). For the examples and problems in this book we use the simple table of temperature-dependent properties of air (table A.1) and the correlations of the properties of air and of combustion products given in appendix A.

Energy equation for liquid flow

For machinery working with liquids, such as hydraulic turbines and pumps, the SFEE form of equation 2.11 is not useful because the enthalpy cannot easily be measured. It is more useful to return to the original form of the SFEE (equation 2.4), and to form different groups of the energy terms. Typically for machines working with liquids the stagnation pressure p_0 is derived from equation 2.9 for incompressible (constant specific volume v or density ρ) fluids only, and for lossless deceleration of the flow to stagnation conditions:

$$p_0 = p_{st} + \frac{C^2}{2vg_c} = p_{st} + \frac{\rho C^2}{2g_c} \tag{2.12}$$

Note that this equation relating static to stagnation pressure is valid for liquids only (see example on pump losses in section 2.4) and cannot be used for gases with acceptable accuracy except at low Mach numbers. For high Mach numbers the compressible-flow functions developed below (equation 2.61-2 developed in section 2.5) must be used (see example on compressor losses in section 2.4). Also, incompressible fluid means that there is only one value of density or specific volume, corresponding to static properties.

For liquid machines the total pressure p_T is defined as the stagnation pressure p_0 plus the elevation head:

$$p_T \equiv p_{st} + \frac{\rho C^2}{2g_c} + \frac{\rho g z}{g_c} = p_0 + \frac{\rho g z}{g_c} \tag{2.13}$$

The SFEE (equation 2.4) can be written as:

$$\frac{\dot{Q}_{in} + \dot{W}_{in} - \dot{Q}_{ex} - \dot{W}_{ex}}{\dot{m}} = \Delta_1^2 (u_{st}) + \Delta_1^2 \left(\frac{p_0}{\rho} \right) + \Delta_1^2 \left(\frac{g z}{g_c} \right) \tag{2.14}$$

For a given left-hand side the losses in a pump result in unwanted internal-energy increases (Δu_{st}) due to friction that reduce the total-pressure increase on the right-hand side. Similarly for a given total-pressure drop the losses in a water turbine result in unwanted internal-energy increases (Δu_{st}) due to friction that reduce the work output of the turbine. Because of the large specific heat capacity of most liquids, these internal-energy increases result in small but finite temperature increases in both pumps and liquid turbines.

It is usually more convenient to express the pressures as heads of liquid (from $p = \rho g H/g_c = (H/v)(g/g_c)$):

$$\frac{\dot{Q}_{in} + \dot{W}_{in} - \dot{Q}_{ex} - \dot{W}_{ex}}{\dot{m}} = \Delta_1^2\,(u_{st}) + \Delta_1^2\left(\frac{H_0 g}{g_c}\right) + \Delta_1^2\left(\frac{gz}{g_c}\right) \qquad (2.15)$$

Consider a hydraulic turbine working between two reservoirs, with the head measurements shown in figure 2.4. In applying this form of the SFEE, we can assume that the heat transfer will be negligible. But we cannot assume that the change of internal energy will be negligible. In fact, the change of temperature of the water will be extremely small (usually measured in hundredths of a degree Celsius) but this change of internal energy is a consequence of the frictional losses in the feedpipe and the turbine itself. In large hydraulic turbines, where the measurement of the mass flow and power output is difficult, efficiency measurements are often based on the precise measurement of these very small increases in water temperature.

Therefore the change of total head across a hydraulic turbine is not analogous to the change of stagnation enthalpy. The change of stagnation enthalpy (together with the change of potential energy, if applicable) represents the actual work output of the turbine rotor in all cases, however poor the efficiency of the machine. In the absence of friction the change of total head gives the ideal work output per unit mass.

This concept may be better understood by reducing the case to the limiting one of a turbine so lossy that it gives no work at all. It would in this case be precisely similar to a throttle valve. There would be no change of enthalpy across it, but there would be a large change in total head.

In the general case the work output of the hydraulic turbine is then

$$\frac{\dot{W}_{ex}}{\dot{m}} = \left[(H_{0,1} + z_1) - (H_{0,2} + z_2)\right]\frac{g}{g_c} - (u_{st,2} - u_{st,1}) \qquad (2.16)$$

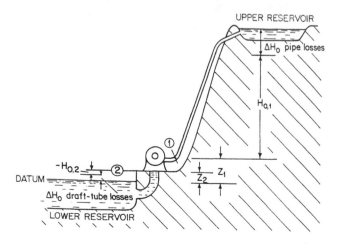

Figure 2.4. Hydraulic-turbine heads

or, in words: [actual specific work] = [ideal specific work] - [losses]. The units of specific work are in SI units (and British-U.S. units) Joules per kilogram (and British thermal units per pound-mass) or kilowatts per kilogram per second (and horsepower per pound-mass per second): kJ/kg (and Btu/lbm); or kW/(kg/s) (and hp/(lbm/s)).

2.2 Examples of the use of the SFEE

Gas-turbine expander

A gas-turbine expander is measured on test to be passing 84 kg/s of combustion products of known mean specific heat capacity at constant pressure 1,130 J/(kg · K). What is the shaft power output if the mean inlet temperature is 1,250 °C and the mean exhaust temperature is 550 °C, both measured with stagnation probes?

For gases, the changes in enthalpy between two temperatures can be calculated from the change in temperature multiplied by the mean specific heat at constant pressure between those two temperatures (see equation 2.44 below). Therefore

$$\dot{W}_{ex} = \dot{m}\Delta_2^1 h_0 = \dot{m}\overline{C_p}\Delta_2^1 T_0$$

$$\dot{W}_{ex} = 84\frac{\text{kg}}{\text{s}} \times 1,130\frac{\text{J}}{\text{kg} \cdot \text{K}} \times (1,250 - 550)\text{K}$$

$$= 66.44 \times 10^6 \text{ J/s} = 66.44 \text{ MW}$$

(degrees Kelvin [K] = degrees Celsius [°C] for temperature differences).

Air compressor

We are giving two examples of the use of enthalpy tables in British-U.S. units. Most other examples, and the property data in appendix A, tables A.1 and A.2, are given in SI units.

What will be the outlet stagnation temperature of an air compressor taking in 25 lbm/s of atmospheric air at 72.3 °F and absorbing 1, 950 hp?

From the 1948 version of the Gas Tables, table 1, the inlet stagnation enthalpy is 127.14 Btu/lbm. The work input per pound is

$$1,950\text{hp} \times \frac{550 \text{ ft} \cdot \text{lbf}}{\text{s}} \cdot hp \times \frac{\text{Btu}}{778 \text{ ft} \cdot \text{lbf}} \times \frac{s}{25 \text{ lbm}} = 55.14\frac{\text{Btu}}{\text{lbm}}$$

Then the outlet stagnation enthalpy is (127.14 + 55.14) = 182.28 Btu/lbm. This corresponds in table 1 of the Gas Tables to an outlet stagnation temperature of 760.8 °R = 301.1 °F.

In these two examples of the application of the SFEE, we are unconcerned with the efficiency of the internal processes, except that we should know the destination of the bearing, seal, and windage (disk friction, etc.) losses. If the power is measured at the machine shaft, and if the machine is sufficiently well insulated so that the friction losses go to increasing the enthalpy of the working fluid, the use of the SFEE is correct. It is also correct if the power is that at the rotor blading and if the friction losses are ducted away in oil drains and ventilation ducts, external to the working fluid. When none of these situations applies, a correction must be made for the additional energy transfer across the control-volume boundaries.

Intercooler

Find the energy loss in the air passing into an air-compressor intercooler at 50 psia, 650 °R, and leaving at 550 °R. There is a stagnation-pressure loss in the air of 10 psi. The air mass flow is 10 lbm/s.

Application of the SFEE, equation 2.11, gives:

$$\dot{Q}_{ex} = \dot{m}\left(h_{0,1} - h_{0,2}\right)$$

The mass flow rate is $\dot{m} = 10$ lbm/s, and from the Gas Tables $h_{0,1} = 155.50$ Btu/lbm and $h_{0,2} = 131.46$ Btu/lbm. Therefore the energy loss, or simply the heat transferred out, $\dot{Q}_{ex}$ is 240.4 Btu/s.

The pressure losses do not enter the energy-loss calculation. The pressure drop does not constitute a loss of energy. It produces a drop in thermodynamic availability (meaning that the air would be able to give less work in expansion to ambient pressure through a thermodynamically ideal engine than if there were no pressure drop). Available work is a second-law concern, and the SFEE is strictly a first-law, conservation-of-energy, expression. Pressure losses, even to the extent of throttling of the flow, do not produce energy losses.

2.3 The second law of thermodynamics

The second law of thermodynamics is the second principal postulate upon which the science is based. It is known in many forms that are corollaries of one another. The most popular, perhaps, is, *Heat cannot pass from a cooler to a warmer body without the expenditure of work.*

Although this seems a straightforward statement, it requires an understanding of temperature. In the early days of modern science there was much confusion between temperature and heat. As we have not yet defined temperature, we shall start with what might seem a more abstruse statement, yet which in fact requires a smaller degree of intellectual faith: that interacting bodies can approach a state of equilibrium. We shall present a simplified and compressed treatment of the statistical development given in such texts as Reynolds' *Thermodynamics* (1968).

Matter consists of molecules and atoms, which are themselves composed of many elementary particles. The particles have various forms of energy, and continually exchange energy in quanta with interacting particles. In gases, these interactions occur principally through collisions.

Consider an assembly of particles, each of which can exist in any of several energy states. Suppose that somehow (the physicist James Clerk Maxwell conceived of a "demon" that could manipulate molecules) one could force them all into their mean energy states and then release them to interact. Rapidly the assembly would progress from a state of complete order (uniformity) to a state of maximum disorder. The second law states simply that this process from order to disorder, from disequilibrium to equilibrium, occurs. (As a simple analogy, consider an assembly of dice, each with faces of six different colors. At the start, all show the same color uppermost. Then the dice are thrown, one at a time, in a random sequence. The state of the assembly—the "macro-state"—would proceed from order to disorder, from maximum disequilibrium to maximum equilibrium.)

We could conceptually quantify the approach to equilibrium by calculating the number of ways the particles could be arranged among the "microstates" to give the "macrostate" of the assembly. Let us call this number Ω. When the particles are all at the identical (mean-energy) microstate, there is only one way the particles can be arranged, since they are all identical, and that number is 1. As equilibrium is approached, the number becomes extremely large, even for a very small collection of particles (figure 2.5).

If two collections or assemblies of particles, A and B, are allowed to interact, the new number of ways the combined macrostate C could be arranged is the product of the number of ways A and B could individually be arranged:

$$\Omega_C = \Omega_A \times \Omega_B \tag{2.17}$$

It is more appropriate that this measure of disorder of the combined system should be the sum, rather than the product, of the "disorders" of the constituents, and that the numbers be smaller than they would otherwise be for realistic collections of molecules. We therefore define an "extensive" property (one that is proportional to the mass of the system) S, which we call entropy, for reasons which will be more obvious later, and which is proportional to the logarithm of Ω:

$$S \equiv K \ln\Omega \tag{2.18}$$

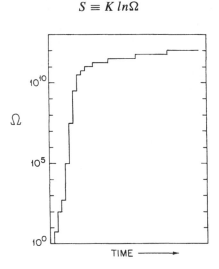

Figure 2.5. Approach to equilibrium

Therefore

$$K \, ln\Omega_C = K \, ln\Omega_A + K \, ln\Omega_B \qquad (2.19)$$

and

$$S_C = S_A + S_B \qquad (2.20)$$

It is more rigorous to define entropy as the time average of $K \, ln\Omega$. This overcomes to some extent the major objection to this model, which is that it could allow the entropy of an isolated assembly to decrease occasionally even though the general direction of the approach to equilibrium is for the entropy to increase. But to decrease through even one interaction would contravene the second law, which can be restated: *The entropy of an isolated assembly must increase or in the limit remain constant.*

Derivation of Gibbs' equation

We use the model just described to consider two bodies, or assemblies of particles, brought into thermal contact and allowed to reach equilibrium by exchange of heat (figure 2.6). They are isolated from the surroundings, and they are not allowed to perform work upon each other. We define (or choose) these assemblies to be "simple" substances, which by definition means that the state is fixed by any two independent, intensive properties. (Thus dry air at normal temperatures and pressures can be modelled as a simple substance, while an ionized gas or a mixture of reacting substances is not. Gibbs' equation requires additional terms for increased degrees of freedom.)

Because entropy is a property the entropy of each assembly A and B and of the combined assembly C, it can be stated as a function of any two other properties. It is convenient to use the internal energy, u, and the specific volume, v. We shall deal with quasi-stationary states and omit subscripts differentiating "static" from "stagnation" because (at negligible velocity) they will be identical:

$$S = S(u, v) \qquad (2.21)$$

Therefore

$$dS = \left(\frac{\partial S}{\partial u}\right)_v du + \left(\frac{\partial S}{\partial v}\right)_u dv \qquad (2.22)$$

Also

$$S_C = S_A + S_B \qquad (2.23)$$
$$dS_C = dS_A + dS_B \qquad (2.24)$$
$$= \left(\frac{\partial S_A}{\partial u_A}\right)_v du_A + \left(\frac{\partial S_A}{\partial v_A}\right)_u dv_A + \left(\frac{\partial S_B}{\partial u_B}\right)_v du_B + \left(\frac{\partial S_B}{\partial v_B}\right)_u dv_B \qquad (2.25)$$

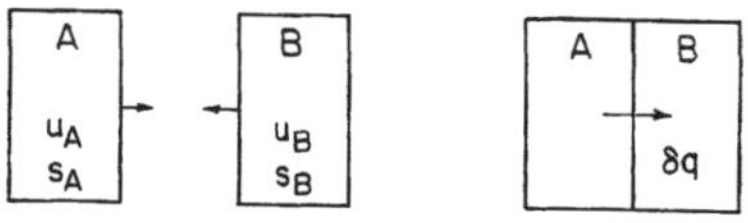

Figure 2.6. A thermal interaction to equilibrium

At equilibrium, $dS_C = 0$. Because no work interaction has been allowed, $dv_A = dv_B = 0$. A and B will have increased and decreased their internal energies by the amount of heat transferred δq, so that

$$du_A = -du_B = \delta q$$

Equation 2.25 then simplifies to

$$\left(\frac{\partial s_A}{\partial u_A}\right)_v = \left(\frac{\partial s_B}{\partial u_B}\right)_v \qquad (2.26)$$

where we have used lower-case s for "specific entropy", or entropy per unit mass.

This, then, is the condition for thermal equilibrium. We can define this condition as one in which the temperatures of the two assemblies are equal. We could in fact define the quantities in equation 2.26 as thermodynamic temperature. The result, however, would be the inverse of what we commonly call temperature. Therefore we make the following definition:

$$T \equiv \frac{1}{(\partial s/\partial u)_v} = \left(\frac{\partial u}{\partial s}\right)_v \qquad (2.27)$$

We now incorporate this definition into equation 2.26 and apply it to a second interaction between the assemblies A and B in which we allow the common boundary to move by an increase in the volume of, say, A and an equal decrease in the volume of B, while maintaining thermal equilibrium, until pressure equilibrium has been attained:

$$dS_C = \frac{1}{T_A}du_A + \left(\frac{\partial S_A}{\partial v_A}\right)_u dv_A + \frac{1}{T_B}du_B + \left(\frac{\partial S_B}{\partial v_B}\right)_u dv_B \qquad (2.28)$$

At equilibrium, by arguments similar to those already made,

$$\begin{aligned} T_A &= T_B \\ du_A &= -du_B \\ dv_A &= -dv_B \\ dS_C &= 0 \end{aligned}$$

Then equation 2.28 reduces to

$$\left(\frac{\partial S_A}{\partial v_A}\right)_u = \left(\frac{\partial S_B}{\partial v_B}\right)_u \qquad (2.29)$$

This condition applies for equality of pressure, and of temperature, for the two assemblies. The quantities in equation 2.29 have the dimensions of pressure divided by temperature and are given that definition:

$$\frac{p}{T} \equiv \left(\frac{\partial S}{\partial v}\right)_u \qquad (2.30)$$

Incorporating this definition, equation 2.28 becomes the Gibbs' equation for a simple substance in the absence of energy storage due to motion, gravity, electricity, magnetism, and capillarity:

$$T\,ds = du + p\,dv \qquad (2.31)$$

We shall refer to these restrictions as "simple circumstances".

Comparison of Gibbs' equation with the first law

Gibbs' equation has some similarities with the first law written for the same circumstances as equation 2.31 and is often confused for it. Gibbs' equation:

$$T\,ds = du + p\,dv$$

First law:

$$\delta q = du + \delta w$$

Gibbs' equation applies to any change, ideal or not: its only restriction is that it must apply to a simple substance. It is composed entirely of properties or changes in property values.

The first-law equation applies to a defined collection of material but is otherwise generally applicable. It includes only one property, du. The quantities δq and δw are dependent on the type of process undergone.

If one considers a thermodynamically reversible process in simple circumstances, one can write for δw the work of a reversible process, equal to $p\,dv$. Then, by comparison with Gibbs' equation, δq for a reversible process is $T\,ds$, and the equations are identical. This discussion underscores the essential difference in application of the first and second laws. The first law makes no condition about the type of process undergone, only about the end states. The second law is concerned about the process, and allows the end state to be predicted, given the initial state, or vice versa.

We examine process descriptions in the following derivations from Gibbs' equation.

The perfect- and semi-perfect-gas models

By definition a perfect gas[2] has constant values of specific heat capacities at constant volume and constant pressure, C_v and C_p respectively. By definition a semi-perfect gas has specific heat capacities that are functions only of temperature, $C_p = C_p(T)$ and $C_v = C_v(T)$. Both perfect and semi-perfect gases by definition obey the equation of

---

[2]Some thermodynamics texts define a perfect gas as one with constant specific-heat capacities, and an ideal gas as one with heat capacities as functions of temperature. Other thermodynamics texts define perfect and ideal gases exactly inversely. Both expressions, "perfect" and "ideal", imply perfection in some way. In this text we define as perfect gas one with constant specific-heat capacities, and as semi-perfect gas (a "less than perfect" gas) one with specific-heat capacities that are functions of temperature. Equation $C_p - C_v = R$ and definition $\gamma \equiv C_p/C_v$ apply to both perfect and semi-perfect gases.

state:

$$pV = n\Re T = \frac{m}{M_w}\Re T = mRT$$

$$pv = RT \qquad (2.32)$$

$$p = \rho RT$$

where n is the number of *kmoles* of the substance, $\Re$ is the "universal gas constant", M_w is the molecular weight of the gas, and $R = \Re/M_w$ is the "gas constant".

The universal gas constant is:

$$\Re = 8,313.219 \text{ J/kmole/K} \qquad (2.33)$$
$$= 1.98587 \text{ Btu/lbmole/}^\circ\text{R}$$
$$= 1,545.32 \text{ ft.lbf/lbmole/}^\circ\text{R}$$

The molecular weight for dry air is $M_w = 28.970$ kg/kmole. Therefore the gas constant for dry air is

$$R = 286.96 \text{ J/kg/K} \qquad (2.34)$$
$$= 0.068549 \text{ Btu/lbm/}^\circ\text{R}$$
$$= 53.32 \text{ ft.lbf/lbm/}^\circ\text{R}$$

The above equation of state dictates that the specific internal energy and specific enthalpy of perfect and semi-perfect gases also obey the following relations:

$$dh = C_p dT \qquad (2.35)$$
$$du = C_v dT \qquad (2.36)$$

where the ratio of specific heats, γ, is defined by:

$$\gamma \equiv \frac{C_p}{C_v} \qquad (2.37)$$

The definition of enthalpy $h \equiv u + pv$ implies $dh = du + pdv + vdp$, and the equation of state $pv = RT$ implies $pdv + vdp = RdT$. Gibbs' equation then leads to:

$$du + pdv = Tds = dh - vdp$$
$$C_v dT + RdT = C_p dT$$
$$C_v + R = C_p \qquad (2.38)$$

Perfect and semi-perfect gases have a value of $R \equiv \Re/M_w$. For perfect gases R and one other value of either C_p, C_v, or γ specify the remaining two using $C_v = C_p - R$ and $\gamma \equiv C_p/C_v$. For semi-perfect gases $C_p(T)$ is evaluated from temperature, and R and $C_p(T)$ are used to evaluate $C_v(T) = C_p(T) - R$ and $\gamma \equiv C_p(T)/C_v(T)$.

Equations 2.31 and 2.33, and the the relation for the velocity of propagation of a small pressure wave (a sound wave), a, through a fluid, are used (Keenan, 1970, pp. 96–105) to derive:

$$a^2 = -g_c v^2 \left(\frac{\partial p}{\partial v}\right)_s \qquad (2.39)$$

For a perfect gas $vdp = RdT + C_v dT - Tds$, so that

$$v\left(\frac{\partial p}{\partial v}\right)_s = -p\frac{C_p}{C_v} = -\gamma\frac{RT}{v}$$

$$a_{st} = \sqrt{g_c \gamma R T_{st}} = \sqrt{\frac{g_c C_p T_{st}}{(C_p/R) - 1}} \qquad (2.40)$$

where we have noted that the appropriate temperature is T_{st} for the actual speed of sound a_{st}.

Models for semi-perfect gases usually give mathematical expressions for C_p as a function of temperature ($C_p(T)$). For example such expressions for air and products of combustion modeled as semi-perfect gases are given in Appendix A. All of the above expressions are valid for both perfect and semi-perfect gases. For perfect gases one can evaluate changes between two states as:

$$u_2 - u_1 = C_v(T_2 - T_1) \qquad (2.41)$$

$$h_2 - h_1 = C_p(T_2 - T_1) \qquad (2.42)$$

For semi-perfect gases the expressions are modified to:

$$u_2 - u_1 = \int_{T_1}^{T_2} C_v dT \approx \overline{C_v}(T_2 - T_1) \qquad (2.43)$$

$$h_2 - h_1 = \int_{T_1}^{T_2} C_p dT \approx \overline{C_p}(T_2 - T_1) \qquad (2.44)$$

where the integrals are exact. The expressions with $\overline{C_v}$ and $\overline{C_p}$ can also be exact with appropriate choice of the values of the mean specific heats. However, it is convenient to calculate the mean specific heat as the value at the mean temperature, given by $\overline{C_p} = C_p((T_1 + T_2)/2)$ and $\overline{C_v} = \overline{C_p} - R$, which results in a small error, hence the approximations. In the remainder of the text we use this approximation: whenever C_p is used for semi-perfect gases it implies the value of $\overline{C_p}$ evaluated at the average of the two end-temperatures of the process.

Perfect-gas T-s and h-s diagrams

The enthalpy-entropy diagram is the most-used vehicle for illustrating and analyzing turbomachinery. One reason for this is that the high-efficiency streamline flow and the short fluid residence times reduce heat transfer to the level at which it can usually be neglected. The work transfers are then given by the change of enthalpy from the steady-flow energy equation (equation 2.4).

The ideal adiabatic process is isentropic, as can be seen from Gibbs' equation 2.31 and the first law written for an ideal, reversible process, as described earlier. Then, by definition, $Q = T ds = 0$ for a reversible adiabatic process. Therefore $ds = 0$.

In a perfect gas, enthalpy changes are proportional to temperature changes (equation 2.44), so that a temperature-entropy diagram may be used interchangeably with an enthalpy-entropy (Mollier) diagram with only a scale change for the temperature axis.

Constant-pressure lines on the T-s diagram

The shape of the constant-pressure (and constant volume) lines on a temperature-entropy diagram can be found from the definition of enthalpy $h \equiv u + pv$, the perfect-gas law (equation 2.33) and Gibbs' equation:

$$dh - vdp = Tds = du + pdv$$
$$C_p dT - vdp = Tds = C_v dT + pdv \tag{2.45}$$

Therefore the left- and right-hand sides of the above equation respectively lead to:

$$\left(\frac{\partial T}{\partial s}\right)_p = \frac{T}{C_p} \tag{2.46}$$

$$\left(\frac{\partial T}{\partial s}\right)_v = \frac{T}{C_v} \tag{2.47}$$

Equation 2.46 states that the slope of any constant-pressure line (isobar) is a function only of temperature. (C_p for a perfect gas is constant. For a real gas it is a function only of temperature for all normal pressures, table A.1.) The constant-pressure lines on a $T - s$ diagram are therefore identical, have an increasing slope with temperature, and are displaced horizontally from one another (figure 2.7). They do not "diverge", as is often stated; the $T - s$ diagram of figure 2.8 (used in some textbooks) is wrong and misleading. Similarly (equation 2.47) the slope of the constant-volume lines (isochors) is a function only of temperature, but it is higher than that of the constant pressure lines (because $C_p > C_v$).

The apparent virtue of assuming diverging rather than parallel constant-pressure lines is that the temperature drop in a high-temperature expansion process between two pressures can be larger than the temperature (and therefore the enthalpy) rise in the compression process between the same pressures. This inexactitude is not necessary, as will be seen.

Isentropic work for compression and expansion processes

We wish to find the temperature rise for an isentropic process over a small pressure rise. We use the same relations as before:

$$Tds = C_p dT - vdp$$
$$\left(\frac{\partial T}{\partial p}\right)_s = \frac{v}{C_p} = \frac{RT}{pC_p} \tag{2.48}$$

The specific work done in a steady-flow compression or expansion process is dh, which for a perfect gas is $C_p dT$. The change in pressure δp can be expressed as a normalized pressure change $\delta p / p$. Then

$$\left(\frac{c_p \partial T}{\partial p/p}\right)_s = RT \tag{2.49}$$

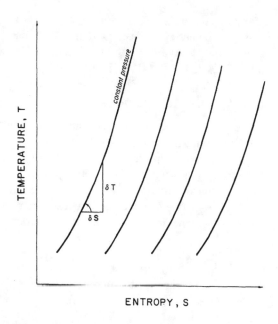

Figure 2.7. Slope of constant pressure lines on the $T - s$ diagram

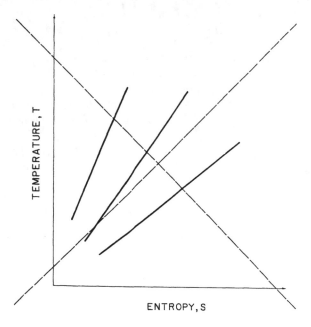

Figure 2.8. Incorrect representation of constant-pressure lines

The work required for an isentropic process between two pressure levels is therefore proportional to the absolute temperature. An isentropic turbine expanding from $1,500$ K $(2,700\ °\text{R},\ 2,240\ °\text{F})$ would produce five times the work required by an isentropic compressor compressing from 300 K $(540\ °\text{R},\ 80\ °\text{F})$ between the same pressure levels.

The second law derived for flow processes

The above statement of the second law "entropy of an isolated system must increase or remain constant" applied to bulk flow systems such as the one shown in figure 2.2 is expressed as: [change of entropy within the control volume] = [entropy coming in] - [entropy going out] + [entropy generated by irreversibility within the control volume]. The entropy flows in and out can be from flow through ports or from heat transfers at corresponding temperatures. Losses due to imperfect processes (e.g., friction) within the control volume appear as entropy generated by irreversibility within the control volume. Therefore:

$$\frac{dS_A}{dt} = \int \frac{dq_{in}}{T_{in}dt} - \int \frac{dq_{ex}}{T_{ex}dt} + \frac{dm_1}{dt}s_1 - \frac{dm_2}{dt}s_2 + \frac{dS_{ir}}{dt} \qquad (2.50)$$

where T_{in} and T_{ex} are the local temperatures at which heat transfers dq_{in} and dq_{ex} occur, and dS_{ir} is the entropy generated by irreversibility inside the control volume due to losses.

For steady-flow and quasi-steady flow the corresponding "steady-flow entropy equation" (SFSE) is:

$$\dot{m}s_2 + \int_T \frac{\dot{Q}_{ex}}{T_{ex}} = \dot{m}s_1 + \int_T \frac{\dot{Q}_{in}}{T_{in}} + \dot{S}_{ir} \qquad (2.51)$$

which, in words, states that the entropy transferred out of the control volume via bulk flow through the ports or heat transfers is equal to the entropy transferred into the control volume via bulk flow through the ports or heat transfers plus the entropy generated within the control volume due to irreversibilities (losses).

The relationship between stagnation and static condition at any flow station is by definition isentropic. Therefore changes in entropy $\dot{m}(s_2 - s_1)$ may be evaluated: either by using stagnation properties at both 1 and 2; or by using static properties at both 1 and 2. For simple substances $s_2 - s_1$ may be evaluated from tables. The entropy change for an incompressible substance with constant density ρ and specific heat capacity C is given by:

$$s_2 - s_1 = C\,ln\frac{T_2}{T_1} \qquad (2.52)$$

Gibbs' equation 2.31 and the equation of state for perfect and semi-perfect gases lead to:

$$
\begin{aligned}
ds &= \frac{dh - vdp}{T} = C_p\frac{dT}{T} - R\frac{dp}{p} \\[2mm]
ds &= \frac{du + pdv}{T} = C_v\frac{dT}{T} + R\frac{dv}{v} \\[2mm]
ds &= \frac{dh - vdp}{T} = C_p\frac{dv}{v} + C_v\frac{dp}{p}
\end{aligned}
\qquad (2.53)
$$

Therefore for a perfect gas the entropy change between any two states 1 and 2 is given by any one of:

$$
\begin{aligned}
s_2 - s_1 &= C_p \ln\frac{T_2}{T_1} - R\ln\frac{p_2}{p_1} \\
s_2 - s_1 &= C_v \ln\frac{T_2}{T_1} + R\ln\frac{v_2}{v_1} \\
s_2 - s_1 &= C_p \ln\frac{v_2}{v_1} + C_v \ln\frac{p_2}{p_1}
\end{aligned}
\tag{2.54}
$$

For a semi-perfect gas one may use equations 2.53 and integrate, or approximate the entropy change with equations 2.54 with a suitably averaged value of C_p as described above. (C_v is given by $C_v = C_p - R$.)

If two states 1 and 2 are connected via an isentropic process, or if the flow between two flow stations 1 and 2 is isentropic, then $s_2 - s_1 = 0$ and equations 2.54 lead to:

$$
\begin{aligned}
\frac{T_2}{T_1} &= \left(\frac{p_2}{p_1}\right)^{\frac{R}{C_p}} - \left(\frac{\rho_2}{\rho_1}\right)^{\frac{1}{[(C_p/R)-1]}} = \left(\frac{v_1}{v_2}\right)^{\frac{1}{[(C_p/R)-1]}} \\
\frac{T_2}{T_1} &= \left(\frac{p_2}{p_1}\right)^{\frac{R}{C_p}} = \left(\frac{\rho_2}{\rho_1}\right)^{\frac{R}{C_v}} = \left(\frac{v_1}{v_2}\right)^{\frac{R}{C_v}} \\
\frac{T_2}{T_1} &= \left(\frac{p_2}{p_1}\right)^{\frac{\gamma-1}{\gamma}} = \left(\frac{\rho_2}{\rho_1}\right)^{\gamma-1} = \left(\frac{v_1}{v_2}\right)^{\gamma-1}
\end{aligned}
\tag{2.55}
$$

Since the relationship between stagnation and static conditions at any flow station is by definition isentropic, the above equations also relate stagnation to static properties at any location for perfect gases. For semi-perfect gases the above equation can be used as an approximation, with average values of C_p and C_v evaluated as described above. Therefore for semi-perfect gases γ is also a function of temperature ($\gamma = C_p/C_v = C_p/(C_p - R)$).

2.4 Examples of the use of the SFEE and the SFSE

Example 1 Compressor losses

At the inlet of an air compressor the area-averaged velocity is $C_1 = 300$ m/s, the flow area is $A_1 = 0.08\ m^2$, and the environmental temperature and pressure are atmospheric at $T_{en} = 300\ K$ and $p_{en} = 100$ kPa. The compressor blading delivers 3.0 MW of work to the working fluid. Evaluate: (a) the stagnation and static pressures, temperatures, and densities at the inlet; (b) the mass flow rate at the inlet; (c) the maximum stagnation pressure $P_{0,2,mx}$ that can be reached at the outlet; and (d) the entropy generated by irreversibility, if the actual stagnation pressure reached at the outlet is $P_{0,2} = 0.97 \times P_{0,2,mx}$. Solve the problem in two ways: assuming that the air is a perfect gas with gas constant $R = 286.96$ J/kg/K and specific heat capacity at constant pressure $C_p = 1010$ J/kg/K; and by modeling air as a semi-perfect gas using the data in Appendix A.

The SFEE is used to evaluate conditions at various flow stations, and then the SFSE is used to evaluate the entropy increase. The numerical answers for this example appear underlined, in parentheses, and in square brackets, representing respectively: perfect gas, first iteration for semi-perfect gas between temperatures at the end states of the process, and converged solutions after three to eight iterations (using progressively updated values of $\overline{C_p}$) for semi-perfect gas.

We consider the air at two flow stations of the same elevation: far away from the inlet of the compressor, denoted by subscript en, where the air is at rest, and static and stagnation conditions are identical; and at station 1 at the inlet of the compressor, where the air has acquired velocity C_1 and the static and stagnation conditions have different values. Since there is no mechanism for work or heat to be put into or taken out of the air stream from station en to station 1, the SFEE (equation 2.10 or 2.11) applied between these two stations gives:

$$\frac{\dot{Q}_{in} + \dot{W}_{in} - \dot{Q}_{ex} - \dot{W}_{ex}}{\dot{m}} = 0 = \left(h_{0,1} - h_{0,en}\right)$$

$$C_p(T_{0,1} - T_{en}) = 0 \qquad \Rightarrow \qquad T_{0,1} = T_{en} = 300 \text{ K}$$

The friction associated with accelerating the air particles from zero velocity at station en to velocity C_1 at station 1 is negligible. There is no mechanism to increase or decrease the stagnation pressure from station 1 to station en:

$$p_{0,1} = p_{en} = 100 \text{ kPa}$$

Therefore

$$\rho_{0,1} = \frac{p_{0,1}}{RT_{0,1}} = 1.1616 \text{ kg/m}^3$$

At compressor inlet:

$$h_{0,1} = h_{st,1} + \frac{C_1^2}{2} \qquad \Rightarrow \qquad T_{0,1} - T_{st,1} = \frac{C_1^2}{2C_p}$$

$$T_{st,1} = \underline{255.45 \text{ K}} \quad (255.12 \text{ K}) \quad [255.12 \text{ K}]$$

$$C_p = \underline{1010 \text{ J/kg/K}} \quad (1002.78 \text{ J/kg/K}) \quad [1002.11 \text{ J/kg/K}]$$

The corresponding $p_{st,1}$ is found by applying equation 2.55 between stagnation and static conditions at station 1:

$$p_{st,1} = p_{0,1} \left(\frac{T_{st,1}}{T_{0,1}}\right)^{\frac{C_p}{R}}$$

$$p_{st,1} = \underline{56.787 \text{ kPa}} \quad [56.763 \text{ kPa}]$$

The static density at the inlet can now be found either from equation 2.55 or from the equation of state for the gas (equation 2.33):

$$\rho_{st,1} = \frac{p_{st,1}}{RT_{st,1}} = \underline{0.7747 \text{ kg/m}^3} \quad [0.7754 \text{ kg/m}^3]$$

The mass flow rate through the compressor, $\dot{m}_c$, is given by:

$$\dot{m}_c = \rho_{st,1} A_1 C_1 = \underline{18.5926 \text{ kg/s}} \quad [18.6102 \text{ kg/s}]$$

We assume that the compressor is adiabatic and receives work W_{in}. Application of the SFEE between compressor inlet and outlet, flow stations 1 and 2, gives:

$$\dot{W}_{in}/\dot{m}_c = \left(h_{0,2} - h_{0,1}\right) = C_p(T_{0,2} - T_{0,1})$$
$$\Rightarrow \quad T_{0,2} = \underline{459.76 \text{ K}} \quad (460.86 \text{ K})$$
$$C_p = \underline{1010 \text{ J/kg/K}} \quad (1002.78 \text{ J/kg/K}) \quad [1002.11 \text{ J/kg/K}]$$

The maximum stagnation pressure at compressor outlet is attained by an isentropic compressor, and it is found by applying equation 2.55 between stagnation conditions at compressor inlet and outlet:

$$\frac{p_{0,2,mx}}{p_{0,1}} = \left(\frac{T_{0,2}}{T_{0,1}}\right)^{\frac{C_p}{R}}$$
$$\Rightarrow \quad p_{0,2,mx} = \underline{449.33 \text{ kPa}} \; [447.83 \text{ kPa}]$$

If the actual stagnation pressure at compressor outlet is

$$p_{0,2} = 0.97 p_{0,2,mx} = \underline{435.85 \text{ kPa}} \; [434.40 \text{ kPa}],$$

then the entropy generated by irreversibility in the compressor is given by the top expression of equation 2.54 applied between stagnation conditions at compressor inlet and outlet:

$$\dot{S}_{ir}/\dot{m}_c = s_{0,2} - s_{0,1} = C_p \ln\frac{T_{0,2}}{T_{0,1}} - R \ln\frac{p_{0,2}}{p_{0,1}}$$
$$\Rightarrow \quad \dot{S}_{ir} = \underline{162.5 \text{ W/K}} \; [162.51 \text{ W/K}]$$

The stagnation conditions at the inlet are environmental, but the corresponding static temperature and static pressure are lower than the environmental levels. The stagnation temperature at the outlet of an adiabatic compressor is determined by the power input into the compressor and the mass-flow rate, and it is not affected by compressor inefficiencies. Given the amount of power input into the compressor, the maximum stagnation pressure at the outlet is achieved by an isentropic compressor. Compressor inefficiencies reduce the maximum attainable pressure at compressor outlet, and generate entropy by irreversibility inside the compressor control volume. This increases the entropy of the outgoing stream from $s_{0,1}$ to $s_{0,2}$ (the entropy generated by irreversibility inside the compressor leaves the control volume with the outgoing air stream).

Example 2   Pump losses

An agricultural pump is operating in an environmental temperature $T_{en} = 300$ K, pressure $p_{en} = 100$ kPa, and gravitational acceleration $g = 9.8$ m/s^2. The pump is pumping water from near the surface of a large lake (station 1) which is at atmospheric temperature. The pump outlet (station 2) is at the same elevation as that of the lake water. A pipe carries the water from station 2 to station 3, the beginning of an open conduit 10 m above the lake level. At the outlet of the pipe the effective cross-sectional area is $A_3 = 0.1$ m^2, and the corresponding mean velocity of the water is $C_3 = 2$ m/s. The pump inputs 187.16 kW to the water flow, and the total pressure at the pump outlet (station 2) is

$p_{T,2} = 417.9$ kPa. Model the water as an incompressible fluid with constant specific-heat capacity $C_w = 4,179$ J/(kg · K) and density $\rho = 1,000$ kg/m³. What is the static pressure at the pump inlet if the effective cross-sectional area is $A_1 = 0.15$ m²? Evaluate the water temperature at the pump outlet and pipe outlet (stations 2 and 3), and the stagnation and total pressure at the pipe outlet (station 3). Explain the reasons for any temperature changes in the water. What are the entropy increases from pump inlet to pump outlet to pipe outlet?

Continuity is used to evaluate mass flow, the SFEE is used between flow stations to evaluate properties at the stations, and then the SFSE is used to evaluate entropy increases. Continuity is used to evaluate the mass flow rate through the pump, $\dot{m}_{pu}$, and the flow velocity at the inlet, C_1:

$$\dot{m}_{pu} = \rho A_3 C_3 = 1,000 \frac{\text{kg}}{\text{m}^3} \times 0.1 \text{ m}^2 \times 2 \frac{\text{m}}{\text{s}} = 200 \frac{\text{kg}}{\text{s}}$$

$$C_1 = \frac{\dot{m}_{pu}}{\rho A_1} = \frac{200 \text{ kg/s}}{1,000 \text{ kg/m}^3 \times 0.15 \text{ m}^2} = 1.333 \text{ m/s}$$

In perfect (isentropic) incompressible-fluid flow the temperature of the working fluid does not change (equation 2.52) and there is no difference between the static, stagnation, and total temperatures. The work input (pumps) or output (turbines) is used to change the total head or total pressure of the working fluid, and the losses appear as (static) internal-energy increases; i.e. losses result in (static) temperature increases (equations 2.14, 2.15, and 2.16).

Let's assign the zero elevation level at the surface of the lake ($z_1 = 0$). Similarly to the example on compressor losses above, there is no mechanism to increase the stagnation pressure from environmental pressure at pump inlet, ie $p_{T,1} = p_{0,1} = p_{en} = 100$ kPa. Friction in accelerating the water from environmental conditions at rest to C_1 at the inlet is negligible; therefore the temperature at pump inlet is environmental, $T_1 = T_{st,1} = T_{en} = 300$ K. The static pressure at the inlet is:

$$p_{st,1} = p_{0,1} - \frac{\rho C_1^2}{2} = 99.11 \text{ kPa}$$

Since $z_2 - z_1 = 0$, the total and stagnation pressure at station 2 are equal:

$$p_{T,2} = p_{0,2} = 417.9 \text{ kPa}$$

At station 3 the static pressure is atmospheric, $p_{st,3} = 100$ kPa. The stagnation and total pressures, respectively, are:

$$p_{0,3} = p_{st,3} + \frac{\rho C_3^2}{2} = 102 \text{ kPa}$$

$$p_{T,3} = p_{0,3} + \frac{\rho g z_3}{g_c} = 200 \text{ kPa}$$

Assuming that the pump and pipe are adiabatic ($\dot{Q}_{in} = \dot{Q}_{ex} = 0$) the SFEE (equation 2.14) applied from pump inlet to pump outlet ($z_2 = z_1$) gives:

$$\frac{\dot{W}_{in}}{\dot{m}_{pu}} = \left(u_{st,2} - u_{st,1} \right) + \left(\frac{p_{0,2} - p_{0,1}}{\rho} \right) = C_w \left(T_2 - T_1 \right) + \left(\frac{p_{0,2} - p_{0,1}}{\rho} \right)$$

$$\Rightarrow \quad (T_2 - T_1) = 0.15 \text{ K} \qquad \Rightarrow \qquad T_2 = 300.15 \text{ K}$$

The SFEE (equation 2.14) applied from pump inlet to pipe outlet ($z_3 = z_1 + 10$ m) gives:

$$\frac{\dot{W}_{in}}{\dot{m}_{pu}} = \left(u_{st,3} - u_{st,1}\right) + \left(\frac{p_{0,3} - p_{0,1}}{\rho}\right) + \left(\frac{g(z_3 - z_1)}{g_c}\right)$$

$$= C_w\,(T_2 - T_1) + \left(\frac{p_{T,3} - p_{T,1}}{\rho}\right)$$

$$\Rightarrow \quad (T_3 - T_1) = 0.20 \text{ K} \qquad \Rightarrow \qquad T_3 = 300.20 \text{ K}$$

If the pump was frictionless, then the water temperatures at pump inlet and outlet would be equal. Similarly, if the pipe was frictionless, then the water temperatures at pipe inlet and outlet would be equal. In this case losses due to friction in the pump and pipe increase the internal energy from $u_{st,1}$ to $u_{st,2}$ to $u_{st,3}$ and the corresponding temperature from T_1 to T_2 to T_3, and reduce the total outlet pressure that can be achieved with the given pump work. The entropy generated by irreversibility in each segment of the process is evaluated using equations 2.51 and 2.52. Since the pump and pipe are adiabatic $\dot{Q}_{in} = \dot{Q}_{ex} = 0$.

$$\Delta_1^2(\dot{S}_{ir}) = \dot{m}_{pu}(s_2 - s_1) = \dot{m}_{pu}\,C_w\,ln\frac{T_2}{T_1} = 417.80 \text{ W/K}$$

$$\Delta_2^3(\dot{S}_{ir}) = \dot{m}_{pu}(s_3 - s_2) = \dot{m}_{pu}\,C_w\,ln\frac{T_3}{T_2} = 139.22 \text{ W/K}$$

2.5 Compressible-flow functions for a perfect gas

Many flow problems in the design of turbomachinery involve the specification of the stagnation pressure and temperature, p_0 and T_0, and either the absolute velocity C, or the mass-flow rate $\dot{m}$ and characteristic area A. If the Mach number can be found, all the actual, static, properties are immediately calculable.

We can apply equation 2.9, relating stagnation and static conditions at a point or along an adiabatic stream tube in a moving fluid, to a perfect gas, and then substitute from equation 2.40 to obtain:

$$h_0 - h_{st} = \frac{C^2}{2g_c}$$

$$C_p(T_0 - T_{st}) = \frac{C^2}{2g_c}$$

$$\frac{T_0}{T_{st}} = 1 + \frac{C^2}{2g_c C_p T_{st}} = 1 + \frac{C^2}{2\left(\frac{C_p}{R} - 1\right)a_{st}^2}$$

Substituting in the last equation the definition of Mach number,

$$M \equiv \frac{C}{a_{st}} \tag{2.56}$$

we obtain

$$\frac{T_0}{T_{st}} = 1 + \frac{M^2}{2\left(\frac{C_p}{R} - 1\right)} = 1 + \left(\frac{\gamma - 1}{2}\right) M^2 \qquad (2.57)$$

Equation 2.57 is the fundamental equation of one-dimensional compressible flow.
    Equation 2.55 coupled with equation 2.57 gives:

$$\frac{p_0}{p_{st}} = \left[1 + \frac{M^2}{2\left(\frac{C_p}{R} - 1\right)}\right]^{(C_p/R)} = \left[1 + \left(\frac{\gamma - 1}{2}\right) M^2\right]^{\frac{\gamma}{\gamma - 1}} \qquad (2.58)$$

$$\frac{\rho_0}{\rho_{st}} = \frac{v_{st}}{v_0} = \left[1 + \frac{M^2}{2\left(\frac{C_p}{R} - 1\right)}\right]^{[(C_p/R)-1]} = \left[1 + \left(\frac{\gamma - 1}{2}\right) M^2\right]^{\frac{1}{\gamma - 1}} \qquad (2.59)$$

Equation 2.57 has a significant difference from equations 2.58 and 2.59. Equation 2.57 is based on the first law only, and on adiabatic (no-heat-transfer), but not necessarily isentropic (not loss-less) flow. However, equations 2.58 and 2.59 are valid only for isentropic flows.

Continuity in terms of Mach number

The Mach number $M \equiv C/a_{st}$ can be introduced into the continuity equation $\dot{m} = \rho_{st} A C$:

$$\frac{\dot{m}}{A} = \rho_{st} a_{st} M$$

We substitute for $\rho = 1/v$ and for a, from equation 2.40. Then:

$$\frac{\dot{m}}{A} = \frac{p_{st}}{R T_{st}} M \sqrt{\frac{g_c C_p T_{st}}{\left(\frac{C_p}{R} - 1\right)}}$$

$$\dot{m} \frac{\sqrt{R T_{st}/g_c}}{A p_{st}} = M \left[1 - \frac{R}{C_p}\right]^{-\frac{1}{2}} \qquad (2.60)$$

We usually know the stagnation, rather than the static conditions, and we therefore convert this relation into a more useful form using equations 2.57 and 2.58:

$$\frac{\dot{m}\sqrt{R T_0/g_c}}{A p_0} = \frac{M}{\sqrt{1 - \frac{R}{C_p}}} \left[1 + \frac{M^2}{2\left(\frac{C_p}{R} - 1\right)}\right]^{-\left(\frac{C_p}{R} - \frac{1}{2}\right)} \qquad (2.61)$$

$$= M\sqrt{\gamma}\left[1+\left(\frac{\gamma-1}{2}\right)M^2\right]^{-\frac{\gamma+1}{2(\gamma-1)}}$$

The gas constant R can be found from the universal gas constant $\Re$ and the molecular weight M_w (values given by equations 2.33 and 2.34). A chart of equation 2.61 is shown in figure 2.9.

Substituting equation 2.57 into equation 2.61 gives:

$$\frac{C}{\sqrt{g_c R T_0}} = \sqrt{2\frac{C_p}{R}\left[1-\left(1+\frac{M^2}{2\left(\frac{C_p}{R}-1\right)}\right)^{-1}\right]} \tag{2.62}$$

$$= \sqrt{2\left(\frac{\gamma}{\gamma-1}\right)\left[1-\left(1+\frac{\gamma-1}{2}M^2\right)^{-1}\right]} \tag{2.63}$$

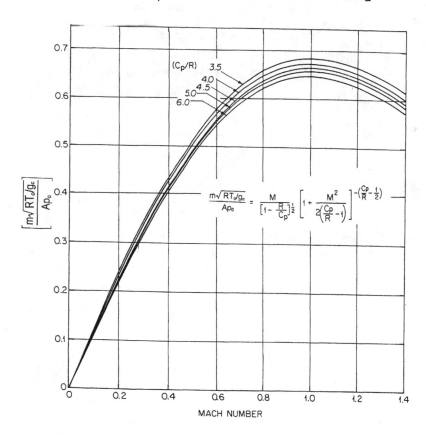

Figure 2.9. Universal flow-function plot

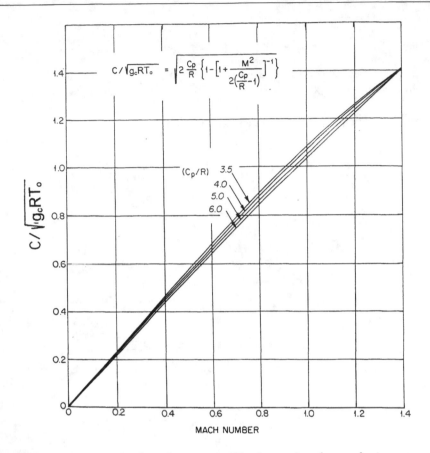

Figure 2.10. Velocity function versus Mach number for perfect gases

Equation 2.62 indicates that the left-hand side, a velocity function, is a function of M and C_p/R. A chart of equation 2.62 is shown in figure 2.10.

The use of the above equations and of figures 2.9 and 2.10 is illustrated in the following examples.

Example 3   Use of the compressible-flow functions and charts

Calculate the turbine-nozzle throat area for the following specifications:

$$
\begin{aligned}
T_0 &= 1500 \text{ K} \\
p_0 &= 10^6 \text{ N/m}^2 \\
\dot{m} &= 15 \text{ kg/s} \\
C_1 &= 650 \text{ m/s} \\
M_w &= 29.0 \\
R &= 8313.219/29.0 = 286.66 \text{ J/kg/K}
\end{aligned}
$$

where C_p/R can be taken as equal to the values in table A.1.
(i) Calculate $C_1/\sqrt{g_c R T_0} = 0.9912$.
From table A.1, C_p/R at 1500 K = 4.2173 and at 1000 K = 3.9741.
Let us start with a guess at $C_p/R = 4.0$
(ii) Read throat Mach number from figure 2.10 $M_1 \approx 0.911$.
(iii) Calculate T_0/T_{st} from equation 2.57:

$$\frac{T_0}{T_{st}} = 1 + \frac{0.911^2}{2(4-1)} = 1.1383$$

Therefore $T_{st,1} \approx 1318\ K$.
Interpolating in table A.1 gives $C_p/R = 4.182$.
(iv) A second iteration through steps (i) to (iii) gives: $M_1 = 0.921$; $T_{st,1} = 1323.6$ K; and $C_p/R = 4.188$.
(v) Enter figure 2.9 or equation 2.61 with M_1 and C_p/R.

$$\frac{\dot{m}\sqrt{R T_0/g_c}}{A p_0} = 0.666$$

Therefore $A = 0.01477\ m^2$.

Appropriate areas and velocities

All the relationships developed apply to the actual velocities of the streams relative to the observer, and the areas are those actually available. The static density must always be found using the actual (inclined) value of the velocity C.

In turbomachines, velocities are often in directions far from that parallel to the axis of the duct or passage containing the flow. If the angle the flow makes with the axial direction is α, then the available flow area is $A_{an} \cos \alpha$, where A_{an} is the annulus (normal) area.

Alternatively, one can take the axial component of velocity, $C_z = C \cos \alpha$, in combination with the annulus area A_{an} in the continuity equation.

Example 4   Calculation of flow area required for specified velocity

Find the annulus diameter required at the outlet of the last rotor-blade row in a high-pressure-ratio air compressor. The conditions specified are as follows:
Mass flow, $\dot{m} = 22$ kg/s.
Stagnation-pressure ratio $r \equiv p_{0,2}/p_{0,1} = 25.4$.
Inlet stagnation pressure, $p_{0,1} = 8.3 \times 10^4$ N/m^2.
Inlet stagnation temperature, $T_{0,1} = 283.2$ K.
Axial velocity at rotor outlet $C_{z,2} = 194$ m/s.
Mean flow direction relative to axial, $\alpha_{C,2} = 41°$.
Hub-shroud diameter ratio $\Lambda \equiv d_{hb,2}/d_{sh,2} = 0.85$.
Estimated polytropic efficiency, $\eta_{p,c} = 0.89$ in the expression (developed later in the chapter):

$$\frac{T_{0,2}}{T_{0,1}} = \left(\frac{p_{0,2}}{p_{0,1}}\right)^{\left[\left(\frac{R}{C_p}\right)\frac{1}{\eta_{p,c}}\right]}$$

We calculate first the stagnation conditions at rotor outlet. Then we calculate the absolute outlet velocity, and use figure 2.10 to find the Mach number. From this we calculate the static density and the flow area required.

Using the given relationship for polytropic efficiency, we find the stagnation rotor-outlet conditions:

$$\frac{T_{0,2}}{T_{0,1}} = \left(\frac{p_{0,2}}{p_{0,1}}\right)^{\left[\left(\frac{R}{C_p}\right)\frac{1}{\eta_{p,c}}\right]} = (25.4)^{\left(\frac{R}{C_p}\frac{1}{0.89}\right)}$$

We need to iterate to arrive at a reasonable value of (C_p/R). From table A.1 guess $C_p/R = 3.6$. Then $T_{0,2} = 777.2$ K, and the mean temperature across the compressor is 530 K. Interpolating between 500 K and 550 K in table A.1 gives $C_p/R = 3.601$, and $T_{0,2} = 777.0$ K.

The actual velocity, C_2, is found from the axial velocity, from $C_2 = C_{z,2}/\cos\alpha_{C,2} = 257.05$ m/s.

We can use these values to enter figure 2.10. However, here the appropriate value of (C_p/R) is that at the outlet static temperature. We will guess $(C_p/R) = 3.8$.

$$\frac{C_2}{\sqrt{g_c R T_{0,2}}} = \frac{257.05}{\sqrt{286.96 \times 777.0}} = 0.5444$$

The Mach number is approximately $M_2 = 0.477$.
The static temperature is calculated from equation 2.57: $T_{st,2} = 746.5$ K.
Interpolating in table A.1 we find that $(C_p/R) = 3.78$, which means that our guess was good enough.

The stagnation density is,

$$\rho_{0,2} = \frac{p_{0,2}}{R T_{0,2}} = \frac{25.4 \times 8.3 \times 10^4}{286.96 \times 777} = 9.455 \text{ kg/m}^3$$

The static density is calculated using equation 2.59: $\rho_{st,2} = 8.458$ kg/m^3.
The annulus area, A_{an}, can now be calculated using the axial component of the velocity and the continuity equation:

$$\begin{aligned}
A_{an} &= 22/(8.458 \times 194) = 0.0134 \ m^2 \\
&= \pi d_{sh,2}^2 (1 - \Lambda^2)/4 \\
d_{sh,2} &= 248 \text{ mm} \qquad d_{hb,2} = 211 \text{ mm}
\end{aligned}$$

Speed of sound in terms of stagnation conditions

Another useful compressible-flow relation is that giving the "true" speed of sound, a_{st}, in terms of the stagnation temperature, T_0, and the Mach number, M. It is obtained by combining equations 2.40 and 2.57:

$$a_{st} = \sqrt{\frac{g_c C_p T_{st}}{\frac{C_p}{R} - 1}} = \sqrt{\frac{g_c C_p T_0}{\frac{M^2}{2} + \frac{C_p}{R} - 1}} = \sqrt{\frac{g_c R T_0 \frac{\gamma}{\gamma - 1}}{\frac{M^2}{2} + \frac{1}{\gamma - 1}}} \qquad (2.64)$$

2.6 Turbomachine-efficiency definitions

The overall energy efficiencies of turbomachines compare the actual work transfer with that which would occur in an ideal process. In an energy-using machine, such as a pump or compressor, the work transfer in the ideal process is in the numerator. In an energy-producing machine, the ideal-process work transfer is in the denominator.

Efficiencies are seldom defined with sufficient precision. The statement that "we can deliver a compressor with an efficiency of 95 percent" is so undefined as to be almost useless for analysis. In fact, competition in machine efficiencies occasionally leads to intentional or unintentional failure to define which of many efficiencies is being used. To take an extreme example, an intercooled compressor may have at the same time, and at the same operating point, an isothermal efficiency of 75 percent and an isentropic efficiency of over 100 percent. Or a turbine may simultaneously have isentropic stagnation-to-static efficiencies of 90 percent and 80 percent, depending on what assumption or specification is made about the outlet plane.

General efficiency definitions

The efficiency of a pump η_{pu} or compressor η_c is defined as follows: "[power transfer in ideal process from an inlet stagnation pressure and temperature to a defined outlet stagnation pressure] divided by [the actual compressor or pump power]". The efficiency of an expander or turbine η_e is defined as the inverse of this ratio.

$$\eta_{pu} \equiv \frac{\dot{W}_{in,ie}}{\dot{W}_{in,ac}} \qquad \eta_c \equiv \frac{\dot{W}_{in,ie}}{\dot{W}_{in,ac}} \qquad \eta_e \equiv \frac{\dot{W}_{ex,ac}}{\dot{W}_{ex,ie}} \qquad (2.65)$$

These definitions require more precision, particularly in four aspects.

1. The ideal process must be identified. (The principal choices are isentropic, polytropic, and isothermal.)

2. The inlet and, more significantly, the outlet plane must be identified.

3. It should be stated whether the "defined outlet stagnation pressure" is the actual stagnation pressure or the "useful" stagnation pressure (which might in fact be the static pressure).

4. The actual work transfer may be just that in the machine blading (sometimes called the "internal" or "blading" work, and hence efficiency), or it may also include disk and seal friction and bearing losses; which of these is to be used should be specified.

Even in scientific papers it is common to identify only the ideal process. Let us discuss the implications of these aspects.

Ideal processes

The two principal ideal processes used are the isentropic for adiabatic processes and the isothermal for gas compression when intercooling is (or can be) employed. In addition, a variation of the isentropic efficiency is to take the limiting value at unity pressure ratio. This is termed the "small-stage" or "polytropic" efficiency, for reasons that will become obvious later. In machines using liquids, such as pumps and hydraulic turbines, the ideal isentropic process is also isothermal (equation 2.52). The ideal work transfer in these various cases is found as follows.

Isentropic process

Consider the SFEE for gas turbomachinery (equation 2.11) for an adiabatic process ($\dot{Q}_{in} = \dot{Q}_{ex} = 0$). If the machine is a compressor, then $\dot{W}_{ex} = 0$ and $\dot{W}_{in} > 0$. If the machine is a turbine, then $\dot{W}_{in} = 0$ and $\dot{W}_{ex} > 0$. In either case, the actual and isentropic processes result in:

$$\left(\frac{\dot{W}_{in} - \dot{W}_{ex}}{\dot{m}}\right)_{ac} = \Delta_{in}^{ex} h_0 = h_{0,ex} - h_{0,in} \tag{2.66}$$

$$\left(\frac{\dot{W}_{in} - \dot{W}_{ex}}{\dot{m}}\right)_{s} = \Delta_{in}^{ex} h_{0,s} = h_{0,ex,s}^{\otimes} - h_{0,in} \tag{2.67}$$

where $h_{0,ex,s}^{\otimes}$ is defined as the enthalpy after an isentropic process from $h_{0,in}$ and $p_{0,in}$ to $p_{0,ex}^{\otimes}$, the defined outlet pressure (to be discussed below). These are the general expressions.

In the special case of a perfect gas,

$$\left(\frac{\dot{W}_{in} - \dot{W}_{ex}}{\dot{m}}\right)_{ac} = h_{0,ex} - h_{0,in} = C_p(T_{0,ex} - T_{0,in})$$

$$\left(\frac{\dot{W}_{in} - \dot{W}_{ex}}{\dot{m}}\right)_{s} = h_{0,ex,s}^{\otimes} - h_{0,in} = (T_{0,ex,s}^{\otimes} - T_{0,in}) = C_p T_{0,in}\left(\frac{T_{0,ex,s}^{\otimes}}{T_{0,in}} - 1\right)$$

$$\left(\frac{\dot{W}_{in} - \dot{W}_{ex}}{\dot{m}}\right)_{s} = C_p T_{0,in}\left[\left(\frac{p_{0,ex}^{\otimes}}{p_{0,in}}\right)^{R/C_p} - 1\right] = C_p T_{0,in}\left(r^{R/C_p} - 1\right) \tag{2.68}$$

$$= \left(\frac{\gamma}{\gamma - 1}\right) R T_{0,in}\left\{r^{[(\gamma-1)/\gamma]} - 1\right\}$$

$$r \equiv \frac{p_{0,ex}^{\otimes}}{p_{0,in}}$$

where r is the pressure ratio, representing the compressor pressure ratio $r_c > 1$, or the turbine pressure ratio $r_e < 1$. Equation 2.68 illustrates why the authors are keen on the "(C_p/R)" rather than the "γ" form of the compressible flow and isentropic equations. This equation, and others like it, are usually given as

$$\left(\frac{\dot{W}_{in} - \dot{W}_{ex}}{\dot{m}}\right)_{s} = C_p T_{0,in}\left[r^{[(\gamma-1)/\gamma]} - 1\right]$$

leading users to find a value of C_p appropriate for temperature but to take γ as constant with temperature. This approach leads to errors and inconsistencies in precise calculations.

Substituting equations 2.66, 2.67, and 2.68 in equations 2.65, the "internal" isentropic efficiencies for a compressor and turbine, $\eta_{s,c}$ and $\eta_{s,e}$, respectively, are:

$$\eta_{s,c} = \frac{h_{0,ex,s}^{\otimes} - h_{0,in}}{h_{0,ex} - h_{0,in}} = \frac{T_{0,ex,s}^{\otimes} - T_{0,in}}{T_{0,ex} - T_{0,in}} = \frac{r_c^{R/C_p} - 1}{\dfrac{T_{0,ex}}{T_{0,in}} - 1} \qquad (2.69)$$

$$\eta_{s,e} = \frac{h_{0,in} - h_{0,ex}}{h_{0,in} - h_{0,ex,s}^{\otimes}} = \frac{T_{0,in} - T_{0,ex}}{T_{0,in} - T_{0,ex,s}^{\otimes}} = \frac{1 - \dfrac{T_{0,ex}}{T_{0,in}}}{1 - r_e^{R/C_p}} \qquad (2.70)$$

The above equations in terms of temperature differences and r are exact for perfect gases, and approximations (with appropriate values of $C_p = \overline{C_p}$) for semi-perfect gases.

In the special case of an adiabatic liquid machine with equal inlet and outlet elevations the SFEE (equation 2.4) reduces to:

$$\left(\frac{\dot{W}_{in} - \dot{W}_{ex}}{\dot{m}}\right)_s = \Delta_{in}^{ex} h_{0,s} = \Delta_{in}^{ex}(u + p_0 v)_s$$

If the liquid can be treated as incompressible (it is a reasonable approximation for water in most pressure ranges for pumps and water turbines), the specific volume, $v = 1/\rho$, is constant. For an isentropic process in an incompressible liquid (equation 2.52) $\Delta T = \Delta u = 0$ (hence the subscript it for isothermal as well as s for isentropic). Therefore:

$$\left(\frac{\dot{W}_{in} - \dot{W}_{ex}}{\dot{m}}\right)_s = \left(\frac{\dot{W}_{in} - \dot{W}_{ex}}{\dot{m}}\right)_{it} = v\Delta_{in}^{ex} p_{0,s} = \frac{p_{0,ex,s}^{\otimes} - p_{0,in}}{\rho} \qquad (2.71)$$

For very-high-pressure-ratio pumps where the liquid is compressible, the general expressions, equations 2.66 and 2.67, must be used.

Isothermal process

For incompressible liquids, equation 2.71 gives the isentropic and isothermal ideal power required to attain $p_{0,ex}^{\otimes}$.

For any simple substance, let $T = T_{0,in} = T_{0,ex}$ be the temperature of isothermal compression from inlet pressure $p_{0,in}$ to outlet pressure $p_{0,ex}^{\otimes}$. Assuming inlet and outlet are at the same elevation, and starting from the SFEE (equation 2.4),

$$\begin{aligned}
\left(\frac{\dot{W}_{in} - \dot{W}_{ex}}{\dot{m}}\right)_{it} &= \Delta_{in}^{ex} h_0 - \left(\frac{\dot{Q}_{in} - \dot{Q}_{ex}}{\dot{m}}\right) = \\
&= \Delta_{in}^{ex} h_0 - T\Delta_{in}^{ex} s_0 = (h_{0,ex} - h_{0,in}) - T(s_{0,ex} - s_{0,in}) \quad (2.72)
\end{aligned}$$

where we used $\dot{Q}_{in}/\dot{m} = T\Delta s_{st} = T\Delta s_0$ for the isothermal process. Equation 2.72 and property tables can be used to evaluate the entropy change from inlet to outlet for most substances.

The isothermal power required for perfect and semi-perfect gases is obtained by substituting in the general equation 2.72 $h_{0,ex} - h_{0,in} = C_p(T_{0,ex} - T_{0,in}) = 0$ and Δs_0 from the first of equations 2.54, where $T = T_{0,in} = T_{0,ex}$:

$$\left(\frac{\dot{W}_{in} - \dot{W}_{ex}}{\dot{m}}\right)_{it} = TR\ln\frac{p_{0,ex}^{\otimes}}{p_{0,in}} \tag{2.73}$$

Actual work transfer

The actual work transfer in the process is given by the SFEE:

$$\left(\frac{\dot{W}_{in} - \dot{W}_{ex}}{\dot{m}}\right) + \left(\frac{\dot{Q}_{in} - \dot{Q}_{ex}}{\dot{m}}\right) = \Delta_{in}^{ex}h_0 = (h_{0,ex} - h_{0,in}) \tag{2.74}$$

Turbomachines are generally adiabatic, in which case $\dot{Q}_{in} = \dot{Q}_{ex} = 0$. In some cases, however, heat transfer is significant. The two principal examples are intercooled compressors and cooled turbine expanders. Another case, a rare one, is the multistage turbine with reheat combustion or heat exchange between one or more stages. In these cases the heat transfer, or its equivalent in the case of combustion of fuel, must be included.

Equation 2.74 is the general and exact expression for the actual work transfer in a process not involving significant changes in fluid height. In other words, this will be the work transfer between the fluid and the machine's rotor(s), stator(s), and ducts. In many uses of energy efficiencies, this "process" work transfer is employed in the evaluation of efficiency, and it gives a measure of the quality of the aerodynamic and thermodynamic design.

The purchaser or user of a turbomachine is less concerned with aerodynamics than with actual energy cost, however. The purchaser wants to know how much work input is required or is produced at the shaft. Therefore, "external" energy losses, those external to the working fluid, must be added to or subtracted from the process energy.

External energy losses result from friction in bearings, labyrinths, and possibly on the faces of disks ("windage"). We say "possibly" because, if the fluid causing the disk friction is bled from the working fluid and passes back into the working fluid, as is frequently the case, the energy losses appear as an increase in enthalpy (and entropy) in the working fluid, and therefore are measured as an increase in the outlet enthalpy and a loss in work. The same could be true of working-fluid-lubricated bearings, for example gas bearings, that exhaust into the machine discharge. The same is occasionally true of labyrinth fluid.

Therefore it is not possible to identify the external energy losses in a general way. We shall simply term them Σ[external losses] for identification in each individual case (see table 2.2). The questions of using appropriate reference planes and defining the useful outlet pressure will be discussed below.

Table 2.2. Types of turbomachine efficiencies

Efficiency	Compressor or pump	Expander or turbine
Isentropic, general fluid, η_s	$\dfrac{h^{\otimes}_{0,ex,s} - h_{0,in}}{h_{0,ex} - h_{0,in} + \dot{W}_{g3}/\dot{m}}$	$\dfrac{h_{0,in} - h_{0,ex} - \dot{W}_{g3}/\dot{m}}{h_{0,in} - h^{\otimes}_{0,ex,s}}$
Isentropic, perfect gas, η_s	$\dfrac{(p^{\otimes}_{0,ex}/p_{0,in})^{(R/C_p)} - 1}{(p^{\otimes}_{0,ex}/p_{0,in})^{n'} - 1 + \frac{\dot{W}_{g3}}{(\dot{m}h_{0,in})}}$	$\dfrac{1 - (p^{\otimes}_{0,ex}/p_{0,in})^{n'} - \frac{\dot{W}_{g3}}{(\dot{m}C_p T_{0,in})}}{1 - (p^{\otimes}_{0,ex}/p_{0,in})^{(R/C_p)}}$
Isentropic, incompr. liquid, η_s	$\dfrac{(H^{\otimes}_{0,ex} - H_{0,in})(g/g_c)}{\dot{W}_{in}/\dot{m}}$	$\dfrac{\dot{W}_{in}/\dot{m}}{v(p_{0,in} - p^{\otimes}_{0,ex}) + (z_{in} - z_{ex})(g/g_c)}$
Isothermal, general gas, η_{it}	$\dfrac{h^{\otimes}_{0,ex} - h_{0,in}}{\dot{W}_{in}/\dot{m}}$	$\dfrac{\Delta^{ex}_{in}h_0 - T\Delta^{ex}_{in}s_0 - \dot{W}_{g3}/\dot{m}}{\Delta^{ex}_{in}h_0 - T\Delta^{ex}_{in}s_0}$
Isothermal, perfect gas, η_{it}	$\dfrac{RT\ln(p^{\otimes}_{0,ex}/p_{0,in})}{C_p\Delta^{ex}_{in}T_0 - T\Delta^{ex}_{in}s_0 + \dot{W}_{g3}/\dot{m}}$	Not used for expanders

Notes: The exponent n' is defined in section 2.6. W_{g3} indicates group 3 losses as defined in chapters 7, 8, and 9, the Σ[external losses]. $\Delta^{ex}_{in}h_0 = h_{0,ex} - h_{0,in}$ and $\Delta^{ex}_{in}s_0 = s_{0,ex} - s_{0,in}$.

Polytropic efficiency for perfect gas

The isentropic efficiency has a serious disadvantage if it is used as a measure of the quality of the aerodynamic design (or as a measure of the losses) in an adiabatic machine. The value of the isentropic efficiency is a function of pressure ratio as well as of losses, for the following reasons.

Suppose that we combine two compressors of equal isentropic efficiency and equal enthalpy rise to make a compressor of higher pressure ratio (figure 2.11). The combined actual enthalpy rise will be the sum of the individual enthalpy rises. But the sum of the isentropic enthalpy rises will be larger than the isentropic enthalpy rise for the combined compressor, as explained below. The enthalpy rises are equal:

$$\Delta h_{0,1} = \Delta h_{0,2}$$

For each stage

$$\eta_{s,c} = \frac{\Delta h_{0,1,s}}{\Delta h_{0,1}} = \frac{\Delta h_{0,2,s}}{\Delta h_{0,2}}$$

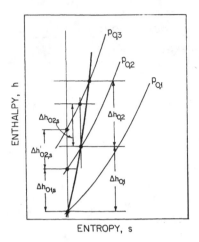

Figure 2.11. Isentropic efficiency of a two-stage compressor

Therefore $\Delta h_{0,1,s} = \Delta h_{0,2,s}$. But $\Delta h_{0,2,s} > \Delta h'_{0,2,s}$ (see equations 2.46 and 2.68). For the two stages

$$\eta_{s,c} = \frac{\Delta h_{0,1,s} + \Delta h'_{0,2,s}}{\Delta h_{0,1} + \Delta h_{0,2}}$$

Therefore the isentropic efficiency of the two-stage compressor is less than that of the single-stage compressor.

The losses are indicated by the increase of entropy in the working fluid per unit change of enthalpy: in other words, by the slope of the process line on the $h - s$ chart. This slope is a function of the exponent n in $pv^n = $ constant, which can be used to represent the process. (In fact this relation need connect only the end points of the process. The intermediate points will presumably be close to those given by the relation, but the process does not have to follow the "polytropic" equation for the following to be true.)

To avoid the influence of pressure ratio on the isentropic efficiency, the limiting value of the isentropic efficiency for a given polytropic process can be used as the pressure ratio approaches unity. In steam turbines this efficiency is usually known as the "small-stage" efficiency. In gas-turbine expanders and compressors, it is known as the "polytropic" efficiency.

The combination of the polytropic relation, $pv^n = $ constant, and the equation of state for a perfect gas (equation 2.33) leads to

$$r \equiv \frac{p_{0,ex}^{\otimes}}{p_{0,in}} = \left(\frac{\rho_{0,ex}}{\rho_{0,in}}\right)^n$$

for a polytropic, and

$$r = \frac{\rho_{0,ex}}{\rho_{0,in}}\frac{T_{0,ex}}{T_{0,in}}$$

for any process. Therefore

$$\left(\frac{T_{0,ex}}{T_{0,in}}\right) = r^{\left(\frac{n-1}{n}\right)} = r^{n'}, \qquad \text{where} \qquad n' \equiv \frac{n-1}{n}.$$

The actual enthalpy rise (work input) of an adiabatic compressor is

$$\Delta h_0 = C_p(T_{0,ex} - T_{0,in}) = C_p T_{0,in}\left(\frac{T_{0,ex}}{T_{0,in}} - 1\right) = C_p T_{0,in}(r^{n'} - 1) \qquad (2.75)$$

The isentropic enthalpy rise is

$$\Delta h_{0,s} = C_p T_{0,in}(r^{R/C_p} - 1) \qquad (2.76)$$

Therefore the "internal" isentropic efficiency, $\eta_{s,c}$, of a compressor working on a perfect gas is

$$\eta_{s,c} = \frac{r^{R/C_p} - 1}{r^{n'} - 1} \qquad (2.77)$$

Let the pressure ratio $r = 1 + \zeta$ and let $\zeta \to 0$ so that the pressure ratio approaches unity. The binomial theorem gives:

$$\eta_{s,c} = \frac{(1+\zeta)^{R/C_p} - 1}{(1+\zeta)^{n'} - 1} = \frac{1 + (R/C_p)\zeta + \cdots + \cdots - 1}{1 + n'\zeta + \cdots + \cdots - 1}$$

As $\zeta \to 0$, ζ^2 and higher-order terms can be neglected. Therefore the polytropic efficiency of a compressor, $\eta_{p,c}$, and of a turbine, $\eta_{p,e}$, are defined by:

$$\eta_{p,c} \equiv (\eta_{s,c})_{r \to 1.0} = \frac{R/C_p}{n'} \qquad \eta_{p,e} \equiv (\eta_{s,e})_{r \to 1.0} = \frac{n'}{R/C_p} \qquad (2.78)$$

For perfect gases the polytropic efficiencies defined by equations 2.78 lead to the following useful relations:

$$\frac{T_{0,ex}}{T_{0,in}} = r_c^{\left[\left(\frac{R}{C_p}\right)\frac{1}{\eta_{p,c}}\right]} \qquad (2.79)$$

for compressors, and

$$\frac{T_{0,ex}}{T_{0,in}} = r_e^{\left[\left(\frac{R}{C_p}\right)\eta_{p,e}\right]} \qquad (2.80)$$

for expanders (turbines).

Definition of useful outlet pressure

These equations (2.79 and 2.80) enable the adiabatic work,

$$\left(\frac{\dot{W}_{in} - \dot{W}_{ex}}{\dot{m}}\right)_{\dot{Q}=0} = \Delta h_0 \approx C_p \Delta_{in}^{ex} T_0 \equiv C_p(T_{0,ex} - T_{0,in}) \qquad (2.81)$$

to be obtained directly from the pressure ratio. But which value of $p_{0,ex}^{\otimes}$ should be used? It is the value for which η_p is defined. A stagnation-to-static efficiency at the outlet flange, for instance, defines the useful stagnation pressure at this plane as the actual static pressure there.

Figure 2.12 shows four equivalent ideal polytropic adiabatic compression processes for different definitions of useful outlet pressure. All processes have the same inlet $T_{0,1}$ and outlet $T_{0,2}$ and therefore the same actual work. Their differing values of n' depend only on which of the four possibilities is used for $p_{0,2}^{\otimes}$:

$$p_{0,2}^{\otimes} \equiv p_{st,2} \text{ the rotor-outlet static pressure, or}$$
$$p_{0,2}^{\otimes} \equiv p_{0,2} \text{ the rotor-outlet stagnation pressure, or}$$
$$p_{0,2}^{\otimes} \equiv p_{st,3} \text{ the stator-outlet static pressure, or}$$
$$p_{0,2}^{\otimes} \equiv p_{0,3} \text{ the stator-outlet stagnation pressure.}$$

In a compressor the diffuser outlet and the casing flange may provide other locations for the outlet pressure to be defined. When $p_{0,2}^{\otimes}$ is defined as a static pressure, the efficiency obtained is termed a "stagnation-to-static" efficiency, η_{0s}. When $p_{0,2}^{\otimes}$ is a stagnation pressure, the efficiency obtained is "stagnation to stagnation", η_{00}.

It is probable that turbomachine efficiencies are poorly defined and hence misused more because of confusion over the definition of $p_{0,2}^{\otimes}$ than through any other factor. Here are some comments intended to clear up some of the confusion.

1. There is no "right" definition of efficiency. One may choose to use any definition one wants. It is of course essential that whichever is chosen is exactly identified.

2. There may, however, be a definition that is more appropriate in certain circumstances than others. The reason why it is more appropriate is likely to be that it

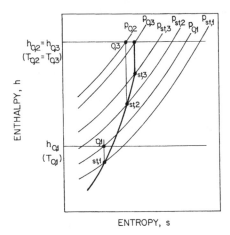

Figure 2.12. Dependence of ideal work transfer of a compressor on the definition of useful outlet pressure

saves some effort in making calculations and in accounting for energy flows and loss locations. For instance, the appropriate definition of $p_{0,2}^{\otimes}$ for the efficiency of a compressor in a turbojet engine is the static pressure at the end of the compressor diffuser, the boundary between the compressor and the combustor, because the dynamic pressure at the point cannot be used by the combustor. There would be nothing thermodynamically wrong in choosing to use the stagnation pressure, but doing so would have undesirable consequences. The combustor would seem to have a higher pressure drop than under the other definition, because the inlet pressure would now be listed as the stagnation, instead of the static, pressure. The compressor designer would then have an incentive to cut back the diffuser, or maybe to eliminate it altogether, which would further increase the numerical value of the compressor efficiency and further increase the allocated combustor pressure drop. At this point we would have gone beyond merely energy accounting. The actual losses would be increased if the diffuser were eliminated, so that defining the compressor efficiency in terms of a stagnation outlet pressure could have wholly undesirable consequences.

Therefore the choice of the "appropriate" value of $p_{0,2}^{\otimes}$ is one that must take incentives into account.

In the case of the turbine of a turbojet engine, the appropriate $p_{0,2}^{\otimes}$ to use in the turbine-efficiency definition is the stagnation pressure upstream of the propulsion nozzle that produces the jet. Obviously in this case the dynamic pressure in the stream is useful, and there is no need for an incentive for the turbine designer to diffuse the leaving flow.

3. Only the ideal work is affected by the choice of definition of useful outlet pressure, $p_{0,2}^{\otimes}$. The actual work is entirely unaffected. One has only one decision to make with regard to the actual work: whether or not to include the "external" friction losses (bearings, windage, and seals).

4. Even if the choice of useful outlet pressure is a static pressure, it is defined as a stagnation pressure, $p_{0,2}^{\otimes}$. The reason is that the ideal work is found from the steady-flow energy equation, equation 2.11, which requires the stipulation of stagnation enthalpies at the beginning and end of the process. We are defining a hypothetical state for a hypothetical (ideal) process. Even though it is confusing, there is nothing illogical about defining a hypothetical stagnation pressure as being numerically equal to a known static pressure.

Shape of polytropic-process lines on the h-s diagram

Just as the lines of constant pressure are often erroneously represented as straight on the $h - s$ diagram for a perfect gas (figure 2.8), so, too, are lines of constant polytropic exponent n. We shall show that, in fact, these lines are curved to maintain a constant differential gradient with the constant-pressure lines.

The equation for a polytropic process is $pv^n = $ constant. Differentiating and dividing give

$$\frac{1}{n}\frac{dp}{p} + \frac{dv}{v} = 0$$

Differentiating the equation of state for a perfect gas $pv = RT$ and dividing gives

$$\frac{dp}{p} + \frac{dv}{v} = \frac{dT}{T}$$

Gibbs' equation (equation 2.31) with equation 2.35 is

$$T\,ds = C_p dT - v\,dp$$

Combining the last three equations gives

$$T\,ds = C_p dT - R\,dT/n'$$

$$\frac{ds}{dT} = \frac{C_p}{T}\left(1 - \frac{R/C_p}{n'}\right) \tag{2.82}$$

Using the definitions of compressor and expander polytropic efficiencies, equations 2.78, we can now derive the slope of the polytropic processes in the following simple forms, for compression and expansion respectively:

$$\frac{dT}{ds} = \left(\frac{T}{C_p}\right)\left(\frac{1}{1 - \eta_{p,c}}\right) \qquad\qquad \frac{dT}{ds} = \left(\frac{T}{C_p}\right)\left(\frac{1}{1 - 1/\eta_{p,e}}\right) \tag{2.83}$$

These differ from the slope of the constant-pressure lines, given in equation 2.46, by a constant: the efficiency function, $(\partial T/\partial s)_p = (T/C_p)$.

The slope of the compression line is positive, and that of the expansion line is negative. For a polytropic efficiency of 100 percent, the slope becomes infinite: the line is vertical and is an isentrope (figure 2.13). At a temperature of absolute zero, both lines have zero slope.

The variation of curvature of the polytropic lines is of little significance for machines of small pressure ratio. However, for large pressure ratios a misrepresentation of the shape of these process lines can lead to the drawing of incorrect conclusions.

Polytropic processes for non-perfect gases

Because non-perfect gases have non-constant values of the isentropic index connecting points on an isentropic process, the relationship just derived for perfect gases cannot apply. Instead, a mean polytropic efficiency is based on a factor that represents the increased value of $(\partial h/(\partial p/p))_s$ with increased temperature. This factor is known as the reheat factor.

We combine Gibbs' equation 2.31 with the definition of enthalpy $h \equiv u + pv$:

$$\begin{aligned} dh &= du + p\,dv + v\,dp \\ dh &= T\,ds + v\,dp \end{aligned}$$

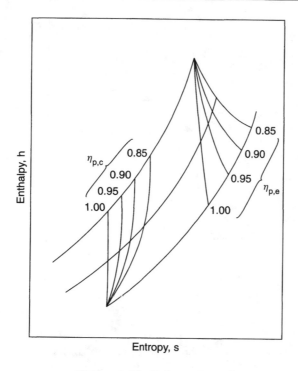

Figure 2.13. Polytropic processes

Therefore

$$\left(\frac{\partial h}{\partial p/p}\right)_s = pv = z_c RT \tag{2.84}$$

The general equation of state, $pv = z_c RT$, applies to non-perfect gases, with the so-called compressibility factor, z_c, being a slowly varying function of (reduced) pressure and temperature. R is the gas constant. Equation 2.84 shows that the isentropic enthalpy drop between lines of constant pressure increases with absolute temperature as with perfect gases, for which $z_c = 1$, and for which the appropriate relation was derived in equation 2.49.

Therefore, when a non-isentropic adiabatic expansion, such as that from 01 to 02 in figure 2.14, is divided into a number of actual or hypothetical small expansions, the isentropic enthalpy drop of each (for example, a', b') will be larger than the corresponding section (a, b) of the overall isentropic enthalpy drop $\Delta h_{0,s}$.

That is $a' > a$, $b' > b$, and $\Sigma \Delta h_{0,s,se} > \Delta h_{0,s}$, where

$\Delta h_{0,s,se} \equiv$ isentropic enthalpy drop of small "stage", and
$\Delta h_{0,se} \equiv$ actual enthalpy drop of small "stage".

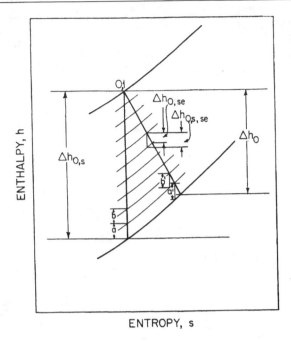

Figure 2.14. Polytropic process for a non-perfect gas

However, $\Delta h_{0,se} = \Delta h_0$. Now the overall isentropic efficiency, η_s, is

$$\eta_s \equiv \frac{\Delta h_0}{\Delta h_{0,s}} = \left(\frac{\Sigma \Delta h_{0,se}}{\Sigma \Delta h_{0,s,se}} \right) \cdot \left(\frac{\Sigma \Delta h_{0,s,se}}{\Delta h_{0,s}} \right)$$

$$\eta_s = \eta_p R_h \tag{2.85}$$

where

$\eta_p \equiv \Delta h_{0,se}/\Delta h_{0,s,se}$, the "small-stage" isentropic efficiency, and
$R_h \equiv \Sigma \Delta h_{0,s,se}/\Delta h_{0,s}$, the "reheat factor".

The reheat factor is obviously a function of pressure ratio, being unity for very low pressure ratios, and of polytropic efficiency, being unity for all pressure ratios when the polytropic efficiency is unity.

Usefulness of polytropic efficiency

In removing the effect of pressure ratio, the polytropic efficiency enables machines of different pressure ratio to be compared with validity. A single-stage compressor may have an isentropic efficiency of 90 percent, stagnation to stagnation. If this stage is combined with others of similar performance to form a high-pressure-ratio compressor, the isentropic stagnation-to-stagnation efficiency of the whole compressor may be only 85 percent, making it appear to be of inferior aerodynamic performance. If, however, the polytropic efficiency is used, both the single-stage and the multistage machines will have the same polytropic efficiency of about 91 percent.

The polytropic efficiency has a double advantage in analysis of Brayton power cycles, particularly when the design pressure ratio is being varied. First, as mentioned, a single value of polytropic efficiency can be entered for each compression and expansion process, instead of isentropic efficiencies that vary with pressure ratio. Second, the equations using polytropic efficiencies are often simpler than those for isentropic efficiencies, leading to a saving in computation time.

In practice, however, the elimination of the reheat factor by the use of polytropic efficiencies does not eliminate all influence of pressure ratio from efficiency. A compressor or turbine of high pressure ratio will have higher Mach numbers, in general, and blading that is relatively longer in the low-pressure region and shorter in the high-pressure area, all factors leading to increased losses. The variation in maximum efficiency with pressure ratio is discussed in chapter 3.

Effect of leaving losses on efficiency values

The stagnation-to-stagnation efficiency (at machine, or diffuser, outlet) has been quoted here because of the complicating effects of the outlet kinetic energy, or "leaving losses". These leaving losses are relatively much more important for a single-stage than for a multistage machine, as will become obvious when velocity diagrams are considered (chapter 5). Therefore, if the more usually quoted stagnation-to-static efficiencies had been used, some of the pressure-ratio effects of isentropic efficiency would have been compensated for by the pressure-ratio effects of leaving losses.

Actual work is unaffected by efficiency definition

From the steady-flow energy equation the actual (internal) work is the difference in stagnation enthalpy from inlet to outlet. This work remains constant at $(h_{0,ex} - h_{0,in})$ regardless of how the outlet pressure is defined.

Relations between polytropic and isentropic efficiencies for perfect gases

For compression

$$\eta_{s,c} = \frac{r^{R/C_p} - 1}{r^{(R/C_p)\eta_{p,c}} - 1} \tag{2.86}$$

$$\eta_{p,c} = \frac{\ln\left(r^{R/C_p}\right)}{\ln\left[\left(r^{R/C_p} - 1\right)/\eta_{s,c} + 1\right]} \tag{2.87}$$

For expansion

$$\eta_{s,e} = \frac{1 - r^{(R/C_p)\eta_{p,e}}}{1 - r^{R/C_p}} \tag{2.88}$$

$$\eta_{p,e} = \frac{\ln\left[1 - \eta_{s,e}\left(1 - r^{R/C_p}\right)\right]}{\ln r^{R/C_p}} \tag{2.89}$$

As before, the outlet stagnation pressure, $p_{0,ex}^{\otimes}$, must be chosen to be consistent with the particular definition of efficiency being used.

References

Gyftopoulos, Elias P., and Beretta, Gian Paolo (1991). Thermodynamics: Foundations and Applications. Macmillan, NY, NY.

Keenan, Joseph H. (1970). Thermodynamics. The MIT Press, Cambridge, MA.

Keenan, Joseph H., Jing Chao, and Joseph Kaye (1983). Gas Tables. Wiley, NY.

Korakianitis, T., Vlachopoulos, N.E., and Zou, D. (1994). Models for the prediction of transients in regenerative gas turbines with centrifugal impellers. ASME paper 94-GT-361, Transactions of the ASME, Journal of Engineering for Gas Turbines and Power (in print).

Reynolds, William C. (1968). Thermodynamics. McGraw-Hill, NY, NY.

Problems

1. Does the stagnation temperature of the working fluid rise or fall in passing through a gas-turbine expander? Why?

2. Does the temperature of the water rise or fall in passing through a water turbine? Why?

3. By how much, and in which direction, does the temperature of water change when it falls over a waterfall 50 m high? What is the mechanism for this change?

4. Sketch in your qualitative estimates of the variation of stagnation and static enthalpy and pressure through the intercooled compressor shown in figure P2.4. Some end points are shown. (Four lines are required.)

5. For the subsonic axial-flow air expander specified, calculate the stagnation and static pressures and temperatures and the Mach number at the rotor-inlet plane (figure P2.5). Also find the rotor rotational speed and the nozzle-inlet blade height for constant-outer-diameter blading. Sketch the form of the complete turbine expansion on an enthalpy-entropy chart. The axial velocity will be constant at this design point.

> Mass flow, $\dot{m} = 2$ kg/s
> Nozzle-inlet stagnation temperature, $T_{0,ni} = 400$ °C.
> Absolute nozzle-inlet stagnation pressure, $p_{0,ni} = 3$ bars($= 3 \times 10^5$ N/m^2)
> Mean diameter, $d_m = 0.25$ m.
> Blade height at rotor entry, $l = 0.1 d_m (= 1/2[d_s - d_h])$.
> Flow angle at nozzle exit, $\alpha_1 = 70°$ to axial direction.
> Drop in stagnation pressure in nozzle, $\Delta p_0 = 0.05$ bar
> Rotor peripheral speed at mean diameter, $u_m = 0.5 \times$ (component of nozzle outlet velocity, $C_{\theta,1}$).

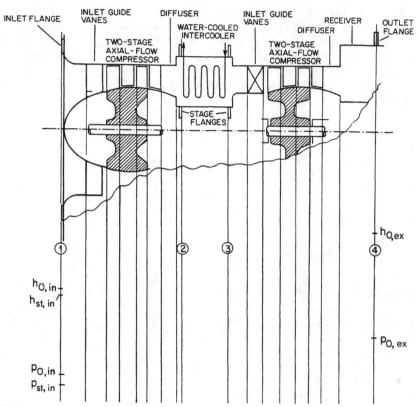

Figure P2.4

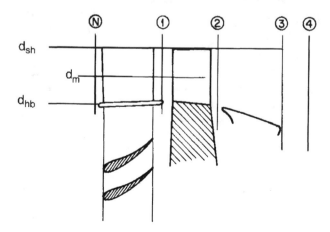

Figure P2.5

6. Complete the table of efficiencies, table P2.6, for the turbine expander of problem 2.5, for inlet conditions $1,500$ K and 8×10^5 N/m^2. The turbine stagnation enthalpy drop is twice the mean blade speed squared ($\Delta h_0 = 2u_m^2$) and the mean blade speed is 500 m/s. The flow velocity at rotor exit is axial and is 0.728 of the mean blade speed, and it is reduced by 50 percent in the diffuser. The rise in static pressure in the diffuser is 75 percent of the value it would have if the diffuser flow were isentropic. Also calculate the power, in kW, delivered by the blading. (Do not count the "external" losses—bearing friction and so forth—here or in the efficiencies.) The inlet conditions to be used for the efficiencies are the stagnation conditions at plane N.

Table P2.6

Efficiency	Outlet 2	plane 3
$\eta_{s,tt}$	0.95	
$\eta_{s,ts}$		
$\eta_{p,tt}$		
$\eta_{p,ts}$		

7. Given that an axial-compressor stage is a row of rotor blades which act as diffusers, followed by a row of stator blades which also act as diffusers, explain (from first-law and second-law considerations) why a high Mach number is necessary if the compressor is to give a high pressure ratio.

8. Calculate the isentropic stagnation-to-static internal efficiency of an intercooled air compressor (rather similar to that shown in problem 2.4) from inlet flange to outlet flange. Comment on your findings. Suppose that instead of the actual arrangement shown, two centrifugal stages are used, one upstream and one downstream of the intercooler. Each stage has a stagnation-to-stagnation flange-to-flange pressure ratio of 4 : 1, and each has a stagnation-to-stagnation polytropic efficiency of 88 percent for the same planes (1 to 2 and 3 to 4). The mean velocity at the outlet flange of each stage is 30 m/s. The intercooler has a stagnation-pressure loss of 6 percent of the stagnation pressure of the flow entering it, and it lowers the stagnation temperature to 1.03 times the first-stage-inlet stagnation temperature of 305 K. Use a constant value of $\gamma = 1.4$. Draw a temperature-entropy diagram.

9. Using figure P2.9, calculate the rotor rotational speed, the static temperature and pressure at nozzle exit, and the (absolute) Mach number at that point, of a radial-inward-flow turbine expander of nozzle-outlet diameter (surface 1) of 250 mm. As an approximation, take this as the rotor-inlet diameter also. The nozzle-exit direction of the flow is 75 degrees from the radial direction, and the axial height of the blade passage is 10 percent of the rotor diameter. The turbine nozzles are supplied with 2 kg/s of air at 2 bars stagnation pressure and 125 °C stagnation temperature. The rotor peripheral speed is 90 percent of the tangential velocity of the air at nozzle exit. The stagnation-pressure losses of the flow through the nozzles are small enough to be neglected. Sketch the nozzle expansion on a temperature-entropy diagram.

10. Draw lines approximating the changes of stagnation and static enthalpy and pressure through the compressor test rig shown diagrammatically below (figure P2.10). The centrifugal compressor is driven by a motor and by the energy-recovery turbine, which expands the flow

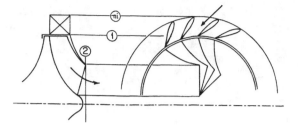

Figure P2.9

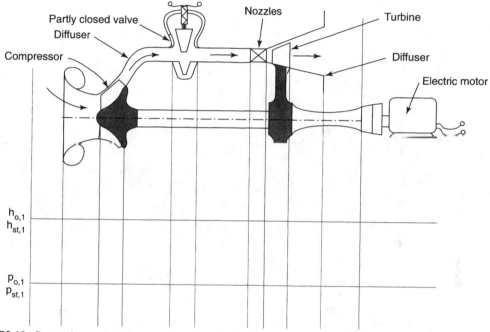

Figure P2.10. Stagnation and static pressure and enthalpy through a compressor test rig

back to atmospheric pressure (as static pressure). A throttle valve is used to produce different back pressures on the compressor.

11. Figure P2.11 shows a block diagram of the compressor test rig of problem 2.10. Your task is to draw control volumes around each of the three components, and then one around the three as an assembly, excluding the motor. Calculate and discuss the external energy flows and changes of enthalpy using the steady-flow energy equation. Do this for two different cases. In the first case, the throttle valve is wide open, so that there is no loss of stagnation pressure between the outlet of the compressor diffuser and the inlet of the turbine bell mouth. In the second case, the throttle reduces the turbine expansion ratio from 10 to 5 to 1.

In both cases the compressor takes in 25 kg/s of air at 288 K and compresses it through a stagnation-to-static pressure ratio (to diffuser exit) of 10 : 1 and a polytropic efficiency (same conditions) of 0.80. The turbine takes the flow from the throttle valve and expands

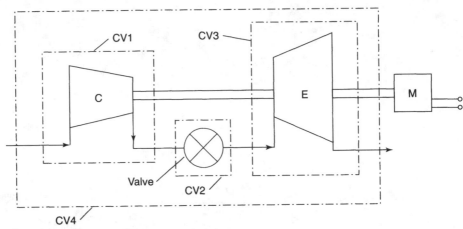

Figure P2.11. Block diagram of compressor test rig

it to atmospheric pressure with a stagnation-to-static polytropic efficiency of 0.90. A mean specific heat of 1020 J/(kg · K) and a gas constant of 286.96 J/(kg · K) may be used throughout.

12. If a centrifugal air compressor has a stagnation-to-stagnation isentropic efficiency to diffuser outlet of 0.80, would the isothermal efficiency for the same conditions be higher? Or lower? Why?

    Would the stagnation-to-stagnation polytropic efficiency from rotor (impeller) inlet to rotor outlet be higher than 0.80? Or lower?

13. Calculate the power output in kW of a radial-inflow turbine for a Diesel-engine turbocharger. The exhaust gas has a molecular weight of 28.5 and a mass flow of 0.25 kg/s. The gas enters the turbine at a stagnation temperature of 675 K and a stagnation enthalpy of 695 kJ/kg. The stagnation pressure at inlet is 280, 000 Pa. The rotor is 150 mm diameter and runs at 50, 000 rpm. The gas enters the rotor with a velocity of 400 m/s at 70° to the radial direction, and leaves without swirl at a Mach number of 0.22. The static pressure at diffuser outlet is 100, 000 Pa.

14. Suppose that you are a manufacturer of high-speed boats, wanting to specify and purchase the most effective axial-flow water-pump jet-propulsion units to be connected to the boat's engine (see figure P2.14). Write down the four factors needed to define turbomachine efficiency. Then choose your specifications for each factor to enable you to choose the best pump for your application.

15. Figure P2.15 is a diagram of a test setup of a device with which an inventor wants to heat buildings under construction. Plot the qualitative variations of four air properties: the stagnation and static pressures and temperatures. Assume that the motor waste heat (from inefficiencies) is conducted into the airflow.

16. At what flow angle will the flow become just sonic downstream of the variable-setting-angle nozzles of a high-pressure turbine? The annulus into which the nozzles discharge has a shroud diameter of 283 mm, and a hub/shroud diameter ratio of 0.85. The mass flow is 8 kg/s; the

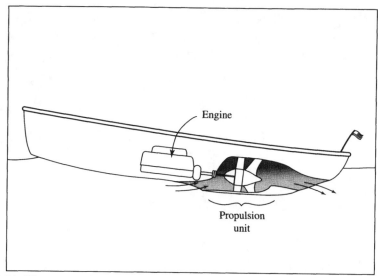

Figure P2.14. Water-pump jet propulsion

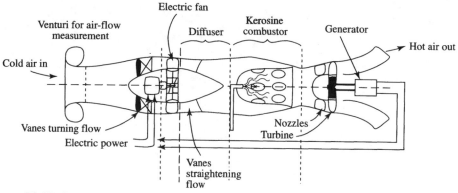

Figure P2.15. Stagnation and static pressure and temperatures in building-heating device

stagnation temperature and pressure at nozzle exit are 1550 K and 15 bar; R is 286.96 J/(kg · K); and C_p/R is 4.1. The flow enters the nozzles without swirl, and an impulse turbine is downstream of the nozzles.

17. Does the steady-flow energy equation apply precisely to a turbine in which the blades are cooled? Explain.

Chapter 3

Thermodynamics of gas-turbine cycles

Gas-turbine engines are rotors and stators with blading, combustors, casings, heat exchangers and so forth. Each molecule of gas that enters the engine undergoes a sequence of processes that is called a "cycle". In a theoretical cycle, and in actual so-called "closed-cycle" gas turbines, the gas molecule returns to its original state. Most gas turbines work on the "open" cycle, in which ambient air enters the compressor, fuel is burned in the air, and the products of combustion emerge from the engine exhaust at above-atmospheric temperature. The regeneration of this gas by cooling and by the conversion of carbon dioxide to oxygen is then carried out in the atmosphere by natural processes.

The cycle to be used for a gas-turbine engine must be chosen before any component design can be started. The thermodynamic analysis of the cycle will yield the potential efficiency, power output and approximate size of the engine.

Gas turbines, in contrast to steam turbines, spark-ignition and compression-ignition engines, can operate on a wide variety of different cycles. In this chapter we develop the performance of "simple" (compressor, burner, expander; or "CBE") cycles; similar cycles that incorporate an exhaust-gas heat exchanger ("CBEX"); and heat-exchanger cycles that use an intercooled, instead of a nonintercooled, compressor (e.g., "CICBEX" cycles)[1]. We will also describe and briefly discuss some other cycles (for instance, turbojet and turbofan cycles for jet propulsion, and combined and steam-injection cycles for industrial power generation) among the many that have been designed or considered.

This chapter is concerned with the start of the design process. The first decisions the designer must make before the start of a stand-alone piece of turbomachinery or of a gas-turbine engine is the choice of the thermodynamic conditions, including the temperatures, pressures, pressure losses, component efficiencies and so forth. The designer might, in fact, have only limited freedom in many of these respects. The customer or someone at a more-senior level might already have chosen the overall component specifications or the engine cycle. The component efficiencies, pressure losses and so on may be very

---

[1]This is a slightly modified version of the cycle-designation system apparently developed by E. S. Taylor at MIT. We have substituted the generic "expander, E" for "turbine, T".

restricted choices. The designer is more likely to choose a type of compressor, for instance (from the principal types: axial-flow, radial-flow or axial-centrifugal combinations) and accept whatever efficiency is produced by either a very conservative or a very aggressive approach to the type chosen. If you, the reader, are being exposed to turbomachinery for the first time you will not have the experience to choose appropriate values, and you will need to be guided by the typical numbers used in the examples in this chapter. You will be able to choose some values for yourself after going through chapters four and five, which cover the performance of diffusing components and the preliminary design of fans, compressors, pumps, and turbines. Much more precise estimates of component efficiencies, including those of heat exchangers, can be made after the material in chapters 7 through 10 is absorbed.

3.1   Temperature-entropy diagrams

By definition, a gas turbine uses a gas as a working fluid. In the great majority of gas-turbine applications, the working fluid is air, or is a gas that is well removed from its liquefaction temperature in the cycle conditions chosen (for instance, hydrogen or helium). Under these conditions, it is a good approximation to treat the working fluid as a perfect or semi-perfect gas, defined as a substance that obeys the equation of state:

$$pv = RT$$

where p is the pressure, v is the specific volume, R is a constant (the "gas constant"), and T is the (absolute) temperature. It is convenient to perform initial analyses of cycles assuming that the working fluid is a perfect gas simply because the calculations are greatly simplified, and because the resulting ease of analysis makes it possible to gain a deeper insight into the variations that may be expected from changes in cycle conditions. Final calculations may then be made using real-gas properties in the knowledge that conditions will not be greatly changed. We shall give examples of calculations made using different assumptions for working-fluid properties.

Property diagrams are particularly useful for giving the conditions of, and relationships among, the end points of processes making up gas-turbine cycles. Perfect gases are simple substances, for which the state can be found from the value of any two independent properties. Therefore we could make cycle diagrams on charts with axes of p and v, or v and h, for instance.

More suitable choices for the axes of a diagram to represent the ideal gas-turbine cycle, which is known as the Brayton or Joule cycle in one form, and the Ericsson cycle in another form, are T and s or h and s. The fluid stagnation temperatures at compressor and turbine inlets are normally part of the cycle specifications. The ideal compression and expansion processes are isentropes in the Brayton cycle, and isothermals in the ideal Ericsson cycle (for compression which in practice could be approached by using many intercoolers and expansion by using many reheat combustors). Both are easily drawn on $T - s$ diagrams. The thermodynamic or material limits for gas-turbine cycles are lines of constant temperature, again easily drawn. Atmospheric temperature is one limit for open-cycle engines, and the maximum gas temperature that can be tolerated by the expander is the other.

Atmospheric pressure is a limit for open cycles. Heat addition is carried out at a higher, constant, pressure in an ideal Brayton cycle. Constant-pressure lines have a distinct shape on a temperature-entropy diagram, a shape that helps us understand how gas-turbine cycles produce positive net power. The shape is shown in figure 2.7, and the reason why a turbine expander can give surplus power beyond that needed by the compressor is explained in section 2.3 (subsection "isentropic work for compression and expansion processes").

3.2   Actual processes

Compression processes in gas turbines are normally virtually adiabatic, or are occasionally adiabatic compression processes separated by intercoolers. Expansion processes for "uncooled" turbines are adiabatic, but turbines cooled with air or water are not adiabatic. The expansion process can also be split by one or more "reheat" combustors.

In real adiabatic processes the entropy must increase. The work required for compression between two pressure levels increases for an entropy-increasing process compared with that for an isentropic process. Conversely, the work obtained from a real expansion process, involving an increase in entropy, decreases from that given by an isentropic process.

Whether or not there is net power available in a real gas-turbine cycle depends on the efficiency of the nonisentropic processes (represented by the slope of the process line on the $T-s$ or $h-s$ diagram, as shown in figures 2.13 and 3.1, and on the ratio of the turbine-inlet temperature to the compressor-inlet temperature (figure 3.2).

In a real cycle there will also be pressure losses, which means that the compressor pressure ratio will be larger than the turbine expansion ratio. There may also be a leakage of mass out of the compressed-air side of the cycle, some unwanted, and some scheduled for cooling and sealing; in open-cycle engines the fuel mass flow is added to the burner and expander flow, and in high-temperature engines at least some of the cooling air is mixed with the main flow in the expander. All these effects make the choice of parameters for a real gas-turbine cycle challenging and interesting.

3.3   Choice of the pressure ratio for maximum power

It is easy to show that the thermal efficiency of an ideal Brayton cycle is a function only of pressure ratio. We shall not derive this result here, because it is somewhat trivial and can be misleading. One can still read claims in the technical literature that the ability to design higher-pressure-ratio compressors has opened the way to higher-efficiency gas turbines. This is true for aircraft engines, but false for engine cycles incorporating heat exchangers. Such cycles have maximum efficiencies at low pressure ratios, as will be shown. In fact, in a real gas-turbine cycle, the higher the turbine-inlet temperature, the greater in general are the net output and the cycle efficiency. The higher pressure ratios used in CBE cycles (composed of compressor

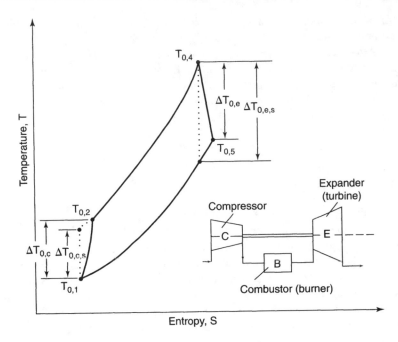

Figure 3.1. Gas-turbine cycle with nonisentropic compression and expansion

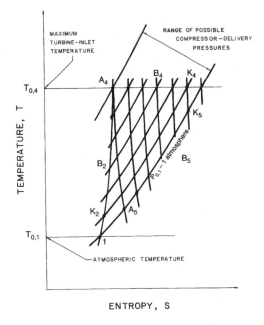

Figure 3.2. Constraints on gas-turbine cycle parameters

"C", combustor or burner "B", and expander "E") are more a consequence of the high temperature ratios than a determinant of the high efficiency and specific output. Also, in real cycles, component efficiencies have a crucial effect on gas-turbine-engine viability. It will be shown that for every combination of cycle temperature ratio and component efficiencies, there exists an optimum pressure ratio for maximum thermodynamic cycle efficiency, and another optimum pressure ratio for maximum specific power.

Consider the choices open to the designer of a gas-turbine cycle within the constraints shown in figure 3.2. For an open cycle the start of compression is at atmospheric temperature and pressure. A metallurgical limit fixes the expander-inlet temperature[2]. The end of expansion is also at or close to atmospheric pressure. The best compressor technology available produces, it is supposed, the compression-process line shown. Another polytropic-process line represents the best available expansion process, which may start from different points along the $T_{0,4}$ line.

Stated in these terms, the only freedom left to the designer is to choose the pressure ratio. A range of choices is shown. Let us look first at the extreme choices.

At pressure ratios close to unity, typified by the cycle 1, K_2, K_4, K_5, the compressor absorbs little work but the turbine expander also produces little work. Net output is vanishingly small.

The other extreme is the pressure ratio so high that compressor-delivery temperature is equal to the turbine-inlet temperature. No heat can be added, and the flow is immediately expanded to atmospheric pressure - cycle 1, A_4, A_5. Turbine work (which is proportional to the change of enthalpy, and therefore a function of the change of temperature) during expansion is obviously less than compressor work, so that net output is negative.

In between these two unattractive extremes there are pressure ratios at which the cycle produces net work. Therefore there is an optimum pressure ratio, perhaps in the approximate region of cycle 1, B_2, B_4, B_5, at which the cycle produces maximum net power (see figure 3.3).

This same optimum pressure ratio will also apply to a cycle incorporating a heat exchanger (figure 1.7) if, as a first approximation, we specify that the added heat-exchanger pressure losses will be zero. The heat-exchanger cycle is shown on a $T - s$ diagram in figure 3.4.

At this point we would like to work an example to illustrate the existence of an optimum pressure ratio, even though we have not yet developed the tools to do so easily. We resolve this minor dilemma by using a formula derived later in this chapter. Readers who prefer to wait for the formula to be developed may wish to pass over this example.

---

[2]In practice, the expander-inlet temperature may be influenced partly by the pressure ratio, in addition to the strong influence of the metallurgical limit. The reason is that, when an engine has a very low pressure ratio, which would be the case, for instance, if a high-effectiveness heat exchanger is used, a low blade speed may be used in the turbine expander, leading to low blade stresses and to the possibility of using a higher gas temperature for a given blade life. Also, the cooling air, coming principally from the compressor outlet, will be considerably cooler than for an engine with a high pressure ratio, so that the proportion of cooling air required at a given turbine-inlet temperature should be less.

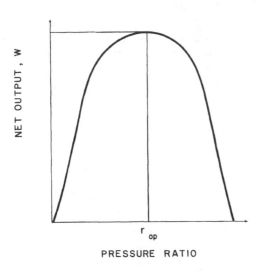

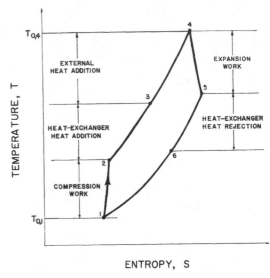

Figure 3.3. Optimum pressure ratio for maximum specific power

Figure 3.4. Energy transfers in a gas-turbine-engine cycle on the $T - s$ diagram

---

Example 1   Calculation of pressure ratio for maximum power

   Calculate the pressure ratio, r_c that gives maximum specific power in a gas-turbine engine with the following specifications.

$$\text{Compressor-inlet temperature, } T_{0,1} = 310\text{K}$$
$$\text{Expander-inlet temperature, } T_{0,4} = 1300\text{K}$$
$$\text{Compressor polytropic efficiency, } \eta_{p,c} = 0.85$$
$$\text{Expander polytropic efficiency, } \eta_{p,e} = 0.875$$
$$\text{Stagnation relative-pressure losses, } \Sigma(\Delta p_0/p_0) = 0.06$$
$$(R/\overline{C_{p,c}}) = 0.240$$
$$(R/\overline{C_{p,e}}) = 0.284$$
$$\dot{m}_e \overline{C_{p,e}}/\dot{m}_c \overline{C_{p,c}} = 1.150$$

(Taking the last three parameters as constants is an approximation. In fact, the mean specific heats would vary with pressure ratio, as would the rate of fuel addition. The effects of this approximation, however, are minor for present purposes.)

Solution.
   We use equation 3.1, given and derived subsequently, defining the specific power:

$$\dot{W}' \equiv \frac{\dot{W}}{\dot{m}_c \overline{C_{p,c}} T_{0,1}} = \left[\frac{\dot{m}_e \overline{C_{p,e}}}{\dot{m}_c \overline{C_{p,c}}} \right] E_1 T' - C$$

with the following definitions:

$$E_1 \equiv (T_{0,4} - T_{0,5})/T_{0,4}$$

$$= 1 - \left\{ \left[1 - \Sigma(\Delta p_0/p_0) \right] r \right\}^{(R/\overline{C_{p,e}})\eta_{p,e}}$$

$$T' \equiv (T_{0,4}/T_{0,1})$$

$$C \equiv (T_{0,2} - T_{0,1})/T_{0,1} = \left[r^{(R/\overline{C_{p,c}})/\eta_{p,c}} - 1 \right]$$

Inserting the specified values we obtain the specific power in terms of the pressure ratio:

$$\dot{W}' = 1.15\{1 - ((1 - 0.06)r)^{(-0.24 \times 0.875)}\}(1300/310 - (r^{(0.284/0.85)} - 1)$$

$$= 4.8226\{1 - (0.94r)^{-0.21}\} - r^{0.334} + 1$$

At maximum specific power:

$$\frac{d\dot{W}'}{dr} = 0 = -4.7603 \times (-0.21)r^{-1.21} - 0.334r^{-0.666}$$

from which the optimum pressure ratio is found to be 7.5.

Comments

1. This optimum pressure ratio for maximum power output is unaffected by the presence or absence of a heat exchanger if we neglect, as an approximation, the effects of the additional heat-exchanger pressure drops.

2. The optimum pressure ratio for maximum efficiency is, however, different for simple and for heat-exchanger cycles.

3. Table 3.1 shows the effect of cycle temperature ratio on optimum pressure ratio.

Table 3.1. Effect of cycle temperature ratio on optimum pressure ratios in the CBEX cycle (typical)

Temperature ratio, T'	4.0	5.0
Pressure ratio for maximum specific power	7.0	10.0
Maximum specific power	0.62	1.02
Pressure ratio for maximum thermal efficiency	3.5	4.5
Maximum thermal efficiency	0.445	0.525

3.4 Choice of optimum pressure ratio for peak efficiency

Suppose that the pressure ratio for maximum output has been selected in the way suggested in section 3.3. One can consider making small changes in pressure ratio without changing the net output, because it is at the maximum shown in figure 3.3. Suppose first that we are concerned with a simple (no-heat-exchanger) cycle. When the pressure ratio is reduced a small amount from the optimum for peak net power, the compressor-outlet temperature, $T_{0,2}$ (figure 3.1), is reduced, and additional fuel must be supplied to make up the increased temperature rise $(T_{0,4} - T_{0,2})$. With the net output staying constant or falling slightly, and the heat input to the cycle increasing, the efficiency falls. Conversely, if the pressure ratio is raised slightly, the compressor-outlet temperature is higher, and less fuel need be burned to bring the working fluid up to turbine-inlet temperature, $T_{0,4}$. The thermal efficiency is therefore increased by raising the pressure ratio above that which gives maximum output.

The maximum cycle thermal efficiency is reached at a pressure ratio only a little higher than that for maximum power. At higher pressure ratios the effect on the thermal efficiency of the fall in power output is greater than the effect of the reduced heat-addition required.

Now suppose that we are concerned with a heat-exchanger cycle, figure 3.4. For simplicity in presenting the argument, we suppose that the heat exchanger has an effectiveness of unity, which means that the compressor-discharge air will be brought up to the temperature of the turbine discharge, i.e., $T_{0,3} = T_{0,5}$. In addition, as before, we assume that the heat-exchanger pressure losses are zero.

Now when the pressure ratio is raised above its maximum-output level, the turbine-discharge temperature $T_{0,5}$, falls because of the larger expansion ratio. Therefore the temperature of the compressed air leaving the heat exchanger, $T_{0,3}$, also falls, and more fuel must be burned to keep the turbine-inlet temperature at the desired value. When, on the other hand, the pressure ratio is dropped below that for maximum output, the expander-outlet temperature $T_{0,5}$, increases, and with it the temperature of the compressed air leaving the heat exchanger. Less fuel need be burned, and the thermal efficiency increases. The optimum pressure ratio for maximum thermal efficiency is therefore lower than the pressure ratio for maximum output. It may be considerably lower for a cycle with high-efficiency components.

All these effects can be seen in the charts of cycle thermal efficiency against specific power, with temperature ratio (turbine inlet to compressor inlet) and compressor pressure ratio as parameters, shown later in this chapter (figures 3.19, 3.23, et seq.). Each plot is made for a set of values of component efficiencies and pressure losses. They will be discussed below.

3.5   Cycle designation—fuller specification

The types of cycles are indicated on the performance charts and state diagrams by symbols for the components, in the order in which they are encountered by the working fluid:

$C \equiv$ compressor,

$I \equiv$ intercooler,

$B \equiv$ heat addition from external source (e.g. a "burner"),

E ≡ expander, and

X ≡ heat exchanger from turbine-outlet to compressor-outlet flows.

Although the working fluid passes twice through the heat exchanger, the symbol X is used in the designation only in the expander-exhaust position. Thus a simple cycle is designated CBE. A heat-exchanger cycle is CBEX (not CXBEX). A cycle with a single intercooler in the compressor and an exhaust heat exchanger is CICBEX, and so forth. The thermodynamic-cycle designation is unaffected by whether or not the components have more than one stage, or even more than one shaft. Thus a two-shaft multistage engine is designated simply as working on a CBE cycle.

There are, naturally, more-complex cycles in which gas-turbine engines can be configured. Two of much current interest are the combined cycles, in which the exhaust from a CBE-cycle engine passes to a steam generator, which in turn supplies a steam turbine; and the steam-injected cycle, similar to the combined cycle except that the steam raised is mixed with the hot gas and passed through the turbine of the gas-turbine engine. The analysis of these two cycles is beyond the scope of this text, but they are described a little more fully in a later section.

3.6   Cycle performance calculations

Simple and heat-exchanger cycles

The two measures of Brayton-cycle performance usually wanted in preliminary design are the specific power, sometimes expressed as the power per unit mass flow, but better expressed nondimensionally as $\dot{W}' \equiv (\dot{W}/(\dot{m}\,\overline{C_{p,c}}T_{0,1})$, and the thermal efficiency η_{th}. The two most important cycle characteristics are the compressor pressure ratio, $r_c \equiv (p_{0,2}/p_{0,1})$, and the turbine-inlet-to-compressor-inlet temperature ratio $T' \equiv (T_{0,4}/T_{0,1})$.

In a real, as distinct from an ideal, cycle there are many other variables that affect the specific power and the efficiency in addition to the pressure and temperature ratios. We shall account for eight: the compressor and expander polytropic efficiencies; the leakage; the sum of the losses in relative stagnation pressure; the heat-exchanger effectiveness; and the mean specific heats for compression, expansion, and heat addition. Rather than combine all these variables into one large equation for each performance measure, we shall introduce some new variables that will simplify the task of dealing with changes in the variables.

The relations that can be used for both the simple, CBE, cycle and the heat-exchanger, CBEX, cycle are given first, and the derivations are made subsequently. More-complex cycles are treated in later sections.

Specific power

$$\dot{W}' \equiv \frac{\dot{W}}{\dot{m}_c \overline{C_{p,c}} T_{0,1}} = \left[\frac{\dot{m}_e \overline{C_{p,e}}}{\dot{m}_c \overline{C_{p,c}}} \right] E_1 T' - C \tag{3.1}$$

Cycle thermal efficiency

$$\eta_{th} = \frac{[\dot{m}_e \overline{C_{p,e}}/\dot{m}_c \overline{C_{p,c}}]E_1 T' - C}{[\dot{m}_b \overline{C_{p,b}}/\dot{m}_c \overline{C_{p,c}}][T'(1 - \epsilon_{hx}(1 - E_1)) - (1 + C)(1 - \epsilon_{hx})]} \tag{3.2}$$

where

$$
\begin{aligned}
\dot{W} &\equiv \text{power output} \\
\dot{m}_c &\equiv \text{mass flow through compressor} \\
\dot{m}_e &\equiv \text{mass flow through expander (including effects of fuel addition} \\
&\qquad \text{and working-fluid leakage)} \\
\dot{m}_b &\equiv \text{mass flow through burner or heat-addition device} \\
\overline{C_{p,c}} &\equiv \text{mean specific heat during compression} \\
\overline{C_{p,e}} &\equiv \text{mean specific heat during expansion} \\
\overline{C_{p,b}} &\equiv \text{mean specific heat during heat addition} \\
E_1 &\equiv (T_{0,4} - T_{0,5})/T_{0,4} \\
&= 1 - \left\{ \left[(1 - \Sigma(\Delta p_0/p_0)) \, r_c \right]^{-(R/\overline{C_{p,e}})\eta_{p,e}} \right\}
\end{aligned}
\tag{3.3}
$$

$$
\begin{aligned}
\Sigma(\Delta p_0/p_0) &\equiv \text{sum of losses of relative stagnation pressure} \\
r_c &\equiv (p_{0,2}/p_{0,1}) \\
R &\equiv \text{gas constant} \\
\eta_{p,e} &\equiv \text{expander polytropic efficiency} \\
T' &\equiv (T_{0,4}/T_{0,1}) \\
C &\equiv (T_{0,2} - T_{0,1})/T_{0,1} = \left[r_c^{(R/\overline{C_{p,c}})/\eta_{p,c}} - 1 \right]
\end{aligned}
\tag{3.4}
$$

$$
\begin{aligned}
\eta_{p,c} &\equiv \text{compressor polytropic efficiency and} \\
\epsilon_{hx} &\equiv \text{heat-exchanger effectiveness} = (T_{0,3} - T_{0,2})/(T_{0,5} - T_{0,2}) \tag{3.5}
\end{aligned}
$$

For the simple CBE cycle the specific power is given by equation 3.1, as before. In the efficiency relation the heat-exchanger effectiveness, ϵ_{hx}, vanishes, and the expression becomes

$$\eta_{th} = \frac{[\dot{m}_e \overline{C_{p,e}}/\dot{m}_c \overline{C_{p,c}}]E_1 T' - C}{[\dot{m}_b \overline{C_{p,b}}/\dot{m}_c \overline{C_{p,c}}][T' - C - 1]} \tag{3.6}$$

In preliminary design, fuel addition and working-fluid leakage (e.g., for cooling) are often ignored, and calculations are made with a single value of mean specific heat. Then equation 3.6 becomes

$$\eta_{th,CBE} \approx \frac{E_1 T' - C}{T' - C - 1} \tag{3.7}$$

Equations 3.1 and 3.2 can be derived as follows.

The steady-flow energy equation (equation 2.11), $(\dot{Q}_{in} - \dot{Q}_{ex} + \dot{W}_{in} - \dot{W}_{ex})/\dot{m} = h_{0,2} - h_{0,1}$, can be used to find the work transfer in the (adiabatic) compressor and expander, and the heat transfer in the heat exchanger and burner:

$$\text{Net power} \equiv \dot{W} \equiv \text{expander power} - \text{compressor (positive) power}$$

$$= \dot{m}_e(h_{0,4} - h_{0,5}) - \dot{m}_c(h_{0,2} - h_{0,1}) \tag{3.8}$$

$$\equiv \dot{m}_e \overline{C_{p,e}}(T_{0,4} - T_{0,5}) - \dot{m}_c \overline{C_{p,c}}(T_{0,2} - T_{0,1}) \tag{3.9}$$

$$\text{Specific power} \equiv \frac{\dot{W}}{\dot{m}_c \overline{C_{p,c}} T_{0,1}}$$

$$= \left[\frac{\dot{m}_e \overline{C_{p,e}}}{\dot{m}_c \overline{C_{p,c}}}\right]\left[\frac{T_{0,4} - T_{0,5}}{T_{0,4}}\right]\left[\frac{T_{0,4}}{T_{0,1}}\right] - \left[\frac{T_{0,2} - T_{0,1}}{T_{0,1}}\right]$$

$$= \left[\frac{\dot{m}_e \overline{C_{p,e}}}{\dot{m}_c \overline{C_{p,c}}}\right] E_1 T' - C \tag{3.10}$$

which is equation 3.1. The parameters E_1, T' and C were defined earlier. The parameter C can be obtained from the pressure ratio using equation 2.78:

$$\left(\frac{T_{0,2}}{T_{0,1}}\right) = \left(\frac{p_{0,2}}{p_{0,1}}\right)^{(R/\overline{C_{p,c}})/\eta_{p,c}}$$

Therefore,

$$\left(\frac{T_{0,2} - T_{0,1}}{T_{0,1}}\right) \equiv C = \left(r_c^{(R/\overline{C_{p,c}})/\eta_{p,c}} - 1\right)$$

Similarly, the parameter E_1 can be obtained using equation 2.79:

$$\left(\frac{T_{0,4}}{T_{0,5}}\right) = \left(\frac{p_{0,4}}{p_{0,5}}\right)^{(R/\overline{C_{p,e}})\eta_{p,e}}$$

But the expansion ratio $p_{0,4}/p_{0,5} = (p_{0,2}/p_{0,2})(1 - \Sigma(\Delta p_0/p_0))$; therefore,

$$\frac{T_{0,4} - T_{0,5}}{T_{0,4}} \equiv E_1 = 1 - \frac{1}{(T_{0,4}/T_{0,5})} = 1 - \left[r_c\left(1 - \Sigma(\Delta p_0/p_0)\right)\right]^{(-R/\overline{C_{p,e}})\eta_{p,e}}$$

The specific power is the numerator of the expression for thermal efficiency, equation 3.2. It is unaffected by the presence or absence of a heat exchanger except for the influence of any additional pressure drop on E_1. The denominator is the heat addition, derived as follows for CBE and CBEX cycles:

$$\text{Heat addition} = \dot{m}_b(h_{0,4} - h_{0,3}) \tag{3.11}$$

$$= \dot{m}_b \overline{C_{p,b}}(T_{0,4} - T_{0,3}) \tag{3.12}$$

We substitute for $T_{0,3}$ from the definition of heat-exchanger effectiveness, and we divide by the same terms used to make the specific power nondimensional. We call this the specific heat addition, $\dot{Q}'$:

$$
\begin{aligned}
\dot{Q}' &\equiv \left[\frac{\text{Heat addition}}{\dot{m}_c \overline{C_{p,c}} T_{0,1}} \right] = \left[\frac{\dot{m}_b \overline{C_{p,b}}}{\dot{m}_c \overline{C_{p,c}}} \right] \left[\frac{T_{0,4}}{T_{0,1}} - \epsilon_{hx} \frac{T_{0,5} - T_{0,2}}{T_{0,1}} - \frac{T_{0,2}}{T_{0,1}} \right] \\[2ex]
&= \left[\frac{\dot{m}_b \overline{C_{p,b}}}{\dot{m}_c \overline{C_{p,c}}} \right] T' - \epsilon_{hx} \left[\frac{T_{0,5} - T_{0,4}}{T_{0,4}} \right] \left[\frac{T_{0,4}}{T_{0,1}} \right] - \epsilon_{hx} \left[\frac{T_{0,4}}{T_{0,1}} \right] + (\epsilon_{hx} - 1) \left[\frac{T_{0,2}}{T_{0,1}} \right] \\[2ex]
&= \left[\frac{\dot{m}_b \overline{C_{p,b}}}{\dot{m}_c \overline{C_{p,c}}} \right] \{ T' + \epsilon_{hx} E_1 T' - \epsilon_{hx} T' + \epsilon_{hx}(1 + C) - (1 + C) \}
\end{aligned}
$$

Therefore,

$$
\left[\frac{\text{Heat addition}}{\dot{m}_c \overline{C_{p,c}} T_{0,1}} \right] = \left[\frac{\dot{m}_b \overline{C_{p,b}}}{\dot{m}_c \overline{C_{p,c}}} \right] \{ T'[1 - \epsilon_{hx}(1 - E_1)] - (1 + C)(1 - \epsilon_{hx}) \} \qquad (3.13)
$$

Some influence coefficients for heat-exchanger (CBEX) cycles calculated using these relations are shown in table 3.2.

Intercooled and reheat cycles

Intercooled cycles

We showed in section 2.6 that compression work is directly proportional to absolute temperature at compressor inlet. A large pressure ratio will, in an adiabatic machine, produce a relatively large increase in temperature, so that the work required to accomplish an incremental pressure ratio in the latter part of the compression will be much larger than for the same incremental pressure ratio in the low-pressure part.

It is therefore attractive to consider the possibility of breaking the compression process into several parts and cooling the compressed gas between the stages or groups of stages. The power required can, even with quite inefficient compressor stages, be less than for an ideal adiabatic process, so that polytropic and isentropic efficiencies of well over 100 percent can be obtained. Accordingly, the efficiencies of intercooled compressors for the process industries are normally given in terms of the ideal isothermal process, figures 3.5a and 3.5b.

When the compressor power is reduced to such an extent in a gas-turbine cycle, the net output increases to high levels. Without a heat exchanger, however, the thermal efficiency would be low because of the large amount of additional fuel required to heat up the compressed gas from the low compressor-delivery temperature. This disadvantage is completely overcome when a heat exchanger is incorporated.

Even for a high-temperature-ratio highly regenerated CBEX cycle having a low optimum pressure ratio of 3.5, the addition of one intercooler could increase the compressor efficiency by about seven points (figure 3.5b), which would produce a significant increase in thermal efficiency and specific power.

Table 3.2. Influence coefficients for heat-exchanger cycles

Compressor pressure ratio, r_c	3.0	3.5	4.0
Specification of base cases			
Compressor polytropic efficiency	0.9	0.9	0.9
Expander polytropic efficiency	0.9	0.9	0.9
Compressor-inlet total temperature, K	300	300	300
Turbine-inlet total temperature, K	1200	1200	1200
Heat-exchanger effectiveness	0.9	0.9	0.9
Sum of stagnation-pressure losses	0.10	0.10	0.10
Leakage	0.05	0.05	0.05
Mechanical efficiency	0.98	0.98	0.98
Base-case performance:			
Thermal efficiency, η_{th}	0.414	0.420	0.488
Specific power (non-dimensional)	0.452	0.501	0.818
Effect on thermal efficiency of:			
Leakage increasing from 0.05 to 0.06, $-\Delta\eta_{th}$, %	0.39	0.43	0.34
$-(\Delta\eta_{th}/\eta_{th})$, %	0.94	1.02	0.70
Effectiveness falling from 0.90 to 0.89, $-\Delta\eta_{th}$, %	0.71	0.60	0.71
$-(\Delta\eta_{th}/\eta_{th})$, %	1.71	1.43	1.45
Expander efficiency falling from 0.90 to 0.89, $-\Delta\eta_{th}$, %	0.49	0.51	0.44
$-(\Delta\eta_{th}/\eta_{th})$, %	1.18	1.21	0.90
Compressor efficiency falling from 0.90 to 0.89, $-\Delta\eta_{th}$, %	0.47	0.53	0.42
$-(\Delta\eta_{th}/\eta_{th})$, %	1.13	1.26	0.86
Pressure losses increasing from 0.10 to 0.11, $-\Delta\eta_{th}$, %	0.50	0.44	0.35
$-(\Delta\eta_{th}/\eta_{th})$, %	1.21	1.05	0.72
Turbine-inlet temperature falling by 100 K, $-\Delta\eta_{th}$, %	2.77	3.08	1.80
$-(\Delta\eta_{th}/\eta_{th})$, %	6.69	7.33	3.69

There have been several attempts to harvest the benefits of the intercooled cycle. Rolls Royce produced the RM 60, with two intercoolers and three separate rotating shafts (figure 3.6.) This engine was used to power naval vessels, in which the ready availability of cooling water made it advantageous. Ford developed an automotive engine with a rather similar, but additionally reheated cycle (figure 3.6). In this case the intercoolers had to be air-cooled, giving higher losses and reduced intercooling. Both engines were too complex for the then-current state of the art. With US Navy funding in the 1990s Westinghouse (later Northrop-Grumman) and Rolls Royce developed the WR-21, an intercooled regenerative gas-turbine engine for marine propulsion (figure 3.22). The intercooled cycle will undoubtedly find a place in the future.

Performance calculations for intercooled and reheat cycles

The formulas for calculating the specific power and the thermal efficiency of an intercooled reheated heat-exchange cycle are given in this section. The derivation is similar to that for equations 3.1 and 3.2 and will not be given in detail.

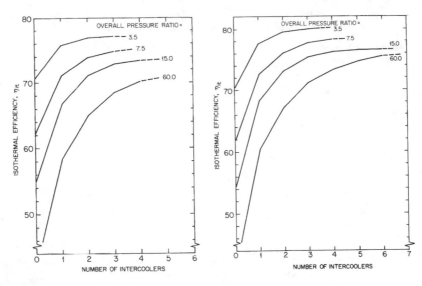

Figure 3.5. Isothermal efficiency of intercooled (a) centrifugal and (b) axial compressors. From Wilson (1965)

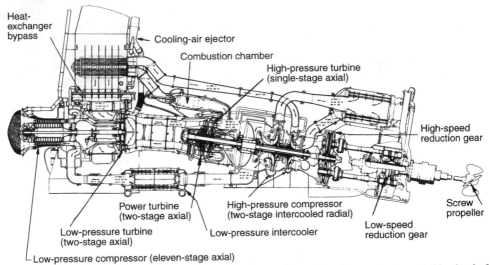

This compact complex-cycle engine powered the Royal Navy frigate Grey Goose. It was probably ahead of its time: in the 1950s and 1960s most complex-cycle engines ended up with component mismatches through the inability to predict component performance maps with sufficient accuracy. The design of this engine was somewhat unusual in that the power was extracted through the intermediate-pressure turbine. The long ducts were made in the form of narrow-angle diffusing cones to reduce pressure losses.

Figure 3.6. Rolls-Royce RM 60 marine engine working on the CICICBEX cycle. Courtesy Rolls-Royce, plc

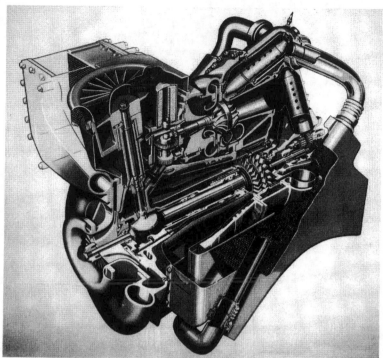

This is a complex (three-shaft intercooled-reheated-recuperated) 225-kW (300-bhp) engine developed in the late 1950s. The low-pressure compressor at lower left is driven at 46,500 rpm by the two-stage low-pressure turbine at right-center. The air is delivered to an intercooler, through which ambient air is drawn by the axial fan. This is driven from the low-pressure-compressor shaft at about 18,000 rpm by bevel gears. The cooled compressed air then passes to the high-pressure compressor, top center, formed as a "monorotor" with the high-pressure radial-inflow turbine to its right. This is free-running except for its bevel-geared connection to the starter-generator to its left. The high-pressure air, at about 16 bar, passes through twin recuperators to the high-pressure combustor, angled down at the top to the high-pressure radial-inflow turbine. The exhaust gases, still rich in oxygen, from this turbine go to the intermediate-pressure reheat combustor at the right. This feeds the intermediate-pressure single-stage power turbine that provides the output power, and then the two low-pressure turbines. The turbine outlet flow finally passes to the twin recuperators and then to the exhaust duct. The engine developed maximum efficiency at 60% power, and consumption was below 85 g/MJ (0.5 lbm/hp-h) from 38% to 82% power. The success of this engine led Ford to start development of an engine of double the power on a similar cycle (the "705"), but the program was stopped in 1964 when the decision was made to change to a much simpler engine because of the high estimated production costs of the 704 and 705 engines.

Figure 3.7. Ford 704 truck engine working on the CICICBEBEX cycle. Courtesy of The Ford Motor Company and A. F. McClean

The cycle considered has three compressors of equal pressure ratio and three expanders also of equal pressure ratio. The pressure ratio of each expander will then equal the compressor pressure ratio less one third of the sum of the relative-stagnation-pressure losses $\Sigma(\Delta p_0/p_0)$. The first heat-addition combustor, 3-4a in figure 3.8, and the two reheat combustors, 5a-4b and 5b-4c, take the working fluid to the same maximum temperature: $T_{0,4a} = T_{0,4b} = T_{0,4c}$.

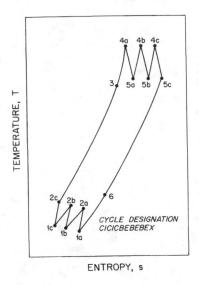

Figure 3.8. Diagram of a multiple intercooler reheated heat-exchanger cycle

The compressors, however, have different inlet and outlet temperatures. The two intercoolers have been specified to have the same effectiveness, ϵ_{ic}, and to reject heat to the ambient at $T_{0,1a}$. Therefore $T_{0,1c} > T_{0,1b} > T_{0,1a}$ and $T_{0,2c} > T_{0,2b} > T_{0,2a}$.

Specific power

$$W' \equiv \dot{W}/(\dot{m}\overline{C_{p,c}}T_{0,1a})$$

where $\dot{W} \equiv$ work from three expanders + (negative) work to three compressors

$$\frac{\dot{W}}{\dot{m}\overline{C_{p,c}}T_{0,1a}} = \left[\frac{\dot{m}_e\overline{C_{p,e}}}{\dot{m}_c\overline{C_{p,c}}}\right] T' \times 3E_3$$

$$- C\{1 \qquad \text{first compressor}$$

$$+ [C(1 - \epsilon_{ic}) + 1] \qquad \text{second compressor}$$

$$+ [(C(1 - \epsilon_{ic}) + 1)(1 + C)(1 - \epsilon_{ic}) + \epsilon_{ic}]\} \quad \text{third compressor} \qquad (3.14)$$

Specific heat addition

$$\dot{Q}' \equiv \dot{Q}'/\dot{m}\overline{C_{p,c}}T_{0,1}$$

$$= \frac{\dot{Q}}{\dot{m}\overline{C_{p,c}}T_{0,1a}} = \left[\frac{\dot{m}_b\overline{C_{p,b}}}{\dot{m}_c\overline{C_{p,c}}}\right] T'$$

$$\times \left\{\left[1 - \epsilon_{hx}(1 - E_3) + (1 - \epsilon_{hx})(C(1 - \epsilon_{ic}) + 1)(1 + C)^2(1 - \epsilon_{ic})\frac{1}{T'}\right]\right.$$

$$\text{first combustor}$$

$$\left. +2E_3\right\} \qquad \text{second and third combustors} \qquad (3.15)$$

The cycle thermal efficiency is the ratio of the specific power (equation 3.14) to the specific heat addition (equation 3.15):

$$\eta_{th} = \frac{[\dot{W}/(\dot{m}\,\overline{C_{p,c}}T_{0,1a})]}{[\dot{Q}/(\dot{m}\,\overline{C_{p,c}}T_{0,1a})]}$$
(3.16)

The symbols are as defined above except that here

$$E_3 \equiv \frac{T_{0,4c} - T_{0,5c}}{T_{T_{0,4c}}} = \left\{ 1 - \left[\left(1 - \frac{1}{3}\Sigma\left(\frac{\Delta p_0}{p_0}\right) \right) r_c \right]^{-(R/\overline{C_{p,e}})\eta_{p,e}} \right\}$$

$$r_c \equiv \frac{p_{0,2a}}{p_{0,1a}} = \frac{p_{0,2b}}{p_{0,1b}} = \frac{p_{0,2c}}{p_{0,1c}}$$

$$C \equiv \frac{T_{0,2a} - T_{0,1a}}{T_{0,1a}} = \frac{T_{0,2b} - T_{0,1b}}{T_{0,1b}} = \frac{T_{0,2c} - T_{0,1c}}{T_{0,1c}} = \left[r_c^{R/(\overline{C_{p,c}}\eta_{p,c})} - 1 \right]$$

$$\epsilon_{hx} \equiv \frac{T_{0,3} - T_{0,2c}}{T_{0,5} - T_{0,2c}}$$

$$\epsilon_{ic} \equiv \frac{T_{0,2a} - T_{0,1b}}{T_{0,2a} - T_{0,1a}} = \frac{T_{0,2b} - T_{0,1c}}{T_{0,2b} - T_{0,1a}}$$

$$T' = \frac{T_{0,4}}{T_{0,1}}$$

The relative magnitudes of these terms may be seen from an example.

Example 2 Double-intercooled reheated heat-exchanger cycle

Find the specific power, the thermal efficiency, and the individual contributions of the different compressors, expanders, and combustors to the performance characteristics of a double-intercooled-reheated heat-exchanger cycle as shown in figure 3.8, with the following specifications:

$$T_{0,1a} = 300\text{K}$$
$$T_{0,4a} = T_{0,4b} = T_{0,4c} = 1,500\text{ K}$$
$$p_{0,1a} = 0.95\text{ bar}$$
$$\frac{p_{0,2a}}{p_{0,1a}} = \frac{p_{0,2b}}{p_{0,1b}} = \frac{p_{0,2c}}{p_{0,1c}} = 2.90$$
$$\eta_{p,e} = \eta_{p,c} = 0.9$$
$$\Sigma\left(\frac{\Delta p_0}{p_0}\right) = 0.15$$
$$\epsilon_{hx} = \epsilon_{ic} = 0.9$$

The leakage and fuel addition are such that

$$\left[\frac{(\dot{m}_e\,\overline{C_{p,e}})}{(\dot{m}_c\,\overline{C_{p,c}})} \right] = 1.0$$

Solution.

From the above specifications:

$$T' = 5.0, \qquad C = 0.40, \quad \text{and} \quad E_3 = 0.1933 \text{ (after iterating for } \overline{C_{p,e}}),$$

These parameters are inserted into equations 3.14 through 3.16. The contribution of:

the first expander to the specific power $= 0.1933 \times 5 = +0.967$
the second expander to the specific power $= 0.1933 \times 5 = +0.967$
the third expander to the specific power $= 0.1933 \times 5 = +0.967$
the first compressor to the specific power $= 1.0 \times (-0.4) = -0.40$
the second compressor the specific power $= 1.04 \times (-0.4) = -0.416$, and of
the third compressor to the specific power $= 1.0456 \times (-0.4) = -0.418$
Net specific power $= \dot{W}/(\dot{m}_c \overline{C_{p,c}} T_{0,1}) = 1.661$

The contribution of:

the first combustor to the heat added $= 0.278 \times 5 = 1.390$
the second combustor the heat added $= 0.1933 \times 5 = 0.967$, and of
the third combustor to the heat added $= 0.1933 \times 5 = 0.967$
Total heat added $= \dot{Q}(\dot{m}_c \overline{C_{p,c}} T_{0,1a}) = 3.324$
Thermal efficiency $= 0.500$

Performance calculations for other cycles

With appropriate changes, equations 3.14 and 3.15 may be used for other cycles.

Single-intercooled-reheated heat-exchanger cycle (CICBEBEX cycle, figure 3.9)

$$\text{Specific power} \equiv \frac{\dot{W}}{\dot{m}_c \overline{C_{p,c}} T_{0,1a}}$$

$$= \left[\frac{\dot{m}_e \overline{C_{p,e}}}{\dot{m}_c \overline{C_{p,c}}} \right] T' \times 2E_2 - C[2 + C(1 - \epsilon_{ic})] \qquad (3.17)$$

where the symbols are as defined above, except that

$$E_2 \equiv 1 - \frac{T_{0,5}}{T_{0,4}} = \left\{ 1 - \left[\left(1 - \frac{1}{2}\Sigma \left(\frac{\Delta p_0}{p_0} \right) \right) r_c \right]^{(-R/\overline{C_{p,e}})\eta_{p,e}} \right\}$$

Specific heat addition

$$\frac{\dot{Q}}{\dot{m}_c \overline{C_{p,c}} T_{0,1a}} = \left[\frac{\dot{m}_b \overline{C_{p,b}}}{\dot{m}_c \overline{C_{p,c}}} \right] \times T' \{(1 + E_2) + \epsilon_{hx}(1 - E_2) -$$

$$\frac{(1 + C)}{T'}(1 - \epsilon_{ic})[(1 + C)(1 - \epsilon_{ic}) + \epsilon_{ic}] \right\} \qquad (3.18)$$

The cycle thermal efficiency is the ratio (specific power/specific heat added).

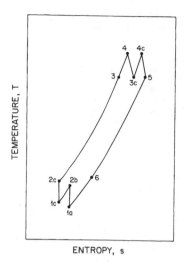

Figure 3.9. Intercooled-reheated (CICBEBEX) cycle

Single-intercooled heat-exchanger cycle CICBEX (figure 3.10)

$$\text{Specific power} \equiv \frac{\dot{W}}{\dot{m}_c \overline{C_{p,c}} T_{0,1a}}$$

$$= \left[\frac{\dot{m}_e \overline{C_{p,e}}}{\dot{m}_c \overline{C_{p,c}}} \right] T' E_3 - C[2 + C(1 - \epsilon_{ic})] \tag{3.19}$$

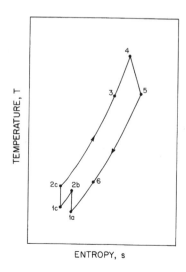

Figure 3.10. Intercooled heat-exchanger (CICBEX) cycle

where the symbols are as defined above except that

$$E_2 \equiv \frac{T_{0,4} - T_{0,5}}{T_{0,4}} = \left\{ 1 - \left[\left(1 - \Sigma \left(\frac{\Delta p_0}{p_0} \right) \right) r_c \right]^{(-R/\overline{C_{p,e}})\eta_{p,e}} \right\}$$

$$\text{Specific heat addition} \equiv \frac{Q'}{\dot{m}_c \overline{C_{p,c}} T_{0,1a}} = \left[\frac{\dot{m}_b \overline{C_{p,b}}}{\dot{m}_c \overline{C_{p,c}}} \right]$$

$$\times \, T' \left\{ 1 - (1 - E_3)\epsilon_{hx} - \frac{(1 + C)}{T'}(1 - \epsilon_{hx}) \times [1 + C(1 - \epsilon_{ic})] \right\} \quad (3.20)$$

The cycle thermal efficiency is the ratio of equations 3.19 and 3.20 in the same way as before.

3.7 Efficiency versus specific power for shaft-power engines

While the performance of shaft-power cycles can be calculated to a high degree of accuracy using the foregoing relations (extended in later sections to more-complex cycles), even greater realism can be introduced by using computer codes that allow for all possible leakage paths, for different systems of flow mixing, and for different blade-cooling arrangements. In the following we are giving some performance plots using the computer programs presented by Korakianitis and Wilson (1994). Besides accounting for a network of possible leakage paths (figure 3.11), the programs incorporate alternative blade-cooling flows and compressor and turbine peak polytropic efficiencies that are functions of the design-point pressure ratio, r_c, mass flow $\dot{m}$ and/or component diameter, as explained in the following.

Compressor and turbine efficiencies

As mentioned in section 2.6, the use of isentropic efficiency incorporates an unwanted "reheat" effect that decreases the value of compressor efficiency as the pressure ratio is raised, even if losses (e.g., the rate of entropy production) are not increased. The use of polytropic efficiencies avoids this problem. There are, however, other effects that reduce efficiency as the pressure ratio is increased.

The major effect is that the passage size must be greatly reduced at the high-pressure end of a turbine or compressor, producing small blades for which the tip-clearance losses become relatively large. Also, the low-pressure blades become long and highly twisted, with high aerodynamic loading at the hub, again decreasing efficiency.

Axial compressors and turbines would tend to have higher efficiencies than radial-flow machines, except in small sizes, where the boundary layers would dominate the area available to the flow. In that case radial components would be more efficient. In addition to boundary-layer effects, the efficiency of components is also a function of specific speed (chapter 5). Open-cycle engines with power output in excess of about 500 kW are likely to have axial compressors and turbines, while smaller engines (typically those with power output under 200 kW) are likely to have radial (centrifugal) compressors and

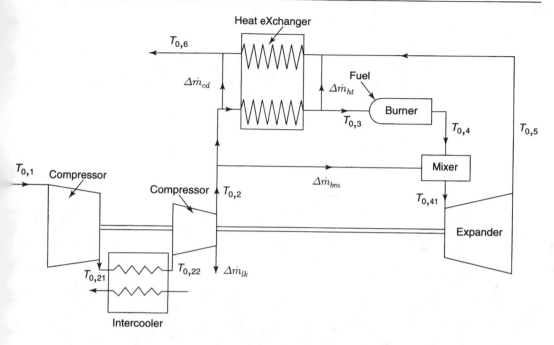

Figure 3.11. Network of leakage paths used in computing performance charts. From Korakianitis and Wilson (1994)

radial turbines. In the 200-kW to 500-kW range the turbines are likely to be axial, while the compressors may be axial or radial. (Examples of these axial and radial components are shown in various figures throughout the book, starting from the introductory historical chapter.) Figures 3.12 to 3.15 show estimates of best published polytropic efficiencies (defined using stagnation conditions at inlet to static pressure at diffuser outlet; see chapter 2[3]) for axial and radial compressors and turbines as a function of component size. The axial-component-efficiency lines can be modeled by:

$$\eta_{p,c,ts} = 0.862 + 0.015 \ln(\dot{m}) - 0.0053 \ln(r_c) \tag{3.21}$$

$$\eta_{p,e,ts} = 0.7127 + 0.03 \ln(d_m) - 0.0093 \ln(r_e^{-1}) \tag{3.22}$$

where $\dot{m}$ is the mass flow rate in kg/s and should not exceed the value of 90 kg/s in these equations, d_m is the rotor mean blading diameter in mm, and r_c and r_e are the compressor and turbine pressure ratios. The radial-component efficiency lines can be modeled by:

$$\eta_{p,c,ts} = 0.878 + 0.030 \ln(\dot{m}) - 0.0037 (r_c) \tag{3.23}$$

$$\eta_{p,e,ts} = 0.6984(d_{sh})^{0.0449} - \frac{r_e^{-1} - 1}{200} \tag{3.24}$$

where d_{sh} is the rotor shroud (outside) diameter in mm.

---

[3] Therefore the outlet static pressures are defined as the useful stagnation outlet pressures, as in section 2.6.

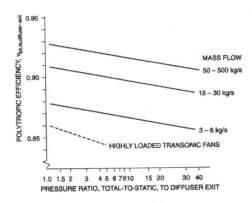

Figure 3.12. Polytropic stagnation-to-static efficiencies of axial compressors as function of pressure ratio

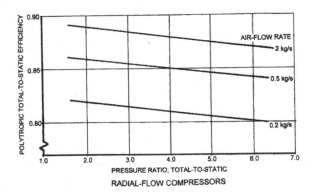

Figure 3.13. Polytropic stagnation-to-static efficiencies of radial compressors as function of pressure ratio

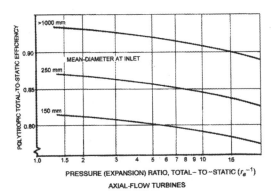

Figure 3.14. Polytropic stagnation-to-static efficiencies of axial turbines as function of pressure (expansion) ratio

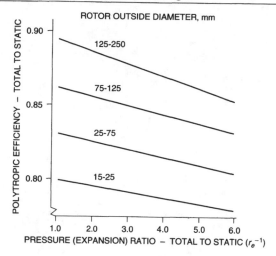

Figure 3.15. Polytropic stagnation-to-static efficiencies of radial turbines as function of pressure (expansion) ratio

Cooling flows

In advanced gas-turbine engines the turbine-inlet temperature is lower than the maximum temperature of the flame, and higher than the maximum temperature of the blading material T_{bm}, or the maximum temperature modern metal alloys for the hot sections of gas turbines can withstand. This is accomplished with cooling. A portion of the compressor exit flow is used to dilute the flow around the flame, thus reducing combustor-exit temperature. A second portion is used to cool the first turbine stator (nozzle). A third portion is used to cool the first turbine rotors, and subsequent stators and rotors downstream, until the gas temperature drops below the blading-material temperature limit. (Blade cooling is treated in chapter 10.) Depending on the engine, one, two, three or more turbine stages may be cooled with compressor-outlet air. The cooling air is ejected around the blades and becomes lower-enthalpy wakes that mix with the main turbine flow while contributing to losses. The frontiers of cooling technology are a trade-off between materials that can withstand hotter temperatures and the development of cooling techniques to minimize the amount of cooling air required by the turbine.

Figure 3.16 shows the amount of blade-cooling flow (percentage of gas-generator flow) as a function of turbine-rotor-inlet temperature for some engines. Figure 3.17, derived from blade-cooling published data (Livingood, 1971) and updated, shows the fraction of compressor-air flow that is used to cool the turbine blading (designated as $\Delta \dot{m}_{bm}$) as function of cooling-technology levels and nondimensional cooling parameter ω_{bm}:

$$\omega_{bm} \equiv \frac{(T_{0,4} - T_{bm})}{(T_{0,4} - T_{0,2})} \tag{3.25}$$

where $T_{0,4}$ and $T_{0,2}$ are the turbine-inlet (first rotor) and compressor-outlet stagnation temperatures respectively. The three lines shown in figure 3.17 correspond to different

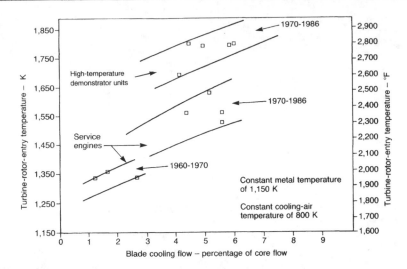

Figure 3.16. Turbine-inlet temperature and blade-cooling flow. Courtesy Rolls Royce plc

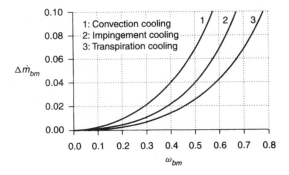

Figure 3.17. Blade-cooling mass flow fraction ($\Delta \dot{m}_{bm}$) as a function of ω_{bm}. From Korakianitis and Wilson (1994)

cooling technology levels, ((1) film convection; (2) film impingement; and (3) transpiration, full-coverage film).

All three cooling flows generate entropy due to mixing of different-temperature gas streams, and this contributes to pressure losses in the cycle from $p_{0,4}$ to $p_{0,41}$. Of these three portions of compressor-outlet air used for cooling flows, only the third one affects the temperature from which turbine expansion starts, dropping the temperature from $T_{0,4}$ to $T_{0,41}$. Traditionally the maximum cycle temperature in a gas-turbine engine is defined as the stagnation temperature at first rotor inlet, $T_{0,4}$. For turbine-expansion calculations this means all flows before $T_{mx} = T_{0,4}$ can be considered adiabatic. In cooled turbines the calculation of expansion work must take into consideration the non-adiabatic effects of the third portion of cooling flow, extracted from compressor outlet to cool the first turbine rotors, and any additional stators and rotors downstream that must be cooled.

Component and process models

We present component and process models that can be used to compute the thermal efficiency and specific power of shaft-power cycles. (The same models can be used to compute the performance of jet-propulsion cycles and combined cycles.) The models are then used to investigate the performance of shaft-power cycles in this section, and of turbojet, turbofan and combined cycles in later sections. The values of: C_p and $\overline{C_p}$ of air; of fuel-air mixture, and of products of combustion, are evaluated as analytic-polynomial functions of temperature and fuel-air mixture as described in appendix A. In the following C_p is $\overline{C_p}$ evaluated at a temperature appropriate to make the equations exact (chapter 2). A close approximation is to evaluate C_p at the mean temperature of the process for the given fuel-air ratio.

The compressor and turbine performance are computed using

$$\left(\frac{T_{0,2}}{T_{0,1}}\right) = \left(\frac{p_{0,2}}{p_{0,1}}\right)^{R/(C_p \eta_{p,c,tt})} \tag{3.26}$$

$$\dot{W}_c = \dot{m}_c(h_{0,2} - h_{0,1}) = \dot{m}_c C_p T_{01} \left(r_c^{R/(C_p \eta_{p,c,tt})} - 1\right)$$

$$\left(\frac{T_{0,5}}{T_{0,41}}\right) = \left(\frac{p_{0,5}}{p_{0,41}}\right)^{(R\eta_{p,e,tt})/C_p} \tag{3.27}$$

$$\dot{W}_e = \dot{m}_e(h_{0,41} - h_{0,5}) = \dot{m}_e C_p T_{0,41} \left(r_e^{-(R\eta_{p,e,tt})/C_p} - 1\right)$$

where $\eta_{p,c,tt}$ and $\eta_{p,e,tt}$ denote the polytropic stagnation-to-stagnation compressor and turbine efficiencies from component inlet to diffuser outlet, excluding bearing losses and where $p_{0,2}$ and $p_{0,5}$ are the "defined outlet stagnation pressures" as discussed in section 2.6. They may therefore be the static pressures at these locations, and the efficiencies to be used would then be the stagnation-to-static values.

The heat exchanger or recuperator and intercooler are modeled by the effectiveness of their respective energy-exchange processes. A formal definition of effectiveness is "actual desirable temperature rise (or drop) divided by the theoretically maximum possible temperature rise (or drop)". For the heat exchanger in CBEX cycles the actual temperature rise is from $T_{0,2}$ to $T_{0,3}$, and the maximum possible temperature rise with an infinitely long counter-flow heat exchanger is from $T_{0,2}$ to $T_{0,5}$, so that the heat-exchanger effectiveness, ϵ_{hx} is defined by:

$$\epsilon_{hx} \equiv \frac{T_{0,3} - T_{0,2}}{T_{0,5} - T_{0,2}} \tag{3.28}$$

Similarly, the intercooler effectiveness in CICBEX cycles ϵ_{ic} is defined by:

$$\epsilon_{ic} \equiv \frac{T_{0,21} - T_{0,22}}{T_{0,21} - T_{0,1}} \tag{3.29}$$

where it is assumed that the cooling medium is available at atmospheric temperature equal to that at compressor inlet, $T_{0,1}$.

All losses modeled as pressure drops reduce r_e from r_c by the sum of the stagnation-pressure-drop fractions in the cycle:

$$r_e^{-1} = r_c \left(1 - \Sigma \frac{\Delta p_0}{p_0} \right) \tag{3.30}$$

Example 3   CBEX-cycle calculations using component models

Estimate the thermal efficiency, specific power, and mass flow rate of a regenerative gas turbine that must produce 500 kW when it operates at sea-level conditions $T_{0,1} = 298$ K and $p_{0,1} = 101.3$ kPa. The turbine is ceramic and uncooled, and $T_{0,4} = 1700$ K. The compressor pressure ratio $r_c = 3.0$. The compressor and turbine stagnation-to-stagnation polytropic efficiencies are 0.88 each, the sum of stagnation-pressure-drop fractions is 0.10, and the heat-exchanger effectiveness is 0.95. For this estimate assume that the specific-heat capacity of the working fluid is $C_p = 1,100$ J/(kg · K) throughout the cycle.

Solution.   Even though this problem can be solved using the equations presented in section 3.6, here we present an alternative solution using the component models presented above.

The compressor outlet temperature is given by equation 3.26,

$$T_{0,2} = T_{0,1} r_c^{R/(C_p \eta_{p,c,tt})} = 298 \times 3^{286.96/(1,100 \times 0.88)} = 412.72 \text{ K}$$

The turbine pressure (expansion) ratio is $r_e = 0.90 r_c = 2.7$, and the turbine outlet temperature is given by equation 3.27

$$T_{0,5} = T_{0,4} r_t^{(R \eta_{p,e,tt})/(C_p)} = 1700 \times 2.7^{(286.96 \times 0.88)/1,100} = 1353.39 \text{ K}$$

The combustor-inlet temperature can now be evaluated from

$$\epsilon_{hx} \equiv \frac{T_{0,3} - T_{0,2}}{T_{0,5} - T_{0,2}} \Rightarrow T_{0,3} = 412.72 + 0.95 \times (1353.39 - 412.72) = 1306.36 \text{ K}$$

Since this is an estimate, we will neglect the increase of turbine mass flow rate due to fuel addition so that $\dot{m} = \dot{m}_c = \dot{m}_t$. Then the thermal efficiency is given by

$$\begin{aligned}
\eta_{th} &= \frac{\dot{W}_e - \dot{W}_c}{\dot{Q}_{in}} = \frac{\dot{m} C_p (T_{0,4} - T_{0,5}) - \dot{m} C_p (T_{0,2} - T_{0,1})}{\dot{m} C_p (T_{0,4} - T_{0,3})} \\
&= \frac{(T_{0,4} - T_{0,5}) - (T_{0,2} - T_{0,1})}{(T_{0,4} - T_{0,3})} = 0.59
\end{aligned}$$

The specific power is given by

$$\begin{aligned}
\dot{W}' &= \frac{\dot{W}_e - \dot{W}_c}{\dot{m} C_p T_{0,1}} = \frac{\dot{m} C_p (T_{0,4} - T_{0,5}) - \dot{m} C_p (T_{0,2} - T_{0,1})}{\dot{m} C_p T_{0,1}} \\
&= \frac{(T_{0,4} - T_{0,5}) - (T_{0,2} - T_{0,1})}{T_{0,1}} = 0.78
\end{aligned}$$

The required mass flow rate is estimated using the specific power

$$\dot{m} = \frac{\dot{W}_e - \dot{W}_c}{\dot{W}' C_p T_{0,1}} = \frac{500 \times 10^3}{0.78 \times 1,100 \times 298} = 1.96 \text{ kg/s}$$

Comments

1. One of the advantages of this type of calculation is that it shows the temperatures at several points in the cycle.

2. An iterative solution (suitable for computer programs) would account for variations of C_p as a function of temperature and fuel-air ratio, and for the non-adiabatic effects of possible cooling flows.

Typical performance charts for CBE, CBEX, and CICBEX cycles

Up to this point thermal efficiency has been evaluated as net power output divided by the enthalpy rise of the working fluid in the combustors or heaters. This form of efficiency is useful when the energy input to the cycle is from means other than fuel, such as solar energy and nuclear energy. Traditionally if the amount of fuel added to the cycle $\dot{m}_F$ is evaluated, then the thermal efficiency is evaluated as:

$$\eta_{th} = \frac{\dot{W}}{\dot{m}_F \cdot L_{HF}}$$

where $\dot{m}_f$ is the mass flow of fuel required at the reduced heating value at combustor-inlet temperature, and L_{HF} is the lower heating value of the fuel (any water produced by the combustion is in the vapor form in the products), evaluated at a reference temperature and pressure. Typically the reference pressure is 1 atmosphere and the reference temperature is 288 K. For the following cycle-performance plots we have used: the above equation to evaluate efficiency; the fuel-air ratio required for combustion products to reach $T_{0,4}$, evaluated as described in appendix A; and $L_{HF} = 43.124$ MJ/kg (evaluated at 288 K, see appendix A).

Using the lower heating value of the fuel results in thermal efficiency figures that are 0.5 points (at lower pressures and temperatures) to 6 points (at higher pressures and temperatures) in efficiency lower than the thermal efficiency evaluated with the enthalpy rise of the working fluid.

Unless otherwise specified in the figure, the performance plots for CBE, CBEX and CICBEX cycles shown in figures 3.18–3.40 have been produced with the following inputs:

> compressor inlet temperature $T_{0,1} = 288$K;
>
> the efficiency of large axial compressors is modeled by the top line of figure 3.12;
>
> the efficiency of large radial compressors is modeled by the top line of figure 3.13;
>
> the efficiency of small radial compressors is modeled by the bottom line of figure 3.13;

the efficiency of large axial turbines is modeled by the top line of figure 3.14;

the efficiency of large radial turbines is modeled by the top line of figure 3.15;

the efficiency of small radial turbines is modeled by the bottom line of figure 3.15;

temperature ratios $T' \equiv T_{0,4}/T_{0,1} = 4.5, 5.5, 6.5, 7.5$;

turbine blading material temperature T_{bm} (equation 3.25) and cooling technology levels and flows as described in the next paragraph;

mass-flow fraction from compressor outlet to overboard leakage or auxiliary uses 0.02;

burner efficiency 0.995;

cycle stagnation-pressure losses $\Sigma \Delta_{p0/p0}$ as prescribed in the figure captions;

compressor pressure ratios r_c as shown in the figures;

regenerator effectiveness as prescribed in the figure captions;

for CBEX and CICBEX cycles regenerator hot-side and cold-side leakage mass-flow fractions 0.025; and

for CICBEX cycles cold-side intercooler inlet temperature $T_{0,1} = 288$ K and intercooler effectiveness 0.90.

A portion of the cooling flow ($\Delta \dot{m}_{bm}$ in figure 3.17) is used in the rotors, and a portion is used in the stators of cooled stages. The cooling flow mixes with the main flow and lowers the average enthalpy for downstream stages. With the present practice of using the turbine first-rotor inlet temperature to define $T_{0,4}$, the first stator cooling flow is "not chargeable" to the cooling flow. Different numbers of cooled stages are used in different engines. We have assumed that cooling flow $\Delta \dot{m}_{bm}$ (figures 3.11 and 3.17) is used to cool the chargeable blade rows, thus lowering turbine inlet temperature from $T_{0,4}$ to $T_{0,41}$ in figure 3.11. For T' = 4.5 we used convection cooling (line 1 in figure 3.17) and $T_{bm} = 1120$ K, for T' = 5.5 we used impingement cooling (line 3 in figure 3.17) and $T_{bm} = 1140$ K, for T' = 6.5 we used transpiration cooling (line 3 in figure 3.17) and $T_{bm} = 1160$ K, and for T' = 7.5 we used future cooling technology (represented by a line 3.2, linearly interpolated "between" lines 2 and 3, below line 3, in figure 3.17); and $T_{bm} = 1200$ K. In addition, the maximum value of $\Delta \dot{m}_{bm} = 0.10$, or ten percent of the core flow, even if the cooling lines indicate higher values.

Figures 3.18, 3.19, and 3.21 are representative of the performance characteristics of CBE, CBEX, and CICBEX cycles. Later figures investigate details of the performance of each one of these cycles.

Figure 3.18 has been computed using the above component models and other cycle-input parameters shown in the table. The specific work is a function of r_c and is equal for the CBE and CBEX cycles. With $T_{mx} = T_{0,4} = T' \cdot T_{0,1} = 1584$ K the maximum pressure ratio for the combination of inputs is about 175:1. Specific power reaches a maximum at $r_c \approx 15$, where η_{th} for the CBE cycle is about 0.38. The corresponding CBEX cycle gives the same amount of work as the CBE cycle, but the heat exchanger reduces the required energy input such that the corresponding η_{th} for the CBEX cycle

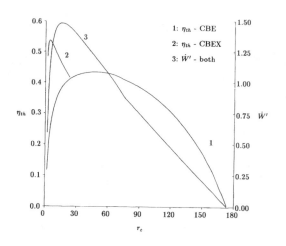

Cycle	CBE	CBEX
T_{01} (K)	288	288
T'	5.5	5.5
$\eta_{p,c,tt}$	large axial	
$\eta_{p,e,tt}$	large axial	
η_b	0.995	0.995
$\Sigma(\Delta p_0/p_0)$	0.07	0.07
T_{bm} (K)	1160	1160
$\Delta\dot{m}_{bm}$	line 3	line 3
$\Delta\dot{m}_{lk}$	0.02	0.02
ϵ_{hx}	-	0.95
$\Delta\dot{m}_{cd}$	-	0.025
$\Delta\dot{m}_{ht}$	-	0.025

Figure 3.18. Thermal efficiency and specific work as functions of pressure ratio for simple and regenerative cycles (input parameters shown in the table).

is about 0.46. The CBE cycle reaches its maximum thermal efficiency $\eta_{th,mx} \approx 0.43$ at $15 < r_c \approx 48$. The CBEX cycle reaches its maximum thermal efficiency $\eta_{th,mx} \approx 0.54$ at $15 > r_c \approx 4$. The heat exchanger becomes infeasible ($T_{05} < T_{02} + 20$ K) for pressure ratios above 25:1. The figure illustrates that there are three "optimum" pressure ratios: for maximum $\dot{W}'$ $r_c \approx 15$; for maximum $\eta_{th,CBE} \approx 0.40$ $r_c \approx 48$; and for maximum $\eta_{th,CBEX} \approx 0.55$ $r_c \approx 4$. This figure indicates that designers should choose different pressure ratios depending on cycle and intended application. It also explains why the thermal efficiency versus specific power lines (for a given T') turn counterclockwise for a CBE cycle, and clockwise for CBEX and CICBEX cycles (figure 3.19).

The cycle-parameter inputs for the performance of typical CBE, CBEX and CICBEX cycles shown in figure 3.19 are listed in the table next to the figure. Each cycle is shown for $T' = 5.5$. Each point plotted in figure 3.19 represents the design-point performance of a different engine. Some additional preliminary choices have been made for the types of components in the cycles. It has been specified that the heat exchanger is a high-effectiveness (0.95) rotary ceramic regenerator (see chapter 10). Such regenerators are better suited to low-pressure-ratio regenerative cycles, which result in relatively higher $T_{0,5}$. Conventional metal heat exchangers (such as the one used in the Textron Lycoming AGT 1500 engine) can be used only with relatively lower $T_{0,5}$ and are better suited to higher-pressure-ratio cycles. The leakage flows past the wiping seals of rotary regenerators ($\Delta\dot{m}_{cd}$ and $\Delta\dot{m}_{ht}$) increase as r_c increases, and they become excessive for $r_c > 6$ (McDonald, 1978). Values of $\Delta\dot{m}_{cd} = \Delta\dot{m}_{ht} = 0.025$ have been specified, based on the experiments reported by Helms et al. (1984). These suggest that the measured effectiveness of rotary ceramic regenerators was between 93.3% and 98.7%, the measured pressure drops were between 2.5% and 5.4%, and the measured leakages were 3.4% with NiO/3O; CaF$_2$ wearface and 4.5% with cooled seals.

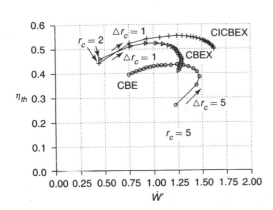

Cycle	CBE	CBEX	CICBEX
T_{01} (K)	288	288	288
T'	5.5	5.5	5.5
$\eta_{p,e,tt}$		large axial	
$\eta_{p,t,tt}$		large axial	
η_b	0.995	0.995	0.995
first r_c	5	2	2
r_c increment	5	1	1
$\Sigma(\Delta p_0/p_0)$	0.04	0.12	0.15
$\Delta \dot{m}_{bm}$	line 2	line 2	line 2
T_{bm} (K)	1140	1140	1140
$\Delta \dot{m}_{lk}$	0.02	0.02	0.02
ϵ_{hx}	-	0.95	0.95
ϵ_{ic}	-	-	0.90
$\Delta \dot{m}_{cd}$	-	0.025	0.025
$\Delta \dot{m}_{ht}$	-	0.025	0.025

Figure 3.19. Thermal efficiency versus specific power for the CBE, CBEX, and CICBEX cycles (input parameters shown in the table).

Figure 3.19 illustrates the advantages and disadvantages of the simple, regenerative, and intercooled-regenerative cycle for each application. In all cycles, as r_c increases η_{th} reaches a maximum value and then it declines. In general, the simple cycle has higher $\dot{W}'$ and lower η_{th} than the other cycles. Intercooled-regenerative cycles have higher η_{th} and higher $\dot{W}'$ than regenerative cycles. For a given T' the points of increasing r_c move counterclockwise for the simple cycle, and clockwise for the regenerative and intercooled-regenerative cycles.

The effects of ϵ_{hx} on CBEX-cycle performance are illustrated in figure 3.20. In this figure $T' = 5.5$, and the remaining inputs are identical to those of figure 3.19. As expected (figure 3.18) the maximum specific work is obtained at $r_c \approx 15$. The figure illustrates that for a given T' and a given r_c the specific work is not affected by ϵ_{hx}. This is because the heat exchanger does not affect work output. However, it does affect the required energy input, and the cycles with lower ϵ_{hx} have lower efficiencies. Higher values of ϵ_{hx} mean larger and heavier heat exchangers, and higher values of thermal efficiency mean less fuel is required for a given task (or mission profile for a military engine). The design of CBEX-cycle engines involves balancing thermal efficiency, regenerator effectiveness, and specific power. The Textron Lycoming AGT1500 CBEX-cycle engine illustrated in figure 3.21 has $r_c \approx 13.8$ and $\epsilon_{hx} \approx 0.68$, and a consequently fairly low average thermal efficiency (the major criticism against the Abrams M1 tank, powered by this engine, is that the amount of fuel the tank can carry limits its range).

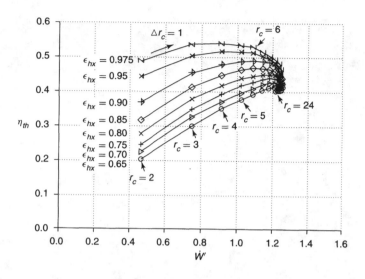

Figure 3.20. Fffect of regenerator effectiveness on CBEX-cycle performance for $T' = 5.0$. Other inputs as shown in figure 3.19.

This 1.1-MW (1500-shp) three-shaft recuperated (CBEX) engine drives the Abrams tank. The high-pressure cooled turbine drives the high-pressure axial-radial compressor. The intermediate-pressure turbine drives the axial low-pressure compressor. The two-stage low-pressure turbine supplies the output power to the epicyclic gearbox. The cylindrical recuperator surrounds the output shaft. The great majority of gas-turbine engines running on the CBEX cycle have been this engine.

Figure 3.21. Textron-Lycoming AGT-1500 CBEX cycle engine. Courtesy Allied-Signal Aerospace

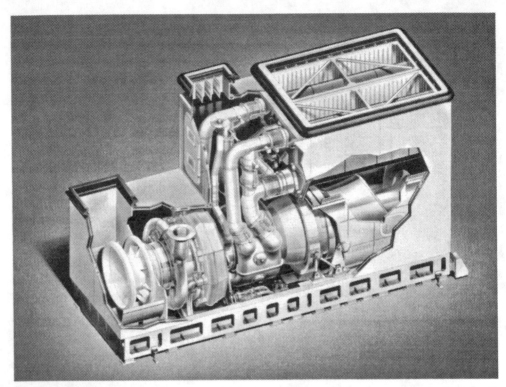

The CICBEX cycle has several advantages for marine propulsion. It promises a thermal efficiency of over 50%, similar to that of the highest-efficiency Diesel engine, while being far lighter and more compact. (Major servicing of marine gas turbines is normally carried out on land, the engine being withdrawn through the stack and replaced by a spare.) The sea provides a heat sink for the intercooler. The higher overall pressure ratio leads to a high turbine expansion ratio, giving turbine-outlet temperatures low enough for the use of a metal recuperator. The turbomachinery in the WR-21 engine is derived from different members of the family of Rolls-Royce RB-211 fan-jets. The intercooler and recuperator are made by Allied-Signal (see chapter 10). The overall engineering, initially by Westinghouse, is by Northrop-Grumman.

Figure 3.22. U.S.Navy WR-21 intercooled-recuperative gas turbine (CICBEX). Courtesy Northrop-Grumman

Efficiency versus specific power for CBE, CBEX, and CICBEX cycles

The first plots (figures 3.23-3.26) show CBE cycles with efficiencies typical of large axial-flow and large- and small-radial-flow compressors and turbines, and total losses of stagnation pressure of 0.04-0.07.

Heat-exchanger cycles are treated in figures 3.27 to 3.35 with various combinations of component efficiencies, heat exchanger effectiveness, and pressure drops.

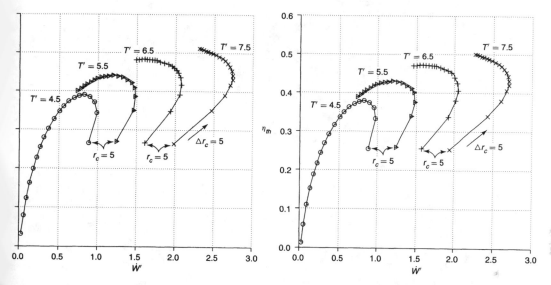

Figure 3.23. CBE-cycle performance with 4% stagnation-pressure losses and large-axial component efficiencies

Figure 3.24. CBE-cycle performance with 7% stagnation-pressure losses and large-axial component efficiencies

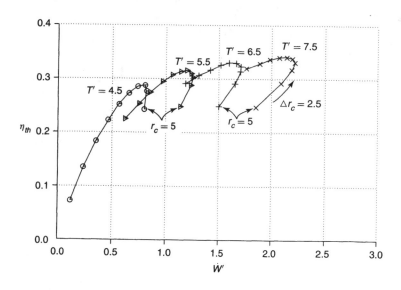

Figure 3.25. CBE-cycle performance with 4% stagnation-pressure losses and large-radial component efficiencies

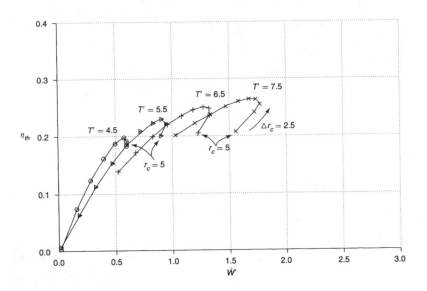

Figure 3.26. CBE-cycle performance with 4% stagnation-pressure losses and small-radial component efficiencies

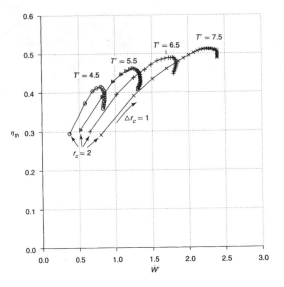

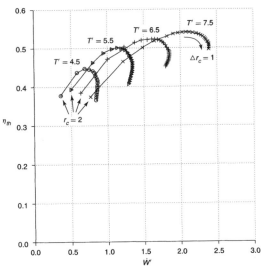

Figure 3.27. CBEX-cycle performance with 8% stagnation-pressure losses, large-axial component efficiencies, and 80%-effectiveness heat exchanger

Figure 3.28. CBEX-cycle performance with 10% stagnation-pressure losses, large-axial component efficiencies, and 90%-effectiveness heat exchanger

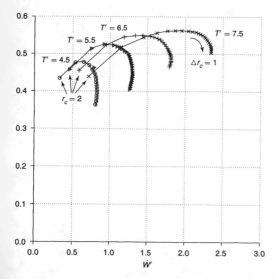

igure 3.29. CBEX-cycle performance 'ith 12% stagnation-pressure losses, large-xial component efficiencies, and 95%-ffectiveness heat exchanger

Figure 3.30. CBEX-cycle performance with 14% stagnation-pressure losses, large-axial component efficiencies, and 97.5%-effectiveness heat exchanger

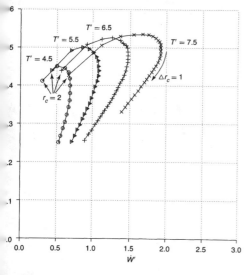

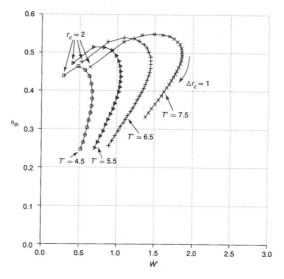

gure 3.31. CBEX-cycle performance th 12% stagnation-pressure losses, large-dial component efficiencies, and 95%-fectiveness heat exchanger

Figure 3.32. CBEX-cycle performance with 14% stagnation-pressure losses, large-radial component efficiencies, and 97.5%-effectiveness heat exchanger

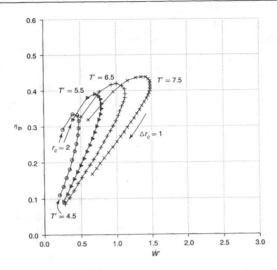

Figure 3.33. CBEX-cycle performance with 10% stagnation-pressure losses, small-radial component efficiencies, and 90%-effectiveness heat exchanger

The performance of intercooled heat-exchanger cycles is given in figures 3.36 to 3.40. As in the previous series, the first plot of the group shows a cycle that should be easily attainable, with fourteen-percent stagnation pressure losses, large-axial component

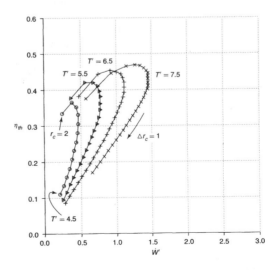

Figure 3.34. CBEX-cycle performance with 12% stagnation-pressure losses, small-radial component efficiencies, and 95%-effectiveness heat exchanger

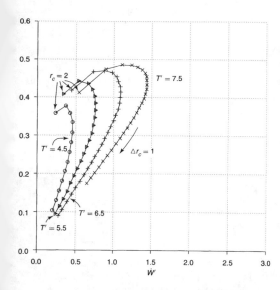

Figure 3.35. CBEX-cycle performance with 14% stagnation-pressure losses, small-radial component efficiencies, and 97.5%-effectiveness heat exchanger

Figure 3.36. CICBEX-cycle performance with 14% stagnation-pressure losses, large-axial component efficiencies, and 90%-effectiveness heat exchanger

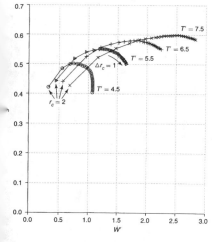

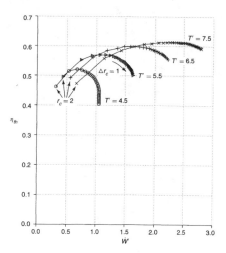

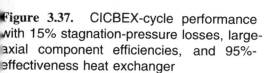

Figure 3.37. CICBEX-cycle performance with 15% stagnation-pressure losses, large-axial component efficiencies, and 95%-effectiveness heat exchanger

Figure 3.38. CICBEX-cycle performance with 16% stagnation-pressure losses, large-axial component efficiencies, and 97.5%-effectiveness heat exchanger

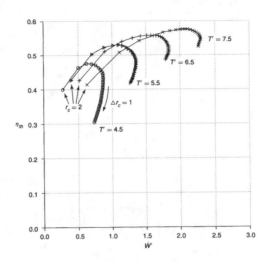

Figure 3.39. CICBEX-cycle performance with 15% stagnation-pressure losses, large-radial component efficiencies, and 95%-effectiveness heat exchanger

efficiencies, and a heat exchanger of 90-percent effectiveness. The plot in figure 3.38 shows the large increase in thermal efficiency (reaching over 60 percent) resulting from an increase in the heat-exchanger effectiveness to 97.5 percent.

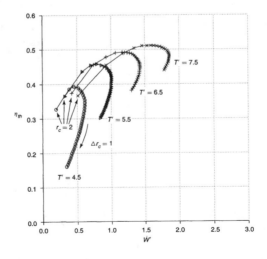

Figure 3.40. CICBEX-cycle performance with 15% stagnation-pressure losses, small-radial component efficiencies, and 95%-effectiveness heat exchanger

3.8   Jet-propulsion cycles

In all types of gas-turbine engines a portion of the power available at turbine inlet is absorbed to drive the compressor. There is a lot of power remaining after the compressor demand is met in the form of high-temperature high-pressure gas, and for this reason the combination of compressor, combustor, and compressor-driving portion of the turbine is frequently called the "gas generator" or "gasifier". If the power is delivered via an output shaft the engine is called a turboshaft. These engines use a "power turbine" after the gas generator. The power turbine can be on the same shaft with the compressor and compressor-driving turbine, or on a separate shaft, thus separating the operating speed of the gas generator from that of the load driven by the power turbine. In turbojet applications the flow produced by the gas generator is expanded in a nozzle, sometimes after the burning of additional fuel in an "afterburner", thus producing thrust via the momentum of this flow at nozzle outlet. In turboprop engines the gas-generator flow drives a turbine which is used to power a propeller. In turbofan engines the gas-generator flow is partly used to drive a turbine that in turn drives a fan (the fan-outlet flow acting as a nozzle with cold air flowing through it), and the remaining gas-generator flow is further expanded to atmospheric pressure in a nozzle with hot exhaust gas flowing through it. If the cold and hot flows are not mixed in the engine ducts then the turbofan is called a separate-exhaust turbofan, otherwise it is called a mixed-exhaust turbofan. It is expensive to develop new components for different types of engines, so engine companies frequently rely on developing several types of engines with one common or nearly common gas generator. For example, figure 3.41 shows the same gas-generator "core" being used in a turbojet and an industrial shaft-power engine.

The main jet-propulsion cycles are turbojets, turbofans, and turboprops. Turbojets are mainly used by military planes at higher altitudes and flight-Mach numbers. Commercial aviation uses high-bypass turbofans at just-subsonic flight-Mach numbers. High-bypass turbofans are more efficient than turbojets, but they also have a larger cross-sectional area. As a compromise between turbojets and turbofans some military planes use low-bypass-ratio turbofans with moderate cross-sectional areas. Turboprops can be considered as very-high-bypass turbofans that usually operate at lower subsonic flight-Mach numbers and altitudes than commercial turbofans. The altitude and flight-Mach-number operating regions of these types of engines are illustrated in figure 3.42.

The turbojet cycle (TBJET)

This cycle (figure 3.44) represents an engine producing thrust by expanding a stream of working fluid in a nozzle. It would normally power an aircraft flying at altitude A_L with velocity V_a, corresponding to Mach number M:

$$M \equiv \frac{V_a}{\sqrt{C_p \cdot R \cdot T_{st,0}/(C_p - R)}} \tag{3.31}$$

In figure 3.44 the static conditions at inlet correspond to point $st, 0$. These are evaluated as functions of altitude A_L, based on Shepherd (1972).

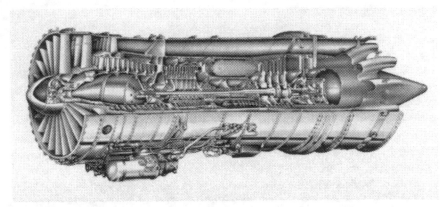

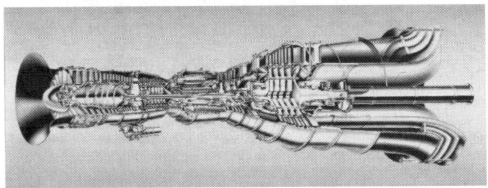

Figure 3.41. The same "gas generator" or "core" is used in Pratt & Whitney's JT8D turbofan and its FT8 industrial and marine shaft-power engine. Courtesy Pratt & Whitney, United Technologies

For $0 \leq A_L \leq 11,000$ m $T_{st,0}$ in K and $p_{st,0}$ in N / m^2 are given by:

$$T_{st,0} = 288.16 - 0.0065 \cdot A_L \tag{3.32}$$

$$p_{st,0} = 101,325 \cdot \left(\frac{T_{st,0}}{288.16}\right)^{5.25757}$$

For $A_L > 11,000$ m $T_{st,0}$ in K and $p_{st,0}$ in N / m^2 are given by:

$$T_{st,0} = 216.66 \text{ K}$$

$$p_{st,0} = 22,622.50 \cdot \exp\left(\frac{11,000 - A_L}{6339.87}\right) \tag{3.33}$$

The stagnation conditions for the flight-Mach number M correspond to point $0,0$. The diffuser at the engine inlet recovers part of the dynamic head at the inlet (ram static-pressure recovery), with a small loss in stagnation pressure from $p_{0,0}$ to $p_{0,1}$. This process can be modeled using the typical efficiency curve for ram pressure recovery shown in figure 3.43. With the exception of ram pressure recovery other cycle inputs are similar to those for the CBE cycle. Most turbojet cycles after the turbine have an afterburner (which is "on" only when the pilot demands excess thrust, because with the afterburner "on" the engine consumes more fuel). If the afterburner is on, the flow expands from

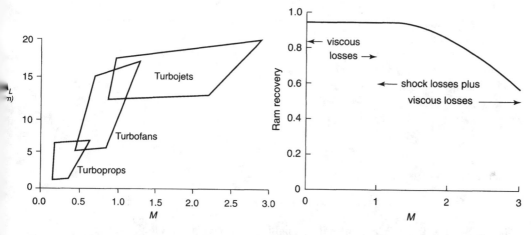

gure 3.42. Altitude and Mach-number nges of turbojet, turbofan, and turboprop eration

Figure 3.43. Ram-diffuser pressure-recovery factor $(p_{0,1} - p_{st,0})/(p_{0,0} - p_{st,0})$ as a function of flight Mach number. From Kerrebrock (1992)

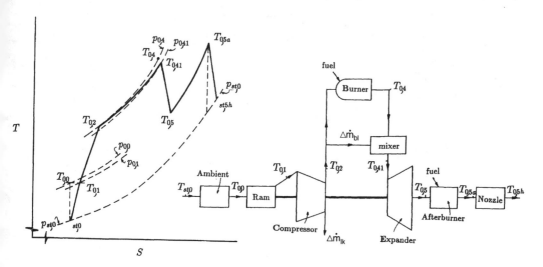

Figure 3.44. Temperature-entropy and block diagram of the TBJET cycle. From Korakianitis and Wilson (1994)

$0, 5a$ to $p_{st,0}$; otherwise it expands from $0, 5$ to $p_{st,0}$. The nozzle pressure (expansion) ratio corresponds to an isentropic temperature ratio. This can be combined with the isentropic efficiency to evaluate a (non-isentropic) enthalpy drop though the nozzle Δh_N, producing a gas stream of velocity $C_{5,ht}$:

$$C_{5,ht} = \sqrt{2(\Delta h_N)} \tag{3.34}$$

The cycle is (practically) infeasible if the pressure ratio available for the nozzle is less than 2% of the cycle pressure ratio.

In the following jet-propulsion cycle-performance plots, the thrust developed by the engine, the power output, and the thrust specific fuel consumption have been calculated for unit mass-flow rate through the engine inlet ($\dot{m}_a = 1.0$). Let $\dot{m}_{ht}$ be the mass flow through the (hot) nozzle, $\dot{m}_F$ be the fuel-flow rate to the cycle and F the jet thrust developed by the engine, given by:

$$F = \dot{m}_{ht} \cdot C_{5,ht} - \dot{m}_a \cdot V_a \tag{3.35}$$

The equivalent useful jet power output $\dot{W}_j$ is given by:

$$\dot{W}_j = F \cdot V_a \tag{3.36}$$

The propulsive, thermal, and overall (availability) efficiencies are defined by:

$$
\begin{aligned}
\eta_{pr} &\equiv \frac{F \cdot V_a}{(1/2) \cdot (\dot{m}_{ht} \cdot C_{5,ht}^2 - \dot{m}_a \cdot V_a^2)} \\
\eta_{th} &\equiv \frac{F \cdot V_a + (1/2) \cdot \dot{m}_{ht}(C_{5,ht} - V_a)^2}{\dot{m}_F L_{HF}} \\
\eta_{ov} &\equiv \frac{F \cdot V_a}{\dot{m}_F L_{HF}}
\end{aligned}
\tag{3.37}
$$

The specific power $\dot{W}'$ is defined by:

$$\dot{W}' \equiv \frac{F \cdot V_a}{\dot{m}_a \cdot C_p(T_{0,1}) \cdot T_{0,1}} \tag{3.38}$$

The thrust specific fuel consumption (for unit mass flow through the engine inlet) is given by

$$t_F = \frac{\dot{m}_F}{F} \tag{3.39}$$

(In the following performance plots this number is multiplied by 1000, so that the units of t_F in the figures are [(kg fuel)/(s.kN)] per unit mass flow rate at the inlet ($\dot{m}_a$)).

The turbofan cycle (TBFANS)

In this cycle (figure 3.45) the engine core is a turbojet engine whose turbine drives an external fan as well as the compressor. The fan flow expands through a "cold" nozzle, as

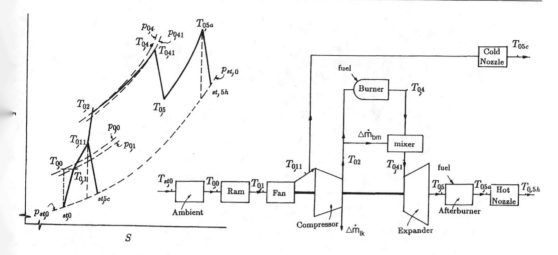

Figure 3.45. Temperature-entropy and block diagram of the TBFANS cycle. From Korakianitis and Wilson (1994)

opposed to the "hot" nozzle of the core. The hot and cold nozzles expand separately to atmospheric conditions. In some turbofan cycles the hot and cold nozzle flows are either fully or partially mixed. The bypass ratio, b_r, is defined by:

$$b_r \equiv \frac{\dot{m}_{cd}}{\dot{m}_{ht}} \qquad (3.40)$$

where $\dot{m}_{cd}$ is the mass-flow rate through the cold nozzle. The following performance plots have been obtained for $\dot{m}_{ht} = 1.0$ (non-dimensionalized with the mass-flow rate through the core (gas generator) of the engine).

Much of the cycle is similar to the TBJET cycle. The fan after the ram inlet raises the pressure and the temperature to point $0, 11$, which corresponds to the condition at compressor inlet and at cold-nozzle inlet. The stagnation conditions at afterburner exit correspond to point $0, 5a$. The hot nozzle expands the flow to $p_{st,0}$ producing a gas stream of velocity $C_{5,ht}$. The cold nozzle expands the flow to $p_{st,0}$ producing a gas stream of velocity $C_{5,cd}$. The cycle is (practically) infeasible if the pressure ratio available for the hot nozzle is less than 2% of the cycle pressure ratio. The thrust developed by the engine, the power output, and the thrust specific fuel consumption are calculated for unit mass-flow rate through the core (hot part) of the engine. For unit mass-flow rate through the core of the engine, the mass-flow rate through the inlet of the engine, $\dot{m}_A$ is:

$$\dot{m}_A = 1.0 + \dot{m}_{cd} \qquad (3.41)$$

The thrust F is given by:

$$F = \dot{m}_{ht} \cdot C_{5,ht} + \dot{m}_{cd} \cdot C_{5,cd} - \dot{m}_a \cdot V_a \qquad (3.42)$$

The takeoff thrust of this engine is from 334 kN to 445 kN—over 100,000 lbf. The bypass ratio (fan flow divided by core-engine flow) is 8.4, believed to be the highest of any engine. The pressure ratio is 39.3. Despite this high pressure ratio the engine incorporates sufficient flexibility with a two-spool design and an ungeared very-wide-chord fan (3.12m diameter). The small combustor resulting from this high compression ratio has staged combustion giving, GE claims, a one-third reduction in NO_x emissions over current high-bypass engines. The specific fuel consumption should also show a 10% reduction.

Illustration. GE90 high-bypass fan-jet engine. Courtesy GE Aircraft Engines

The propulsive, thermal, and overall (availability) efficiencies are defined by:

$$\eta_{pr} \equiv \frac{F \cdot V_a}{(1/2) \cdot (\dot{m}_{ht} \cdot C_{5,ht}^2 + \dot{m}_{cd} \cdot C_{5,cd}^2 - \dot{m}_a \cdot V_a^2)}$$

$$\eta_{th} \equiv \frac{F \cdot V_a + (1/2) \cdot (\dot{m}_{ht}(C_{5,ht} - V_a)^2 + \dot{m}_{cd}(C_{5,cd} - V_a)^2)}{\dot{m}_F L_{HF}}$$

$$\eta_{ov} \equiv \frac{F \cdot V_a}{\dot{m}_F L_{HF}} \tag{3.43}$$

The equivalent $\dot{W}_j$, $\dot{W}'$, and t_F for unit mass flow through the engine inlet are given by equations 3.36, 3.38 and 3.39.

3.9 Performance characteristics of jet-propulsion cycles

The thrust specific fuel consumption of representative turbojet and turbofan cycles with afterburners are shown in figure 3.46. The symbols indicate jet-cycle pressure ratio r_j, where $r_j = r_r \times r_c$ for turbojets, $r_j = r_r \times r_f \times r_c$ for turbofans, r_c is compressor pressure ratio, r_r is ram pressure ratio, and r_f is fan pressure ratio.

The turbojet cycles, (shown with the relatively larger symbols), can be considered as the limiting case of turbofans with zero bypass ratio ($b_r = 0.0$). The other cases correspond to turbofans of bypass ratios 0.4, 4.0 and 8.0. Four combinations of flight

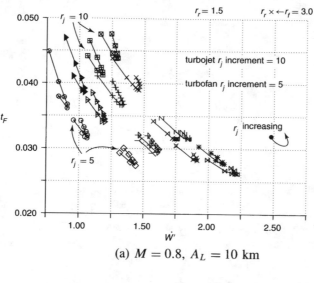

(a) $M = 0.8$, $A_L = 10$ km

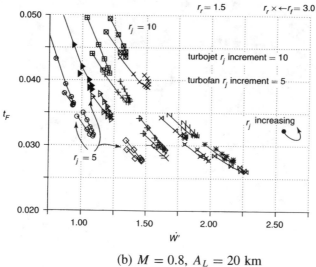

(b) $M = 0.8$, $A_L = 20$ km

Symbols used in the figure

⊘	turbojet, b = 0.0,	$T_{04} = 1400$K	◇	turbofan, b = 4.0,	$T_{04} = 1400$K
▶	turbojet, b = 0.0,	$T_{04} = 1600$K	⊕➙	turbofan, b = 4.0,	$T_{04} = 1600$K
⊞	turbojet, b = 0.0,	$T_{04} = 1800$K	⋈	turbofan, b = 4.0,	$T_{04} = 1800$K
⊠	turbojet, b = 0.0,	$T_{04} = 2000$K	N	turbofan, b = 4.0,	$T_{04} = 2000$K

Figure 3.46. Thrust specific fuel consumption versus specific power for typical jet-propulsion cycles. From Korakianitis and Wilson (1994)

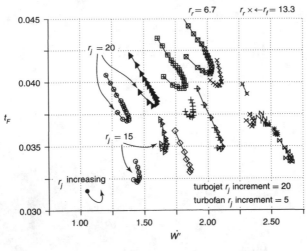

(c) $M = 2.0$, $A_L = 10$ km

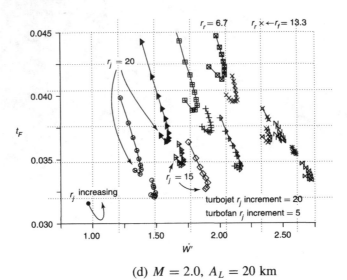

(d) $M = 2.0$, $A_L = 20$ km

Symbols used in the figure

⊙	turbofan, b = 0.4,	$T_{04} = 1400$K	≺	turbofan, b = 8.0,	$T_{04} = 1400$K
▷	turbofan, b = 0.4,	$T_{04} = 1600$K	⋈	turbofan, b = 8.0,	$T_{04} = 1600$K
+	turbofan, b = 0.4,	$T_{04} = 1800$K	✱	turbofan, b = 8.0,	$T_{04} = 1800$K
×	turbofan, b = 0.4,	$T_{04} = 2000$K	⋈	turbofan, b = 8.0,	$T_{04} = 2000$K

Figure 3.46. Continued

Mach number and altitude are included in figure 3.46: the top left (figure 3.46a) is for M=0.8 and A_L=10 km; the top right (figure 3.46b) is for M=0.8 and A_L=20 km; the bottom left (figure 3.46c) is for M=2.0 and A_L=10 km; and the bottom right (figure 3.46d) is for M=2.0 and A_L=20 km.

Other inputs common to the jet-power cycles shown in figure 3.46 are: overboard leakage $\Delta \dot{m}_{lk} = 0.02$; compressor and turbine polytropic stagnation-to-stagnation efficiencies 0.90; nozzle isentropic efficiencies of 0.96; cooling-technology level 2.0 and $T_{bm} = 1300$ K; burner and afterburner efficiencies 0.996; afterburner-outlet temperatures $T_{05,a} = 1.05 \cdot T_{0,4}$; fan pressure ratio 2.0:1; fan polytropic stagnation-to-stagnation efficiency 0.89; pressure-loss fraction 0.08 for the turbojet cycle; and pressure-loss fraction 0.10 for the turbofan cycle.

For $M = 0.8$ the ram pressure ratio $r_r = 1.5$ and for $M = 2.0$ it is $r_r = 6.7$. The ambient conditions are a function of altitude. Thus for a given engine configuration, r_c, and $T_{0,4}$, the values of A_L and M affect the jet-cycle temperature ratio (corresponding to T' in the shaft-power cycles) and the jet-cycle pressure ratio. Each of these cycles is suitable to distinct ranges of flight Mach number and altitude, and has different performance with an afterburner.

In figures 3.46a and 3.46b for the turbojet cycle the values of r_j start from $r_j = 10$ for each $T_{0,4}$ and they move in increments of 10 counter-clockwise tracing letters "U" as r_j increases. For every combination of other inputs there is an optimum cycle (and compressor) pressure ratio for maximum $\dot{W}'$, and a different optimum pressure ratio for minimum t_F. Similar conclusions can be reached for η_{th}, η_{pr}, and η_{ov}. The optimum pressure ratios are finite. The turbofan cycle r_j start from $r_j = 5$ for each $T_{0,4}$ and they move in increments of 5 counter-clockwise. With other jet-cycle inputs kept constant, as b_r increases t_F decreases and $\dot{W}'$ increases. (Matching engines with airframes for different applications is complex. These figures do not include the effect of the increased drag on the airframe from the larger inlets and engine diameters as the bypass ratio increases.) Figures 3.46a and 3.46b indicate that the effect of altitude on cycle performance is small, but comparison of cycles should be performed with extreme care. The same r_j in figures 3.46a and 3.46b correspond to different ambient conditions, and therefore to different cycle temperature ratios (T').

In figures 3.46c and 3.46d the turbojet-cycle pressure ratios r_j start from $r_j = 20$ for each $T_{0,4}$ and they move in increments of 20 counter-clockwise. The turbofan-cycle pressure ratios r_j start from $r_j = 15$ for each $T_{0,4}$ (ram times fan pressure ratio is 13.3:1 in this case) and they move in increments of 5 counter-clockwise.

Comparison of figures 3.46a and 3.46b with figures 3.46c and 3.46d should incorporate the different flight-Mach numbers corresponding to different ram pressure ratios, and therefore the same r_j corresponding to different compressor pressure ratios r_c. Other conclusions are similar to those of the last paragraph.

Higher bypass ratios are more fuel-efficient than lower ones, illustrating the reasons for the move to higher-bypass turbofans, unducted fans and propfans. As flight-Mach numbers increase larger engine diameters cause substantial increases in weight and drag from the engine and its inlet. Current fighter planes use turbofan engines with low bypass ratios. The design matrix for jet-power cycles covers different combinations of cycle variations and cycle parameters that are advantageous for different reasons (fuel consumption, propulsive efficiency, etc.) in different regimes. The off-design performance of the cycles in different operating regimes also plays a significant role in the choice of cycle and engine pressure ratios.

3.10   Descriptions and performance of alternative cycles

In addition to the cycles just described, there is a huge variety of alternative cycles on which gas-turbine engines could run. We shall describe only a few of the more prominent variations here.

Cryogenic cycles

A gas-turbine engine normally produces power from fuel heat while rejecting waste heat to a cold sink. Engines can also be designed to produce power from cold, while absorbing heat ("rejecting cold") to a (relatively) warm sink. Figure 3.47 shows, for instance, a temperature-entropy diagram of a hydrogen closed cycle that could be used for evaporating liquefied natural gas and simultaneously recovering a proportion of the power used at the gas field in liquefying it. Currently, natural gas is usually "deliquefied" by the expenditure of heat, often from combustion of the fuel itself, an obviously wasteful process.

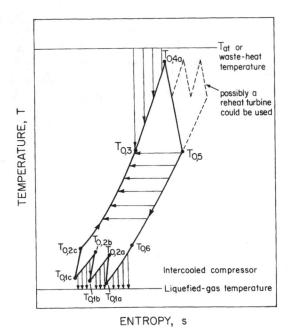

Figure 3.47. Cryogenic intercooled-reheated cycle

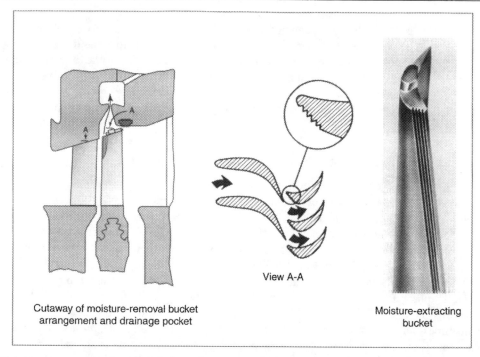

Cutaway of moisture-removal bucket
arrangement and drainage pocket

View A-A

Moisture-extracting
bucket

Wet steam is encountered particularly in nuclear steam turbines. Moisture removal is essential because the tip speed of the longest low-pressure blades is often 700 m/s and above, and large drops of condensed water impacting at relative velocities that can be much higher than the peripheral speed can erode blades relatively quickly. The leading edges are often made of a very hard material such as Stellite, brazed in. In the example above, extraordinary measures such as grooving the leading edges are taken to encourage water to flow radially outward to circumferential drainage pockets.

Illustration.   Moisture removal in wet-steam turbines.   Courtesy GE Aircraft Engines

In the cycle shown, which is similar to the cycle of figure 3.8, the latent heat of the liquefied gas would be given up in cooling the hydrogen to compressor-inlet temperature after leaving the heat exchanger, from $T_{0,6}$ to $T_{0,1a}$, and in recooling the hydrogen in one or more intercoolers, $T_{0,2a}$ to $T_{0,1b}$ and $T_{0,2b}$ to $T_{0,1c}$. After the final compression stage the hydrogen would be heated in the high-pressure side of the heat exchanger, $T_{0,2c} \rightarrow T_{0,3}$, by internal heat transfer, and would receive its final increment of temperature $T_{0,3} \rightarrow T_{0,4a}$, by absorbing heat from the atmosphere or from, for instance, a hot-water power-plant cooling-water discharge.

The intercooled cycle could also be used for this duty. The charts of section 3.7 can be used as a guide to the efficiency and specific power. They have been calculated on the basis of air properties, but the similar value of (C_p/R) for hydrogen should produce similar optimum pressure ratios and specific-power values. Equations 3.18 and 3.19 can be used for more exact values for the intercooled heat-exchanger cycle. More complex cycles may also be used at low temperatures.

This cryogenic application is one case where the use of reheaters between expansion stages would be easily practicable and would increase the power output and the efficiency. Many stages of compression and expansion are needed to produce a change of temperature sufficient for the use of intercoolers or reheaters when hydrogen is used, because of the very high specific heat. Whether or not the high additional cost of intercoolers and reheaters would be justified would depend on the value of the additional power produced.

Cycles that incorporate water or steam

The combined cycle is the most-used variation of the basic gas-turbine cycle in the last part of the twentieth century. The simplest form is the combined-heat-and-power plant, or CHP. A gas-turbine engine, usually one working on the "simple cycle" (CBE), exhausts hot gas into a heat-recovery steam generator (HRSG). In the case of CHP, the steam from the HRSG is led to a process application (for instance, a paper-making plant), or to building or district heating (figure 3.48). In a true combined-cycle plant the steam operates a steam-turbine plant (figure 3.49), and the plant is sometimes called a "CCGT" plant, for "combined-cycle gas turbine", although manufacturers like to devise their own names for their particular offerings. GE uses "STAG", for "steam and gas". Sometimes the gas-turbine part is called the "topping cycle" and the steam-turbine portion the "bottoming cycle". Most of the new generating plants being built around the world are designed to this cycle. Efficiencies of the small plants are in the range of 50%, while for the larger plants it can go as high as 60% (figure 3.50). (This is forecast for the GE Power Systems "H" technology, which uses steam in another way, blade cooling, to allow turbine-inlet temperatures of 1430 °C (2600 °F), to be reached in heavy-duty gas turbines. The 60-percent figure is for the so-called STAG 109H, a 480-MW combined-cycle plant.) The efficiencies rise with power output partly because Reynolds-number effects and tip-clearance losses become relatively smaller as gas-turbine plants become larger, and partly because the incorporation of efficiency-improving measures in steam-turbine plant (feed-water heating for instance) is economic only for the largest plants.

There is sufficient oxygen in the exhaust of a simple-cycle gas turbine to support additional combustion. However, most combined-cycle plants do not have supplementary

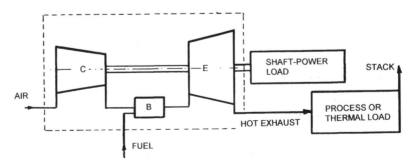

Figure 3.48. Combined heat and power (CHP) plant

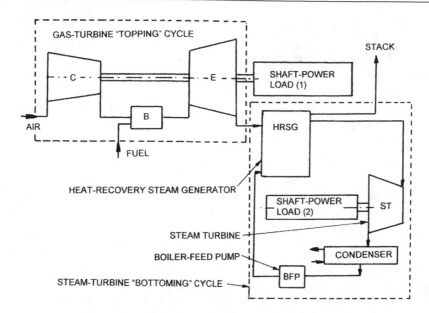

Figure 3.49. Combined-cycle plant

firing. The temperature of the steam at the stop-valve of large steam turbines is around 566 °C (1050 °F), a temperature limit set by hydrogen embrittlement of superheater tubes (at high pressures and temperatures water dissociates to OH^- and H^+). It is desirable that the steam reach, but not exceed, this temperature. The increasing turbine-inlet temperatures of modern gas-turbine plant match the required steam conditions without the need for further combustion. There is also benefit in increasing the output of the gas-turbine by incorporating intercooling and reheat.

Another variation is the integrated-gasification combined cycle (IGCC) that incorporates a system producing gas from coal. Where the gasifier is oxygen-fed, the system must include an oxygen plant in addition to the gasification plant, leading to a capital cost reported as approximately three times that of a CCGT fired by natural gas. The ability to use a low-cost fuel, coal, in an environmentally benign manner will justify the additional capital cost in certain circumstances today, and presumably in more circumstances in the future when natural-gas prices are certain to rise. The 250-MWe Demkolec plant in the Netherlands started trial operation in 1994, and the Wabash River plant in Indiana started trials in 1995. The capital cost of larger plants is estimated at about $1600/kW; several other IGCC plants are in the advanced planning stage (Stambler, 1996).

Coal is also being used to power combined-cycle gas turbines by using pressurized fluidized beds for combustion, initially in Spain, Japan, and the US. The beds contain limestone and other sorbents that, together with slag-melting on the walls and base of the bed, produce a hot gas that can pass through a gas turbine expander without causing more than minor erosion, corrosion, or deposition. The prices forecast for the plants are 75 percent of those for the IGCC plants.

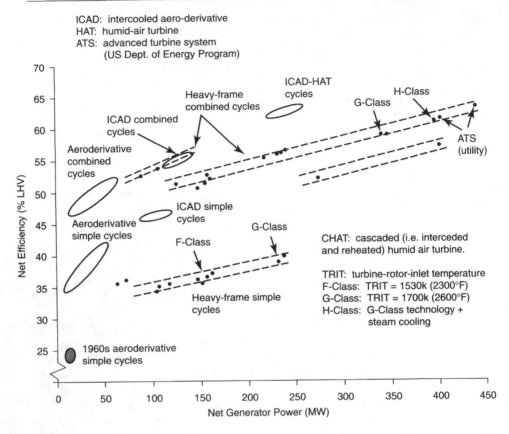

ICAD: intercooled aero-derivative
HAT: humid-air turbine
ATS: advanced turbine system
     (US Dept. of Energy Program)

Figure 3.50. Thermal efficiency versus power output and type. From Touchton (1996)

Steam injection in a location where it will expand through the turbine blading with the combustion gases is a third use of the steam generated in a HRSG (figure 3.51). It is a modern version of the system pioneered by Lemale and Armengaud in the first decade of this century (see the historical introduction). Steam may be injected upstream of or into the combustion chamber, or into the turbine nozzles anywhere along the expansion. The steam does less work the further along the expansion it is injected. In comparison with the combined cycle, the steam-injected cycle has the following advantages. A substantial increase in power can be obtained from the gas-turbine engine with no modification in the configuration of the expansion turbine itself. The part-load efficiency is improved. The production of NO_x is reduced. In a review of the status of steam-injected gas turbines, Tuzson (1992) states that combined-cycle turbines have demonstrated the highest power-generation efficiencies and the lowest cost in sizes above 50 MW (although he also quotes a study giving the power level below which steam-injection systems become more attractive than combined cycles as 150 MW). At lower power levels the steam-injected

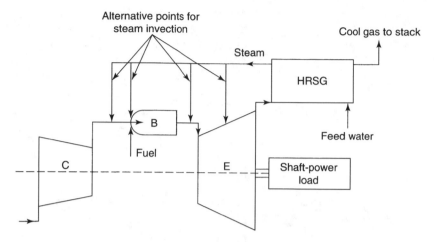

Figure 3.51. Steam-injection gas turbine

gas turbine becomes attractive because of the avoidance of the large cost of the steam turbine. A typical power gain from steam injection for a GE LM5000 gas-turbine engine was quoted as increasing the engine output from 34 MW to 49 MW, together with an efficiency increase from 37 percent (simple cycle) to 41 percent. GE analyzed the gains that would be obtained from a combination of intercooling and steam injection for the LM5000: a power increase from 34 MW to 110 MW and an efficiency improvement from 37 to 55 percent. The water-purification requirements are more demanding for steam injection than for the combined cycle because virtually all the water is normally lost in the exhaust rather than being circulated in a closed system, and because the specifications are more stringent. Any dissolved solids that become deposited on the turbine blades or elsewhere could form corrosion sites or potential blockages. However, Tuzson states that water-purification cost is of the order of five percent of the fuel cost and is not, therefore, a decisive factor. The reliability of early steam-injected units has been high, for instance 99.5 percent. Rather surprisingly, combustor-liner durability has been found to increase.

One of the advanced gas-turbine systems being developed in Japan uses an intercooled-reheated gas turbine (the intercooler is a water-spray direct-contact type) in which the steam raised in the HRSG can power a conventional steam turbine, or the steam can be injected into the gas turbine (Takeya and Yashui, 1988). The configuration shown in figure 3.52 is for the steam-injection system. The output, 400 MW, and the predicted efficiency, 54.3 percent, place it outside Tuzson's guidelines above.

A gas turbine is a good candidate for steam injection if the compressor has a wide range of operation (in particular, a good "surge margin"—see chapter 8) because the increased flow creates a higher back pressure. A high pressure ratio and a high turbine-inlet temperature are also desirable. These conditions seem to favor the aircraft-derivative turbine. However, Tuzson (1992) points out that heavy-duty industrial turbines can accommodate concentrations of contaminants about five times higher than can the aircraft-derivative turbines.

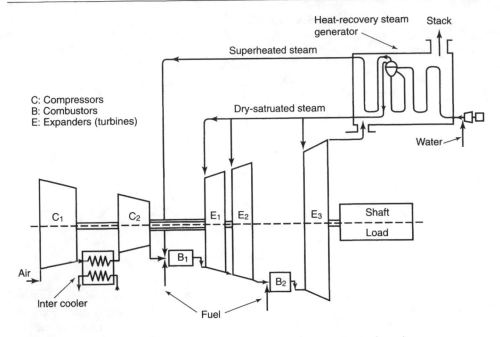

Figure 3.52. Intercooled-reheated steam-injected cycle

There are many variations of these relatively simple forms of water/steam injection. El-Masri (1988) proposes an intercooled-recuperative cycle in which the intercooler and an aftercooler are direct-contact water-injected evaporative units and there is subsequent water injection into the recuperator (figure 3.53). There is no steam generator. The results of his analysis show considerably higher efficiencies over the conventional intercooled-recuperative cycle and over steam-injected cycles. The humid-air cycle proposed by D.D. Rao in 1990 (figure 3.54) was analyzed by Stecco et al. (1993): it incorporates a water-cooled "surface" intercooler and a similar aftercooler, followed by a water-injection evaporator, a recuperator, and combustor. The authors believe that the cycle has advantages over the steam-injected cycles. Bolland and Stadaas (1993), working with El-Masri, propose an intercooled-recuperated cycle in which the recuperator is coupled to the HRSG with water injection in the intercooler and aftercooler and steam injection in and after the combustor (figure 3.55). Their analysis persuades them that this cycle, which they termed the DRIASI cycle, is superior to both the combined cycle and the steam-injected cycle for low-power applications.

The estimated efficiencies of some of these cycles have been incorporated into figure 3.50.

Performance charts of cogeneration plants

Cogeneration power plants are already in use in northern Europe. In a typical cogeneration plant there is a topping CBE gas-turbine cycle, a bottoming steam-turbine

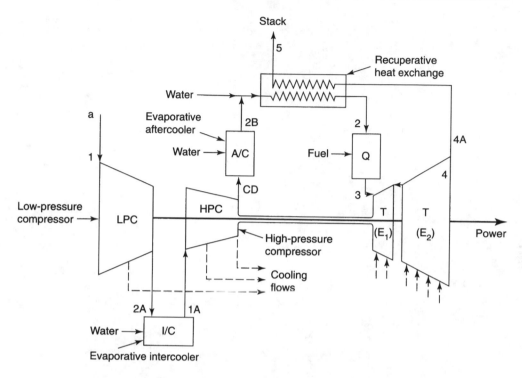

Figure 3.53. Water-injected recuperated cycle. From El-Masri (1988)

cycle, and part of the condenser flow provides district hot-water heating. Tsatsaronis and Chen (1995) and Agazzani and Massardo (1996) have proposed methods for the combined thermodynamic and economic optimization of simpler combined power plants. In some of these power plants the gas turbine is steam injected. In other power plants there is supplementary firing. Biomass (which produces large amounts of water upon burning) is abundant in Scandinavian countries, so in yet other power plants some of the water available from the flue gas in the exhaust is condensed to provide additional energy to the steam cycle, or to district heating. In the United States utility companies are facing the problem of meeting peaking electricity demand (with peaking CBE units or energy-storage schemes) without purchasing additional power plants. In other parts of the world power-plant managers are faced with complex economic problems: for example, electricity and hot water are already traded in the spot market in some countries (e.g., Sweden). This situation raises the question of analysing the thermodynamic performance of combined power plants with performance figures similar to those used in shaft-power gas turbines.

Korakianitis et al. (1997) address this question on a typical cogeneration plant shown in figure 3.56. This power plant is powered by natural gas (energy input F1) and has supplementary biomass firing (energy input F2). The gas turbine is steam injected ($\dot{m}_5$, station 5). The water available in the exhaust gas, ($\dot{m}_{17}$, station 17), is condensed (satu-

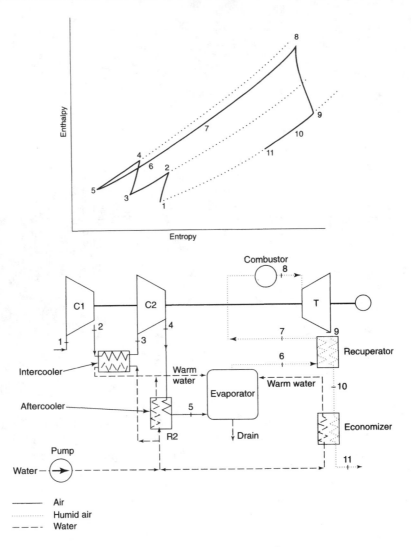

Figure 3.54. Humid-air cycle

rated water at exhaust temperature and at its partial pressure in the exhaust products of combustion is returned to the steam-turbine cycle). The flow rate of air at the gas-turbine inlet is ($\dot{m}_{15}$, station 15), and the flow rate of steam at the HP steam-turbine inlet is ($\dot{m}_3$, station 3). Energy available from condensation of the flue gas and from the steam-turbine condenser provide district hot-water heating $\Delta \dot{E}$ at 110 °C. For typical values of CBE-cycle compressor pressure ratio and turbine-inlet temperature, they obtained figures such as 3.57 showing power-plant thermal efficiency versus specific power and effectiveness versus specific energy of district heating as functions of F2/F1, ratios of steam plant to

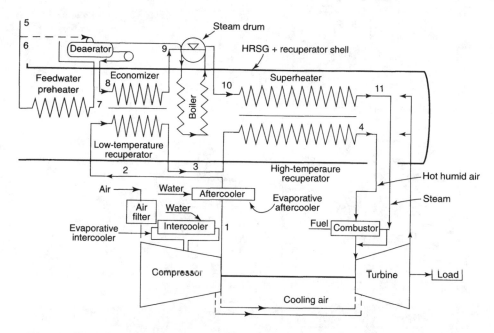

Figure 3.55. Dual-recuperated intercooled-aftercooled steam-injected cycle (DRI-ASI)

gas plant mass flow rates, steam-injection rates, etc. A typical set of performance figures is shown in figure 3.57. For these figures the following definitions were used:

$$\eta_{th} = \frac{\dot{W}_S}{\dot{E}_{F1} + \dot{E}_{F2}}$$

$$e_{ff} \equiv \frac{\dot{W}_S + \Delta \dot{\Omega}_{hw}}{\dot{E}_{F1} + \dot{E}_{F2}}$$

$$\dot{W}' \equiv \frac{\dot{W}_S}{\dot{m}_c C_p T_{0,1}}$$

$$\Delta \dot{E}'_{hw} \equiv \frac{\Delta \dot{E}_{hw}}{\dot{m}_c C_p T_{0,1}}$$

where $\dot{W}_S$ is the sum of the electrical work from the gas and steam turbines, $\dot{\Omega}_{hw}$ is the thermodynamic availability of the energy to district hot-water heating $\dot{E}_{hw}$, and the specific power and specific energy are nondimensionalized with the same term as the shaft-power cycles in earlier sections.

Figures 3.57a and 3.57b show regions of performance for different combined power plants with gas-turbine-inlet temperatures (TIT) of 1500 K, 1600 K and 1700 K and

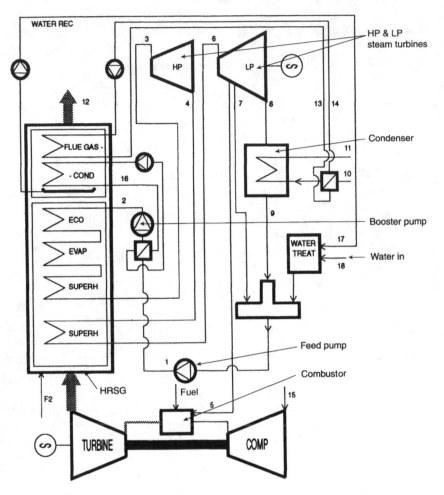

Figure 3.56. Cogeneration power plant with STIG, supplementary firing, and exhaust gas condensation. From Korakianitis et al. (1997)

gas-turbine compressor pressure ratios $r_c = 10, 20, 30, 40$. (The regions for 1500 K are shaded.) Within each of these regions one can observe further details of the effect of different amounts of supplementary firing, steam injection, exhaust-gas condensation, etc., as will be explained below. Similarly to the earlier performance figures of the CBE cycles, for a given TIT the regions move counterclockwise with increasing r_c. Although there is some overlap between regions, it is clear that the TIT and r_c of the gas turbine dominate the overall performance of the power plant.

Figures 3.57c and 3.57d are representative of the details in each one of the above regions.

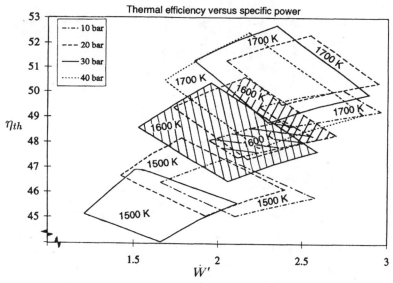

(a) Regions of η_{th} versus $\dot{W}'$ as functions of TIT and r_c

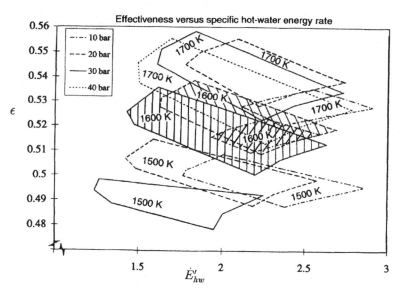

(b) Regions of e_{ff} versus $\Delta\dot{E}'_{hw}$ as functions of TIT and r_c

Figure 3.57. Typical performance of cogeneration power plant in figure 3.56. From Korakianitis et al. (1997)

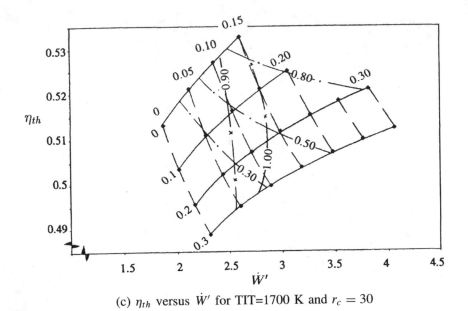

(c) η_{th} versus $\dot{W}'$ for TIT=1700 K and $r_c = 30$

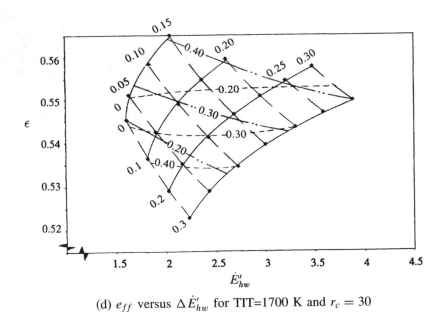

(d) e_{ff} versus $\Delta \dot{E}'_{hw}$ for TIT=1700 K and $r_c = 30$

Figure 3.57. Continued

The solid lines in figures 3.57c and 3.57d correspond to the fraction of energy supplied from supplementary firing ($\dot{E}_{F2}/\dot{E}_{F1} = 0.0, 0.1, 0.2, 0.3$), indicating that as supplementary firing increases η_{th} and e_{ff} decrease while $\dot{W}'$ and $\Delta\dot{E}'_{hw}$ increase.

The long dashed lines in figures 3.57c and 3.57d correspond to the ratio of mass-flow rate of steam injection in the gas turbine to mass-flow rate of air at the gas-turbine inlet ($\dot{m}_5/\dot{m}_{15} = 0.00, 0.05, 0.10, 0.15, 0.20, 0.25, 0.30$), indicating that as the rate of steam injection increases η_{th}, e_{ff}, $\dot{W}'$, and $\Delta\dot{E}'_{hw}$ increase.

The long dashed lines with one dot in figure 3.57c correspond to the ratio of mass-flow rate of steam injection in the gas turbine to mass-flow rate of steam at the steam-turbine inlet ($\dot{m}_5/\dot{m}_3 =$), indicating that as the rate of steam injection increases η_{th}, e_{ff}, $\dot{W}'$, and $\Delta\dot{E}'_{hw}$ increase. As $\dot{m}_5/\dot{m}_3 =$ approaches 1.0 the steam-turbine plant becomes smaller and all the steam at the HP turbine outlet is injected in the gas turbine.

The short dashed lines in figure 3.57d correspond to the power produced by the steam-turbine plant ($\dot{W}_{sm}$ divided by the power produced by the gas-turbine plant $\dot{W}_{gt}$), indicating that most of the power is produced by the gas turbine.

The long dashed lines with an "x" in figure 3.57d correspond to ($\dot{m}_{17}$ divided by $\dot{m}_5$). When this ratio is greater than 1.0, there is more water condensed in the exhaust boiler than needed for steam injection, and when this ratio is less than 1.0 the plant needs make-up water.

The efficiencies shown in these figures are slightly lower than those that might be expected because: the demand for hot water heating at 110 °C dictates a higher condenser pressure than combined power plants that do not involve cogeneration; and the steam power plant has only one open and one closed feed-water heaters (it is intended to be a representative power plant).

Based on these Korakianitis et al. (1997) concluded that the performance of these plants should be "optimized" by the following sequence: (a) maximizing turbine-inlet temperature in the gas turbine; (b) optimizing the gas-turbine pressure ratio for gas-turbine performance; (c) optimizing steam-turbine boiler pressure; and (d) maximizing steam-injection in the gas turbine. Most of the power is produced by the gas turbine. Supplementary firing should be considered only as a power-enhancing scheme; it lowers plant thermal efficiency because it occurs at boiler temperatures, which are lower than the maximum cycle temperature.

Chemical recuperation*

Energy can be recovered from the turbine exhaust by means other than heat transfer to the compressed-air flow. A chemically recuperated gas turbine (CRGT) has been proposed by Kesser, Hoffman, and Baughn (1994) in which a simple-cycle gas turbine passes its exhaust to a heat-recovery methane-steam reformer (figure 3.58). This takes the place of a HRSG in a steam-injected plant. "The tubes in the reformer (unlike the tubes in the HRSG superheater) are filled with a nickel-based catalyst that promotes a chemical reaction between steam and methane". The gaseous fuel that results is a mixture of CO, H_2, and CO_2, and is reckoned to produce NO_x emissions as low as 1 ppm (vs. 25 ppm for a steam-injected system). The authors calculated the thermal efficiency of the cycle

---

*The work cited here was predated by a feasibility study in 1975, followed by the construction of a plant in 1980.

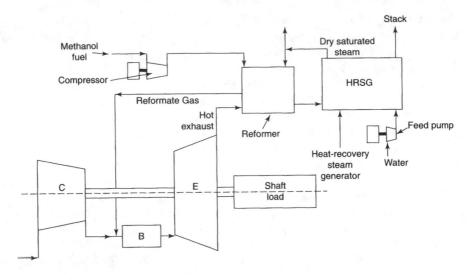

Figure 3.58. Chemically-recuperated gas turbine

(around 47 percent) as less than the combined cycle but better than the steam-injected cycles for the power level of interest.

Inlet-cooling cycles

A somewhat similar approach to chemical recuperation is to use the exhaust heat via a HRSG to operate an absorption chiller that can be used to cool the compressor-inlet air (Ebeling et al., 1992) (figure 3.59). However, a more widespread practice is to use power off-peak-load (in the case of the Butler-Warner plant in Fayetteville, NC, having eight GE Frame-5 units, 4.3 MWe is used off-peak, 148 hours per week) to operate a vapor-compression ice-making system (Ebeling, 1994). Ice is melted during power peaks for four hours per day, five days per week. An increase of hot-weather capacity of 29 percent was given for the plant quoted. The modification paid for itself in under a year.

Other forms of energy storage

A different form of energy storage to meet daily demands is to compress air and to store it in underground caverns using off-peak night-time electrical energy driving a gas-turbine plant's alternator as a motor. During high-load periods the air is taken from the cavern to the combustor and from then to the turbine, with the compressor uncoupled (figure 3.60). The output is much higher than would be normal for the size of the plant because of the absence of compressor power demand, and the alternator must be sized appropriately. The first major air-storage plant was built over a salt deposit in Huntorf, near Bremen, Germany (Ruch, 1977). The electrical output was 290 MW, the pressure ratio was 4:1, and the gas-turbine plant was supplied by what was then the Brown Boveri Company.

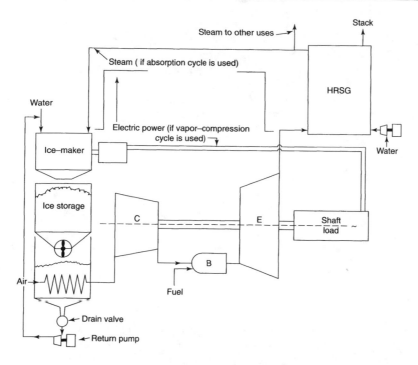

Figure 3.59. Inlet-air cooling

In another energy-storage plant currently used in southeast Missouri, excess power at off-peak hours is used to pump water to a lake at a higher elevation. Peaking demand is met by running the stored water through a hydroelectric power plant.

Closed-cycle applications

There is no change in the thermodynamics of cycles that are closed rather than open (figure 3.61). An open cycle uses air that is normally taken from the atmosphere, compressed, combined with fuel in a combustion process, expanded, and discharged. In a closed-cycle engine the "working fluid" could be air, but is usually a gas with characteristics that are more favorable to the application (helium for nuclear turbines, for instance). Combustion must be external rather than internal, and the combustion heat must be transferred to the fluid, usually through heat-exchanger surfaces like tubes. These surfaces cannot be cooled in the way that other surfaces in the gas stream, such as turbine blades, are cooled, so that closed-cycle gas turbines necessarily use a lower turbine-inlet temperature than do internal-combustion open-cycle turbines. (There is much activity to develop ceramic and other nonmetallic heat exchangers that would allow a much higher gas temperature to be reached than can be attained with metal surfaces.) Open-cycle engines normally discharge hot gases from the turbine or heat-exchanger outlet to the atmosphere, whereas the gases in a closed cycle must be cooled back to compressor-inlet temperature. This temperature will, therefore, be a little higher than ambient.

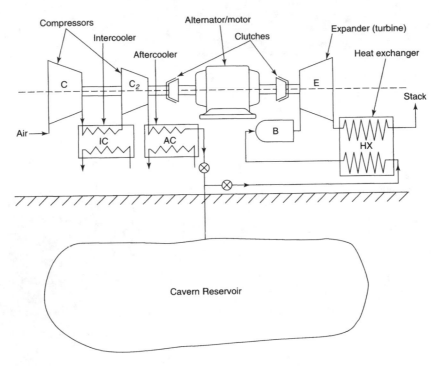

Figure 3.60. Air-storage plant

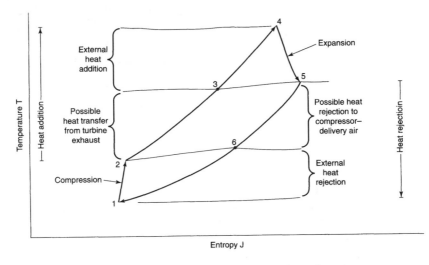

Figure 3.61. $T - s$ diagram of a closed cycle

Closed-cycle engines can use a base pressure—e.g., that at the compressor inlet that is far higher—or lower—than atmospheric, allowing units to be made smaller, or larger, than would be the case for an open-cycle unit with atmospheric-pressure entry. The engine output can be varied by removing or adding working fluid to change the base pressure, giving potentially almost constant part-power efficiencies (if the plant runs at constant shaft speed). A gas of high molecular weight can be chosen as the working fluid if there is a requirement for low blade speeds and a small number of stages; or a gas like hydrogen or helium can be used to run at low Mach numbers and to use the favorable thermal and/or nuclear properties of these gases, and if the resultant large number of stages is not penalizing. (A low molecular weight gives a high speed of sound in a gas. The maximum peripheral speed of rotors is, in general, stress-limited, and rotor blades in low-molecular-weight gases will normally run at low Mach numbers and hence produce only small pressure ratios per stage.)

Closed-cycle gas turbines have been used successfully on dirty fuels like coal with which the sensitive compressor and turbine blading and other internal surfaces stay virtually as new. This is in sharp contrast to the many experimental open-cycle coal-burning turbines where rapid degradation of the blading is the principal concern. The thermal efficiency is limited by the temperature limits mentioned above, but can approach or possibly exceed 50 percent. The use of closed cycles was pioneered by Ackeret and Keller (1970), and the Swiss company Escher-Wyss (now part of ASEA-Brown-Boveri), working with GHH Sterkrade in Germany.

GHH has also advocated the use of closed-cycle gas turbines for the recovery of some of the energy put into compressing and liquefying natural gas (LNG) (Krey, 1980), as discussed under "cryogenic cycles" at the beginning of this section. Krey proposes to use nitrogen as the working fluid in the closed-cycle engine (figure 3.62), with about 0.5 percent oxygen by volume added to prevent nitration embrittlement of the high-temperature materials. He quotes a design study in which 78.5 kg/s of LNG would be vaporized by cooling the inlet flow to the compressor of a closed-cycle gas turbine to 144 K. Choosing a conservative turbine-inlet temperature of 993 K, the plant output would be 90 MW and the thermal efficiency 54.4 percent. The compression ratio would be about 6:1.

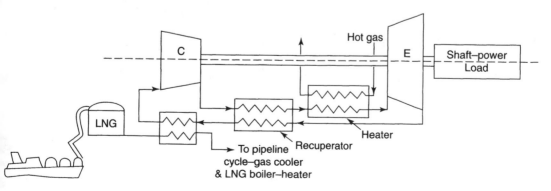

Figure 3.62. Closed-cycle LNG gasifier

The supplementary-fired exhaust-heated cycle

The exhaust-heated cycle is a semi-closed cycle invented by Stolze in the last century and further analyzed by Jendrassik in the 1930s and 1940s (see the historical introduction). It was developed by Donald L. Mordell at McGill University in the 1950s (Mordell, 1955) to burn coal (figure 3.63). He did this in a combustor downstream of the turbine, and heated the compressed air in a metal-tube heat exchanger. The plant worked well, but the economics were unattractive, given the very low turbine-inlet temperature that were possible with this arrangement. Mordell showed his plant in 1953 to one of the authors (DGW) who was visiting Montreal as a very junior engine-room officer on a cargo boat. Thirty-five years later, MIT and Wilson were funded by the US Department of Energy to bring about an improvement to the Stolze-Mordell cycle, principally through using a ceramic rotary regenerator. Subsequently, two developments further improved the attractiveness of this cycle: the suggestion by Janos Beer of MIT that if the coal were fractionated so that a clean gas could be burned ahead of the turbine and the char burned downstream, a large improvement in efficiency and specific power would be realized; and the granting of a patent to Wilson and MIT for a new form of ceramic regenerator, particularly suitable for this type of cycle (see chapter 10). Some aspects of what is sometimes called the "inverted cycle" (see below) were also incorporated. The cycle is shown in figure 3.64, and some predicted relations between thermal efficiency and specific work are in figure 3.65.

The incoming air passes through an intercooled compressor, through one "side" of the regenerator, and into the high-temperature combustor in which is burned clean fuel such as natural gas or oil. The hot gas expands through the turbine and passes to the low-temperature combustor (a fluidized bed being one candidate), fed with coal, coal char, refuse-derived fuel (RDF), etc. One of the several methods of hot-gas cleanup being developed for the Department of Energy could be used at combustor outlet if found necessary (RDF being a relatively benign fuel, it may not be needed in this case). The hot gases then pass through the hot side of the regenerator to a waste-heat boiler (WHB) producing hot water or low-temperature steam. The emerging gases go to a counter-flow water washer where the temperature is reduced to near ambient. Finally the cold, saturated gases are compressed in an induced-draft fan (IDF) and discharged to the stack.

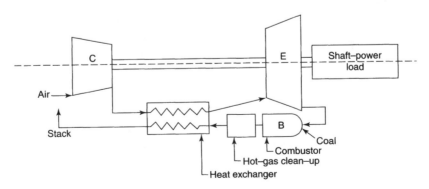

Figure 3.63. Stolze-Mordell exhaust-heated cycle

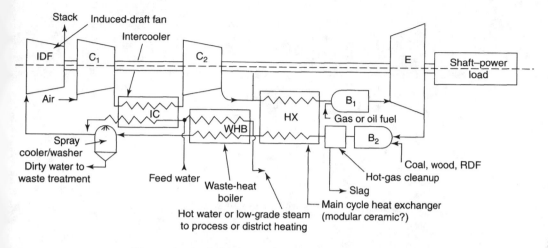

Figure 3.64. Supplementary-fired exhaust-heated cycle

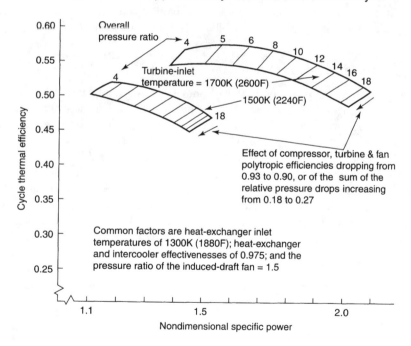

Figure 3.65. Predicted performance of supplementary-fired exhaust-heated cycle

The IDF adds to the expansion pressure ratio of hot gases through the turbine, producing more work, while requiring the input of a smaller quantity of work through the compression of cold gas. This cycle, when used alone for the production of work from a flow of hot gas, is called the "inverted cycle" (figure 3.66). Besides increasing the thermal

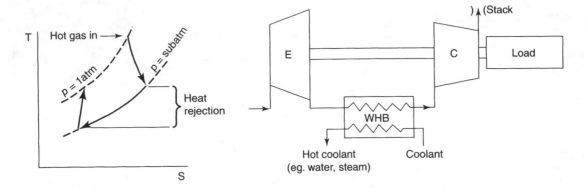

Figure 3.66. The simple inverted cycle

efficiency and specific work of the present cycle, it has the virtue of putting the solid-fuel combustor at just below atmospheric pressure, thus increasing plant safety and facilitating fuel feeding. It can be seen from figure 3.65 that the predicted efficiencies are very high for a plant burning 50-percent solid fuel. A fuller description is given by Wilson (1993).

Pressure-gain-combustion cycles[*]

In piston-type internal-combustion engines, combustion is accompanied by a pressure rise, whereas in gas-turbine engines combustors typically have a 3-to-6-percent pressure loss. There have been many attempts to produce pressure-rise combustors for turbines. The most promising approach appears to be to use wave-engine technology (Weber, 1995). The flow from the compressor—which does not have to be diffused—would be ducted into the relatively long channels of a so-called "wave rotor" (figure 3.67). The channels pass ports on both the intake and the outlet ends. When the compressor flow is admitted, the outlet is closed by a wall, and the flow comes to rest, reaching approximately its stagnation pressure, and a compression wave is reflected. By the time this reaches the intake end this is also closed. Fuel is injected and ignited (several alternative methods are available) and the pressure rises in constant-volume combustion. The rotor reaches a port opening to the turbine and the port "empties" (not, of course, to a vacuum) at a higher stagnation pressure than before. An expansion wave travels back to the inlet end, reaching it just as the channel opens to the compressor-delivery flow.

A wave rotor working on this arrangement would probably not need cooling, as the channel walls would take up a mean temperature between that of the compressor-delivery flow and that of the turbine inlet. The main problems to be solved are to minimize the losses, particularly those connected with the various clearances between rotor and stators.

Closed-cycle plants for nuclear and other applications

Running a gas-turbine engine on a closed cycle has several pronounced advantages and disadvantages.

---

[*]A cycle of this type was proposed by P. Meyer in 1935.

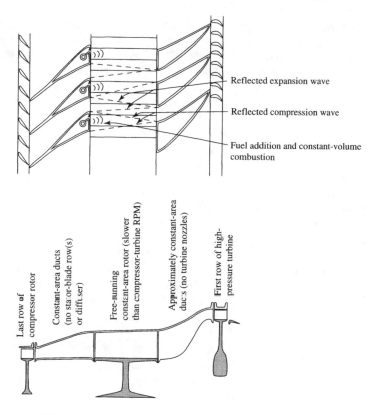

Figure 3.67. Wave-rotor pressure-rise combustor

Advantages

1. Gases other than air can be used. Hydrogen has the advantage of outstanding heat-transfer properties and low pumping-power requirements, and helium the advantage of a low neutron-capture cross section, chemical inertness, and nontoxicity.

2. A wide range of fuels can be employed: coal, peat, and nuclear fuels are examples.

3. The gas path remains extremely clean and noncorrosive. Blade paths retain their as-manufactured efficiencies.

4. Power output at constant output-shaft speed can be varied by varying the loop pressure (or the mass of gas in the circuit) while all temperatures and velocity-diagram operating points can be maintained. Thus part-load efficiencies are high.

Disadvantages

1. An externally fired or heated gas heater must be used instead of the open-cycle compact combustion chamber. A gas heater must use large quantities of heat-resisting materials, and becomes large and expensive, with long transient times.

2. Despite the heat-resisting materials used in the heater, the turbine-inlet temperature must necessarily be at or below those used for uncooled blades in open-cycle engines, e.g., 1000°C instead of 1500°C. These lower temperatures greatly reduce the power density and cycle efficiency.

3. A gas cooler must be used, resulting in a considerable increase in the bulk mass and cost of the engine, a requirement for water or air cooling, and an increase (above coolant temperature) of the gas temperature at compressor inlet. This higher temperature further decreases the specific power and the thermal efficiency.

The performance charts can be used for closed cycles: one would take those with higher stagnation-pressure losses and lower temperature ratios. The gases used as working fluids should have similar values of (C_p/R) to those for air for the approximations to be close.

The Swiss company Escher-Wyss has been responsible for most of the closed-cycle plants built (Zenker, 1976), starting with a 2-MW unit in 1939 and increasing in output to a 50-MW helium-gas plant at Oberhausen, West Germany in 1974. Helium closed-cycle gas turbines are being proposed as fail-safe high-efficiency means of utilizing nuclear energy for power production (McDonald, 1986).

References

Agazzani, A. and Massardo, A.F. (1996). A tool for thermoeconomic analysis and optimization of gas, steam, and combined plants. ASME Journal of Energy for Gas Turbines and Power, and ASME paper 96-GT-479, ASME, New York, NY.

Bolland, Olav, and Stadaas, Jan Fredrik (1993). Comparative evaluation of combined cycles and gas-turbine systems with water injection, steam injection, steam injection, and recuperation, ASME paper 93-GT-57, ASME, New York, NY.

Chappell, M.S., and Cockshutt. E.P. (1974). Gas-turbine cycle calculations: thermodynamic data tables for air and combustion products for three systems of units. National Research Council of Canada Aeronautical Report LR-579, NRC no. 14300. Ottawa.

Dunteman, N. Richard (1970). A new look at the competitive position of the inverted-cycle gas turbine for waste-heat utilization and other applications. MSME thesis, MIT, Cambridge, Mass.

Ebeling, J.E., Halil, R., Bantam, D., Bakenhus, B., Schreiber, H., and Wendland, R. (1992). Peaking-gas-turbine capacity enhancement using ice storage for compressor-inlet-air cooling, ASME paper 92-GT-265, New York, NY.

Ebeling, Jerry, et al. (1994). Thermal-energy storage and inlet-air cooling for combined cycle, ASME paper 94-GT-310, ASNE, New York, NY.

El-Masri, M.A. (1988). A modified high-efficiency recuperated gas-turbine cycle, J. of Engg. for Gas Turbines and Power vol. 110, pp. 233–242, ASME, New York, NY.

Forbes, S.M. (1956). TA turbine performance data. T.D. report 125. Ruston & Hornsby Ltd., Lincoln, England.

Helms, H.E., Heitman, P.W., Lindgren, L.C., and Thrasher, S.R. (1984). Ceramic applications in turbine engines. NASA CR 174715, October.

Hodge, James. (1955). Gas-turbine cycles and performance estimation. Butterworths, London.

Keller, C. and Schmidt, D. (1970). Industrial closed-cycle gas turbines for conventional and nuclear fuel, ASME paper 67-GT-10, ASME, New York, NY.

Kerrebrock, J.L. (1992). Aircraft engines and gas turbines, 2nd edition. The MIT Press, Cambridge, MA.

Kesser, K.F., Hoffman, M.A., and Baughn J.W. (1994). Analysis of a basic chemically recuperated gas-turbine power plant, J. of Engg. for Gas Turbines and Power, vol. 116, pp. 277–284, ASME, New York, NY.

Korakianitis, T. and Wilson, D.G. (1994). Models for predicting the performance of Brayton-cycle engines. ASME J. of Engineering for Gas Turbines and Power, vol. 116, no. 2, pp. 381–388, ASME, New York, NY.

Korakianitis, T. and Beier, K.J. (1994). Investigation of the part-load performance of two 1.12 MW regenerative marine gas turbines. ASME J. of Engineering for Gas Turbines and Power, vol. 116, no. 2, pp. 418–423, ASME, New York, NY.

Korakianitis, T., Grantstrom, J., Wassingbo, P., and Massardo, A. (1997). Parametric performance of combined-cogeneration power plants with various power and efficiency enhancements. ASME J. of Engineering for Gas Turbines and Power, (in print). ASME paper 97-GT-285, ASME, New York, NY.

Krey, G. (1980). Utilization of the cold by LNG vaporization with closed-cycle gas turbine, J. of Engg. for Power, vol. 102, pp. 225–230, ASNE, New York, NY.

Livingood, J.N.B., Ellerbrock, H.H. and Kaufman, A. (1971). NASA turbine-cooling research. NASA TM X-2384.

McDonald, C.F. (1978). The role of the recuperator in high-performance gas-turbine applications. ASME paper 78-GT-46, ASME, New York, NY.

McDonald, Colin F. (1986). Performance potential of a future advanced nuclear gas-turbine concept. GA Technologies report GA-A18576, San Diego, CA.

Mordell, D.L. (1955). An experimental coal-burning gas turbine, Proc. I. Mech. Engrs., vol. 169 no. 7, Instn. Mech. Engrs., London, UK.

Ruch, Skip (1977). Storing energy for peak demand, Turbomachinery International, September–October, pp. 22–26. Norwalk, CT.

Shepherd, D.G. (1972). Aerospace propulsion. American Elsevier Publishing Co.

Stambler, Irwin (1966). in IGCC and advanced cycles outlined at EPRI meeting, Gas Turbine World, January-February, pp. 16–33, Stamford, CT.

Stecco, S.S., et al. (1993). The humid-air cycle: some thermodynamic considerations, ASME paper no. 93-GT-77, ASME, New York, NY.

Takeya, K., and H. Yasui (1988). Performance of the integrated gas and steam cycle (IGSC) for reheat gas turbines, J. of Engg. for Gas Turbines and Power, vol. 110 pp. 220–232, ASME, New York, NY.

Touchton, G. (1996). Gas turbines: leading technology for competitive markets. Global Gas Turbine News vol. 36 no. 1, ASME IGTI, Atlanta, GA.

Tsatsaronis, G., and Chen, Y. (1995). Exergoeconomic evaluation and optimization of cogeneration systems. Internatioonal Journal of Global Energy Issues, vol. 7, no. 3/4, pp. 148–161.

Tuzson, J. (1992). Status of steam-injected gas turbines, J. of Engg. for Gas Turbines and Power, vol. 114, pp. 682–686, ASME, New York, NY.

Weber, H.E. (1995). Shock-wave-engine design, John Wiley, New York, NY.

Wilson, D.G. (1965). The specification of high-mass-flow intercooled air compressors for process applications. Paper 65-WA/FE-25. ASME. New York, NY.

Wilson, D.G. (1993). The supplementary-fired exhaust-heated cycle for coal, wood, and refuse-derived fuel, Proc. I. Mech. Engrs., vol. 207, pp. 203–208, Instn. Mech. Engrs., London, UK.

Zenker, P. (1976). The Oberhausen 50-MW helium turbine plant. Combustion, pp. 2-1-41, April. New York, NY.

Problems

1. The diagram in figure P3.1-P3.2 is of a pumped-storage scheme in which an air compressor can be electrically driven during the night to pump air into an underground cavern. What is the isentropic efficiency of the air compressor, stagnation to stagnation, from the stagnation inlet conditions shown at 1 to the stagnation conditions at outlet from the cavern, shown at 3? The stagnation-to-stagnation polytropic efficiency from the inlet of the cavern, 1, to the outlet of the machine, 2, is 90 percent. The conditions at the various stations are shown on the diagram, and the air may be assumed to be a perfect gas of $\gamma = 1.4$, $C_p = 1,005$ J/(kg $\cdot$ K). State other assumptions you need to make.

2. For the complete cycle shown in figure P3.1-P3.2 the air from the underground cavern can be heated in a combustor B and expanded through a turbine E, and this can be done during the day with the compressor declutched (not driven).

   (a) What is the qualitative effect on the cycle efficiency of the pressure loss from point 2 to point 3?
   (b) What is the qualitative effect on the cycle thermal efficiency and the specific work (per kg of air compressed) of the cooling of the compressed air to 15 °C in the cavern?
   (c) What would be the effect on the cycle thermal efficiency and the specific power of adding a heat exchanger (turbine exhaust to burner inlet)?

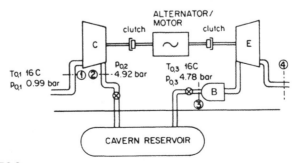

Figure P3.1 and P3.2

(d) How does the size of the alternator/motor compare with an alternator for a single-shaft simple-cycle gas-turbine engine (CBE) using the same compressor and turbine?

(e) What is the approximate effect of a 1 percent drop in alternator/motor efficiency on the overall cycle thermal efficiency?

(f) Would it be useful to make a second cavern into which the air from an intermediate stage of the compressor could be delivered and returned? Should the walls of this cavern and of the storage cavern after the last stage be good insulators or good conductors?

3. Find the maximum percentage of air that must be taken from the compressor delivery to cool a high-temperature turbine expander for the overall CBE cycle efficiency to drop not more than five percentage points from the value calculated for no usage of compressed air. Here are the CBE cycle specifications:

Compressor pressure ratio, 20 : 1.
Compressor inlet temperature, 38 °C.
Turbine inlet temperature, 1,094 °C.
Turbine exhaust pressure $p_{0,4} = p_{0,1} =$ atmospheric.
Combustor pressure drop, 5 percent of inlet p_0
Compressor and turbine polytropic efficiencies, $\eta_{p,c} = \eta_{p,e} = 0.90$ (based on given pressures).

Assume that neither the compressed air added to the turbine for cooling nor the mass of the fuel added in the burner affect the turbine power. For simplicity in calculation use $c_p = 1,005$ J/(kg·k) throughout the cycle.

4. (a) Find the pressure ratio for maximum efficiency for a nonregenerative (CBE) simple-cycle gas turbine by calculating the overall efficiency at three pressure ratios and attempting to draw a curve through the efficiencies obtained. The cycle specifications are as follows:

Inlet temperature to compressor, $T_{0,1} = 38$ °C.
Inlet temperature to the turbine, $T_{0,4} = 983$ °C.
Compressor and turbine polytropic efficiencies, $\eta_{p,c} = \eta_{p,e} = 0.88$.
Combustor pressure drop 5 percent of the inlet stagnation pressure $= \Sigma(\Delta p_0/p_0)$
$(\dot{m}_e/\dot{m}_c) = 1.025$.

Calculations using mean air tables of constant specific heats can be used and the effects of fuel addition on the gas properties can be ignored for simplicity.

(b) Then determine how the optimum pressure ratio, and the efficiency at that optimum, change when intercooling is added. The intercooling should divide the compressor into two compressors having equal pressure ratios and should cool the air to 36 °C. There will be a 4 percent loss in stagnation pressure through the intercooler.

5. (a) Find the cycle thermal efficiency for an open-cycle gas-turbine engine with a heat exchanger (CBEX cycle) having an inlet temperature of 27 °C, and the turbine-inlet temperature of 927 °C, a pressure ratio of 4:1 in the compressor; a compressor polytropic efficiency of 92.5 percent; a turbine polytropic efficiency of 92.5 percent; a heat-exchanger effectiveness of 98 percent; and pressure drops in the air and gas streams through each side of the heat exchanger and through the combustor of $5-\frac{1}{3}$ percent of the inlet pressure in each case.

(b) Calculate the power output per kg per second air flow and compare your values of power output and efficiency with those given by the performance curves in the chapter. Then assume that there is a leakage of 2 percent of the air entering the heat exchanger, going directly to atmosphere; or that there is a 6 percent pressure drop in the combustor; or that there is both leakage and increased combustor pressure drop together. Hence discuss the sensitivity of this cycle to leakage and pressure drop.

(c) These calculations can be made using the data in appendix A for air and gas enthalpies, or mean values of specific heats. However, make one complete cycle calculation using gas tables and compare the results.

6. (a) The temperature ratio of a simple-cycle (CBE) gas-turbine engine could be increased with blade cooling, but compressed air from the compressor delivery would be needed to cool the turbine blades; this cooling air would produce no useful work in the turbine expansion. Also it is possible that the turbine itself would have a lower polytropic efficiency because of the thicker blade profiles necessary. In the two cases of blade cooling given in table P3.6 find the maximum amount of compressor air that may be used for cooling if the specific power of the gas turbine is to be the same as for the base case.

(b) The mean specific heat in compression is 1005 J/(kg·K) and in expansion 1160 J/(kg·K). The universal gas constant is 8313 J/(kmole · K); the molecular weight is 28.9; and the inlet stagnation temperature and pressure are 5 °C and 0.95 bar. Effects of fuel mass addition may be neglected. Will the cycle thermodynamic efficiencies be higher or lower than for the base case?

Table P3.6

	Base case (BC)	Blade cooling (A)	Blade cooling (B)
Temp. ratio, T'	4	5	5
Pressure ratio, r_c	36	36	36
Compressor $\eta_{p,c}$	0.90	0.90	0.90
Turbine $\eta_{p,e}$	0.90	0.90	0.87
Cooling air ($\delta\dot{m}/\dot{m}_c$)	0	?	?
$\Sigma(\Delta p_0/p_0)$	0.14	0.14	0.14

7. Investigate the thermal efficiency and the power output per kg per second of the following heat-exchanger open-cycle (CBEX) gas-turbine engines and comment on their possible attractiveness for automotive, merchant-ship, and naval use. Compare the results for the new cycles with the nearest equivalents in the performance charts shown in the chapter.
Common values:

Inlet temperature, $T_{0,1} = 15$ °C.
Inlet pressure, $p_{0,1} = 0.98$ bar.
Compressor and turbine polytropic efficiencies $= 0.90$.
Regenerator thermal effectiveness, $\epsilon_{hx} = 0.90$.
Sum of stagnation-pressure losses, $\Sigma(\Delta p_0/p_0) = 0.12$.

Expander-inlet temperature, $T_{0,4}$, and pressure ratio, r_c (compressor), varying values as follows:

$$T_{0,4} = 1,227 \,°C, \; r_c = 6,$$
$$T_{0,4} = 1,077 \,°C, \; r_c = 5,$$
$$T_{0,4} = 927 \,°C, \; r_c = 4.$$

8. Find the cycle thermal efficiency and the hydrogen mass flow in kg per second for the cycle shown, if it is to evaporate 30 kg/s of liquid methane. The heat absorption is from furnace waste heat. The hydrogen may be treated as a perfect gas. Property values and cycle conditions are shown on the $h - s$ diagram in figure P3.8. Make and state such other assumptions or approximations as seem necessary or desirable.

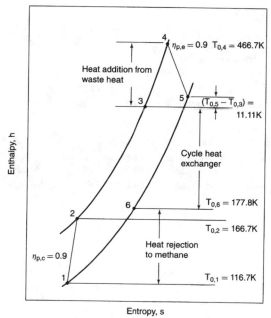

Entropy, s

Figure P3.8 In this figure $p_{0,4}/p_{0,5} = 0.85(p_{0,2}/p_{0,1})$. Hydrogen properties: $M_w = 2.0$, $C_p = 14,317 \text{ J/(kg} \cdot \text{K)}$, $\gamma = 1.41$, $R = 4,125.6 \text{ J/(kg} \cdot \text{K)}$. Liquid methane properties: $M_w = 16.0$, Boiling point at atmospheric pressure $= 111.11$ K. Latent heat of evaporation $= 514.52$ kJ/kg.

9. We want to find the effects of various component losses on the net output (specific power) and the thermal efficiency of simple (CBE) and of heat-exchanger (CBEX) gas turbine cycles. Choose one pressure ratio for the CBE cycle from those listed and another pressure ratio for the CBEX cycle. Calculate the thermal efficiency in each case. Then successively make the compressor and turbine efficiencies to be 100 percent, and then eliminate the pressure drops. Then, for the heat-exchanger cycle, make the heat-exchanger effectiveness 100 percent. Find the thermal efficiency at each step. We want (from a group of students) to produce a plot of base-cycle thermal efficiencies against pressure ratio and to give the effects of component losses.

Simple-cycle pressure-ratio choices: 5, 10, 24, 48, 64.

Heat-exchanger-cycle pressure-ratio choices: 2.5, 5, 10, 20.

Stagnation-to-stagnation (effective) polytropic efficiencies, compression: 0.87; expansion: 0.85.

Compressor-inlet temperature 288.2 K; pressure 1.0 atm.

Expander-inlet temperature 1,673.0 K.

Pressure losses (reducing the expansion ratio): 10 percent for the CBE cycle; 15 percent for the CBEX cycle.

Heat-exchanger effectiveness: 75 percent.

Ignore leakage. Preferably the correlations in appendix A should be used to calculate fuel mass flow, and air, and gas properties.

10. (a) Find the effect on the overall thermal efficiency, the waste heat recovered, and the cycle mass flow of designing a heat-exchanger for waste-heat recovery (or "waste-heat boiler", WHB) on a CBE gas-turbine engine with and without an induced-draft fan or compressor. This is the component between points 5 and 6 in figure P3.10. If employed, it would have a pressure ratio of 1.8 and an isentropic efficiency of 85 percent and would thereby increase both the pressure ratio to be designed for in the expander and the power extracted by the compressor-drive shaft.

The design shaft power for this marine power unit is to be 15,000 kW. The main compressor pressure ratio is 4.5 and the turbine-inlet temperature is 1,027 °C. The compressor-inlet temperature is 16 °C, and the atmospheric pressure at the inlet (upper-deck level) is 1 bar. The pressure losses as proportions of local stagnation pressure are as follows:

Inlet duct, 0.03,

Burner, 0.06

Turbine duct, 0.02,

Waste-heat boiler, 0.04,

Exhaust duct, 0.03.

In both cases the waste-heat boiler is designed for an exhaust-gas outlet temperature $T_{0,5} = 150$ °C. The expander isentropic efficiency, which includes the effects of leakage, is 86 percent, and that of the compressor is 87 percent.

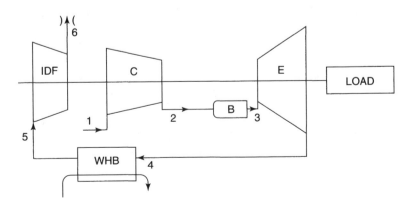

Figure P3.10.

(b) Assume that the expander and WHB mass flow is increased by 1.05 percent in both cases to account for fuel addition (even though it will differ somewhat in the two cases). Use Keenan et al., Gas Tables, table 1, for the air side and table 4 (400 percent excess air) for the products of combustion, taking the molecular weight of the products to be 28.9 kg/kmole, or appendix A. Give your calculations in the form of a table. Briefly discuss your results. Indicate how you forecast your results would change (qualitatively) if there were a heat exchanger (with the turbine exhaust heating the compressor-delivery air) before the waste-heat boiler.

11. Which of the following two modifications to a simple-cycle gas-turbine engine has the thermodynamic potential for the larger improvement in efficiency and for the larger increase in power output? Why?

In both cases the turbine exhaust passes to a waste-heat boiler fed with high-pressure water and producing dry steam. In one case this steam is led to a steam turbine that exhausts into a condenser, and the water is recycled. In the second case the steam is fed into the combustor of the gas turbine, and expands with the combustion gas, through the existing turbine. More fuel is burned in the combustor to maintain the turbine-inlet temperature constant. If a larger turbine is required, it will be supplied.

12. Calculate the (mixed-mean) stagnation and static pressures and temperatures at the exit of the first-stage turbine nozzles of a gas-turbine engine given the following specifications.

The compressor-inlet stagnation temperature and pressure are 300 K and 100, 000 Pa. The compressor pressure ratio is 3.5, and the associated polytropic efficiency is 0.90. These are based on the "effective" stagnation pressure at compressor discharge, although it is actually the static pressure at outlet (the dynamic pressure can't be utilized in the combustor). There is a five-percent drop in stagnation pressure from the effective compressor-outlet value to the turbine-nozzle inlet. One percent of the compressor air is "bled" off for cooling the first-stage nozzle row, and is discharged into the nozzle-exit flow. Guess that this mixing contributes to a drop in stagnation pressure of one percent from nozzle entrance to exit. Calculate the nozzle-exit conditions as if this cooling flow had become fully mixed, and as if the nozzle blades were otherwise adiabatic. The cycle temperature ratio (turbine-nozzle inlet over compressor inlet) is 6.0. The nozzle-exit Mach number is 0.90. Ignore both leakage and the effects of fuel-mass addition.

13. Calculate the thermal efficiency and specific power of a "simple-cycle" shaft-power engine using liquid hydrogen as the fuel, in two alternative configurations. In the first, the liquid hydrogen is injected direct into the combustion chamber as a cryogenic liquid at its boiling point. In the second configuration, the liquid hydrogen is used to cool the air going into the compressor to the point where the two temperatures are within 25 K of each other. For both cases the cycle pressure ratio is 20 : 1; the polytropic efficiencies of compressor and turbine are 0.9; and the sum of the stagnation-pressure drops is 0.05. For the first case the compressor-inlet temperature and pressure are 310 K and 120, 000 Pa and the turbine-inlet temperature is six times that of the compressor inlet. For the second case the turbine-inlet temperature stays at its first-case value; the compressor-inlet temperature drops. The lower calorific value of hydrogen may be taken as 120 MJ/kg. Approximate the heat required to vaporize and heat the hydrogen as 5 MJ/kg in both cases. A constant specific heat of 1100 J/(kg · K) may be used. Leakage and added fuel mass may be ignored. You may take the effects of the increased cycle temperature ratio in the second case to only one iteration.

Chapter 4

Diffusion and diffusers

The design of turbomachinery is dominated by diffusion, which is the conversion of velocity or "dynamic head" (or "dynamic pressure") into stream static pressure. Every blade row in a typical axial compressor is a collection of parallel diffusers. In most centrifugal compressors, the performance capabilities of both the rotor and the radial diffuser are limited by the diffusion capabilities of the flow channels. The performance of turbines of all types is enhanced, sometimes markedly, by an efficient exhaust diffuser. A large proportion of the aerodynamic and hydrodynamic losses (apart from kinetic-energy leaving losses) in turbomachines is due to local or general areas of boundary-layer separation resulting from a local or general degree of diffusion that is too large for the boundary layer to overcome. It was principally the lack of understanding of diffusion that delayed the arrival of efficient dynamic pumps and compressors until some decades into this century, and of viable gas-turbine engines until the mid-1930s.

This chapter contributes to the design of turbomachinery by giving the reader the ability to specify diffuser performance directly, useful where the diffusers will be used after single-stage or multistage compressors and turbines, and by setting the groundwork for the use of diffusion limitations in the choice of velocity diagrams in radial-flow and axial-flow blade rows (in chapter 5). This groundwork will be useful for preliminary design. It can be used to refine our estimates of stagnation-to-static efficiencies for compressors and turbines, covering cases where there is an outlet diffuser and where no benefit can be won for the kinetic energy in the flow leaving the diffuser. It will be further used for the more-detailed performance predictions arrived at in chapters 7 through 9, in which the design of axial- and radial-flow turbines and compressors is treated.

The analysis of diffusion is extremely complicated. Only a small part of the multi-dimensional matrix of diffusion parameters has been investigated experimentally. Nevertheless, we shall recommend a very useful rule to be employed in preliminary design: that a simple mainstream velocity ratio, W_2/W_1, should be kept above a minimum value if losses are to be low.

Diffusion can occur on an isolated surface (figure 4.1a) or in a duct (figure 4.1b). In either case, for the desired velocity reduction from W_1 to W_2 to occur, the boundary layers must remain attached. Where separation occurs, the main flow usually forms a jet that dissipates into turbulence. Occasionally separation can be local: see the discussion of "separation bubbles" below.

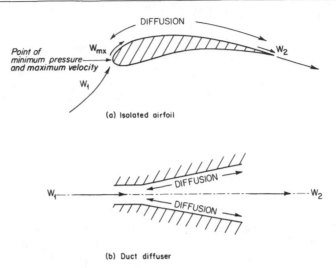

Figure 4.1. Examples of diffusion

The data quoted and analyzed here are for ducts, but we shall apply the findings to diffusion in general, including that found on isolated or nearly isolated bodies (such as axial-compressor blades).

4.1   Diffusion in ducts

Diffusion can occur because of area change or because of flow curvature. Subsonic diffusers have increasing area. Figure 4.2 shows some different types of symmetrical, straight-axis diffusers.

Curved diffusers

Diffusion occurs locally when flow approaches a bend in a constant-area duct (figure 4.3). Potential (incompressible, inviscid) flow, which is a good model of flow away from areas dominated by compressibility or viscosity, requires that the flow speed up on the inside of a bend and slow down on the outside. Therefore diffusion occurs approaching the outside of a bend and leaving the inside of a bend, and these are locations where separation is possible. In general boundary layers are less stable on convex than on plane surfaces.

When the axis of a diffuser is curved, the overall diffusion is added to the local diffusion. For separation not to occur, less overall diffusion must obviously be attempted than could be allowed in a straight diffuser. In most cases separation has to occur in only one place for complete flow breakdown to follow.

Boundary-layer considerations

Since static pressure is increased and velocity is decreased along the length of diffusers, let us consider viscous-fluid flow over a segment of flat plate along the length

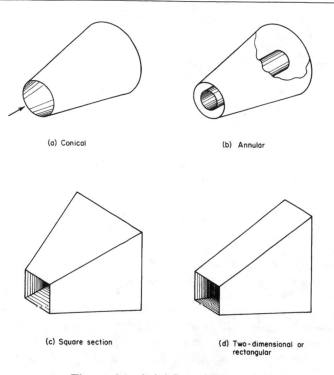

(a) Conical   (b) Annular

(c) Square section   (d) Two-dimensional or rectangular

Figure 4.2. Axial-flow diffusers

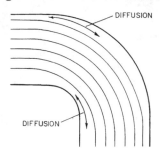

Figure 4.3. Diffusion in a constant-area bend

of which we gradually increase the static pressure. Boundary-layer theory (Schlichting, 1987) indicates that the velocity profile of the boundary layer will be as illustrated in figure 4.4. Let's assume that the static pressure increases from left to right, so that $p_{st,2} > p_{st,1}$. In the left diagram $p_{st,1}$ is low enough so that the velocity profile is relatively full; and τ_w, the shear stress of the fluid on the plate surface, is a function of fluid viscosity μ and the gradient of the velocity profile:

$$\tau_w = \mu \left(\frac{\partial C}{\partial y} \right)_w \tag{4.1}$$

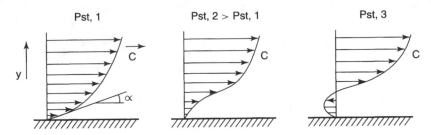

Figure 4.4. Boundary-layer velocity profiles in increasing-pressure regions (the static pressure is increasing from left to right)

$$\tan \alpha_w = \left(\frac{\partial C}{\partial y} \right)_w \tag{4.2}$$

As the static pressure is increased the velocity profile becomes less full (angle α increases) and the boundary layer becomes thicker (boundary-layer displacement thickness δ^* increases), so that where the static pressure is $p_{st,2} > p_{st,1}$ the gradient of the velocity profile is $\alpha_2 > \alpha_1$. In theory the static pressure can increase up to a point where $\alpha = 90°$. Any further attempt to increase the pressure will result in flow reversal in the wall region, a phenomenon called "separation" and explained further below.

This indicates that there is a limit of how much flow area can be increased and pressure can be decreased in diffusers. Consider a subsonic diffuser of rectangular cross section in which the area ratio can be increased (or decreased) by gradually pulling the walls further apart (or bringing them closer together), as illustrated in figure 4.5. With a small area increase the boundary layer stays attached and a moderate pressure recovery is achieved (figure 4.5a). With a somewhat larger area increase the boundary layer still stays attached, and a higher pressure recovery is achieved (figure 4.5b). The maximum

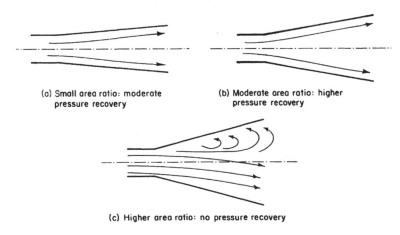

(a) Small area ratio: moderate
pressure recovery

(b) Moderate area ratio: higher
pressure recovery

(c) Higher area ratio: no pressure recovery

Figure 4.5. The risk factor in diffuser design

pressure recovery occurs where the boundary layer nearly separates at a point along the diffuser walls. (Actually maximum pressure recovery occurs when for extremely short periods of time the flow separates and locally reverses direction, and then reattaches for much longer periods of time.) However, if the area is increased further to where the boundary layer separates at one point, then the flow will tend to fully separate from one wall and form a jet against the other wall (figure 4.5c). Experimental data presented below indicate that, if the diffuser angle is now decreased in an attempt to reattach the flow and re-establish pressure recovery, then flow reattachment occurs at a diffuser angle lower than that of the maximum pressure recovery (a "hysteresis" effect). Therefore diffuser design should be reasonably conservative.

Influence of inlet boundary layer

Flows with laminar boundary layers at inlet, or with thick turbulent layers, will not withstand as much diffusion without separation as will thin turbulent layers.

A model or analogy of a flow undergoing diffusion is a line of people on skate-boards on a walled track approaching a hill (figure 4.6). Suppose that the skaters in the middle have enough velocity to get over the hill. The skaters on the outside are brushing against

Figure 4.6. Skate-boarding analogy to diffusion

the walls, and their velocity has been reduced so much that they have insufficient kinetic energy to surmount the hill. If they come to rest halfway up the hill and begin sliding back, any following lines of skate-boarders will have to leave the wall in an analogy of separation.

The only way for these skate-boarders at the sides to get over the hill (if we assume that they are allowed only to roll and are not to use their muscles) is for their neighbors who are going at higher speeds to give them a hand and drag them up the hill. In laminar boundary layers the (viscous) drag forces between layers is small, and this is why laminar flows cannot diffuse far without separation. In turbulent flows the inner strata of the boundary layers are given energy by exchange of "packets" of fluid from the higher-energy regions. In the analogy skate-boarders far from the wall, going at high speed, would exchange places with slow skate-boarders near the wall. The bumping that would occur during these exchanges would ensure that no skate-boarders could go very slowly in conditions of moderate diffusion or moderate slope. The energy losses during this bumping process would, of course, be larger than for the laminar model, but the losses are always small compared with those that occur if separation is allowed to take place.

Thus, the inlet-boundary-layer condition has a very significant influence on the performance of any diffuser.

Separation

The term "separation" is, then, short for "separation of the flow from the guiding surface". It occurs when the main flow is encountering a pressure rise (or rate of diffusion) severe enough to bring the boundary layer to rest. Usually, a vortex forms just downstream of this stationary point, with some of the flow reversing direction against the wall. This vortex effectively presents another bounding surface to the flow. Sometimes the new streamlines of the main flow result in a pressure distribution sufficiently modified (in the direction of reducing the severity of the adverse pressure gradient) for the main flow to "reattach" to the wall. The backward-turning vortex then forms a "separation bubble" which acts rather as does a roller bearing, and overall flow losses are small compared to those that would occur in full separation. We describe this phenomenon, and its effects on the pressure and heat-transfer distributions on a gas-turbine blade, in chapter 10.

Frequently, however, the flow does not reattach but continues as a high-velocity jet flow away from the wall. The jet dissipates in turbulent mixing, and there is little further pressure recovery. This type of separation is to be avoided wherever possible. Where it is unavoidable, it should be limited in extent. For instance, in a typical aircraft-engine layout the compressor will deliver air at around 200 m/s to a coaxial annular diffuser, that can be designed to reduce the velocity efficiently to a little under 100 m/s. The combustion system needs air flowing at a much lower velocity, but to continue to attempt to diffuse will invite a jet-type separation that might then occur further upstream than is necessary. By placing some type of break in the smooth wall, the separation can be located to yield maximum diffusion.

Separation as described is a steady-flow phenomenon. Frequently, however, it is unsteady. Sometimes separation that rapidly travels up and down a wall without ever resulting in gross jet-like separation yields maximum diffusion. On the other hand, the type of unsteadiness where a jet oscillates from one wall of a rectangular diffusing passage to the other will give high losses in the diffuser. Such a flow can also produce losses in, for instance, a downstream blade row, designed to accept a fully attached flow. This problem produces arguments among component designers, with the person responsible for a downstream diffuser or blade row blaming poor performance on the velocity profile delivered by the upstream component. We often cannot resolve such arguments satisfactorily, because we do not know how to characterize an acceptable velocity profile except to state that gross jet-type separation, steady or unsteady, is usually (with combustion systems being necessarily an exception) unacceptable.

Separation that is steady in an upstream component, for instance, a rotor, may appear to be unsteady in the form received by a downstream component. Radial-flow compressors of high pressure ratio generally experience a region of separation near the rotor circumference on the suction side of the blades (chapter 9). In those cases where a circular row of diffuser vanes is used downstream (outside) of the rotor, it appears to be desirable to allow a relatively short radial distance, of about 5 percent of the rotor radius, for the jets and wakes to mix out sufficiently for the diffuser row to be capable of operating efficiently. On the other hand, many studies of the effect of varying the axial distance between rotor and stator rows in axial-flow compressors have failed to show that there is any advantage to increasing the spacing to more than that necessary for purely mechanical reasons. These different experimental conclusions may simply reflect the difference in the extent of the unavoidable separation in radial-flow and axial-flow compressors.

Some of the cases of separation just discussed have been two-dimensional (such as the jet flow oscillating from one side to the other of a rectangular-section passage) and some have been three-dimensional (such as jet flow in a conical or annular diffuser). We do not know enough about the interaction of a separation from one wall of a rectangular diffuser with the flow in the corner and the influence on the flow on the neighboring walls to do more than refer the reader to the experimental results given here and elsewhere. We should also sound a warning about applying results from two-dimensional diffuser tests to three-dimensional cases.

Diffuser flow regimes

Kline and his coworkers at Stanford University have identified the significant flow regimes of diffusers. Figure 4.7 shows these regimes for straight-walled two-dimensional diffusers with thin turbulent inlet boundary layers. They found that maximum pressure recovery occurred when a condition of transitory stall existed. This is a condition where short-duration flow reversal propagates up and down the diffuser walls, but general flow separation does not occur.

4.2   Performance measures

Pressure-rise coefficient

The purpose of diffusers is to convert dynamic pressure at inlet to added static pressure at outlet. Therefore a useful measure of performance is the ratio of the increase in static

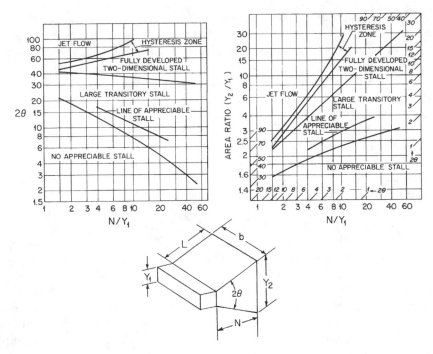

Figure 4.7. Flow regimes in straight-wall two-dimensional diffusers. From Fox and Kline (1962)

pressure to the inlet dynamic pressure: the pressure-rise coefficient

$$C_{pr} \equiv \frac{p_{st,ex} - p_{st,in}}{p_{0,in} - p_{st,in}} \qquad (4.3)$$

The dynamic pressure at diffuser inlet, $p_{0,in} - p_{st,in}$, is also given (later) the symbol $q_{d,in}$. Ideally, the value of C_{pr} could reach unity. In practice, only in very favorable circumstances can a pressure-rise coefficient of over 0.80 be obtained (see for instances figures 4.13 and 4.14).

The pressure-rise coefficient in a compressible-flow lossy diffuser

If flow stations 1 and 2 are defined at diffuser inlet and outlet respectively, the pressure-rise coefficient (equation 4.3) is defined by

$$C_{pr} \equiv \frac{p_{st,2} - p_{st,1}}{p_{0,1} - p_{st,1}}.$$

The pressures, temperatures, and velocities are shown on a T-s diagram in figure 4.8.

$$\frac{p_{st}}{p_0} = \left[\frac{T_{st}}{T_0}\right]^{C_p/R} = \left[\frac{T_0 - C^2/(2g_c C_p)}{T_0}\right]^{C_p/R} = \left[1 - \frac{C^2}{2g_c C_p T_0}\right]^{C_p/R} \qquad (4.4)$$

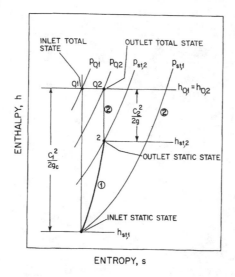

Figure 4.8. Diffusion on the h-s diagram

This is the exact solution, which can also be given in the form of equation 2.58:

$$\frac{p_{st}}{p_0} = \left[1 + \frac{M^2}{2(C_p/R - 1)}\right]^{-(C_p/R)} \tag{4.5}$$

An approximate solution can be obtained by using the first two terms only of the binomial expansion of equation 4.4.

$$\frac{p_{st}}{p_0} \approx 1 - \frac{C^2}{2g_c R T_0} = 1 - \frac{\rho_0 C^2}{2g_c p_0}$$

$$p_{st} \approx p_0 - \frac{\rho_0 C^2}{2g_c} \tag{4.6}$$

This is similar to the frequently used incompressible relation for the dynamic head. For it to be a fair approximation to compressible-flow conditions, the stagnation density, ρ_0 not ρ_{st}, must be used. The term including the velocity, C, in equation 4.4 is a Mach number, as can be seen from the form of the relation given in equation 2.58.

The approximate relation, equation 4.6, can also be obtained in terms of the local Mach number by substitution of equation 2.40:

$$\frac{p_{st}}{p_0} \approx 1 - \frac{C^2}{2g_c R T_0} = 1 - \frac{M^2}{2\left(1 - \frac{R}{C_p}\right)} \tag{4.7}$$

To show the degree of approximation involved in using equation 4.7 instead of the exact form, equation 4.4, values for each, and for the ratio of the approximate to the exact

Table 4.1. Ratio of static pressure to stagnation pressure, p_{st}/p_0, for $C_p/R = 3.5$

	Exact	Approximate	Approx./exact
C relation	$\left[1 - \dfrac{C^2}{2g_c C_p T_0}\right]^{C_p/R}$	$\left[1 - \dfrac{\rho_0 C^2}{2g_c p_0}\right]$	
M relation	$\left[1 + \dfrac{M^2}{2[(C_p/R) - 1]}\right]^{-C_p/R}$	$\left[1 - \dfrac{M^2}{2[1 - (R/C_p)]}\right]$	
M			
0.2	0.97250	0.97222	0.99971
0.4	0.89561	0.89147	0.99538
0.5	0.84302	0.83333	0.98851
0.6	0.78400	0.76493	0.97567
0.7	0.72093	0.68761	0.95379
0.8	0.65602	0.60284	0.91893

values, are listed in table 4.1 for $C_p/R = 3.5$ as a function of Mach number. The errors involved in using the incompressible form for the dynamic head are less than 1 percent up to a Mach number of 0.4, and are still less than 2.5 percent at $M = 0.6$.

The substitution of equation 4.6 into 4.3 leads to

$$C_{pr} \approx \frac{p_{0,2} - p_{0,1}}{\rho_{0,1} C_1^2/(2g_c)} + \left[1 - \frac{\rho_{0,2}}{\rho_{0,1}}\left(\frac{C_2}{C_1}\right)^2\right] \tag{4.8}$$

for low Mach numbers. The stagnation temperature does not change in an adiabatic diffuser, so that we can put $\rho_{0,2}/\rho_{0,1} = p_{0,2}/p_{0,1}$ for perfect-gas flow. For lossless flow,

$$p_{0,2} = p_{0,1} \quad \text{and} \quad C_{pr,tl} = 1 - \left(\frac{C_2}{C_1}\right)^2 \tag{4.9}$$

This is the theoretical pressure-rise coefficient. The velocity ratio (C_2/C_1) is more often given here as (W_{ex}/W_{in}), where W is a relative velocity, so that $C_{pr,tl}$ can be applied to moving as well as to stationary ducts.

The diffuser velocity ratio has been found to be a good guide to permissible diffusion levels in nonisentropic and compressible flows. The ratio W_{ex}/W_{in} is sometimes known as the de Haller number, after the Swiss engineer who suggested that this ratio should be not less than 0.72 for compressor cascades. In some diffusers, such as compressor-outlet diffusers in aircraft engines, the de Haller ratio can be somewhat smaller. The relation between the de Haller ratio and $C_{pr,tl}$ is shown in figure 4.9.

The de Haller ratio of not less than 0.72 for compressor cascades agrees well with the often-used upper limit of 0.5 for the value of $C_{pr,tl}$. If the cascades have to work in conditions where the blade surfaces become roughened, such as by industrial pollution or

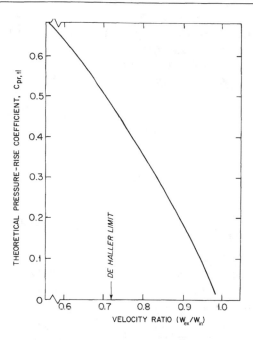

Figure 4.9. Velocity ratio and theoretical pressure-rise coefficient

salt deposits, a lower value of $C_{pr,tl}$ should be used, such as 0.45 or 0.40. A lower value should also be used for blade rows that are operated over a wide range of incidences, such as the first stages in a high-pressure-ratio compressor (see chapter 8). An upper limit of 0.4 would be appropriate for these first stages for many situations.

4.3   Theoretical pressure rise as a design guide

The theoretical value of $C_{pr,tl}$ (or $\Delta p/q$, which is identical, but $C_{pr,tl}$ is more frequently used in blade-row and stage design) chosen for any particular diffuser or blade row may not necessarily be related closely to the C_{pr} actually achieved. It is simply a design guide to the value of outlet relative velocity to be specified. Thus, if the inlet relative velocity is W_{in}, the diffuser outlet relative velocity W_{ex} may be chosen from equation 4.10.

$$W_{ex} = W_{in}\sqrt{1 - C_{pr,tl}} \tag{4.10}$$

4.4   Diffuser effectiveness

There is in fact a performance measure that is defined as the ratio between the actual and the theoretical pressure-rise coefficients. It is called the diffuser effectiveness:

$$\eta_{df} \equiv \frac{(\Delta p/q)_{ac}}{(\Delta p/q)_{tl}} = \frac{C_{pr}}{C_{pr,tl}} \tag{4.11}$$

The effectiveness is sometimes cross-plotted on diffuser-performance charts. It is related to the actual pressure-rise coefficient and to the area ratio, typically as shown in figure 4.10a,b. The variation of the actual to the theoretical pressure-rise coefficients for different inlet-boundary-layer thicknesses is plotted by Nebo in figure 4.10c.

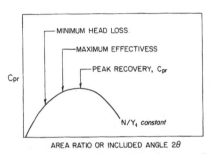

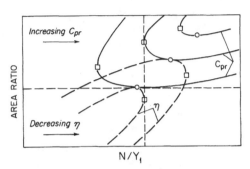

(a) Representative locations of several optima of performance at constant N/Y1. From Reneau, Johnston, and Kline (1967)

(b) Locations of optima of performance at constant length (square symbols) and constant area (round symbols). From Reneau, Johnston, and Kline (1967)

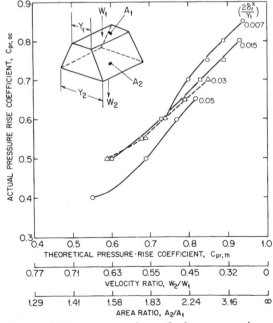

(c) Actual pressure-rise coefficient versus theoretical pressure-rise coefficient, velocity ratio and area ratio for straight two-dimensional diffusers. From Nebo (1975)

Figure 4.10. Relationship of diffuser effectiveness to actual pressure-rise coefficient

Example 1   Axial-compressor diffuser design

Design a conical-annular diffuser, shown diagrammatically in figure 4.11, for an axial compressor to have a theoretical pressure-rise coefficient, $C_{pr,tl}$ of 0.55. Then calculate a one-dimensional value for $C_{pr} = [(p_{st,3} - p_{st,2})/(p_{0,2} - p_{st,2})]$, and compare it with the theoretical value. Also, locate a possible annular diffuser from figure 4.18. The following data are specified:

Mass flow, $\dot{m} = 30$ kg/s
Diffuser-entrance stagnation pressure, $p_{0,2} = 500$ kN/m^2
Diffuser-entrance stagnation temperature, $T_{0,2} = 450$ K
Diffuser-entrance velocity, axial, $C_2 = 200$ m/s
Diffuser stagnation-pressure loss $(p_{0,2} - p_{0,3}) = 0.1(p_{0,2} - p_{st,2})$
Compressor-inlet stagnation temperature, $T_{0,1} = 288$ K
Compressor-inlet stagnation pressure, $p_{0,1} = 100$ kN/m^2
$(d_{hb,2}/d_{sh,2}) = 0.90$; $d_{hb,2} = d_{hb,3}$; $2L/(d_{sh,2} - d_{hb,2}) = 5.0$

Use a mean specific heat based on the mean static temperature of the diffuser. Find $d_{hb,2}, d_{sh,2}$, and $d_{sh,3}$ for one-dimensional-flow conditions.

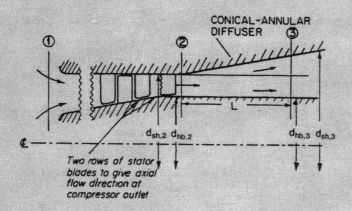

Figure 4.11. Conical-annular diffuser at outlet of axial compressor

Summary of calculations.   The dimensions were calculated to be

$d_{hb,2} = 485$ mm
$d_{sh,2} = 539$ mm
$d_{sh,3} = 560$ mm
$L = 135$ mm

The actual value of the pressure-rise coefficient was calculated to be 0.442, in comparison with the theoretical (incompressible inviscid) value of 0.55. The actual value is lower than the theoretical value, first, because there is a frictional drop in stagnation pressure leading to a smaller increase in static pressure than would otherwise occur and, second, because the calculations include approximations.

Calculations. The theoretical inviscid pressure-rise coefficient is

$$C_{pr,tl} = 1 - \left(\frac{C_3}{C_2}\right)^2$$

where $C_3 = C_{x,3}$ and $C_2 = C_{x,2}$. (The radial components of velocity may be neglected.) For $C_{pr,tl} = 0.55$, $(C_3/C_2) = 0.671$. Therefore $C_3 = 0.671 \times 200 = 134.20$ m/s. To calculate $p_{st,2}$ and $p_{st,3}$, use

$$T_{st,2} = T_{0,2} - \frac{C_2^2}{2g_c C_p},$$

where $C_p = \overline{C_p}$ as described in chapter 2. With a first guess at $C_p = 1,000$ J/(kg $\cdot$ K),

$$T_{st,2} = 450 - \frac{200^2}{2 \times 1,000} = 430 \text{ K}$$
$$T_{st,3} = 450 - 0.45(450 - 430) = 441 \text{ K}$$

Diffuser mean static temperature = 435.5 K. By interpolation in table A.1 (appendix A), $C_p = 1,018.8$ J/(kg $\cdot$ K).

A second iteration with this closer value of C_p leads to

$$T_{st,2} = 430.35 \text{ K}$$
$$T_{st,3} = 441.2 \text{ K}$$

Then

$$\frac{p_{0,2}}{p_{st,2}} = \left(\frac{T_{0,2}}{T_{st,2}}\right)^{C_p/R} = \left(\frac{450}{430.35}\right)^{3.5464} = 1.1716$$

Therefore

$$p_{st,2} = 426.78 \text{ kN/m}^2$$
$$(p_{0,2} - p_{st,2}) = 73.221 \text{ kN/m}^2$$
$$(p_{0,2} - p_{0,3}) = 0.1(p_{0,2} - p_{st,2}) = 7.322 \text{ kN/m}^2$$
$$p_{0,2} = 492.68 \text{ kN/m}^2$$
$$\frac{(p_{st,3} - p_{st,2})}{(p_{0,2} - p_{st,2})} = 0.422$$

And

$$\frac{p_{0,3}}{p_{st,3}} = \left(\frac{T_{0,3}}{T_{st,3}}\right)^{C_p/R} = \left(\frac{450}{441.2}\right)^{3.5464} = 1.07030;$$

therefore

$$p_{st,3} = 459.17 \text{ kN/m}^2$$
$$(p_{st,3} - p_{st,2}) = 32.389 \text{ kN/m}^2$$

Sizing. We must calculate first the stagnation and then the static densities at inlet and outlet of the diffuser:

$$\rho_{0,2} = \frac{p_{0,2}}{RT_{0,2}} = \frac{500,000 \text{ N/m}^2}{286.96 \text{ J/(kg} \cdot \text{K)} \times 450 \text{ K}} = 3.8720 \text{ kg/m}^3$$

$$\frac{\rho_{0,2}}{\rho_{st,2}} = \left(\frac{T_{0,2}}{T_{st,2}}\right)^{C_p/R-1} = \left(\frac{450}{430.35}\right)^{2.550} = 1.1204$$

Therefore,

$$\rho_{st,2} = 3.4558 \text{ kg/m}^3$$

$$\rho_{0,3} = \frac{p_{0,3}}{RT_{0,3}} = \frac{492,680 \text{ N/m}^2}{286.96 \text{ J/(kg} \cdot \text{K)} \times 450 \text{ K}} = 3.8153 \text{ kg/m}^3$$

$$\frac{\rho_{0,3}}{\rho_{st,3}} = \left(\frac{T_{0,3}}{T_{st,3}}\right)^{[(C_p/R)-1]} = \left(\frac{450}{441.15}\right)^{2.550} = 1.0519$$

$$\rho_{st,3} = 3.6271 \text{ kg/m}^3$$

Next we use continuity $\dot{m} = \rho_{st,2} A_2 C_2$ to evaluate areas and diameters:

$$A_2 = \frac{\pi d_{sh,2}^2}{4}[1 - 0.9^2] = \frac{30 \text{ kg/s}}{3.4558 \text{ kg/m}^3 \times 200 \text{ m/s}} = 0.0434 \text{ m}^2$$

Therefore,

$$d_{sh,2} = 0.5393 \text{ m}$$

$$d_{hb,2} = 0.4854 \text{ m}$$

$$A_3 = \frac{\pi}{4}[d_{sh,3}^2 - 0.4854^2] = \frac{30 \text{ kg/s}}{3.6271 \text{ kg/m}^3 \times 134.2 \text{ m/s}} = 0.0616 \text{ m}^2$$

$$d_{sh,3} = 0.5604 \text{ m}$$

$$\frac{d_{sh,2} - d_{hb,2}}{2} = 26.95 \text{ mm} \quad \text{(blade height)}$$

$$L = 5\left(\frac{d_{sh,2} - d_{hb,2}}{2}\right) = 134.75 \text{ mm}$$

Not all the input information was required. (In real-life problems, one usually has to deal with a great deal of superfluous, sometimes conflicting, information, and occasionally with too little.)

4.5 Axial-diffuser performance data

The very simple design rules for selecting values of $(\Delta p/q)_{tl}$ for different diffuser applications are one alternative open to the designer. The other is to find experimental data for the actual configuration and conditions under study. We report some representative test data for different configurations and conditions.

This alternative approach can provide only occasional guidelines in a multidimensional space. There are too many variables for the whole field to have been more than spotted with experiments. These are the principal variables.

Diffuser cross section (circular, rectangular, elliptical, annular, etc.).
Ratio of length to inlet hydraulic diameter[1].
Size and shape of inlet boundary layers.
Mach number of inlet flow.
Direction of inlet flow (axial, at an incidence, swirling, log spiral, etc.).
Steady, periodic, or fluctuating flow.
Straight axis or curved.
Smooth walls or rough, or shaped (e.g., ribbed).
Downstream conditions (plenum, tailpipe, transverse wall, etc.).

With such a vast array of possible combinations of conditions, the designer is unlikely to find test results that exactly match the conditions of interest. The data, however, can be extrapolated or interpolated in many cases, and the area of uncertainty in which the designer's judgment must be relied upon will thereby be reduced.

Conical-diffuser performance

The following performance data for conical diffusers were taken principally from work performed by Dolan and Runstadler (1973) for NASA. (The complete results were later published in a design manual.)

The variables were: length-to-inlet-diameter ratio L/d_{in}; inlet (throat) Mach number M_{in}; inlet boundary-layer blockage B_{in}; inlet Reynolds number $R_{e,in} \equiv W_{in}d_{in}\rho/\mu$; and stagnation-pressure level p_0. The fluid was air; the flow was axial, steady, and non-swirling. The diffusers had straight axes, smooth walls, and discharged into a plenum. The inlet boundary-layer blockage is the ratio of the sum of the boundary-layer displacement thickness (δ^*) (or equivalent stream thickness) to the total inlet width. The performance is given as the actual pressure-recovery coefficient, C_{pr}.

The optimum divergence angle for a fixed L/d_{in} ratio of 8 and midrange values of other conditions is shown to be between 7 and 8 degrees in figure 4.12. This optimum angle gives a maximum pressure-recovery coefficient of 0.65.

At a fixed divergence angle of 4 degrees, the pressure-recovery coefficient C_{pr} rises as the diffuser is made longer, until at $L/d_{in} = 25$, C_{pr} is about 0.76 for the conditions for which figure 4.13 is plotted.

Correlations of C_{pr} against area ratio and L/d_{in} with divergence angle as a parameter are shown in figure 4.14 for different inlet conditions. It can be seen that for all conditions the optimum divergence angle decreases from about 8 degrees for short diffusers (L/d_{in} below 5) to between 4 and 6 degrees for long diffusers (L/d_{in} over 25).

In figure 4.15 the inlet Mach number is shown to have a surprisingly small effect on C_{pr}. The performance improves with increase of Reynolds number at low Mach

---

[1]The hydraulic diameter d_h is used to correlate data from circular and non-circular flow passages: $d_h \equiv 4\times$ (cross sectional area, A) / (passage perimeter, p_e).

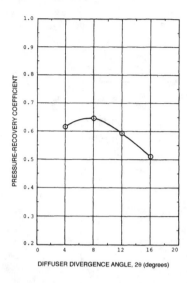

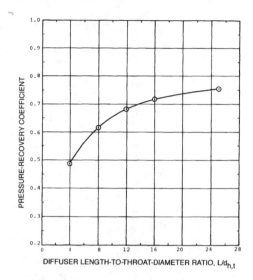

Figure 4.12. Pressure recovery versus divergence angle for conical diffusers $L/d_{in} = 8$, $M_{in} = 0.6$, $B_{in} = 0.06$, $R_{e,in} = 156,000$, $p_{0,in} = 109$ kN/m². From Dolan and Runstadler (1973)

Figure 4.13. Pressure recovery versus diffuser length for annular diffusers. $2\theta = 4°$, $M_{in} = 0.6$, $B_{in} = 0.06$, $R_{e,in} = 156,000$, $p_{0,in} = 109$ kN/m². From Dolan and Runstadler (1973)

numbers (figure 4.14a and 4.14b) but not, apparently, at high Mach numbers (figure 4.14c and 4.14d).

Square-diffuser performance

A comparison of conical- (circular) and rectangular-diffuser (square throat) performance by Dolan and Runstadler (1973) shown in figure 4.16 shows only a slight fall-off for the rectangular diffusers. Therefore the data for conical diffusers may be used for square diffusers, with an appropriate correction for large inlet boundary layers.

Rectangular-diffuser performance

Rectangular diffusers are those with one pair of parallel sidewalls and one pair of diverging sidewalls. Correlations of peak pressure recovery at constant $L/d_{h,t}$ ratios by Reneau, Johnston, and Kline (1967) are shown in figure 4.17. Values of the inlet boundary-layer blockage are given for various test points. The optimum divergence angles (figure 4.17b) can be seen to be about twice those for conical diffusers at equivalent $L/d_{h,t}$ ratios, as would be expected.

Annular-diffuser performance

Figure 4.18a shows that first stall in annular diffusers occurs at a smaller area ratio for a given value of $(N_d/d_{h,t})$ and (N_d/h_b) (where N_d is the diffuser axial length, figure 4.7

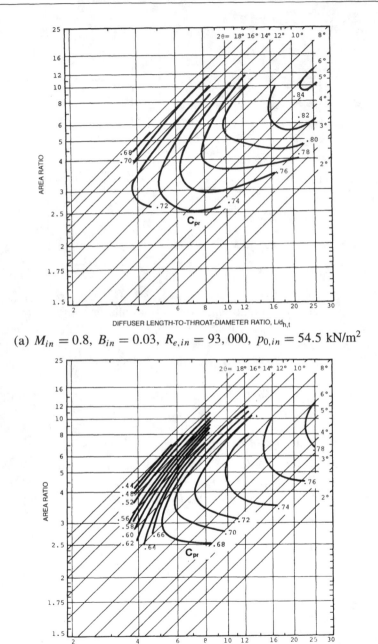

(a) $M_{in} = 0.8$, $B_{in} = 0.03$, $R_{e,in} = 93,000$, $p_{0,in} = 54.5$ kN/m^2

(b) $M_{in} = 0.2$, $B_{in} = 0.03$, $R_{e,in} = 30,000$, $p_{0,in} = 54.5$ kN/m^2

Figure 4.14. Conical-diffuser performance map. From Dolan and Runstadler (1973)

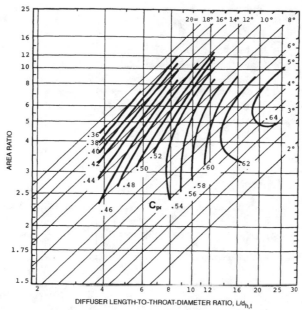

(c) $M_{in} = 0.8$, $B_{in} = 0.12$, $R_{e,in} = 93,000$, $p_{0,in} = 54.5$ kN/m^2

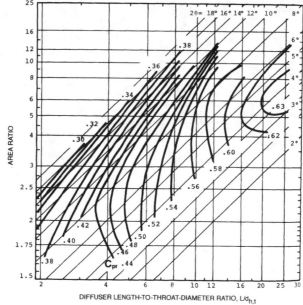

(d) $M_{in} = 1.0$, $B_{in} = 0.12$, $R_{e,in} = 202,000$, $p_{0,in} = 109$ kN/m^2

Figure 4.14. Continued

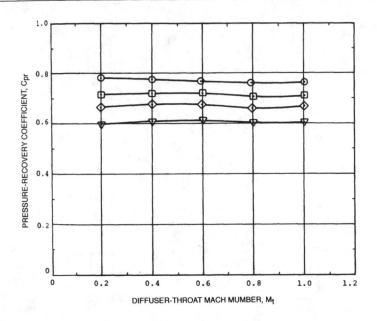

Figure 4.15. Diffuser pressure recovery versus Mach number. $2\theta = 8°$, $L/d_{h,in} = 16$, $p_{0,in} = 218$ kN/m^2. From Dolan and Runstadler (1973)

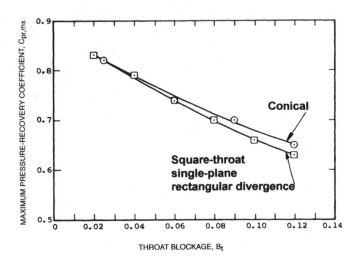

Figure 4.16. Maximum pressure recovery of conical and square diffusers at $M_t = 1.0$. From Dolan and Runstadler (1973)

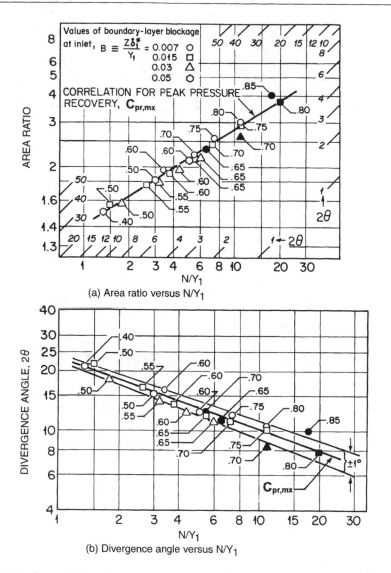

(a) Area ratio versus N/Y_1

(b) Divergence angle versus N/Y_1

Figure 4.17. Correlation of peak pressure recovery at constant N/Y_1. From Reneau, Johnston, and Kline (1967)

and h_b is the annular height) than for a conical diffuser. Values of C_{pr} for three straight-core (constant inner diameter) annular diffusers of divergence angles 2θ 10, 20, and 30 degrees and for a range of Reynolds numbers are shown in figure 4.18b and 4.18c. It is evident that in this configuration (figure 4.18d) the 20-degree angle is optimum for longer diffusers ($N_d/h_b > 2.5$) and the 30-degree angle is better for shorter diffusers. The test points are shown in relation to the C_{pr} contours in figure 4.18e.

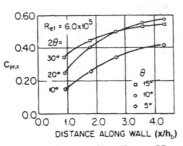

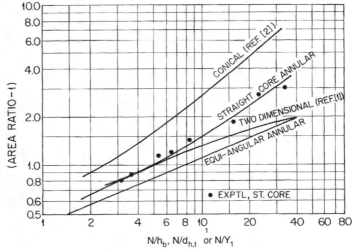

(a) Lines of first stall. From Howard, Henseller, and Thornton-Trump (1967)

(b) Axial static-pressure distribution for three diffusers at $Re_{e,in} = 1.2 \times 10^5$. From Adenubi (1975)

(c) Axial static-pressure distribution for three diffusers at $R_{e,in} = 6.0 \times 10^5$. From Adenubi (1975)

Figure 4.18. Performance of annular diffusers

The performance of annular diffusers with conical centerbodies and constant-diameter outer walls, suitable for turbine exits, has been reported by Adkins et al. (1983). Maximum pressure recovery was produced with cones of 25-35 degrees included angle, of length about three inner-cylinder diameters, with a radius between the parallel inner cylinder and the cone of at least 0.42 of the inner-cylinder diameter, and with a Reynolds number based on diffuser-inlet annulus height of at least 30,000. The highest pressure-recovery coefficient measured with these conditions was about 0.64 for an annulus entering the diffuser with an inner-to-outer-diameter ratio of 0.8. Pressure recovery for inlet annuli with smaller inner diameters naturally lead to lower potential pressure recovery because the area ratios are limited.

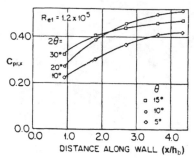

(d) Test geometries in relation to C_{pr} contours of Sovran and Kemp. From Adenubi (1975)

Figure 4.18. Continued

Dump diffusers

A flow along a duct that undergoes a sudden enlargement experiences some diffusion (see Carrotte et al. (1994) for some data and further references). One would expect high losses; however, if a toroidal vortex is stabilized at the sudden enlargement, a fairly efficient diffusion can result.

Griffith diffusers

A faired dump diffuser can be made into a Griffith diffuser, having contoured walls of a shape produced generally by computational fluid dynamics to give boundary layers at the limit of stability (usually with suction) (Yang and Nelson, 1979). The initial rate of area change is very rapid, followed by a much more gradual transition into the tail pipe.

Tailpipe effects

When a rectangular diffuser discharges into a straight constant-area tailpipe, instead of into a plenum as is used for most tests, an increase of pressure recovery normally occurs. Typical tests results of Kelnhofer and Derick are shown in figure 4.19. The parameter for the curves is L_{ex}/Y_{in}, where L_{ex} is the length of added tailpipe and Y_{in} is the inlet or throat width.

Senoo and Kawaguchi (1983) found that a collector or plenum chamber downstream of an annular curved diffuser reduced the pressure. The pressure loss was due to a "cork-screw" vortex, and was reduced by increasing the size of the collector and by incorporating vanes to reduce the swirl.

Wall contouring: splitter vanes

Substitution of a ribbed or finned surface (figure 4.20a) for a smooth surface has been shown by several investigators to give improved pressure recovery. Also the use of a bell-shaped wall contour (figure 4.20b) rather than a straight configuration has been found to give slightly improved performance, or shorter diffusers at the same performance.

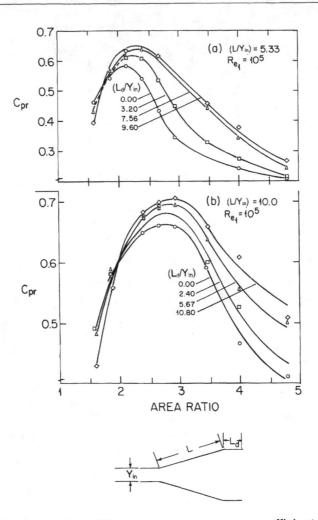

Figure 4.19. Straight-walled diffuser pressure-recovery coefficient with effect of tail-pipe addition. From Kelnhofer and Derick (1971)

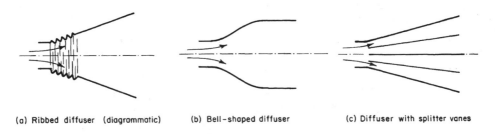

(a) Ribbed diffuser (diagrammatic)   (b) Bell-shaped diffuser   (c) Diffuser with splitter vanes

Figure 4.20. Alternative forms of axial diffuser

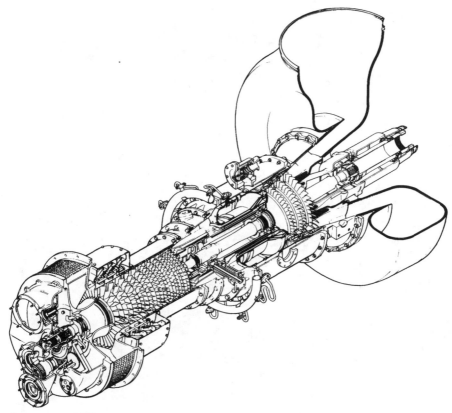

The eleven-stage axial-flow compressor is supplied from a circumferential inlet designed to give axisymmetric flow. The first three rows of stator vanes have controllable setting angles to avoid positive stall during starting. A two-stage axial-flow high-pressure turbine drives the compressor, and a single-stage power turbine is coupled to the output shaft. Both compressor and turbine have what seem to be well-proportioned diffusers.

Illustration. Solar Centaur gas-turbine engine. Courtesy Solar Turbine Company

Shorter diffusers can also be obtained by using splitter vanes to produce a number of small-angle diffusers in parallel (figure 4.20c).

In none of these approaches have design data adequate to cover general cases been produced. Designers wishing to take advantage of potential improvements must resort to test programs.

The effects of swirl on axial-diffuser performance

A small degree of swirl (rotation around the machine or diffuser axis) in the inlet flow of a diffuser with its axis generally in line with the flow (as distinct from a radial diffuser) may cause a sharp drop in performance and perhaps complete flow breakdown. As little as 10 degrees may be sufficient; see figure 4.21. The flow tries to set up

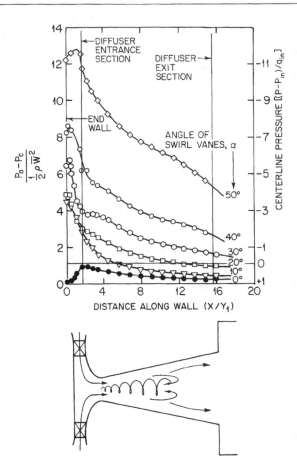

Figure 4.21. Centerline static-pressure distribution for swirling flow in conical diffusers. From So (1967)

a free vortex, with the tangential component of velocity inversely proportional to the radius. Such a flow pattern would result in infinite tangential velocities at the core. Shear forces prevent this occurring, and the central part of such a flow rotates as a solid body. The energy that has been dissipated leaves a stagnation-pressure deficit, and the outer rotating flow produces a low static pressure at the core. (This is similar to the conditions at the center of a hurricane or a tornado.) With only small degrees of swirl, high-pressure downstream flow will reverse to flow into this low-pressure core and to produce general flow breakdown. However, there is some evidence that swirl up to the point of flow breakdown in flow in a conical diffuser actually improves pressure recovery.

As would be expected, having a centerbody in the flow allows a considerably greater degree of swirl before breakdown occurs. Elkersh et al. (1985) tested annular diffusers

A double-sided radial-flow compressor is used for the first stage of this simple-cycle single-shaft engine. (The first Whittle jet also had a double-sided compressor.) The short radial diffuser terminates in several "trumpet" diffusers feeding the second-stage impeller. The air to the "back-side" of the first stage must pass through these diffusing ducts. A single tangential combustor feeds the nozzles to the radial-inflow turbine.

Illustration. Garrett GTCP 85 gas turbine. Courtesy Allied-Signal Aerospace

supplied by a duct with an inner-outer-diameter ratio of 0.5, with varied amounts of inlet swirl, and with various cone angles and lengths of the diffuser. For all tested variations, the minimum loss in stagnation pressure and the maximum pressure-recovery coefficient were found with inlet swirl vanes set at about 30 degrees (the actual swirl angles would probably be a little less than this).

Flow-straightening vanes are desirable upstream of the throat of turbomachine diffusers. Some compressor designs incorporate three rows of straightening vanes. Turbines seldom have true straightening vanes, despite the much larger kinetic energy being discharged. Three to six radial struts are often used to support a bearing or the inner surface of a diffuser. Such struts will remove much of the swirl, but probably not through a diffusion process. It has to be concluded that turbine diffusers seldom recover much pressure over much of their operating ranges, because, in off-design conditions, swirl angles will be high. (One very successful engine that incorporates cambered diffusing straightening vanes in the turbine exhaust is the Pratt & Whitney JT9.)

4.6 Radial-diffuser performance

In contrast to axial diffusers, radial diffusers can diffuse flows with large amounts of swirl. They are used typically downstream of radial-flow compressor and pump rotors. The swirl angles may be 70 degrees to the radial direction, and in compressors the inlet Mach number may be well above unity.

It is useful to consider the two components of the entering flow separately.

The radial-flow component diffuses because the area increases, and because this component has a low Mach number even when the Mach number based on the overall velocity vector is above unity. In axisymmetric flow the pressure gradient is in the radial direction and acts on this radial component.

The tangential-velocity component diffuses because in axisymmetric flow, and in the absence of turning vanes and wall friction, the tangential velocity is inversely proportional to the radius. It is unaffected by area change. It can pass from supersonic to subsonic flow without shock (because it can adjust to direction changes from signals received through the subsonic radial component).

When the radial diffuser is used downstream of a typical compressor or pump, the tangential component of momentum will typically be over twice the radial component (if the flow direction is at more than 63.5 degrees to the radius). The radial component has to surmount the radial pressure gradient, however. When reverse flow of the radial boundary layer occurs, it is not possible for the tangential component to continue diffusing in axisymmetric flow, because this would imply that there would be a pressure increase in one component and not in the other. Therefore breakdown must occur nonsymmetrically. In fact rotating jets are found, producing flow spirals similar to those of spiral nebulae in space. We call this form of radial-flow breakdown a rotating stall. Rotating stall in radial diffusers is different from rotating stall in axial-compressor blade rows, described in chapter 8. The best available guide to the geometric and flow conditions that will cause flow breakdown in straight-walled radial diffusers has been given by Jansen (1964), whose stability charts are reproduced as figure 4.22.

Example 2   Estimation of radial-diffuser stability

Using figure 4.22, estimate whether or not the radial diffuser in figure 4.23 will be stable. The specified data are these.

$p_{0,1} = 100$ kN/m^2
$T_{0,1} = 288.5$ K
$p_{0,2} = 653.7$ kN/m^2
$p_{st,2} = 251.78$ kN/m^2
$\rho_{st,2} = 2.1333$ kg/m^3
$T_{0,2} = 537.8$ K
$C_2 = 508.23$ m/s
$\alpha_{C2} = 65°$ (flow angle of absolute flow C_2 with radial direction)
$p_{0,3} = 641.23$ kN/m^2
$p_{st,3} = 486.16$ kN/m^2
$\alpha_{C3} = 73.72°$ (flow angle leaving diffuser)
$\dot{m} = 5.139$ kg/s
$N = 40,000$ rev/min

Summary of results. Jansen's criterion shown in figure 4.22 indicates this diffuser operates in a stable flow regime.

Calculations. To enter figure 4.22, we need (b_2/r_2), α_{C2}, and $R_{e,2}$ (the Reynolds number). From the geometric data in figure 4.23 $b_2/r_2 = 0.125$, and $\alpha_{C2} = 65°$ as specified.

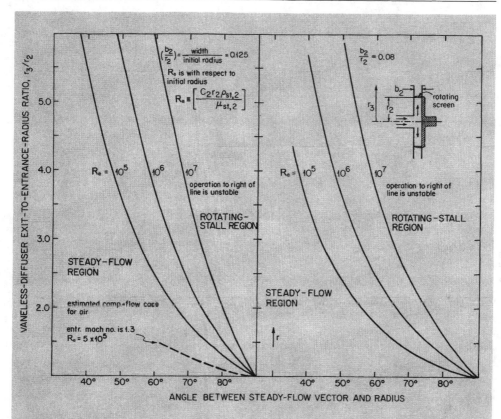

Figure 4.22. Stable operating range of vaneless diffusers. From (Jansen, 1964)

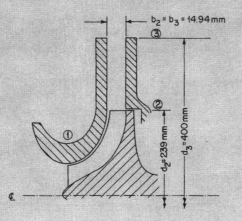

Figure 4.23. Diagram for example on stability analysis of vaneless radial diffuser with plane walls

To evaluate the Reynolds number

$$R_{e,2} \equiv \frac{\rho_{st,2} C_2 d_2}{2\mu_{st,2}}$$

we use table A.1 to first find T_{st} by iterating $C_{p,2}$, which gives $T_{st,2} = 410.6$ K. Then by interpolation $\mu_{st,2} = 2.329 \times 10^{-5}$ kg/ms. Therefore

$$R_{e,2} = \frac{2.1333 \text{ kg/m}^3 \times 508.23 \text{ m/s} \times 0.239/2 \text{ m}}{2.328 \times 10^{-5} \text{kg/ms}} = 5.56 \times 10^6$$

By interpolation on figure 4.22 for the Reynolds number, stability would be given for an inlet flow angle of 65° up to a radius ratio of about 2.75. Here the radius ratio is 1.67; therefore the radial diffuser under consideration appears to have good stability.

Only a small part of the input information provided was needed for this calculation.

Jansen's stability charts are for radial diffusers bounded by plane parallel walls. Some designers have attempted to obtain more diffusion by increasing the axial width between the diffuser walls as the radius is increased. Such wall contouring is likely to have an adverse effect on pressure recovery, because the sluggish radial momentum would then break down earlier. There is much to be said for the opposite approach, that of decreasing the wall spacing with the increase in radius. This is especially the case at the diffuser entrance, where a narrowing of the spacing downstream of the rotor will tend to energize the boundary layers in the radial direction but have no effect on the tangential component wherein is most of the momentum (figure 4.24). Wulf (1992) reports good results with wall contouring giving constant flow area.

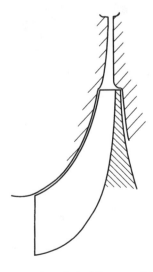

Figure 4.24. Diagram of radial diffuser with contracting walls

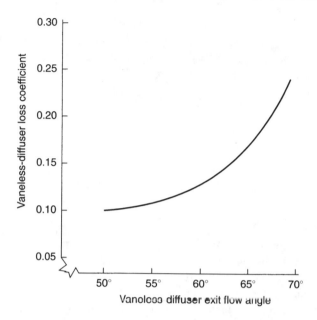

Figure 4.25. Loss data for vaneless diffusers. From Aungier (1993)

There are few data available on vaneless-diffuser performance. Aungier (1993) gives a method for the design of both vaneless diffusers and vaned return channels. A correlation of results from test data and predictions from his paper is given in figure 4.25, showing that the loss coefficient drops to low levels as the exit flow angle decreases below 60° (with the radial direction). These results indicate that the wall boundary layers, which travel at a much larger angle because of the greater effects of the radial pressure gradient on these low-momentum flows, are able to pass out of the diffuser fairly rapidly if the main flow is injected into the vaneless diffuser at a small angle. (The inlet angle will be fairly close to the exit angle for constant-area diffusers with subsonic inlet Mach numbers.) At large angles, e.g., 70°, the boundary-layer flow will have a very long path length because the wall flow will be almost tangential, and will consequently "pile up".

Rotating vaneless diffusers

The concept of attaching the vaneless diffuser to the compressor rotor (figure 4.26) was patented in France by the Neu Company. The principle is that centrifugal action energizes the radial component of the boundary layer and enables it to overcome larger pressure gradients. Accordingly, such compressor rotors were frequently profiled to give increasing diffuser-wall spacing with radius. Critics of the approach maintained that the additional fluid friction on the outside of the rotating walls took away more than was gained in increased pressure recovery. For whatever reason, the concept does not appear to have survived. However, Rodgers and Mnew (1974) at Solar have tested a freely rotating vaneless diffuser with apparently promising results.

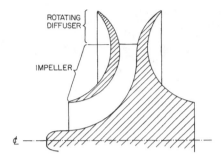

Figure 4.26. Rotating vaneless diffuser

Vaned radial diffusers

To achieve a given pressure rise, a vaneless diffuser requires a considerable radial space, and it incurs fairly high frictional losses because of the long path length of the spiral flow. Vanes can be used to direct the flow more in the radial direction, to increase the rate of diffusion, and to decrease the losses.

Vaned diffusers may be of many forms, some of which are illustrated in figure 4.27. They become similar to axial-flow diffusers with square, rectangular, or occasionally circular cross sections. However, the inlet conditions are very complex, even when the machine is running at its design point (when the flow should enter without incidence). Even in this condition the flow is highly three-dimensional. At off-design conditions incidence is added to complicate further the flow pattern and to increase the likelihood of flow separation and stalling. (The stall is likely to rotate, as it does in vaneless diffusers.) The low-solidity vaned diffuser (figure 4.27f) has been introduced by Senoo et al. (1983) as one that has efficiency gains over vaneless diffusers with little or no loss in range. Aungier (1988) has produced the first known general method for vaned-diffuser design. Hohlweg et al. (1993) compared low-solidity vaned diffusers and concluded that the low-solidity design had better performance with high-Mach-number compressors, but the conventional vaned diffuser was better at low Mach numbers.

Diffuser design data should be used with some caution. Even if the design of a reportedly successful diffuser is scrupulously copied, the use of a different rotor can give quite different inlet conditions and therefore different performance. Development testing must normally be resorted to. However, in general, it can be said that a vaned diffuser can give a higher pressure rise than a vaneless diffuser, but this will be found over a smaller range (figure 4.28).

4.7   Draft tubes for hydraulic turbines

Hydraulic turbines are regarded as too specialized for treatment in this text. They are universally single-stage machines. Diffusers are particularly important for single-stage turbines, as is discussed in chapter 5 and elsewhere. The diffusers after hydraulic turbines are called "draft tubes" or, in German, "Saugrohre". They are often excavated in rock

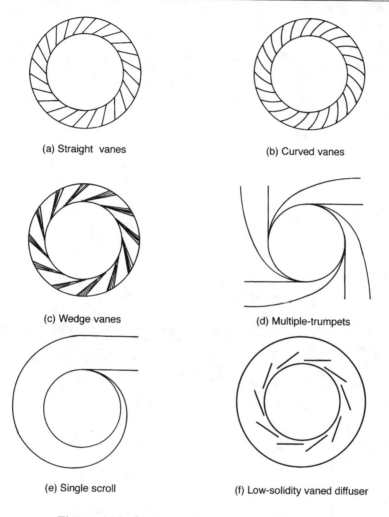

(a) Straight vanes   (b) Curved vanes

(c) Wedge vanes   (d) Multiple-trumpets

(e) Single scroll   (f) Low-solidity vaned diffuser

Figure 4.27. Configuration of vaned diffusers

below the turbine, so that the costs involved in constructing a draft tube are very large. Attempts to save funds by using a sharp bend or an overly flattened cross section can lose several points in efficiency. Deniz et al. (1990, 1991) have made a comprehensive survey of draft tubes and their performance and flow phenomena, and have measured in detail the flow in certain draft tubes over a wide range of turbine operation. Figure 4.29 is from their earlier report showing how draft tubes from vertical-axis turbines (Francis and Kaplan types) have changed their shape over the years. The later versions have required deeper excavation, the additional initial cost being repaid by better performance. "Splitter" guide vanes, figure 4.30, have also been found to be a justifiable additional cost in view of the improved draft-tube performance.

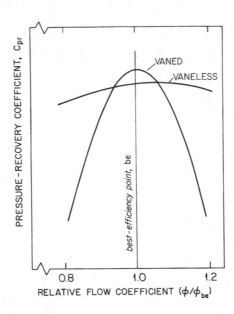

Figure 4.28. Representative pressure rise versus range for vaned and vaneless radial diffusers

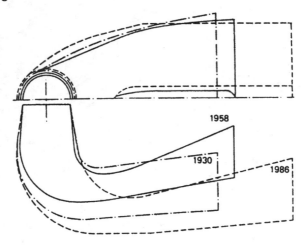

Figure 4.29. Shape of draft tubes in hydraulic turbines. From Deniz et al. (1990)

    The large size of and high fluid density in draft tubes result in very high Reynolds numbers and consequently small boundary layers. The patterns of secondary flow and of stall are, however, similar to those in smaller gas-flow ducts and diffusers, and are well illustrated in these texts.

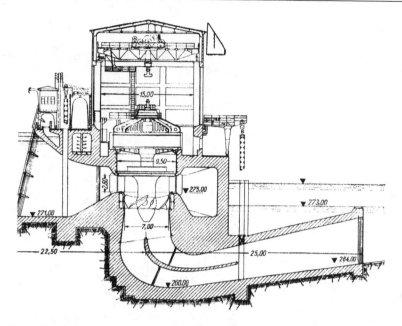

Figure 4.30. Use of splitter guide vanes in draft tubes in hydraulic turbines. From Deniz et al. (1991)

4.8   The risk factor in diffuser design

Large gains in efficiency and in specific power may be made in turbomachinery design by increasing the diffusion of kinetic energy to pressure energy. To be greedy, however, by trying to get too much diffusion is to invite disaster. This is why.

Suppose we have a rectangular diffuser with two diverging walls between two plane walls (figure 4.5). We start with a small amount of diffusion and fully attached flow, and we gradually increase the amount of diffusion either by increasing the length of the diverging walls at constant divergence angle, or by increasing the angle at constant length. The diffusion will increase up to a certain point, when separation will occur. Now, unfortunately, separation in such a case does not take place just at the maximum previous area ratio: it propagates back through the previously attached boundary layer and causes the whole flow to be separated. The last increment of diffusion demanded becomes the straw that breaks the camel's back.

An analogy to the diffuser-design risk is the problem of selling a used car with one advertisement. If one judges that the car is worth at least $500 and possibly $750, one can be absolutely sure of a sale by advertising the car at $450. One can probably make more money by putting the figure at $500, or $600, or even $700, but the chances of no one responding increase. If one advertises at $800, it is almost certain that no one will answer the advertisement. The extra greed has lost the whole enterprise.

References

Adenubi, S.O. (1975). Performance and flow regime of annular diffusers with axial-turbomachine-discharge inlet conditions. ASME paper 75-WA/FE 5, New York, NY.

Adkins, R.C., Jacobsen, O.H., and Chevalier, P. (1983). A preliminary study of annular diffusers with constant-diameter outer walls (suitable for turbine exits). ASME paper 83-GT-218, New York, NY.

Aungier, Ronald H. (1988). A systematic procedure for the aerodynamic design of vaned diffusers. In "Flows in non-rotating turbomachinery components", ASME FED-Vol 69, New York, NY.

Aungier, Ronald H. (1993). Aerodynamic design and analysis of vaneless diffusers and return channels. ASME paper 93-GT-101, New York, NY.

Carrotte, J.F., Denman, P.A., Wray, A.P., and Frey, P. (1994). Detailed performance comparison of a dump and short faired combustor diffuser system. ASME Journal of Engineering for Gas Turbines and Power, vol. 116, pp. 517–526, New York, NY.

Deniz, Sabri, Max Bosshard, Jurg Speerli, and Peter Volkart (1990). "Saugrohre bei flusskraftwerken". Report no. 106, Versuchsanstalt fur Wasserbau, Hydrologie und Glaziologie, ETH, Zurich, Switzerland.

Deniz, Sabri, Jurg Speerli, and Peter Volkart (1991). "Saugrohre, geshwindigkeitsmessungen am saugrohraustritt einer Rohrturbine". Report no 109, Versuchsanstalt fur Wasserbau, Hydrologie und Glaziologie, ETH, Zurich, Switzerland.

Dolan, Francis X., and Runstadler, Peter W., Jr. (1973). Pressure-recovery performance of conical diffusers at high subsonic Mach numbers. Report CR-2299, (July). NASA, Washington, DC.

Elkersh, A.M., Elgammal, A.H., and Maccallum, N.R.L. (1985). An experimental investigation of the performance of equiangular annular diffusers with swirled flow. Proc. I.Mech.E., vol. 199, no. C4, London, UK.

Fox, R.W., and Kline, S.J. (1962). Flow regime data and design methods for curved subsonic diffusers. Trans. ASME J. Basic Eng., D 84 (September), New York, NY.

Hohlweg, William C., Direnzi, Gregory L., and Aungier, Ronald H. (1993). Comparison of conventional and low-solidity vaned diffusers. ASME paper 93-GT-98, New York, NY.

Howard, J.H.G., Henseller, H.J., and Thornton-Trump, A.B. (1967). Performance and flow regimes for annular diffusers. ASME paper 67-WA/FE-21, New York, NY.

Jansen, W. (1964). Rotating stall in a radial vaneless diffuser. Trans. ASME J. Eng. vol. 86, pp. 750–758, New York, NY.

Kelnhofer, William J., and Derick, Charles T. (1971). Tailpipe effects on gas-turbine-diffuser performance with fully devoted inlet conditions. Trans. ASME J. Eng. Power 93 (January): 57–62, New York, NY.

Reneau, L.R., Johnston, J.P., and Kline, S.J. (1967). Performance and design of straight, two-dimensional diffusers. Trans. ASME J. Basic Eng. 89 D, (March); 141–150, New York, NY.

Rodgers, C., and Mnew, H. (1974). Experiments with a model free-rotating vaneless diffuser. Trans. ASME J. Eng. Power (ASME paper 74-GT-58), New York, NY.

Schlichting, H. (1987). Boundary-Layer Theory. (7th ed.). McGraw Hill.

Senoo, Y., and Kawaguchi, N. (1983). Pressure recovery of collectors with annular curved diffusers. ASME paper 83-GT-35, New York, NY.

Senoo, Y., Hayami, H., and Veki, H. (1983) Low-solidity tandem-cascade diffusers for wide-flow-range centrifugal blowers". ASME paper 83-GT-3, New York, NY.

So, Kwan L. (1967). Vortex phenomena in a conical diffuser. AIAA Journal, 5 (June): 1072–1078.

Wulf, James B. (1992). Stage-efficiency effects on vaneless-diffuser wall contours. ASME paper 92-GT-18, New York, NY.

Yang, Tah-teh, and Nelson C.D. (1979). Griffith diffusers. ASME Journal of Fluids Engineering, vol. 101, pp. 473–477, New York, NY.

Problems

1. Write down in equations, relations, or words the steps you would take to find the outlet diameter of a plane-walled, vaneless diffuser in figure P4.1 which is to give a rise in static pressure of 70 kN/m^2 in assumed conditions of isentropic, axisymmetric, steady flow in radial equilibrium. At the inlet the diameter is 300 mm, the air is at 350 kN/m^2, 44.5 °C stagnation pressure and temperature, and flows at 70° to the radial direction with a radial component of velocity of 55 m/s. (These figures are given only to indicate the degree of compressibility, etc., and are not intended to be used for calculations.)

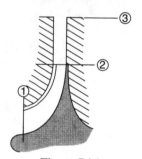

Figure P4.1

2. Determine the rise in static pressure across the blade row shown in figure P4.2 when it is passing 33.93 kg/s air (subsonic flow), assuming uniform one-dimensional flow (ignoring boundary layers, etc). Use the Mach-number charts in chapter 2, or tables.

3. Compare the actual pressure coefficient C_{pr} with the so-called theoretical incompressible $C_{pr,tl}$ as defined here for the row of axial-compressor blades sketched in figure P4.3. Comment on any difference.

$$C_{pr} \equiv \frac{p_{st,2} - p_{st,1}}{p_{0,1} - p_{st,1}}, \qquad C_{pr,tl} \equiv 1 - \left(\frac{C_2}{C_1}\right)^2$$

where C_2 and C_1 are mean velocities at inlet and outlet. Use the Gas Tables or compressible-flow charts for $\gamma = 1.4$, or appendix A data. The loss in stagnation pressure between planes 1 and 2 is 15 percent of the inlet dynamic head $p_{0,1} - p_{st,1}$.

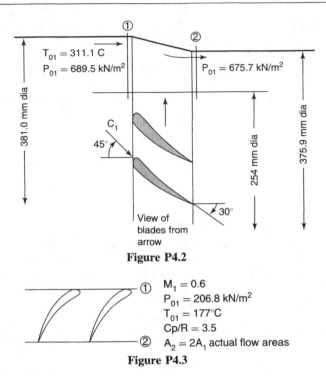

Figure P4.2

$M_1 = 0.6$
$P_{01} = 206.8 \text{ kN/m}^2$
$T_{01} = 177°C$
$Cp/R = 3.5$
$A_2 = 2A_1$ actual flow areas

Figure P4.3

4. If, for experimental convenience, you use air instead of water at similar velocities to develop a diffuser for a water pump, would you expect the actual pressure-rise coefficient to be higher or lower than when the diffuser is run in water? Why?

5. Calculate the tangential component of velocity of the flow leaving the vaneless diffuser of a radial-flow air compressor at its design point (stagnation-pressure ratio 4.2 to diffuser exit; associated polytropic efficiency to the same point 0.85; mass flow 1.7 kg/s; inlet stagnation conditions 300 K and 100, 000 Pa). The impeller has backsweep giving a work coefficient of 0.5; that is, the tangential component of velocity is one-half the blade speed. The enthalpy rise is the product of the blade speed and the tangential component of velocity. The absolute flow direction into the diffuser is 60° from the radial. There is zero swirl in the rotor-inlet flow. Sketch the impeller-outlet velocity diagram and the enthalpy-entropy diagram. You may neglect diffuser-wall friction as an approximation.

Chapter 5

Energy transfer in turbomachines

In this chapter we shall first apply Newton's law to turbomachines and thereby derive Euler's equation governing energy transfer. Then we shall use this to examine the velocity diagrams of the flow into and out of turbine and compressor stationary and rotating passages ("stators" and "rotors"), and we will define and study the various parameters that can be used to define velocity diagrams. Methods by which appropriate mean-diameter velocity diagrams may be chosen for axial-flow machines, and the number of stages determined, will be described. Some correlations of efficiency and other performance data will be given. Thus by the end of the chapter most of the tools that permit the preliminary design of axial and radial compressors, pumps, and turbines will be available.

So far as we know, all other books on turbomachinery deal with velocity diagrams and Euler's equation separately for axial-flow and radial-flow machines, often separately for compressors and turbines, and usually separately for liquid-flow and gas-flow devices. We believe that there is pedagogic value in showing that all types of turbomachine obey the same laws, one result of which value is the ability to carry out general preliminary design once this chapter has been absorbed. However, readers who prefer the more-traditional approach can, for instance, skip the radial-flow sections of this chapter at this point and read them together with chapter 9.

5.1 Euler's equation

Consider the flow of a single streamtube of fluid that enters the rotor of an axial, radial, or mixed-flow turbomachine at one radius with one velocity and leaves at another radius with another velocity (figure 5.1). The change of momentum between the flow entering and leaving the rotor can be used to calculate the force on the rotor and the equal and opposite force on the fluid.

For our purposes it is helpful to consider the three principal components of this force: axial, radial, and tangential. The axial and radial components are important for the design of bearings and for the analysis of vibration excitations, for instances. But these two components cannot contribute to the work transfer between the working fluid and

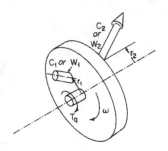

Figure 5.1. Flow through a rotor

the rotor. Only the tangential component of the force can produce a change in enthalpy through a work transfer. This work can be computed from the forces and the torques as follows.

The flow filament enters the rotor at a radius r_1, with a tangential component of the absolute velocity u_1, defined positive in the direction of rotor rotation ω.

Tangential force on rotor from entering fluid $= \dot{m}C_{u,1}/g_c$

Torque on rotor $= \dot{m}r_1 C_{u,1}/g_c$

Net torque, T_q, on rotor from entering and leaving flows, if these are steady,
  is $T_q = \dot{m}(r_1 C_{u,1} - r_2 C_{u,2})/g_c$

Energy transfer $= T_q \omega = \dot{m}w(r_1 C_{u,1} - r_2 C_{u,2})/g_c$

Now $\omega r_1 = u_1$ and $\omega r_2 = u_2$, the rotor peripheral speeds at the corresponding
  radii of entry and exit of the streamtube

Therefore $\dot{W}_{ex} - \dot{W}_{in} = T_q \omega = \dot{m}(u_1 C_{u,1} - u_2 C_{u,2})/g_c$

or

$$\frac{\dot{W}_{ex} - \dot{W}_{in}}{\dot{m}} = \frac{1}{g_c}(u_1 C_{u,1} - u_2 C_{u,2}) \tag{5.1}$$

This is known as Euler's equation. Positive work out $\dot{W}_{ex}$ means that work is delivered from the turbomachine shaft: that is, it is a turbine. A pump, fan, or compressor will have positive $\dot{W}_{in}$.

Euler's equation is almost universally applicable. The flow may be compressible or incompressible, idealized or frictional. The one restriction of steady flow is relatively unimportant, because of the near impossibility of storing fluid within a rotor. Near the rotating impellers or blades at the inlet and outlet of some turbomachines there will be small high-frequency fluctuations of flow angles and velocities. Traditionally flow properties at these locations are measured as averaged properties for steady-flow performance predictions. The effect of averaging this fluctuating flow and using the resultant properties for performance predictions or measurements is negligible; and it is beyond the scope of this book.

No torques other than those given by the fluid during passage through the rotor have been considered. Frictional torques on the disks or bearings or from the casing must be accounted for. Usually this is straightforward. Interaction between the casing (the stationary shroud) and the blade tips in an axial machine (where the flow in a streamtube may flow in a spiral, first performing useful work and then contributing purely to friction losses, or may do both simultaneously) is difficult to account for accurately but is normally relatively small.

For an adiabatic rotor in the absence of external torques, or large changes in elevation, a combination of the steady-flow energy equation (equation 2.6, or 2.11 for gas turbines with no change in elevation) with Euler's equation gives

$$g_c(h_{0,2} - h_{0,1}) = u_2 C_{u,2} - u_1 C_{u,1}$$

or

$$g_c \Delta_1^2 h_0 = \Delta_1^2 (u C_u) \tag{5.2}$$

This is the most useful single relation in turbomachinery design. It can be further simplified in many cases of axial- and radial-flow machines, as follows.

- In the preliminary design of axial-flow machines, the change of radius of the mean flow can often be ignored, so that a more restricted version of Euler's equation becomes

$$g_c \Delta_1^2 h_0 = u \Delta_1^2 C_u \tag{5.3}$$

- In radial-flow compressors and pumps the flow is usually led to the rotor without "swirl" (the tangential component of the inlet velocity, $C_u, 1$, is zero). Therefore, in these cases

$$g_c \Delta_1^2 h_0 = u_2 C_{u,2} \tag{5.4}$$

Example 1   Use of the Euler equation for a gas-turbine stage

What is the power output (kW) of a single-stage axial-flow gas-turbine expander which takes 6 kg/s of gas at 1050 °C and 11 bar stagnation conditions, passes the flow through nozzles, from which it leaves at a direction 70 degrees from that of the axial, at a velocity of 975 m/s, as shown in figure 5.2, and discharges it from the rotor blading without swirl ($C_{u,2} = 0$). The mean diameter of the expander blading is 1 m, and the shaft speed is 10,000 rev/min. The turbine has an isentropic stagnation-to-stagnation stage efficiency, to the rotor outlet, of 75 percent.

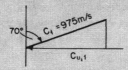

Figure 5.2. Vector diagram of flow leaving nozzles

Solution.

Using equation 5.2, $g_c(h_{0,2} - h_{0,1}) = u_2 C_{u,2} - u_1 C_{u,1} = -u_1 C_{u,1}$,

$u_1 = 10,000 \times \frac{2\pi}{60} \times \frac{1}{2} = 523.60$ m/s and $C_{u,1} = C_1 \sin 70° = 916.20$ m/s

Therefore, $g_c(h_{0,2} - h_{0,1}) = -479,721$ (m/s)^2 where $g_c \equiv 1$ in SI units.

The simple form of the steady-flow energy equation (equation 2.11) can be used to find the power output, $\dot{W}_{ex}$. The turbine can be treated as adiabatic, so that $\dot{Q}_{in} = \dot{Q}_{ex} = 0$. Therefore,

$$\frac{-\dot{W}_{ex}}{\dot{m}} = \Delta_1^2 h_0 = h_{0,2} - h_{0,1}$$

$$\dot{W}_{ex} = \dot{m}(h_{0,1} - h_{0,2}) = 6\frac{\text{kg}}{\text{s}} \times 479,721\frac{\text{m}^2}{\text{s}^2}$$

$$\dot{W}_{ex} = 2,878 \text{ kW}$$

Comments

1. This result gives the true power delivered by the expander-turbine blading. To obtain the shaft power, one would have to deduct only the effect of the various friction torques, principally bearings, seals, and disk friction.

2. The gas conditions were not needed. The same result would have been obtained had the fluid been hydrogen, water, or molasses.

3. We do not need to know the turbine efficiency for this calculation.

Example 2   Application of Euler's equation to a radial-flow air compressor

Calculate the pressure ratio (outlet static pressure/inlet stagnation pressure) of a radial-flow air compressor having a rotor of 300 mm diameter, turning at 20,000 rev/min, and having 15 radial blades. The air (0.9 kg/s) enters the rotor without swirl at 15 °C and 100 kN/m^2, and leaves the rotor with 90 percent of the rotor's tangential velocity. The compressor polytropic efficiency, inlet stagnation to outlet static conditions, is 80 percent.

Solution.

Since $C_{u,1} = 0$ we use the version of Euler's equation for zero inlet swirl (equation 5.4):

$$g_c \Delta_1^2 h_0 = u_2 C_{u,2} = 0.9 u_2^2$$

The peripheral velocity at rotor outlet is $u_2 = 20,000 \times \frac{2\pi}{60} \times \frac{0.3}{2} = 314.61$ m/s

Therefore (recall $g_c \equiv 1$ in SI units) $g_c \Delta_1^2 h_0 = 0.9 u_2^2 = 88,826$ (m/s)^2

We could use tables to find the inlet air enthalpy, $h_{0,1}$, and we could add the isentropic enthalpy rise (which would require converting the polytropic efficiency to the isentropic efficiency) and so find the outlet pressure. Let us use, however, the perfect-gas model, defined by equations 2.32 to 2.44. Recall (see note in chapter 2 after equation 2.44) in this example C_p is $\overline{C_{p,1,2}}$, the value of C_p suitably evaluated between $T_{0,1}$ and $T_{0,2}$.

$$\Delta_1^2 h_0 = \overline{C_{p,1,2}}(T_{0,2} - T_{0,1}) \equiv C_p(T_{0,2} - T_{0,1})$$

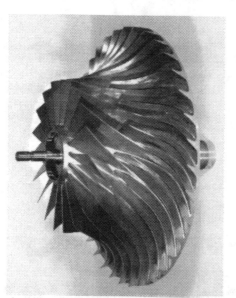

This compressor rotor is from the Allison 404 engine (about 300 kW output). The pressure ratio is about 4:1. Only the outer edges of the blades and the shaft areas have been machined: the rest is as cast. The angle of sweep back at the periphery is about 45 degrees. The blades are considerably "buttressed" at this point to allow for centrifugal loading. At the high-Mach-number inlet tip they are, however, thin and radial. There are fifteen main blades and fifteen "splitters", starting at about one-third of the channel length. The graceful continuous curve of the blades is a result of the design freedom conferred by modern casting methods. The pattern "master" was probably produced from solid material by a five-axis milling machine, a process that is too expensive for production impellers.

Illustration. Allison centrifugal compressor rotor with sweep back. Courtesy Allison Engine Company

We guess C_p and iterate. Guess $C_p = 1,012$ J/(kg · K). Then

$$T_{0,2} = T_{0,1} + \frac{\Delta_1^2 h_0}{C_p} = 288 \text{ K} + \frac{88,826 \ m^2/s^2}{1,012 \text{ J}/(\text{kg} \cdot \text{K})} = 375.78 \text{ K}$$

Then

$$\overline{T_{0,1,2}} = \frac{T_{0,2} + T_{0,1}}{2} = 331.89 \text{ K}$$

At this temperature, from table A.1, $C_p = 1,007.8$ J/(kg · K). We use this value for $\overline{C_{p,1,2}}$ as being close enough (even though the value of C_p at the arithmetic mean temperature is not the mean C_p, because the variation of C_p with T is not linear). Then we use equation 2.78 to find the pressure ratio:

$$\frac{p_{0,2}^{\otimes}}{p_{0,1}} = \left(\frac{T_{0,2}}{T_{0,1}}\right)^{\left[\left(\frac{C_p}{R}\right)\eta_{p,c,ts}\right]} = \left(\frac{375.78}{288}\right)^{\left[\left(\frac{1,007.8}{286.96}\right)0.80\right]} = 2.117$$

Comments

1. The most confusing aspect of this result will undoubtedly be that a ratio of "stagnation" outlet to inlet stagnation pressure, $p_{0,2}^{\otimes}/p_{0,1}$, has apparently been obtained, whereas what was asked for was the ratio of the outlet static pressure to the inlet stagnation pressure, $p_{st,2}/p_{0,1}$. The answer to this difficulty is found in the general definition of a turbomachine efficiency, equation 2.66, where the outlet conditions include the defined outlet stagnation pressure, $p_{0,2}^{\otimes}$. For a stagnation-to-static efficiency, $\eta_{p,c,ts}$, the (useful) outlet stagnation pressure is defined as being equal to the actual outlet static pressure, $p_{0,2}^{\otimes} \equiv p_{st,2}$. Therefore $p_{0,2}^{\otimes}/p_{0,1} \equiv p_{st}, 2/p_{0,1}$, and the answer is as desired.

2. The mass-flow rate is not required in the calculation. Nor is the number of blades or the inlet pressure. Most real-life problems bring with them too much information; only a few have too little.

5.2 Velocity diagrams and the parameters that describe them

The velocity vectors of fluid flow through blade rows of turbomachines can be combined to form velocity diagrams. The velocity vectors shown in such diagrams are conventionally those for the mean flow at a specified radius at entry to, or at exit from, a rotor or a stator, or both. The flow directions within stator and rotor passages and around blades are not shown in velocity diagrams. We shall show that the use of Euler's equation in interpreting velocity diagrams enables the relative enthalpy change obtainable in any given diagram to be estimated on sight.

As an example of the construction of a velocity diagram we show in figure 5.3 the blading cross section and velocity diagrams of simplified typical axial-flow compressor and turbine "stages". A stage is a combination of a stator and a rotor, (not necessarily in that order, although turbine stages are almost inevitably analyzed in this way). The type of velocity-diagram construction that we shall use is based on the rotor peripheral velocity at flow exit, u_2. In this and several later examples of axial-flow machines we show a "simple" velocity diagram, defined as one having a constant axial flow velocity, C_x, from inlet to outlet and a constant rotor peripheral speed, u (constant streamtube diameter), from inlet to outlet.

An example of a more general velocity diagram, in this case for an axial-flow compressor stage, is shown in figure 5.4. Both the axial velocity C_x and the blade speed u are shown varying for the streamsurface under examination.

Absolute and relative velocities

In the velocity diagram we use the convention of giving "absolute" velocities (velocities in the reference frame of the stators or nozzles) the designation C, and of giving "relative" velocities (velocities relative to moving surfaces, the rotor blades) the designation W.

The simple velocity diagrams in figure 5.3 show: the (absolute) flow velocity into the stator is C_2; the (absolute) flow velocity out of the stator is C_1; the relative flow velocity into the rotor is W_1; the relative flow velocity out of the rotor is W_2; the tangential

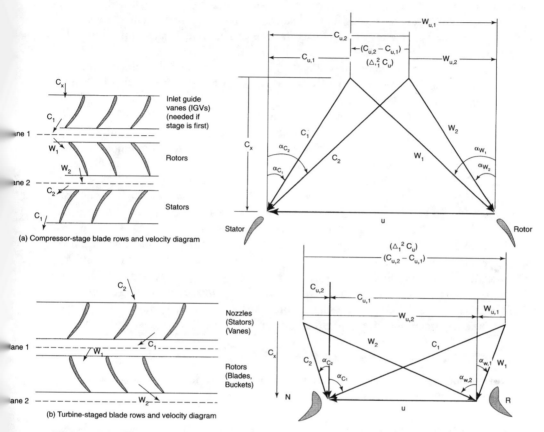

(a) Compressor-stage blade rows and velocity diagram

(b) Turbine-staged blade rows and velocity diagram

Figure 5.3. Representative axial-compressor and axial-turbine stage blade rows and simple (constant u and C_x) velocity diagrams

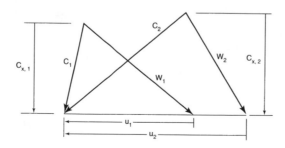

Figure 5.4. General velocity diagram with varying axial velocity and peripheral speed

velocity components $C_{u,1}$, $C_{u,2}$, $W_{u,1}$ and $W_{u,2}$; corresponding flow angles $\alpha_{C,1}$, $\alpha_{C,2}$, $\alpha_{W,1}$ and $\alpha_{W,2}$; the axial flow throughout the stage, C_x, (constant only in simple velocity diagrams); and the peripheral rotor velocity for the stage, u (constant only in simple velocity diagrams).

In figure 5.3 the diagrams are "repeating": the inlet and outlet flow velocities out of the stage are identical, so that a second stage identical to the first could (theoretically) be placed immediately downstream.

In general at the inlet of a compressor the flow would be entering along the axial direction, without a tangential component (without "swirl"), so the flow would not be entering the first stage in the direction of C_2. Therefore, frequently in compressors the first blade row is an "inlet guide vane", usually designated IGV, which turns (accelerates) the flow from the axial direction C_x to the direction of C_1, and it is followed by a rotor blade row with inlet and outlet relative velocities W_1 and W_2 respectively. Similarly, the absolute flow velocity after the last rotor of the compressor would have some swirl. We saw in chapter 4 that it is desirable to remove almost all swirl before entering the diffuser, so after the last rotor blade row most compressors have one or more stator blade rows to remove the swirl (they are called deswirler vanes). Similarly axial-turbine inlet and outlet flows enter and it is desired that they leave the component without swirl, so the considerations at axial-turbine inlets and outlets are similar. Later in this chapter we will show that there exist compressor and turbine stage velocity diagrams that facilitate some of these considerations for tangential components of velocity at inlets and outlets.

Angle and tangential-velocity conventions

We give angles with the axial direction. Another convention, used, for instance, in steam-turbine engineering, gives flow and blade angles relative to the plane of the blade rows.

Absolute tangential components of velocity, and their associated flow angles with the axial direction, are positive in the blade-speed direction.

Relative tangential components of velocity, and their associated flow angles with the axial direction, are negative in the blade-speed direction.

The axial component of velocity is designated C_x.

Flow angles are specified as α subscripted by the flow-vector designation. For instance, the turbine-nozzle angle is named $\alpha_{C,1}$.

Blade angles are specified as β subscripted by the flow-vector designation.

General (not simple) velocity diagrams

The density of the working fluid changes as it flows from stage to stage. The corresponding annulus area through which the flow passes changes accordingly, and the radii at which streamtubes pass through stages change along the length of compressors and turbines change too. Therefore at stator inlet, stator outlet / rotor inlet and at rotor outlet a velocity diagram for a general stage will have different axial velocities, and different radii (and therefore different peripheral rotor velocities). A general velocity diagram is shown in figure 5.4.

Approximate blade shapes

The form of velocity diagram that we use here (there are several alternatives) enables the approximate blade shapes to be sketched. In general, blade angles are not equal to flow angles. The flow usually enters a blade row at an angle of "incidence" and leaves with an angle of "deviation", each usually being a small angle between the flow direction and the tangent to the blade centerline at inlet and exit, respectively. Blades are designed from the velocity diagrams (see chapters 7 and 8), rather than vice versa.

Change in tangential velocity and stage work

The absolute and relative tangential components of velocity C_u and W_u are shown in figure 5.3 as is the change in tangential velocity (also known as the change in "whirl" or "swirl") across the rotor, $\Delta C_u = -\Delta W_u$. For simple diagrams (those with streamtubes or streamsurfaces entering and leaving the rotor at the same diameter, so that $u_1 = u_2$) the specific work done in (by) the stage is equal to $u\Delta_1^2 C_u = g_c\Delta_1^2 h_0$ for that streamtube or streamsurface. The work coefficient, the first of three parameters required to completely specify a stage velocity diagram of the simple kind, is then defined as follows.

The work or loading coefficient, ψ

For an adiabatic stage the work coefficient is defined as:

$$\psi \equiv \frac{-g_c\Delta_1^2 h_0}{u^2} = -\left[\frac{\Delta_1^2(uC_u)}{u^2}\right] \tag{5.5}$$

For "simple" diagrams (constant u from stage inlet to outlet),

$$\psi = \frac{-\Delta_1^2 C_u}{u}$$

The work coefficient ψ is positive for turbines; it is negative for pumps, compressors, fans, and so forth. In some texts, papers, etc. the work coefficient is defined as half this value. It is sometimes known as the "loading" coefficient. In simple diagrams ψ is a non-dimensional measure of: the distance between the apexes of the absolute and relative velocity triangles divided by u. Therefore the approximate value of the work coefficient can be determined immediately from the stage velocity diagram. In the compressor example shown in figure 5.3 $\psi = -0.3$. In the turbine example shown in figure 5.3, $\psi = 1.38$.

In turbines mean-diameter values of $\psi > 1.5$ would lead to them being called "highly-loaded" or "high-work" turbines, or turbine "sections" (if the diagram refers to one radial position along a long blade). Values of $\psi < 1.0$ indicate "low-work" or "lightly-loaded" turbine stages. In compressor stages highly and lowly loaded stages would have values below -0.5 and above -0.3, respectively. (Highly loaded stages have the apexes of the velocity triangles further apart, for both compressors and turbines.) These judgements of "high" and "low" are relative, and apply best to a stage of high hub-shroud ratio.

Stages with low hub-shroud ratios, such as the first stages of fans and compressors and the last stages of axial-flow turbines, normally have very high loadings at the hub and light loadings at the shroud.

The work coefficient alone is insufficient to specify the aerodynamic or boundary-layer loading on the blade rows and inner and outer annulus walls (see chapter 7 for a discussion of high and low aerodynamic loading). With the same value of ψ and a high or low value of C_x, the required flow deflection through any blade row could be low or high, respectively. We therefore define the second velocity-diagram parameter as the flow coefficient.

The flow coefficient, ϕ

$$\phi \equiv \frac{C_x}{u} \tag{5.6}$$

In a simple velocity diagram the flow coefficient is constant and refers to the stage as a whole. In the general case, where both C_x and u vary through the stage for any streamsurface, there will be a different flow coefficient at rotor inlet and at rotor outlet, and the flow coefficient varies with radius.

These two parameters, the work coefficient and the flow coefficient, fix a large part of the velocity diagram. But the geometric relationship of the difference of tangential velocity to the blade peripheral speed has yet to be specified. This is done by a quantity known as the "reaction".

The stage reaction, R_n

The strict definition of reaction is the ratio of the change in static enthalpy to the change in stagnation enthalpy of the flow passing through the rotor:

$$R_n \equiv \left[\frac{\Delta h_{st}}{\Delta h_0} \right]_{rr} = \frac{\Delta h_{st,rr}}{\Delta h_{0,se}} \tag{5.7}$$

(because the change in stagnation enthalpy through a stage, $\Delta h_{0,se}$, is equal to the change through the rotor, $\Delta h_{0,rr}$). The resulting ratio is often referred to as a percentage: for instance, "a 50-percent-reaction turbine". The exact value of reaction is not very significant, however, and turbines with reaction above perhaps 20 percent are often referred to just as "reaction" turbines.

A less-precise definition of reaction substitutes pressures for enthalpies in this relation. It serves as a useful benchmark to remember that zero-reaction (or "pure-impulse") turbines have no change of static pressure through the rotor.

The precise definition of reaction can be developed in terms of velocity-diagram values to show its geometric influence on the final specified form of the diagram.

For either the expansion or compression processes shown in figures 5.5 and 5.6,

$$\Delta h_{st} = \Delta h_0 + \frac{C_1^2}{2g_c} - \frac{C_2^2}{2g_c}$$

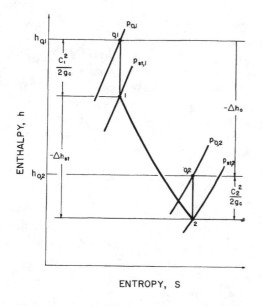

Figure 5.5. Enthalpy-entropy $(h - s)$ diagram for a turbine expansion

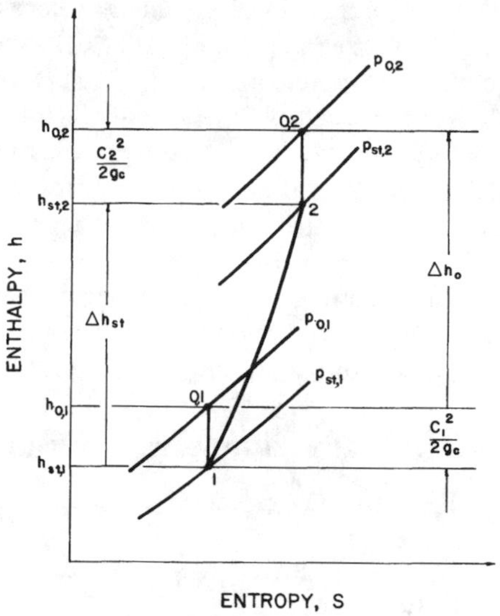

Figure 5.6. Enthalpy-entropy $(h - s)$ diagram for a compression process

where

$$\Delta h_{st} \equiv h_{st,2} - h_{st,1}$$
$$\Delta h_0 \equiv h_{0,2} - h_{0,1} = (u_2 C_{u,2} - u_1 C_{u,1})/g_c$$
$$C_1 \equiv \text{absolute velocity into rotor}$$
$$C_2 \equiv \text{absolute velocity out of rotor}$$

Therefore,

$$R_n \equiv \frac{\Delta h_{st}}{\Delta h_0} = 1 - \frac{(C_2^2 - C_1^2)/2}{u_2 C_{u,2} - u_1 C_{u,1}}$$

This expression is valid for compressors and turbines and for general velocity diagrams (those with u_1 in general not equal to u_2, and $C_{x,1} \neq C_{x,2}$).

For a simple velocity diagram $u_1 = u_2$ and $C_x = C_{x,1} = C_{x,2}$. Also, in general

$$C_1^2 = C_x^2 + C_{u,1}^2 \qquad \text{and} \qquad C_2^2 = C_x^2 + C_{u,2}^2$$

as can be seen from figure 5.3. Therefore for a simple velocity diagram

$$R_n = 1 - \left(\frac{1}{2}\right) \frac{C_x^2 + C_{u,2}^2 - C_x^2 - C_{u,2}^2}{u(C_{u,2} - C_{u,1})}$$

$$R_n = 1 - \left(\frac{1}{2}\right) \frac{(C_{u,2} + C_{u,1})(C_{u,2} - C_{u,1})}{u(C_{u,2} - C_{u,1})}$$

$$R_n = 1 - \frac{C_{u,2} + C_{u,1}}{2u} \tag{5.8}$$

Thus the reaction of a simple velocity diagram is the ratio of the mean tangential-velocity vector, $(C_{u,2} + C_{u,1})/2$, to the peripheral velocity, u, deducted from 1.0. The graphical interpretation of equation 5.8 on velocity diagrams such as figure 5.3 is that R_n is the fraction of vector u at which the bisector of the line joining the apexes of the absolute and relative velocity triangles with base u crosses the vector u. (Assign 0.0 and 1.0 at the tail and point respectively of vector u; then find on u the point where the bisector of the line joining the apexes of the absolute and relative velocity triangles with base u crosses u; that point on u corresponds to a value, usually between 0.0 and 1.0, representing R_n for the velocity diagram of the stage. At design-point operation there are no stage velocity diagrams with negative reaction. We will see later that some compressor stages with R_n just over 1.0 (just over 100 percent) have certain advantages.)

Relations for simple velocity diagrams

The following geometric equations can be derived from a simple velocity diagram of a compressor or turbine:

$$
\begin{aligned}
\tan \alpha_{C,1} &= [\psi/2 + (1 - R_n)]/\phi \\
\tan \alpha_{C,2} &= -[\psi/2 - (1 - R_n)]/\phi \\
\tan \alpha_{W,1} &= -[\psi/2 - R_n]/\phi \\
\tan \alpha_{W,2} &= [\psi/2 + R_n]/\phi \\
(C_1/u)^2 &= [(1 - R_n) + \psi/2]^2 + \phi^2 \\
(C_2/u)^2 &= [(1 - R_n) - \psi/2]^2 + \phi^2 \\
(W_1/u)^2 &= [\psi/2 - R_n]^2 + \phi^2 \\
(W_2/u)^2 &= [\psi/2 + R_n]^2 + \phi^2
\end{aligned}
\tag{5.9}
$$

Zero-reaction diagrams

In zero-reaction turbines the mean tangential-velocity vector $(C_{u,1} + C_{u,2})/2$ is u (the bisector of the line joining the apexes of the absolute and relative velocity triangles falls at the tail of vector u). Zero-reaction diagrams are normally used in turbines, not in compressors, where the flow swirl after rotors and stators would be very high. (The German company Junkers tried zero-reaction compressors at an early phase of development.) A zero-reaction turbine is called an "impulse" turbine because there is no expansion or acceleration of the flow through the rotor blades, and the rotor torque comes wholly from the "impulse" of the nozzle stream. With no pressure drop across the rotor-blade row, pressure seals (which in reaction turbines can take the form of a ring or shroud attached to the blade tips and carrying a labyrinth) are unnecessary.

A frequently used impulse diagram, shown with full lines in figure 5.7, has axial stage entry and exit flows and the reasonably high work coefficient of 2.0. A zero-reaction compressor stage has no intrinsic advantages and is to be avoided. High-reaction (approaching 100 percent) turbine-stage diagrams also have no intrinsic advantages and are also to be avoided. High-reaction compressor-stage velocity diagrams such as the one shown in figure 5.8 do have advantages (discussed below).

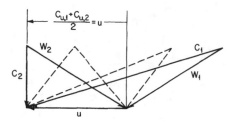

Figure 5.7. Zero-reaction velocity diagrams for turbines

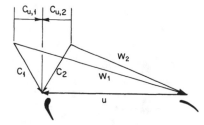

Figure 5.8. One-hundred-percent-reaction velocity diagram for compressors

Fifty-percent-reaction (symmetric) diagrams

In 50-percent-reaction velocity diagrams the bisector of the line joining the apexes of the absolute and relative velocity triangles crosses u in the middle, which is why the diagrams become symmetric. Representative compressor and turbine velocity triangles with 50-percent reaction are shown in figure 5.9.

Such diagrams are frequently favored for turbines because they have accelerating flow to an equal extent in rotor- and stator-blade passages, which leads to low losses. The "rectangular" turbine-stage diagram shown in figure 5.9b has the additional advantage of having axial flow at stage inlet and outlet. Also, tests show that this diagram gives close to the peak polytropic efficiency for turbine stages (see chapter 7).

When fifty-percent reaction is used in compressors, rotors and stators have equal diffusion coefficients $C_{pr,tl}$, and approximately equal (and minimum) relative Mach numbers for a given work coefficient.

High-reaction diagrams

In 100-percent-reaction velocity diagrams $-C_{u,1} = C_{u,2}$, and the bisector of the line joining the apexes of the absolute and relative velocity triangles falls at the point of vector u. For compressors, the small aerodynamic loading of the stator blades makes the machines less sensitive to dirt, which tends to accumulate more on stator blades than on rotor blades. High-reaction diagrams having axial flow at inlet and outlet also have applications.

The rotor-stator (rotor upstream of stator) compressor diagram, figure 5.10a, has the advantage of a low diffusion coefficient $C_{pr,tl}$ for a given work coefficient, and the possible disadvantage of a high rotor Mach number. Conversely, it can produce a desirable

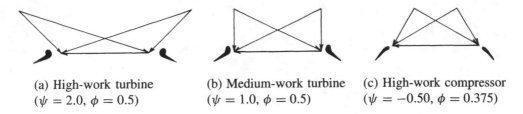

(a) High-work turbine
($\psi = 2.0$, $\phi = 0.5$)

(b) Medium-work turbine
($\psi = 1.0$, $\phi = 0.5$)

(c) High-work compressor
($\psi = -0.50$, $\phi = 0.375$)

Figure 5.9. Symmetric (50-percent-reaction) velocity diagrams

Mach number for a low blade speed, which is an advantage in certain applications. This diagram also applies to fans without stators.

The stator-rotor diagram, figure 5.10b, has over 100-percent reaction, the stators having accelerating flow, and therefore losing some of the static pressure rise produced by the rotor row.

The wind-turbine diagram, figure 5.10c, is similarly over 100-percent reaction, but when applied to a turbine, this becomes the rotor-only diagram of a windmill. In this case C_2 is the vector of the leaving flow.

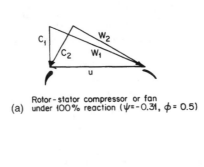

Rotor-stator compressor or fan
(a)   under 100% reaction ($\psi = -0.31$, $\phi = 0.5$)

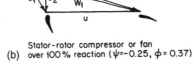

Stator-rotor compressor or fan
(b)   over 100% reaction ($\psi = -0.25$, $\phi = 0.37$)

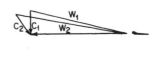

Windmill rotor – no stator
(c)   over 100% reaction ($\psi = 0.1$, $\phi = 0.17$)

Figure 5.10. High-reaction compressor, fan and windmill velocity diagrams

Example 3   Construction and calculation of velocity diagrams

Sketch the velocity diagrams, and calculate the missing data, for the three turbine and three compressor stages with the parameters listed in the table.

Solution.

The velocity diagrams can be drawn from the information given, and are shown in figure 5.11.

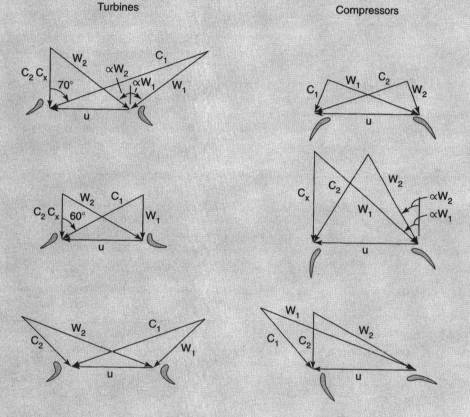

Figure 5.11. Example velocity diagrams

The work coefficient of the first turbine diagram has to be 2.0, because it is an axial-inlet and -outlet impulse diagram. The flow coefficient is obtained by inspection or from the first of equations 5.9, $\phi = 0.73$. The rotor-flow inlet and outlet angles are found from the third and fourth of equations 5.9, $\alpha_{w,1} = \alpha_{w,2} = 53.9°$, and the rotor relative-velocity ratio is $w_2/w_1 = 1.0$.

The second turbine has a rectangular diagram from the two zero angles given. It therefore has fifty-percent reaction. The flow coefficient is $cotan(60°)$, 0.577; the work coefficient is unity, and the rotor relative-velocity ratio is the inverse of $\sin 30°$, 2.0.

For the third turbine the flow coefficient can be found from the first of equations 5.9, $\phi = 0.546$; the flow angles from the third, fourth and second of equations 5.9, $\alpha_{w,1} = 42.5°$,

$\alpha_{w,2} = 70°$, and $\alpha_{c,2} = 42.5°$; and the rotor relative-velocity ratio can be calculated from the seventh and eighth of equations 5.9, $w_2/w_1 = 2.15$.

The same set of equations can be used to calculate the required data for the three compressors. The missing data are, in order across the columns:

1. -0.78, 71.4°, 20.0°, 71.4°, 0.34,

2. 0.845, 75%, 49.8°, 30.6°, 30.6°,

3. 0.55, 125%, 42.5°, 61.4°, 0.714.

The rotor relative-velocity ratio of the first compressor is far too low for it to operate successfully in any circumstances.

	ϕ	ψ	R_n	$\alpha_{C,1}$	$\alpha_{W,1}$	$\alpha_{W,2}$	$\alpha_{C,2}$	W_2/W_1
Turbine	?	?	0	70°	?	?	0°	?
Turbine	?	?	?	60°	0	?	0°	?
Turbine	?	2	50°	70°	?	?	?	?
Compressor	0.3	?	50°	20°	?	?	?	?
Compressor	?	-0.5	?	0°	?	?	?	0.75
Compressor	?	-0.5	?	?	70°	?	0°	?

5.3 Axial-compressor and pump velocity diagrams

Velocity diagrams for axial compressors and pumps can be specified by the three variables described in the previous section: flow coefficient, work coefficient, and reaction. However, other forms of specifications are usually preferred for compressors, for the following reasons.

Axial compressors and pumps are composed of series of rotating and stationary blade rows, each of which has a high relative velocity at entry and a lower relative velocity at exit. (This is strictly true only for machines with reactions between 0 and 100 percent.) Each blade row is therefore a group of parallel diffusers. We should expect the rules governing diffuser design to apply to axial compressors and pumps, and this is the case. The principal rule, as was stated in chapter 4, is that the outlet-to-inlet relative velocity ratio (the de Haller ratio) should be above some minimum value. The range of this ratio should be from 0.7, for machines operating at high Reynolds numbers in clean, near-ideal conditions, to 0.8, for machines that must operate over a wide range of incidence angles (flow coefficients) and/or in dirty, corrosive, or erosive fluids. A typical value for average industrial machines is 0.75.

There are more-sophisticated measures of diffusion in use. We quote one by Lieblein in chapter 8. However, the de Haller ratio has proven to be fully adequate for preliminary design.

In compressor or pump design the relative-velocity ratio, or the lower of the velocity ratios for rotor and stator, is often treated not as a limit but rather as an independent, or input, variable. The designer chooses a value that is estimated to give a safe margin from stall in the normal working range under the conditions the machine is likely to encounter in practice. Thus a value of 0.75 might well be selected.

In figure 5.12 the velocity ratios of the two blade rows is 0.696, which is too low to give efficient, wide-range diffusion. This diagram could be improved either by reducing (increasing) ψ from, say, -0.3 to -0.25, or by increasing ϕ, from, say, 0.4 to 0.5.

In a fifty-percent-reaction diagram the outlet-to-inlet velocity ratios are identical (it is a "symmetric" diagram). At other levels of reaction the velocity ratio will be lower in either the stator or the rotor. No simple rule can indicate which blade row has the limiting velocity ratio (see figure 5.14 and the discussion in "choice of flow coefficient").

The second variable that is significant in certain compressor and pump stages is the inlet relative velocity to either stator or rotor. The first stage of a multistage air compressor is usually designed to a maximum relative Mach number, which is found at the tip section of the rotor blades or possibly at the root or inner-diameter section of the stator blades.

For pumps operating under small suction heads, this same relative velocity governs the cavitation performance.

In neither of these two cases (compressor or pump) is steady blade stress a limiting factor in design (except for compressors of low-molecular-weight gases such as hydrogen or helium). The blade-row velocity ratio and the maximum inlet relative velocity often become the determinants of both the blade speed and the diagram shape. The remaining degree of freedom is expressed alternatively as reaction, or flow coefficient, or blade setting angle ("stagger" angle; see figure 5.13).

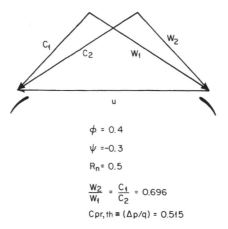

$\phi = 0.4$

$\psi = -0.3$

$R_n = 0.5$

$$\frac{W_2}{W_1} = \frac{C_1}{C_2} = 0.696$$

$C_{pr,th} \equiv (\Delta p/q) = 0.515$

Figure 5.12. Axial compressor or pump velocity diagram

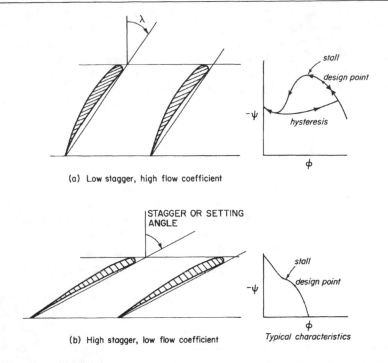

(a) Low stagger, high flow coefficient

(b) High stagger, low flow coefficient

Typical characteristics

Figure 5.13. Blade stagger angle and its effects on flow characteristics

Effect of design flow coefficient

In some cases the axial-compressor or pump designer has considerable freedom to choose the design value of flow coefficient, particularly in those cases where inlet Mach number or cavitation are not problems to be avoided. With the same de Haller ratio but different flow coefficients and work coefficients, blade rows of different stagger or setting angles have not only differing design points but different ψ versus *phi* characteristics in the off-design regions (figure 5.13). The high-work, high-flow, low-stagger blade rows tend to have a sharp stall, with a large fall in pressure rise and possibly a hysteresis effect that makes recovery from stall more difficult (Wilson, 1960). On the other hand, the low-work, low-flow, high-stagger blade rows tend to have an almost imperceptible stall, and a stalled pressure rise higher than at design, with no drop off in pressure differential at the stall itself. These characteristics have been confirmed and quantified by Koch (1981).

The hysteresis in stall characteristics can produce what is known as "nonrecoverable stall" in axial-flow compressors in jet engines. This means in effect that special measures, such as variable-setting stator adjustments, are required to take the machine out of stall. In a single-row fan where one of the authors (DGW) first encountered stall hysteresis, he could "blow" the stall out of the blade row with a hand-held shop-air nozzle.

Choice of flow coefficient

The need to keep the diffusion velocity ratio above some specified limit in both the blade rows of an axial compressor or pump leads to nonintuitive requirements for the minimum flow coefficient for a desired work coefficient in a simple diagram. The results of some calculations for a limiting diffusion ratio of 0.707, approximately the de Haller limit, are shown in figure 5.14. The curves give the lower limit of the flow coefficient (i.e., the flow coefficient that will produce a relative-velocity ratio of 0.707) for each blade row. The lower of the two curves is then the controlling limit; the higher portion of each pair of curves is shown as a broken line. For example, if one wanted a work coefficient of 0.25 in a diagram of reaction 0.75, the flow coefficient would need to be above 0.33, approximately, for the diffusion limit in the stator to be acceptable. At this value of the flow coefficient the rotor diffusion would be well away from the limiting value.

The attractions of high-reaction and 50-percent-reaction designs become clearer from figure 5.14. For practical flow coefficients, which would be those over 0.2, more work can be given by diagrams with reactions near 1.0 (or zero) than by diagrams with intermediate reactions. For instance, for a flow coefficient of 0.3 a 100%-reaction diagram gives a work coefficient of -0.38 without exceeding diffusion limits, whereas at fifty-percent reaction the work coefficient is about -0.24. It is lower at all intermediate reactions.

In practice, compressors and pumps are not designed with constant reaction along the length of the blades, so that the choice of reaction must be influenced by arguments (presented in chapter 6) on some aspects of three-dimensional flow.

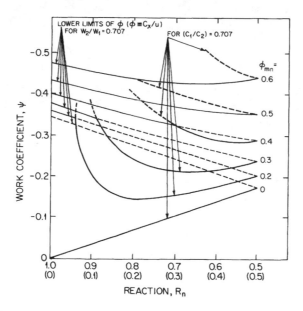

Figure 5.14. Effect of reaction and flow coefficient on stage work for compressor and fan diagrams

Design freedom and the hub-shroud diameter ratio

The degree of freedom there is to choose the shape of the velocity diagrams is a strong function of the hub-shroud diameter ratio, Λ, or the ratio of the inner to the outer diameter of the blade row (figure 5.15):

$$\Lambda \equiv \frac{d_{hb}}{d_{sh}} \tag{5.10}$$

When there is a strong need to keep outside diameter as small as possible, the diameter ratio of the first blade row of an aircraft compressor or fan may be made as small as 0.3. The ratio of the outer-to-inner blade-speed-squared will then be 11.11 : 1. The enthalpy rise is the product of the blade-speed-squared and the work coefficient. The desirability of delivering a similar enthalpy rise along the length of the blades from velocity diagrams having the blade-speed-squared varying more than tenfold puts such constraints on the design choices that there can be little variation from one design to another. In addition, the axial and tangential velocities must satisfy the equations of radial equilibrium, discussed in chapter 6.

In industrial design the optimum configuration will normally be one having a higher hub-shroud ratio at inlet, perhaps of 0.7. Here there can be considerable freedom to choose velocity diagrams. Fashions have developed. In continental Europe, axial-compressor diagrams generally have reactions 30 to 50 percent higher than those used in the United States and Britain. The principal advantages of a high-reaction velocity diagram are the higher (more acceptable) values of the blade-row outlet-to-inlet velocity ratios for the same flow and work coefficients and the lower blade speed that results if the relative Mach number to the rotor blades is to be similar to that for a 50-percent-reaction machine.

A high-reaction machine may be designed with axial inlet and outlet velocity vectors, C_1, which is a major advantage for single-stage units (figure 5.16). The high velocity ratio of the stators makes them less susceptible to flow breakdown when the profiles become roughened by dirt buildup. Normally the stators are affected by dirt more than the rotors, from which at least non-sticky dirt is thrown off.

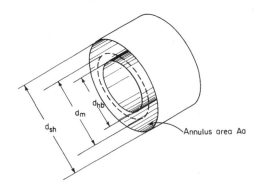

Figure 5.15. Annulus area notation for hub-shroud diameter ratio

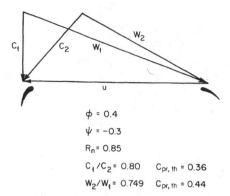

$$\phi = 0.4$$
$$\psi = -0.3$$
$$R_n = 0.85$$
$$C_1/C_2 = 0.80 \qquad C_{pr,th} = 0.36$$
$$W_2/W_1 = 0.749 \qquad C_{pr,th} = 0.44$$

Figure 5.16. High-reaction axial-inlet diagram

This cutaway view of an early Williams automobile engine illustrates the classical design of automotive gas turbines: single-stage centrifugal compressor; twin rotary regenerators; a high-pressure axial turbine driving the compressor and a low-pressure axial turbine driving the output shaft. As is to be expected, the heat exchangers are much larger than the turbomachinery components.

Illustration. Small engine with two ceramic regenerator disks. Courtesy Williams Research Corporation

When severe design constraints are imposed, such as a first-stage hub-shroud diameter ratio of less than 0.5, "simple" velocity diagrams become, for most practical flow distributions, too approximate to be valuable. General velocity diagrams as shown in figure 5.4 should be used. Some guidance for these is given in chapter 6. In addition the streamline curvature produced by the shape of the hub annulus wall in the axial-transverse plane is often used to improve flow distributions and to modify the vector diagrams. That topic, however, is beyond the scope of this book.

5.4   Radial-turbomachine velocity diagrams

Radial-flow pumps, compressors, and turbines cannot have so-called "simple" velocity diagrams because the rotor peripheral velocity is very different at the point the flow leaves the rotor compared with that at which it enters. Often the flow enters centrifugal pumps and compressors without swirl, and at the design point the rotor-outlet flow of radial-inflow turbines is usually nominally without swirl. The rotor-tip velocity triangle is therefore the dominant one for radial-inflow turbines and radial-outflow compressors and pumps.

Radial blades in radial-flow rotors

For stress-reduction reasons, high-pressure-ratio centrifugal compressors usually have rotor blades that are radial at outlet. This shape of blade has high aerodynamic loading at outlet. Circulation, plus apparently inevitable flow separation within the rotor channels, combine to produce a flow deviation, or "slip", with the blade alignment. The combined inlet and outlet velocity diagrams of a radial-blade centrifugal compressor will then approximate figure 5.17. Two hypothetical compressor or pump inlet triangles are shown: one for the hub; and one for the shroud. In comparison to the impeller sketch, in a three-dimensional drawing, the inlet velocity diagrams would be perpendicular to the plane of the page, and the outlet velocity diagram would be in the plane of the page.

Limiting design conditions often are: the shroud relative velocity at rotor inlet, $W_{sh,1}$, because of relative-Mach-number or cavitation considerations; and the relative-velocity ratio, $W_2/W_{sh,1}$ (de Haller ratio in the impeller). Although, as mentioned earlier, it seems to be impossible to prevent separation within the rotor passages of high-pressure-ratio compressors entirely, the proportion of the flow that is separated, and the relative losses thereby caused, seem to be a strong function of the relative-velocity ratio. It is therefore desirable to keep this ratio above, for example, 0.75.

The angle of absolute flow into the diffuser, $\alpha_{C,2}$, has a strong effect on the stability of a radial vaneless diffuser, if used (see chapter 4). In general $\alpha_{C,2}$ should not exceed 70° and should be lower if a large amount of diffusion is to be attempted. As a general guideline for radial-flow compressors with vaneless and vaned diffusers, $\alpha_{C,2}$ should be chosen to be 60° to 65° at design point.

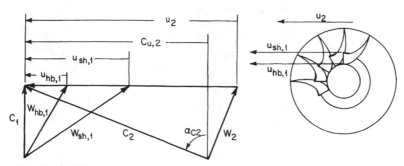

Figure 5.17. Radial-flow velocity diagram and impeller with radial blades. Velocity vectors are shown for compressor or pump operation, and would reverse for turbine flow.

Backward-swept blades for centrifugal machines

The rotor relative-velocity ratio is beneficially increased by using rotor blades that sweep backward relative to the flow direction (figure 5.18).

In addition, a smaller proportion of the stagnation-enthalpy rise is in the form of velocity. For both reasons the polytropic efficiencies of centrifugal machines with swept-back blades are normally considerably higher than those with straight radial blades (the typical range as a function of blade outlet angle β_2 is shown in figure 5.19).

These benefits are won at the cost of a considerable reduction in maximum possible work per stage, for two reasons. First, the work coefficient is reduced (from about 0.9 for radial-blade machines to 0.5 or less for units with swept-back blades). Second, the maximum, stress-limited, peripheral velocity must be lower for swept-back blades than for radial blades of the same relative blade height. In many applications there is no call for maximum work per stage, and blade sweepback angles of up to 50° (and more for pumps) can be advantageously used.

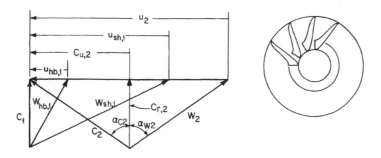

Figure 5.18. Radial-flow velocity diagram and impeller with backswept blades. Velocity vectors are shown for compressor or pump operation, and would reverse for turbine flow.

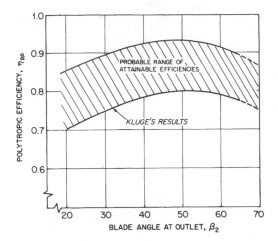

Figure 5.19. Effect of radial-flow blade-outlet angle on polytropic efficiency. From Kluge (1953)

5.5 Correlations of peak stage efficiency with radius ratio and "specific speed"

The hub-shroud (or "hub-tip") ratio Λ (equation 5.10 and figure 5.15), and the equivalent for centrifugal machines, the blade-height-to-diameter ratio, have obvious effects on the peak efficiency that can be expected from any turbomachine stage. At hub-tip ratios near unity the blade annular passage is so small that it may be all taken up with boundary layers ("fully developed flow"), and it will be strongly influenced by tip-leakage flows and other secondary flows. In the limit the efficiency will fall to zero at a hub-tip ratio of unity.

When low hub-shroud ratios are used, the choice of velocity diagrams for the whole radial extent of the stage will be compromised by the need to avoid undue separation or high relative Mach numbers at the blade ends. In the limit the efficiency will not go to zero at zero hub-tip ratio, but for compressors and pumps it will be low. Turbines do not experience so sharp a drop-off in efficiency when areas of separation begin to occur, because the overall pressure gradient is favorable. Therefore there is no danger of the flow reversing through a blade passage, as can occur in compressors.

The hub-shroud ratio can be derived in terms of quantities that appear in the machine specifications, as follows (see figure 5.15):

$$1 - \Lambda^2 = 1 - \left(\frac{d_{hb}}{d_{sh}}\right)^2 \quad = \quad \frac{d_{sh}^2 - d_{hb}^2}{d_{sh}^2} = \frac{A_{an}}{\pi d_{sh}^2 \ / \ 4}$$

$$= \quad \frac{\dot{V}}{C_x \pi d_{sh}^2 \ / \ 4}, \qquad \text{because} \qquad \dot{V} = A_{an} C_x$$

$$= \quad \frac{\dot{V} N^2}{\phi u_2 (\pi d_{sh}^2/4)(60\omega/2\pi)^2}, \qquad \text{because} \qquad \phi = C_x/u_2$$

and because:

$\omega = 2\pi N/60$, where ω is in rad/s and N in rev/min,
$\omega/2 = u_2/d_{sh}$, and
$|\psi| = g_c \Delta h_0 / u_2^2$, we obtain

$$1 - \Lambda^2 = 1 - \left(\frac{d_{hb}}{d_{sh}}\right)^2 = \frac{\pi}{900} \frac{N^2 \dot{V}}{\phi u_2^3}$$

$$1 - \Lambda^2 = 1 - \left(\frac{d_{hb}}{d_{sh}}\right)^2 = \frac{\pi}{900} \frac{|\psi|^{(3/2)} N^2 \dot{V}}{\phi (g_c \Delta h_0)^{(3/2)}}$$

$$\sqrt{1 - \Lambda^2} = \sqrt{1 - \left(\frac{d_{hb}}{d_{sh}}\right)^2} = \frac{\sqrt{\pi}}{30} \frac{|\psi|^{(3/4)}}{\sqrt{\phi}} \frac{N\sqrt{\dot{V}}}{(g_c \Delta h_0)^{(3/4)}}$$

$$\sqrt{1 - \Lambda^2} = \sqrt{1 - \left(\frac{d_{hb}}{d_{sh}}\right)^2} = \frac{|\psi|^{(3/4)}}{\sqrt{\pi \phi}} N_{s1} \qquad (5.11)$$

where the non-dimensional specific speed N_{s1} is defined as

$$N_{s1} \equiv \frac{2\pi N \sqrt{\dot{V}}}{60 |g_c \Delta h_0|^{(3/4)}} \qquad (5.12)$$

Specific speed is therefore principally a function of the relative flow-passage size and of the design-point values of the work coefficient ψ, and of the flow coefficient ϕ.

There are many definitions of specific speed, usually dimensional, but they all have the same relative significance. One used where the old "imperial" system of units is still in force is N_{s3}:

$$N_{s3} \equiv \frac{(N \text{ [rpm]})\sqrt{\dot{V} \text{ [gal/min]}}}{(\Delta H_{T,1} \text{ [ft]})^{(3/4)}} \qquad (5.13)$$

In using the equation for N_{s3}, the "gallon" must be specified as U.S. or "imperial". If the mean diameter d_m is chosen so that

$$d_{sh}^2 - d_m^2 = d_m^2 - d_{hb}^2, \qquad \text{that is} \qquad d_m^2 = \frac{d_{sh}^2 + d_{hb}^2}{2}, \qquad \text{then}$$

$$\frac{d_m}{d_{sh}} = \sqrt{\frac{1 + \Lambda^2}{2}} \qquad \text{and} \qquad \frac{d_m}{d_{hb}} = \sqrt{\frac{1 + \Lambda^{-2}}{2}}$$

Therefore,

$$N_{s1} = \frac{\sqrt{\phi_m}}{|\psi_m|^{3/4}} \sqrt{2\pi} \sqrt{\frac{1 - \Lambda^2}{1 + \Lambda^2}} \qquad (5.14)$$

where

$$\phi_m = \frac{C_x}{u_m} \qquad \text{and} \qquad |\psi_m| = \frac{g_c \Delta h_0}{u_m^2}$$

The derivation of N_{s1} obtainable from equation 5.11 is

$$N_{s1} = \frac{\sqrt{\phi}}{|\psi|^{3/4}} \sqrt{\pi(1 - \Lambda^2)}$$

where ϕ and ψ are based on the peripheral speed, u, at the shroud.

The specific speed of hydraulic machines is usually given with the actual total-head rather than the stagnation-enthalpy change:

$$N_{s2} \equiv \frac{2\pi N \sqrt{\dot{V}}}{60(g\,\Delta H_T)^{3/4}} \tag{5.15}$$

Also a head coefficient, ψ^x, is often used rather than a work coefficient:

$$\psi^x \equiv \frac{g\,\Delta H_T}{u^2} \tag{5.16}$$

In radial-flow machines the rotor velocity, u, is that of the periphery. If b is the axial length of the blades at the outer diameter d, it can be easily shown that

$$N_{s2} = 2\sqrt{\pi}\,\frac{\sqrt{\phi}\sqrt{b/d}}{|\psi^x|^{3/4}} \tag{5.17}$$

Here $\phi \equiv C_r/u$ at the outer diameter.

The relation between N_{s3}, defined by equation 5.13, using U.S. gallons and the nondimensional N_{s2}, equation 5.15, is

$$N_{s3} = 2732\,N_{s2} \tag{5.18}$$

Some charts (from Rodgers (1980) and Rohlik (1975)) showing the influence of specific speed on the maximum efficiencies for compressors and pumps, and the effects of impeller-blade angle and machine size, are shown in figures 5.20, 5.21, and 5.22, and for radial-inflow turbines in figure 5.23. Rodgers' general advice is that for a centrifugal compressor of maximum efficiency the nondimensional specific speed should be near 0.7; for a machine of minimum size the value should be near 1.0. Figure 5.23 shows that this is also good advice for the design of radial-inflow turbines, coupled with an indication that the optimum flow angle at rotor entrance is 70° to 72°.

5.6   Preliminary-design methods for radial-flow turbomachinery

Although the design of radial-flow turbomachinery is based, as is axial machinery, on Euler's equation, the design method differs because the velocity diagram at rotor outlet can be considered separately (but not independently) from that at rotor inlet.

For compressors and pumps with no swirl in the inlet flow, the outlet velocity diagram will completely determine the work, maximum potential pressure ratio, and flow. Similarly, the rotor-inlet velocity diagram will be controlling for radial-inflow turbines designed for zero outlet swirl. In both cases, compressor and turbines, the inducer or exducer (generally axial-flow) section can be designed subsequently (see chapter 9).

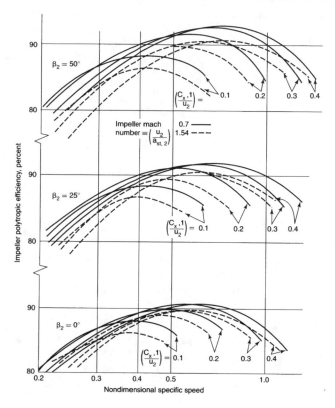

Figure 5.20. Centrifugal-compressor impeller efficiency versus specific speed, flow coefficient and blade angle. From Rodgers (1980)

Compressor-rotor-outlet velocity diagrams

Two outlet diagrams combined with inlet diagrams (one for a radial-blade and one for a backward-swept impeller) are shown in figures 5.17 and 5.18. From the steady-flow energy equation (2.11) and Euler's equation (5.1), the specific work done by the rotor on the flow (in adiabatic conditions) is

$$\frac{\dot{W}_{ex} - \dot{W}_{in}}{\dot{m}} = -\Delta_1^2 h_0 = -(u_2 C_{u,2} - u_1 C_{u,1})/g_c \qquad (5.19)$$

(The rotor will also perform frictional work through the frictional torques exerted by the casing, the fluid scrubbing on the back of the disk and on the back of a rotating shroud, if used, and by bearings and seals. When the frictional work appears as heat in the working fluid, there will be an additional enthalpy rise.)

For a compressor working on a fluid that can be approximated to a semi-perfect gas, equation 5.19 can lead to the following:

$$\frac{\dot{W}_{in}}{\dot{m}} = \Delta_1^2 h_0 = C_p(T_{0,2} - T_{0,1}) = C_p T_{0,1}\left(\frac{T_{0,2}}{T_{0,1}} - 1\right) = (u_2 C_{u,2} - u_1 C_{u,1})/g_c$$

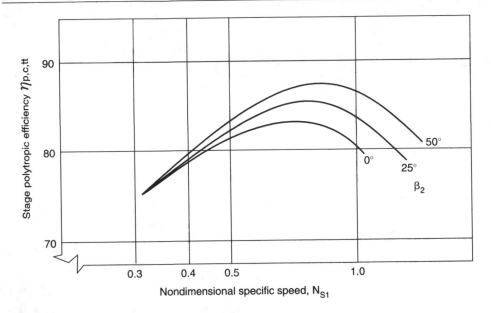

Figure 5.21. Estimated centrifugal-compressor stage efficiency for compressors with rotor-outlet Mach number $(C_{u,2}/a_{st,2}) = 1.54$, and $d_{hb,1}/d_2 = 0.3$. From Rodgers (1980)

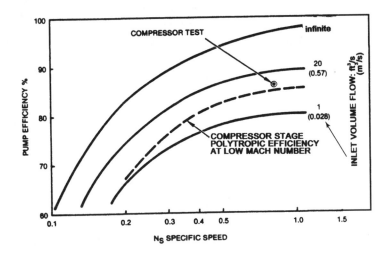

Figure 5.22. Centrifugal-pump efficiency versus specific speed and size. From Rodgers (1980)

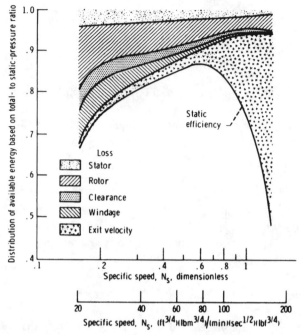

(a) Loss distribution along curve of maximum efficiency

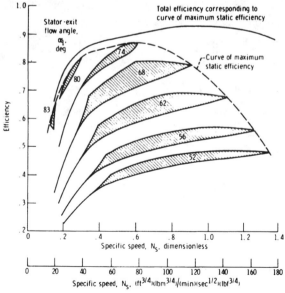

(b) Design-point efficiency versus specific speed

Figure 5.23. Radial-inflow-turbine efficiency versus specific speed. From Rohlik (1975)

Therefore

$$(u_2 C_{u,2} - u_1 C_{u,1}) = g_c C_p T_{0,1} \left\{ \left(\frac{p_{0,2}^\otimes}{p_{0,1}} \right)^{\left[\left(\frac{R}{C_p} \right) \frac{1}{\eta_{p,c}} \right]} - 1 \right\} \tag{5.20}$$

where $C_p \equiv \overline{C_p}$ and $p_{0,2}^\otimes$ is the defined useful stagnation outlet pressure for which $\eta_{p,c}$ is specified.

For a pump working on an incompressible liquid, equation 5.19 may be combined with the appropriate definition of pump efficiency, η_{pu}, to give

$$\frac{\dot{W}_{in}}{\dot{m}} = (u_2 C_{u,2} - u_1 C_{u,1})/g_c = \frac{H_{T,2}^\otimes - H_{T,1}}{\eta_{pu}} \left(\frac{g}{g_c} \right) \tag{5.21}$$

Most compressors and pumps have no swirl in the inlet flow, so that $C_{u,1}$ is zero. To be able to use equations 5.20 and 5.21 we need a relation between U_2 and $C_{u,2}$. This relation is given by Wiesner's correlation (or one of the many other correlations of flow deviation from the rotor-blade direction at outlet).

$$\sigma_w \equiv \frac{C_{u,2,ac}}{C_{u,2,tl}} = 1 - \frac{\sqrt{\cos \beta_2}}{Z^{0.7}} \tag{5.22}$$

where Z is the number of rotor blades, and β_2 is the angle between the radial direction and the tangent to the rotor blade at the periphery. The actual and no-deviation (theoretical) values of $C_{u,2}$ are defined by figure 5.24.

This diagram can be completely solved once a choice has been made for the absolute flow angle of the mean flow leaving the rotor, $\alpha_{C,2}$. It is tempting to make this angle as large as possible, thus reducing the value of C_2 and of the radial component, $C_{r,2}$. However, as pointed out in chapter 4, a choice of a large value of $\alpha_{C,2}$ predisposes the

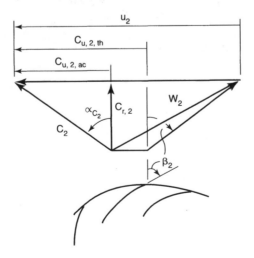

Figure 5.24. Rotor-outlet velocity diagram

downstream diffuser to rotating stall. A value that has been found to be valid for vaneless and vaned diffusers and to be useful for preliminary design is 60°; we use this in the example following.

The most-important property of the rotor-outlet velocity diagram is the ratio of $C_{u,2,ac}/u_2$, which, if there is no swirl in the flow at rotor inlet, is equal to the loading coefficient, ψ. This ratio can be found in terms of known quantities as follows.

$$\tan \beta_2 = \frac{u_2 - C_{u,2,tl}}{C_{r,2}}$$

$$\tan \alpha_{C,2} = \frac{C_{u,2,ac}}{C_{r,2}}$$

$$\sigma_w \equiv \frac{C_{u,2,ac}}{C_{u,2,tl}}$$

$$\frac{\tan \beta_2}{\tan \alpha_{C,2}} = \left[\frac{u_2}{C_{u,2,ac}} \right] - \frac{1}{\sigma_w}$$

$$\left[\frac{C_{u,2,ac}}{u_2} \right] = \left\{ \left[\frac{\tan \beta_2}{\tan \alpha_{C,2}} \right] + \frac{1}{\sigma_w} \right\}^{-1} \tag{5.23}$$

The designer has to make a choice of the number of blades on the rotor periphery and the blade angle there. These are not totally independent choices. The higher the value of β_2, or sweepback, the more blockage is presented by the blades, and therefore the smaller is the number of blades that can be accommodated. The range of possible blade numbers is shown in figure 5.25.

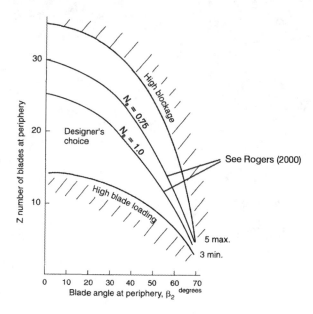

Figure 5.25. Range of number of blades at periphery

We now have enough tools to enable the preliminary design of a radial-outflow compressor to be carried out.

Example 4 Preliminary design of a radial-outflow compressor

Carry out the preliminary design (rotor diameter, blade axial width at outlet, rpm) of a centrifugal compressor to have a stagnation-to-static pressure ratio of 3 : 1 from atmospheric air at 1 bar, 300 K, mass flow 0.75 kg/s. Choose a speed to give maximum efficiency (polytropic, stagnation-to-static).

Solution.

The optimum speed will be that giving the optimum specific speed. Rodgers' data in figure 5.20 for a high sweepback angle (we will choose 45°) suggest a value of about 0.8.

We first need to solve the rotor-outlet velocity diagram. This will need the slip factor. Some choices must be made. We also need to find the enthalpy rise required. These two together will allow us to calculate the blade speed at the rotor periphery. We can use the specific speed to determine the rpm. This will give the rotor diameter. The blade axial width at rotor periphery can be estimated following some approximations and guesses.

Slip factor

We must choose the number of rotor blades at the periphery. It is customary to give an even number of blades so that half may be designated "splitter blades", which means that they do not go all the way to the inducer. We will use an intermediate loading from figure 5.25 and moderate blockage, leading to a choice of 20 blades (10 being "splitter" blades).

The blade angle at the outlet has been specified as 45°, so that Wiesner's correlation (equation 5.22) gives:

$$\sigma_w \equiv \left[\frac{C_{u,2,ac}}{C_{u,2,tl}}\right] = 1 - \frac{\sqrt{\cos 45°}}{20^{0.7}} = 0.8967$$

Rotor-outlet velocity diagram

Everything is specified for this except the direction of the flow leaving the rotor ($\alpha_{C,2}$). We will choose the value suggested above, 60°. Then using equation 5.23 we can find the ratio of the actual outlet tangential velocity to the blade speed:

$$\left[\frac{C_{u,2,ac}}{u_2}\right] = \left\{\left[\frac{\tan \beta_2}{\tan \alpha_{C,2}}\right] + \frac{1}{\sigma_w}\right\}^{-1} = 0.5908$$

With zero swirl at rotor inlet this value is also $-\psi$, the loading coefficient.

Blade peripheral speed

We choose the optimum specific speed, N_s, at 0.8 from figure 5.20. The stage polytropic efficiency is read from figure 5.21 to be 0.854. Then equation 2.78 can be used to find the temperature and the enthalpy rise:

$$\frac{T_{0,2}}{T_{0,1}} = \left(\frac{p_{0,2}}{p_{0,1}}\right)^{\left[\left(\frac{R}{C_p}\right)\frac{1}{\eta_{p,c}}\right]}$$

After iterating on the mean specific heat we obtain:

$$C_p = 1,009.1 \text{ J/(kg} \cdot \text{K)} \quad T_{0,2} = 432.55 \text{ K}$$
$$C_p/R = 3.5165 \quad \Delta h_0 = 133,754 \text{ J/kg}$$
$$-\Delta h_0 = \psi u^2 \quad \text{equation 5.5}$$
$$u_2 = 475.81 \text{ m/s}$$
$$C_{u,2,ac} = 281.11 \text{ m/s}$$
$$C_{r,2} = \frac{C_{u,2,ac}}{\tan 60°} = 162.30 \text{ m/s}$$
$$C_2 = \frac{C_{u,2,ac}}{\sin 60°} = 324.60 \text{ m/s}$$

Specific speed and rotational speed

From figure 5.20 the optimum specific speed is 0.8. We have calculated the enthalpy rise, so that the only other input needed to calculate the shaft speed N is the volume flow at inlet, V_1. For preliminary design it is often sufficient to use the mass flow and the density at stagnation conditions. We will, however, make an approximation to estimate the actual (static) density as follows. We have the freedom to choose the near-optimum parameters for the compressor: we see from figure 5.20 that maximum efficiency was measured by Rodgers for values of $C_{x,1}/u_2$ between 0.3 and 0.4. We will therefore choose the axial velocity at inlet at 185 m/s simply in order to estimate the static density. The Mach number at inlet for this velocity is found from equation 2.62 and/or figure 2.10:

$$\frac{C}{\sqrt{g_c R T_0}}(= 0.6305) = \sqrt{2\frac{C_p}{R}\left[1 - \left(1 + \frac{M^2}{2\left(\frac{C_p}{R} - 1\right)}\right)^{-1}\right]}$$

The inlet Mach number is then 0.55. This is used in equation 2.59 to find the ratio of the stagnation to static density:

$$\frac{\rho_0}{\rho_{st}} = \left[1 + \frac{M^2}{2\left(\frac{C_p}{R} - 1\right)}\right]^{\left[\left(\frac{C_p}{R}\right) - 1\right]} = 1.1577$$

The stagnation density is

$$\rho_{0,1} = \frac{1 \times 10^5}{286.96 \times 300} = 1.1616 \text{ kg/m}^3$$

Therefore the static density is $\rho_{st,1} = 1.0034 \text{ kg/m}^3$.
and the volume flow rate at inlet is $\dot{V}_1 = \frac{0.75 \text{ kg/s}}{1.0034 \text{ kg/m}^3} = 0.7445 \text{ m}^3\text{/s}$.
The shaft speed, N, is found from the specific-speed relation:

$$N = \frac{60 N_s}{2\pi} \frac{(g_c \Delta h_0)^{3/4}}{\sqrt{\dot{V}}} = 61,800 \text{ rpm}$$

The rotor diameter is then

$$d_2 = \frac{u_2 \times 60}{\pi \times 61,800} = 147 \text{ mm}$$

Blade axial width at outlet

We find the rotor stagnation-to-stagnation polytropic efficiency from figure 5.20 to be 0.9 for the chosen conditions. The stagnation pressure at rotor outlet is given, then, by:

$$\frac{p_{0,2}}{p_{0,1}} \left[\frac{432.55}{300} \right]^{3.5165 \times 0.9} = 3.1838 \Rightarrow p_{0,2} = 318,380 \text{ N/m}^2$$

And the stagnation density at rotor outlet is $\rho_{0,2} = p_{0,2}/(RT_{0,2}) = 2.565 \text{ kg/m}^3$.

As before, we find the Mach number at the outlet from figure 2.10 and/or equation 2.62, and we use the results in equation 2.59:

$$\frac{C_2}{\sqrt{g_c R T_{0,2}}} = 0.9213 \qquad\qquad M_2 = 0.8311$$

$$\frac{\rho_{0,2}}{\rho_{st,2}} = 1.3822 \qquad\qquad \Rightarrow \rho_{st,2} = 1.8558 \text{ kg/m}^3$$

$$\dot{m} = C_{r,2}\rho_{st,2}\pi d_2 b_2 \qquad\qquad \Rightarrow b_2 = \frac{0.75}{162.30 \times 1.856 \times \pi \times 0.147}$$

$$b_2 = 5.39 \text{ mm}$$

Comment

The calculated value of b_2 is 5.39 mm. This value ignores blade blockage and the additional blockage due to boundary-layer growth in the rotor passages. The value is, therefore, conservative, in that it is very unlikely to result in flow separation. It could be regarded as the width of the diffuser vaneless space immediately downstream of the rotor. It is good practice to have a narrowing space to this vaneless space, and the value of b_2 calculated above is suitable for the "throat" of this narrowed entrance.

Conditions for flow separation in compressor rotors

The more highly angled the blades of radial-flow compressors are to the radial direction at outlet, the smaller is the extent of flow separation in the impeller. There are some who believe that the flow is so complex, and the requirements of blade loading and changes in direction so severe, that some degree of separation is inevitable. Another point of view is that, if the amount of diffusion required of the most severely loaded streamline (that at the shroud) is limited, separation can be avoided.

We take this latter view. We suggest, more from intuition than from hard evidence, that a lower limit of relative outlet-to-inlet velocity ratio W_2/W_1 of 0.8 be used as a guide to the prevention of separation in rotors with subsonic inlet relative Mach numbers $(W_{sh,1}/a_{st})$ and normal Reynolds numbers. This is a more-conservative value than the 0.71 suggested by Rodgers (1978). This criterion leads to the design rules shown in figure 5.26, derived as follows.

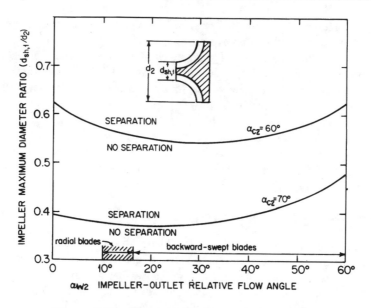

Figure 5.26. Rotor design to avoid separation

The inlet and outlet velocity diagrams, figure 5.18, give the following relationships between $W_2/W_{sh,1}$, and other design parameters:

$$W_{sh,1} = u_{sh,1}/\sin\alpha_{W,sh,1} \tag{5.24}$$

$$W_2 = C_{r,2}/\cos\alpha_{W,2} \tag{5.25}$$

$$C_{r,2} = C_{u,2}/\tan\alpha_{C,2} \tag{5.26}$$

$$C_{u,2} = u_2 - C_{r,2}\tan\alpha_{W,2} \tag{5.27}$$

Therefore,

$$\frac{W_2}{W_{sh,1}} = \frac{\sin\alpha_{W,sh,1}}{(u_{sh,1}/u_2)(\tan\alpha_{C,2}+\tan\alpha_{W,2})\cos\alpha_{W,2}} \tag{5.28}$$

$$\frac{d_{sh,1}}{d_2} = \frac{\sin\alpha_{W,sh,1}}{(W_2/W_{sh,1})(\cos\alpha_{W,2}\tan\alpha_{C,2}+\sin\alpha_{W,2})} \tag{5.29}$$

When the inlet shroud diameter is optimized for minimum relative velocity at the shroud, $W_{sh,1}$, the relative flow angle $\alpha_{W,sh,1}$ is normally found to be close to 60°. Accordingly, figure 5.26 is plotted for $\alpha_{W,sh,1} = 60°$ and for two values of the absolute outlet flow angle, $\alpha_{C,2} = 60°$ and 70°. This range covers the flow angles normally used in design, as implied by the requirements for stability in the vaneless diffuser or vaneless space, figure 4.22. The abscissa is the relative outlet flow angle, $\alpha_{W,2}$, which is a function of, principally, the blade angle at outlet, β_2, and the number of rotor blades. Impellers that have radial blades at the periphery have relative flow angles generally between 10° and 15°, and for the higher outlet flow angle, 70°, the likelihood of separation is at a maximum. Only radial-blade compressors of very low specific speed have the diameter ratio $(d_{sh,1}/d_2)$ below the implied no-separation boundary of 0.38.

The implication of figure 5.26 is that there are twin benefits from decreasing the design value of absolute outlet angle $\alpha_{C,2}$. First, there is a reduced likelihood of separation in the impeller. Second, a smaller flow angle into a vaneless space or vaneless diffuser gives less likelihood of flow instability in that component. A low value of outlet flow angle is given by a small value of axial blade width at the rotor periphery, b_2. Reducing b_2 incurs a disadvantage also: the clearance between blade and casing becomes relatively larger, increasing the losses associated with clearance. We generally use $60°$ as an initial value for $\alpha_{C,2}$ in preliminary design, but there appears to be justification for using lower values (e.g., $55°$) when $d_{sh,1}/d_2$ must be above 0.55.

Velocity diagrams for radial-inflow turbines

All known hot-gas radial-inflow turbines use blades that are aligned along the radii for much of their length. They all use an absolute flow angle into the rotors (the nozzle outlet angle, if a nozzle is used) of close to $70°$. All known designs aim to have zero outlet flow swirl at the design point. Thus there is a great deal of similarity, even commonality, among different hot-gas radial-inflow turbines.

It seems natural to arrange the inlet velocity diagram so that the inlet relative velocity is aligned with the blade direction at inlet: that is, the inlet relative velocity would be radial. However, if the flow were indeed to pass radially into the blading it would immediately have to take up the very high aerodynamic "loading" that would be a consequence of the rapid reduction in momentum. This loading would have to appear as a large pressure difference between the "pressure" and the "suction" sides of the rotor blades. The blade tips are open, and such a sharp pressure difference could not be maintained. It increases from zero at the stagnation point more or less rapidly, depending on the incidence.

It has been found by Wilson and Jansen (1965) that the maximum efficiency in radial inflow turbines is reached for a flow inlet angle corresponding to the inverse of the "slip" or "deviation" if the same rotor were to be driven as a compressor. We have recommended Wiesner's slip-factor correlation for radial-outflow compressors; here is the use of it for the preliminary design of radial-inflow turbines:

$$\sigma_w \equiv \left[\frac{C_{u,1,ac}}{C_{u,1,tl}} \right] = 1 - \frac{\sqrt{\cos \beta_1}}{Z^{0.7}} \tag{5.30}$$

The blade angle β_1 is the angle between the tangent to the blade at the periphery and the radial direction. In hot-gas turbines this angle is usually zero, so that $\cos \beta_1 = 1.0$. The "theoretical" tangential component of velocity at the inlet is the blade speed u. The number of blades at the periphery (i.e., including so called "splitter blades", if used) is Z. The slip factor for radial-bladed rotors then varies with the number of blades as shown in table 5.1.

It is easy in preliminary design to find the approximate blade-tip speed needed in a radial-inflow turbine. The following example will illustrate this.

Table 5.1. Slip factor and number of blades for radial-inflow turbines with radial blades

Number of radial blades on rotor periphery	Slip factor $C_{u,1}/u_1$
9	0.785
11	0.813
13	0.834
15	0.850

Example 5   Preliminary design of a radial-inflow turbine

Calculate the approximate peripheral speed of a radial-bladed rotor of a radial-inflow turbine of the following specifications:

Inlet temperature	1200 K
Inlet pressure	300,000 Pa
Rotor-outlet stagnation pressure	110,000 Pa
Hot-gas inlet mass flow	0.5 kg/s
Fuel/air ratio	0.02
The rotor has	13 radial blades
Nozzle-outlet flow angle	70° to radial direction, at rotor entrance

Solution.

The stagnation-to-stagnation polytropic efficiency of a rotor is high: we will guess 0.92. We use the correlation in appendix A for the properties of combustion air.

For 13 blades $\sigma_w = \psi = 0.834$.

From equation 2.79:

$$\frac{T_{0,2}}{T_{0,1}} = \left(\frac{p_{0,2}}{p_{0,1}}\right)^{\left[\left(\frac{R}{c_p}\right)\eta_{p,e}\right]}$$

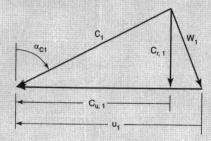

Figure 5.27. Velocity diagram for example of radial-inflow turbine

We guess the mean value of C_p and iterate, arriving at

$$
\begin{aligned}
T_{0,2} &= 961.36 \text{ K} \\
\overline{T} &= 1080.68 \text{ K} \\
C_p &= 1194.64 \text{ J/(kg} \cdot \text{K)} \\
\Delta h_0 &= C_p \Delta T_0 = 285,089 \text{ J/kg} \\
\Delta h_0 &= \psi u_1^2 \quad \Rightarrow u_1 = 584.7 \text{ m/s} \\
C_{u,1} &= 487.6 \text{ m/s} \\
C_1 &= 518.9 \text{ m/s} \\
C_{r,1} &= 181.1 \text{ m/s} \\
W_1 &= 206.4 \text{ m/s.}
\end{aligned}
$$

Use of optimum specific speed to calculate shaft speed, diameter, etc.

The curves of specific speed for radial-inflow turbines (figure 5.23a and b) indicate that the maximum efficiency should be reached at a nondimensional specific speed of about 0.6. We will translate this into a more useful number: the ratio between blade axial width at inlet and the rotor-blade tip diameter at inlet.

$$
N_{s,op} = 0.6 \equiv \frac{2\pi N}{60} \frac{\sqrt{\dot{V}_{in}}}{(\Delta h)^{3/4}}
$$

$$
u_1 = \omega r_1 = \frac{2\pi N}{60} \frac{d_1}{2} = \frac{\pi N d_1}{60}
$$

$$
\dot{V}_{in} = C_{r,1} \pi d_1 b_1 = C_{r,1} \pi d_1^2 \left(\frac{b_1}{d_1} \right)
$$

$$
g_c \Delta h_0 = \psi u_1^2
$$

$$
N_s = \frac{\omega \sqrt{\pi C_{r,1} d_1^2 (b_1/d_1)}}{(\psi u_1^2)^{3/4}}
$$

$$
\Rightarrow \frac{b_1}{d_1} = \frac{N_s^2 \psi^{3/2} u_1^3}{\omega^2 d_1^2 \pi C_{r,1}} = \left[\frac{N_s^2 \psi^{3/2}}{4\pi} \right] \left[\frac{u_1}{C_{r,1}} \right]
$$

$$
\tan \alpha_{C,1} = \psi u_1 / C_{r,1}
$$

$$
\Rightarrow \frac{b_1}{d_1} = \left[\frac{N_s^2 \psi^{1/2} \tan \alpha_{C,1}}{4\pi} \right]
$$

If we insert the appropriate design inputs for the turbine in the example above, we obtain the value of $(b_1/d_1) = 0.072$. We can find both b_1 and d_1 by calculating the actual volume flow at rotor inlet. We use equation 2.62 or figure 2.10 to find the Mach number.

$$
\frac{C_1}{g_c R T_{0,1}} = 0.8843
$$

$$
M_1 \approx 0.815 \quad \text{from equation 2.62 or figure 2.10}
$$

Equation 2.59 is used to find the static density. After iterations at the mean conditions $C_p = 1,194.64$ J/(kg · K) and $C_p/R = 4.163$. Then

$$\frac{\rho_{st,1}}{\rho_{0,1}} = \left[1 + \frac{M_1^2}{2\left(\frac{C_p}{R} - 1\right)} \right]^{-\left[\left(\frac{C_p}{R}\right) - 1\right]} = 0.729$$

The stagnation density at rotor inlet can be taken as that at nozzle inlet for preliminary design. (In fact, there will be a small drop in stagnation pressure and presumably a negligible drop in stagnation temperature.)

The stagnation density at rotor inlet $\rho_{0,1} = p_{0,1}/(RT_{0,1}) = 300 \times 10^3/(286.96 \times 1200) = 0.8712$ kg/m^3.

Therefore the static density at rotor inlet is $\rho_{st,1} = 0.6353$ kg/m^3.

The continuity equation gives $\dot{m} = C_{r,1}\rho_{st,1}\pi d_1 b_1 = C_{r,1}\rho_{st,1}\pi d_1^2 (b_1/d_1)$. Therefore:

$$d_1^2 = \frac{0.5}{181.11 \times 0.6353 \times \pi \times 0.072}$$

$$d_1 = 138.6 \text{ mm}$$

$$b_1 = 9.98 \text{ mm}$$

$$u_1 = \frac{2\pi N}{60}\frac{d_1}{2}$$

$$N = \frac{60 u_1}{\pi d_1} = 80,560 \text{ rpm}$$

This completes the calculation of what is normally needed for preliminary design. We will continue this example in detailed design in chapter 9.

5.7 Choice of number of stages

So far in this chapter we have considered the preliminary design of single stages. There is a wide range of turbomachinery applications that can be covered satisfactorily or ideally by a single stage. A single-stage machine is usually the lowest in first cost and the most compact. However, there may be compelling reasons for using more than one stage.

1. The pressure ratio may be high enough to result in the requirement of a rotor peripheral speed that will lead to unacceptably high blade or disk stresses. (This is a principal reason why large high-pressure-ratio gas turbines use multi-stage axial-flow compressors and turbines.)

2. The high velocities required may lead to Mach numbers that introduce high losses or reduce range to too high a degree. (In industrial gas turbines an increased number of stages in the axial-flow compressor is often accepted as a worthwhile cost in return for subsonic and therefore higher-efficiency initial stages.)

3. If the device works in liquid, the high velocities may lead to cavitation: the problem of vapor bubbles that can reduce output, efficiency and the working life of the unit.

(Boiler feed pumps for modern steam-turbine plants have four to six stages for this reason.)

4. The noise resulting from high relative flow velocities in single-stage machines can be far higher than for multi-stage machines. (Many low-cost vacuum cleaners use two-stage backswept blade centrifugal fans to reduce the noise output.)

5. A single-stage machine will often have a "highly loaded" velocity diagram that intrinsically has a lower stagnation-to-stagnation polytropic efficiency. (See chapter 7 for examples of how high loading in axial-flow turbines reduces the efficiency.)

6. By going to multiple stages of lower aerodynamic loading, the outlet velocity will usually be reduced relative to a more-highly-loaded stage of otherwise identical specifications. (This increases the stagnation-to-static efficiency, which is usually the controlling factor in overall machine efficiency.)

7. By going to multiple stages of identical aerodynamic loading, the outlet velocity and kinetic energy are reduced proportionately: if two stages are used instead of one, the outlet kinetic energy is halved. (Again the stagnation-to-static efficiency is increased.)

The governing efficiency across the total compressor or turbine, including the diffuser, is the stagnation-to-static value for virtually all process compressors and turbines, all compressors in gas-turbine engines, and all turbines in shaft-power engines. Jet engines are almost the only application where the kinetic energy of the leaving flow is useful. If perfect diffusers were available, the stagnation-to-static efficiency would equal the stagnation-to-stagnation value, and the number of stages used would have no influence on the machine efficiency. In practice diffusers used after compressors and turbines have pressure-rise coefficients between 0.1 and 0.7. The effect on stagnation-to-static polytropic efficiency of using from one to four stages of an axial-flow turbine is shown in figure 5.28. In all cases the turbine is represented by the mean-diameter velocity diagram that has 50-percent reaction, a nozzle-outlet angle of 60°, and a work coefficient of 1.8. The stagnation-to-stagnation polytropic efficiency across the blading (excluding the diffuser) is specified to be 0.9. The effects on the stagnation-to-static efficiency of using diffusers with pressure-rise coefficients of 0.0, 0.4, and 0.7 are shown. With an ideal diffuser, the stagnation-to-static efficiency would be constant at 0.9.

Equal enthalpy change per stage is usual but not universal

The simplest method of dividing up the enthalpy rise or fall is to have equal enthalpy changes per stage. However, there are many exceptions.

The first stage of high-temperature gas turbines is often given a large enthalpy drop if by doing so the gas temperature through the stage is reduced to the point where the second and subsequent stages do not need to be cooled.

The first stage in a transonic fan or compressor may be given a large enthalpy rise so that the second and later stages will be subsonic.

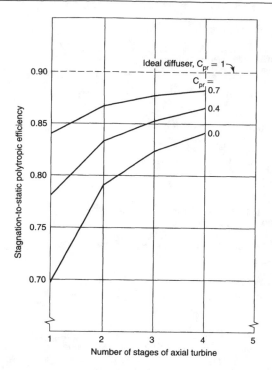

Figure 5.28. Effect of number of stages and diffuser pressure-rise coefficient on stagnation-to-static efficiency of an axial-flow turbine

The first stage, or sometimes the first part of the rotor, of oxygen and hydrogen turbopumps in liquid-fueled rocket engines normally has a very low enthalpy rise so that cavitation resulting from high fluid-dynamic loading is avoided.

Mixed stage types can be used

Although smaller gas turbines may have multistage centrifugal compressors and larger machines have multistage axial-flow compressors and turbines, intermediate-size gas turbines (and other turbomachinery) may have compressors that have several axial-flow stages followed by a centrifugal stage (because if all the stages were axial the high pressure stages would have very small, perhaps delicate, blades of lower efficiency). A few turbines have a radial-inflow stage followed by one or more axial stages. The use of multistage radial-inflow turbines is almost unknown: the general arrangement is difficult to configure neatly, and the inter-stage ducting tends to be expensive and aerodynamically lossy.

A less obvious form of stage mixing has frequently been used in axial-flow turbines: a high-work impulse or low-reaction stage is made the first stage to accomplish the high enthalpy drop mentioned above, so that the following stage or stages can be more-lightly-loaded higher-reaction uncooled stages of high efficiency.

References

Kluge, Friedrich (1953). Kreiselgeblase und Kreisel-verdichter radialer Bauert. Springer-Verlag, Berlin, W. Germany.

Koch, C.C. (1981). Stalling pressure-rise capability of axial-flow compressor stages. Paper 81-GT-3. ASME, New York, NY.

Rodgers, C. (1978). A diffusion-factor correlation for centrifugal-impeller stalling". Trans. ASME, vol. 100, pp. 592–603, New York, NY.

Rodgers, C. (1980). Specific speed and efficiency of centrifugal impellers, In Performance Prediction of Centrifugal Pumps and Compressors. ed. by S. Gopalakrishnan et al. ASME, New York, NY, pp. 592–603.

Rodgers, C. (1991). The efficiencies of single-stage centrifugal compressors for aircraft applications. ASME paper 91-GT-77, New York, NY.

Rodgers, C. (2000). Effects of blade number on the efficiency of centrifugal-compressor impellers. ASME paper 2000-GT-455.

Rohlik, Harold E. (1975). Radial-inflow turbines. In Turbine Design and Application, vol. 3, ed. by Arthur J. Glassman. Special publication SP-290 NASA, Washington, DC, pp. 31–58.

Wiesner, F.J. (1966). A review of slip factors for centrifugal impellers. ASME paper 66-WA/FE-18, New York, NY.

Wilson, David. (1960). Patterning stage characteristics for wide-range axial compressors. Paper 60-WA-113. ASME, New York, NY.

Wilson, D.G., and Jansen, W. (1965). The aerodynamic and thermodynamic design of cryogenic radial-inflow expanders. Paper 65-WA/PID-6, ASME, New York, NY.

Problems

1. Choose a mean-diameter velocity diagram and a blade speed for the single-stage fan delivering cooling air to the condenser of a Rankine-cycle hybrid automotive engine (figure P5.1). Use the following specifications:

    Air flow $\dot{m}$, 2 kg/s.
    Inlet stagnation pressure $p_{0,1}$, 99.285 kN/m^2.
    Inlet stagnation temperature T_{01}, 300 K. Drop in static pressure across condenser
    $p_{st,5} - p_{st,6} = 1.724$ kN/m^2.
    Mean air velocity entering condenser $C_5 = 4.57$ m/s.
    Diffuser pressure-rise coefficient $C_{pr,tl} = 0.75$ (use figure 4.10c to find an average
    $C_{pr,ac}$ at $2\delta^*/Y = 0.03$.)
    Fan hub-tip ratio $\Lambda = 0.75$.
    Maximum pressure-rise coefficient in fan at mean diameter $C_{pr,tl} = 0.4$.
    Fan flow coefficient $C_x/u_m = 0.4$.
    Axial inlet and outlet flow.
    Fan stagnation-to-stagnation polytropic efficiency, rotor inlet to stator outlet $\eta_{p,tt,13} = 0.9$.
    Mean specific heat $C_p = 1,005$ J/(kg · K).
    Other specified conditions are that $p_{st,6} = p_{0,1}$ and $p_{st,4} = p_{0,5}$.

2. Calculate and draw the velocity diagrams suitable for a fire pump to pass 1,250 gpm at a delivery (stagnation) pressure of 250 psi. Find the peripheral velocity u, the degree of reaction

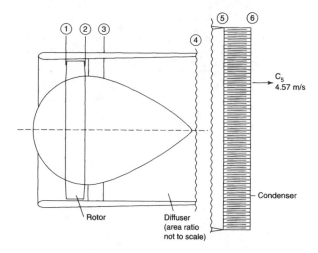

Figure P5.1

R_n, and the rotor and stator diffusion ratio, W_2/W_1, for the axial pumps. Comment on the suitability of each design as a single-stage pump.

(a) Axial pump with work coefficient 0.5, flow coefficient 0.4, and axial outlet flow from the rotor (a stator-rotor combination).

(b) Axial pump with work coefficient 0.4, flow coefficient 0.3, and axial inlet into the rotor (a rotor-stator combination with axial outlet flow from the stator).

(c) A radial-flow (outward) pump with radial (zero-angle) blades giving a relative flow angle at impeller outlet of 12 degrees to the radial direction ("slip"), and an absolute flow angle of 65 degrees at the same point. The diffuser passages have an area ratio of two to one.

(d) A radial-flow (outward) pump with 50-degree swept-back rotor blades giving a relative flow angle at impeller outlet of 55 degrees and an absolute flow angle of 60 degrees, all to the radial direction. The diffuser passages have a two-to-one area ratio.

Assume that the pressure rise required is stagnation-to-stagnation, and make an initial guess at 85 percent for the efficiency (defined as theoretical impeller work over actual blading work) for all machines across the stage.

3. Identify apparently desirable and undesirable features and sketch the approximate rotor and stator configurations of the velocity diagrams in figure P5.3. The following sketch is given as an example of what is wanted:

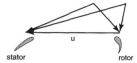

Example diagram P5.3

Desirable features are (W_{ex}/W_{in}) acceptable rotor and stator. Undesirable features are high swirl velocity after both rotor and stator; high relative velocity into stator might lead to high-Mach-number problems.

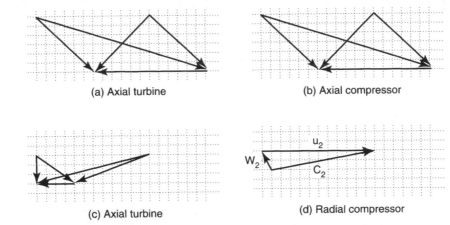

Figure P5.3

4. Calculate the stagnation-to-static polytropic efficiencies from turbine inlet to diffuser outlet of several alternative turbines designed for the same duty. The duty is to expand 12 kg air per second from 4 atmospheres, 650 °C stagnation conditions, to atmospheric pressure. (This will be the static pressure at diffuser outlet.)

(a) Turbine 1 is a single-stage impulse turbine with a gas angle leaving the nozzle of 70 degrees and axial flow at outlet. Diffuser $C_{pr,tl} = 0.65$.

(b) Turbine 2 is a two-stage 50-percent-reaction turbine with axial flow at outlet and a flow coefficient of 0.4. The diffuser $C_{pr,tl} = 0.75$, $\psi = 1.0$.

(c) Turbine 3 is similar to turbine 2 except that it has 4 stages. The diffuser $C_{pr,tl} = 0.75$.

For cases a, b, and c the drop in stagnation pressure in the diffuser is 12.5% of the difference between stagnation and static pressures at inlet.

(d) Turbine 4 is a radial-inflow turbine with an air angle entering the rotor (relative to the radial) of 70 degrees, and a relative flow angle against the direction of rotation of the rotor of 10 degrees. The mean-diameter flow can be assumed to leave the rotor without a swirling component, with a relative flow 50 percent greater than the relative velocity at inlet, and a radius of one-half the inlet radius. The diffuser $C_{pr,tl} = 0.3$. The drop in stagnation pressure in the diffuser is 25 percent of the difference between inlet stagnation and static pressures.

In all cases, the stagnation-to-stagnation polytropic efficiency across the blading (to turbine outlet) can be taken to be 90 percent. Perfect-gas relations or tables may be used, whichever is more convenient. The calculations will involve some iteration. Discuss the reasons for the differences in overall efficiencies.

5. Optimize the design (to the first iteration) of a single-stage axial-flow water pump which has the duty of lifting water from a discharge tank to a supply tank, through a vertical lift of 6 m. There is a frictional head loss of 0.25 m in the pipe additional to this lift, and a kinetic head loss equal to that leaving the pump diffuser. The electrical supply available is enough to power a motor of 25 kW output, and it can be assumed that an ungeared motor of any of the usual standard speeds can be used (approximately 875, 1,150, 1,725, or 3,500 rpm).

(a) As a first approximation, assume that a stagnation-to-stagnation efficiency (ideal power over actual power) across the pump of 0.90 can be used, and that the diffuser C_{pr} is 0.5 (actual static-pressure rise/inlet dynamic head).

(b) Choose a rotor-stator combination and hub-tip ratio, and calculate the conditions for the mean-diameter velocity diagram. Use a flow coefficient of 0.3, axial inlet and outlet flow direction, and a maximum value of $C_{pr,tl}$ of 0.30 at the mean diameter.

(c) Assume that the power dissipated in bearings, couplings and disk friction amounts to one kilowatt, so that the power available at the rotor blading is 24 kW.

6. Suppose that you have a laboratory "work-horse" centrifugal compressor driven by a constant-speed motor, and you use it at different times for compressing hydrogen and for air. For which gas does the compressor produce more work per unit mass, and why?

7. Sketch the velocity diagrams for the constant-axial-velocity, constant-peripheral-velocity turbine and compressor (axial) stages listed in table P5.7. Also sketch in the approximate shapes of the stator and rotor blades. Fill in the missing numbers in table P5.7.

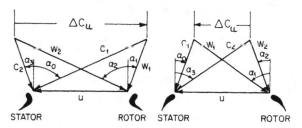

Figure P5.7

Table P5.7

Turbine (T) or Compressor (C) Diagram	ϕ	ψ	R_n	α_0	α_1	α_2	α_3	W_2/W_1
T			0%	70°			0°	
T				60°	0°		0°	
T		2.0	50%	70°				
C	0.3		50%	20°				
C		-0.5		0°				0.75
C		-0.5			70°		0°	

8. Calculate the stagnation-to-stagnation efficiency across the blading of an axial water-turbine stage having the following mean-diameter velocity-diagram characteristics:

R_n (reaction) = 0,
ψ (work coefficient) = 2.0,
ϕ (flow coefficient) = 0.8.

All the stagnation-pressure losses to be considered can be ascribed to the nozzles and to the rotor blade row as follows:
for nozzle row, stagnation-pressure losses

$$\Delta p_0 = 0.03 \frac{\rho C_1^2}{2g_c}$$

for rotor row, stagnation-pressure losses

$$\Delta p_0 = 0.08 \frac{\rho W_2^2}{2g_c}$$

where C_1 and W_2 are the nozzle exit and rotor exit velocities respectively.

9. Why is it unattractive to use a low reaction (having most of the static-pressure rise in the stator) for a single-stage axial-flow cooling fan for electronic equipment? Give two or three reasons if you can.

10. Find the flow coefficient that would be required if an axial-compressor stage of 50-percent reaction must have a theoretical pressure-rise coefficient no greater than 0.5. The work coefficient, $g_c \Delta h_0/u^2$, is to be 1.0.

11. Figure P5.11 shows two alternative diagrams for a high-pressure-ratio, high-hub-tip-ratio, single-stage axial air compressor. Give two advantages of each diagram relative to the other. (Both have equal blade velocity and equal enthalpy rise.)

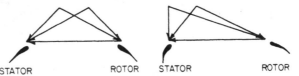

STATOR   ROTOR   STATOR   ROTOR

Figure P5.11

12. Choose the number of stages required and mean-diameter, common, velocity diagrams for a turbine of a gas-turbine engine that has a pressure (or expansion) ratio of 20 to atmospheric pressure, a mass flow of 40 kg/s, a turbine inlet temperature of 1500 K, and an estimated polytropic total-to-static- efficiency overall of 0.87. The mean blade speed should be no more than 550 m/s, and the leaving swirl angle at mean diameter should be no greater than 20°. The reaction should be 0.4 or greater at that point. We would (presumably) want to minimize the number of stages.

This is a design problem and has no one "correct" solution.

13. If diffusers fitted at the outlet of axial-flow compressors can be designed to have an actual pressure-rise coefficient C_{pr} of 0.6 (the actual static-pressure rise divided by the actual dynamic head at inlet), calculate the compressor stagnation-to-static polytropic efficiency and pressure ratio to the end of the diffuser for two alternative compressor designs. Both have a stagnation-to-stagnation polytropic efficiency of 0.9 and a pressure ratio of 10 : 1 to the end of the compressor blading (i.e., to the diffuser entrance). The diffuser-entrance Mach numbers are different for the two designs: 0.2 in one case and 0.4 in the other. The compressor-inlet stagnation conditions are 285 K and 100,000 Pa. A common mean C_p/R of 3.55 may be used. Draw a $T - s$ diagram.

14. To maximize the stagnation-to-static efficiency of a radial-inflow turbine stage, it is desirable to (fill in the blanks):

    a. _____ the rotor-outlet kinetic energy;
    b. _____ the rotor-outlet-flow angular velocity;
    c. make the ratio of the relative velocity at rotor outlet to that at rotor inlet
       _____ along the shroud streamline; and
       _____ along the hub streamline.

    List the implication(s) of these choices for the design of the outlet diffuser.

15. Calculate the hub-shroud diameter ratio at rotor entrance for an axial-flow turbine, based on the following simple mean-diameter diagram parameters and these other inputs.

Reaction	0.5
Work coefficient	1.8
Flow coefficient	0.63
Mass flow	25 kg/s
Stagnation pressure	2,000,000 Pa (nozzle exit)
Stagnation temperature	1600 K
Mean blade speed	575 m/s
Shroud diameter	391 mm
Gas constant R	286.96 J/(kg · K)
Mean specific heat	1235 J/(kg · K)

16. Find the flow angle relative to the axial direction that must be used at the nozzle exit of an axial-flow turbine running on carbon dioxide. The Mach number at nozzle exit is 0.9, and the stagnation temperature and pressure are 800 K and 450 kPa. Carbon dioxide has a molecular weight of 44.01 and a mean specific heat for these conditions of 1137 J/(kg · K). The universal gas constant is 8313.219 J/(kmole · K). The mass flow is 5.5 kg/s. The nozzle-hub diameter is 308 mm and the shroud is 342.5 mm.

17. Calculate and write in the values for the scales of the graph of the performance of a full-size prototype water turbine. The graph and scales shown are for a model turbine having a rotor 300-mm diameter and working under a constant head of 3 m. The full-size turbine is to have

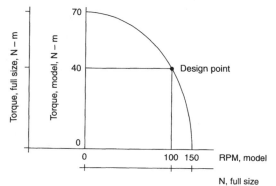

Figure P5.17. Performance curve for a water turbine

a rotor of 6 m in diameter and to work under a head of 48 m. Assume dynamic similarity (identical vector diagrams) and identical efficiencies.

18. Calculate the ratio (b_2/d_2) for a centrifugal blower of the following specifications.

Fluid	air
Nondimensional specific speed	1.5
Inlet flow volume	10 m³/s
Stagnation temperature at inlet	573 K
Stagnation pressure at inlet	100,000 Pa
Stagnation pressure at outlet	150,000 Pa
Stagn.-to-stagn. polytropic eff.	0.85
Mean specific heat	1005 J/(kg · K)
Static density at inlet	1.162 kg/m³
Static density at outlet	1.40 kg/m³
Flow coefficient at outlet	0.35
Work coefficient	0.5

19. How many stages are required in a multistage centrifugal water pump that must produce a total head of 350 m? The rotor peripheral speed must not be greater than 40 m/s; the work coefficient should not be more than 0.5; and the stagnation-to-stagnation efficiency (actual head over theoretical) can be assumed to be 0.86.

20. Calculate the specific enthalpy drop for the air passing through an axial-flow air turbine. The flow leaves the nozzle blades at 70° to the axial direction, with a Mach number of 1.0 and at a stagnation temperature of 600 K. The air leaves the rotor blades axially (i.e., without swirl), and at a static pressure of 120 kPa. The rotor-blade inlet angle is 42°. The rotor mean diameter (which can be used for calculations) is 100 mm and the rotational speed is 50,000 rpm. Use the specific heat at the nozzle-exit static temperature as an approximation to the mean value. Calculate the "simple-diagram" reaction. How is the stagnation enthalpy drop divided between nozzles and rotor? Calculate the apparent incidence angle of the flow entering the rotor blades. What approximations do you have to make in your calculations?

21. Calculate the design-point nondimensional specific speed, defined in equation 5.15 for the centrifugal water pump shown. At design point the rotation speed, N, is 1000 rpm, and the inlet stagnation pressure and temperature are 210 kPa and 285 K. The flow enters the rotor without swirl and leaves with an absolute swirl of 65° to the radial direction. The sweepback angle of the rotor blades at the periphery is 60°. The absolute tangential velocity of the flow leaving the rotor is 90 percent of the value it would have were the flow to follow the rotor blades exactly, for the same radial velocity. The hydraulic efficiency is 0.9 (the ratio between the actual and the theoretical head).

22. Calculate the blading power output in watts of the radial-inflow turbine of a Diesel-engine turbocharger that takes 200 g/s of gas, molal mass 28.5, at 723 K and 1.6 bar stagnation conditions and exhausts it to the atmosphere at 1 bar static pressure (1 bar is 100,000 Pa) without swirl. The mean specific heat can be taken as 1227J/(kg · K). The gas leaves the nozzles at 400 m/s at an angle of 70° to the radial direction. The rotor speed is 40,000 rpm and the rotor outside diameter is 186.2 mm.

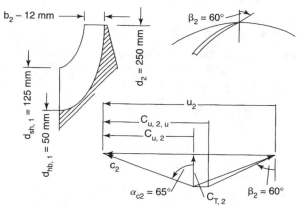

Figure P5.21. Centrifugal water pump

23. Draw the velocity diagram for the Kaplan water turbine illustrated, assuming that the relative flow into the rotor blades is aligned with the blade direction, and that there is only a small difference between the flow direction and the blade mean line at exit. (The absolute leaving flow into the diffuser (the "draft tube") must be without swirl.)

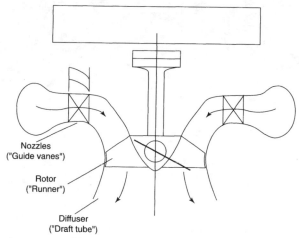

Figure P5.23 Kaplan water turbine

Comment on the approximate value of the loading coefficient that your diagram indicates.

Kaplan turbines are used for low-head high-flow applications. The alternators are always directly coupled (i.e., without gearing). What are the implications with regard to blade speed and shaft speed that the value of the loading coefficient leads to?

24. Draw the velocity diagram of a Pelton wheel (see figure 1.4b), used for high-head low-flow applications. In the sketch there is just one nozzle (sometimes several are used). The jet issues horizontally just above the outlet flow stream (the "tail race"). The "buckets" (blades) cut the jet, dividing it into two equal streams, turning the relative flow through 165° and, at

design point, allowing the flow to fall into the tail race without swirl. Neglect friction on the relative flow in the buckets.

What is the reaction? What is the value of the loading coefficient? If the nozzle is isentropic, what is the value of the mean blade speed if the head is 1150 m? What implications does this value of blade speed have for directly coupled alternators?

25. The sketch is of a "squirrel-cage" radial-outflow blower. Calculate the value of b_2/d_2 that must be used if the following specifications apply.

$$d_{sh,1}/d_2 = 0.7$$
$$C_{r,2}/C_{x,1} = 1.2$$
$$C_{r,2}/u_2 = 0.25$$

The absolute flow angle leaving the rotor is $60°$ from the radial. The flow is essentially incompressible, and uniform leaving the rotor (plane 2), and uniform and without swirl entering the rotor (plane 1).

Draw the outlet velocity diagram and calculate the nondimensional specific speed.

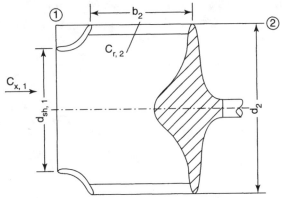

Figure P5.25 Squirrel-cage blower

26. Calculate the specific speed of a radial-outflow air compressor, the dimensions of which are shown in the sketch. The absolute angle of air flow leaving the rotor is $60°$. The blade angle at outlet (sweepback) is also $60°$, measured again from the radial direction. The work coefficient is 90 percent of the value that would be given if the flow followed the blade direction exactly, with the same value of radial velocity in the theoretical and actual vector diagrams. The rotor peripheral speed is 100 m/s, and at this condition the static density of the air at rotor outlet is ten percent higher than at rotor inlet. There is no swirl at rotor inlet.

27. Give two reasons why a multistage turbomachine is likely to have a higher overall efficiency than a single-stage machine designed for the same duty.

28. Why are stages with inlet and outlet flow vectors that are axial in direction used only infrequently in multistage turbomachines?

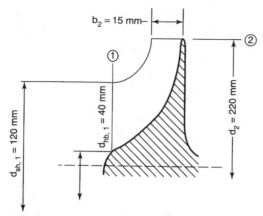

Figure P5.26 Radial-outflow air compressor

29. Find how many jet engines acting as hot-gas generators or "gasifiers" are required to produce a minimum of 44 MW for a naval boost engine. The jet engines have their exhausts ducted individually to segments of the nozzle ring of a single-stage impulse turbine. Each jet engine supplies 75 kg/s of exhaust at 2.0 bar and 850 K stagnation conditions, and the turbine exhausts to a static pressure (effective total pressure) of 1.1 bar. The stagnation-to-static polytropic efficiency for these conditions is 0.87. At design point the turbine has axial inlet and outlet flow. The gas has a specific heat of 1.11 kJ/(kg · K) and a gas constant of 286.96 J/(kg · K). There are 97 nozzle vanes and 83 rotor blades, and the design speed of the (impulse) power turbine is 4, 500 rpm.

Then calculate the required blade speed of the turbine at the reference diameter at which the impulse diagram applies. If the flow leaves the turbine nozzles at 70° to the axial direction, find the enthalpy of the kinetic energy in the flow leaving the rotor and express this as a percentage of the stagnation enthalpy drop through the turbine stage.

Chapter 6

Three-dimensional velocity diagrams for axial turbomachines

In the last chapter several degrees of freedom were shown to be available for the choice of velocity diagrams for axial-flow turbomachines. Some of these degrees of freedom are appropriated by the necessary preliminary selection of change in stagnation enthalpy per stage, and mean or tip blade speed. Both of these quantities are usually decided upon, perhaps by comparison with previous machines, in the early stage of making preliminary specifications.

There usually remain at least three degrees of freedom. This chapter is concerned with describing how one of these freedoms must be given up, at least in machines with low hub-shroud diameter ratios (relatively long blades and small hubs), to satisfy, or partially satisfy, the requirement that the flow approach radial equilibrium. That is, if radial-flow accelerations are to be avoided, as they usually are to a large degree, the radial distribution of pressure must balance (or provide centripetal acceleration for) the flow "swirl".

That being so, the choice of velocity diagrams reduces usually to two degrees of freedom. The shape of the velocity diagram may be selected, often within fairly narrow limits, at one radius. And the variation of either the axial velocity or the tangential velocity, which are coupled by the equations of radial equilibrium for an interstage plane, may be chosen, again within sometimes narrow limits. Whether these limits are indeed narrow or broad depends primarily on the value of the hub-shroud ratio (the ratio of the inner to the outer diameter of the flow annulus).

A large hub-shroud ratio, perhaps defined as between 0.85 and 1.0, can involve only small changes of the velocity diagram with radius. For preliminary design it is usually sufficient to characterize the machine by the velocity diagram at mean diameter. Thus we can usefully refer to a "50-percent-reaction turbine", or to a "low-stagger compressor stage", even though there will be some radial variation of reaction and stagger.

When the hub-shroud ratio is less than, say, 0.75, the necessary variation in velocity diagram from hub to shroud is large enough for the characterization of the machine by the mean-diameter velocity diagram to be no longer useful. This is not obvious at this point. Having at least partial freedom to choose one velocity diagram and the variation of the diagram with radius, it is in fact possible to choose to have, for instance, constant reaction from hub to shroud radius. But it turns out that such a choice produces an unacceptable design for various reasons.

In sections 6.1 to 6.6 we present a simplified method for the design of three-dimensional velocity diagrams one stage at a time (with the use of the simple radial-equilibrium equation, SRE). Design guidelines for the use of this equation are given. In section 6.7 we present the elements of a streamline-curvature computational method for three-dimensional velocity diagrams. This is an iterative method that can be used for the design of all stages of an axial component at once, while taking into consideration the axial change of radius of streamlines along all stages of the component.

Lakshminarayana (1996) covers flow in turbomachines in considerable depth, including radial-equilibrium flows (relevant to this chapter); flow in blade rows, impellers and diffusers (chapters 7, 8, and 9, in addition to chapter 4); and heat transfer to and from turbine blades (chapter 10).

6.1   The constant-work stage

It is normally desirable to design for a constant change in stagnation enthalpy, Δh_0, for the stage at every radius. The occasions when one may not follow this rule are those where there are strong radial pressure gradients due to flow curvature in the axial-meridional plane. The treatment of this flow curvature is beyond the scope of preliminary design, and design methods accounting for it (such as the one described in section 6.7) can be used to design for constant or not constant Δh_0. These streamline-curvature methods must run on computers. In sections 6.1 to 6.5 we shall restrict ourselves to the case of constant Δh_0.

The variation of work coefficient, ψ, with radius is therefore prescribed:

$$\psi \equiv \frac{-g_c \Delta_1^2 h_0}{u^2}$$

Since $u = \omega r$,

$$\psi = \frac{-g_c \Delta_1^2 h_T}{\omega^2 r^2} \tag{6.1}$$

One design limit stems from this rapid increase of work coefficient with reduction in radius. The maximum permissible work coefficient is chosen at the hub radius (usually coupled with other limits such as of Mach number or reaction). Long blades are necessarily highly "loaded" (that is, ψ is high) at the hub and lightly loaded at the shroud.

6.2   Conditions for radial equilibrium

When the fluid flowing through a duct or annulus has a tangential component of velocity, a pressure gradient with radius is naturally set up. The combination of the radial variation of static pressure, the variation of tangential velocity, and the variation of axial velocity, produces the radial variation of stagnation enthalpy and stagnation pressure. In preliminary design we shall deal exclusively with designs having constant-enthalpy conditions at all transverse planes. This requirement then sets the coupling relation between axial- and tangential-velocity variations.

In figure 6.1 the pressure gradient in the elemental annulus due to the tangential velocity C_u alone is

$$\delta p_{st} r \delta \theta \delta l = \frac{r \delta \theta \delta r \delta l \rho_{st} C_u^2}{r g_c}$$

Therefore,

$$\left(\frac{\partial p_{st}}{\partial r} \right)_{C_r} = \frac{\rho_{st} C_u^2}{r g_c} \tag{6.2}$$

Specifying that the radial component of velocity, C_r, be constant, means that we shall not include the effects of flow curvature in the axial-meridional plane on the radial pressure gradient. We want to find how this radial pressure gradient influences the axial-velocity component and the other fluid properties.

By definition,

$$h_0 \;\equiv\; h_{st} + \frac{C^2}{2 g_c} \qquad \text{(equation 2.9)}$$

$$C^2 \;=\; C_x^2 + C_u^2 + C_r^2$$

$$h_{st} \;\equiv\; u_{st} + p_{st} v_{st} \qquad \text{(equation 2.5)}$$

Therefore,

$$dh_{st} = du_{st} + p_{st} dv_{st} + v_{st} dp_{st}$$

Gibbs' equation, equation 2.31, is

$$T_{st} ds_{st} = du_{st} + p_{st} dv_{st}$$

Therefore,

$$dh_{st} = T_{st} ds_{st} + v_{st} dp_{st}$$

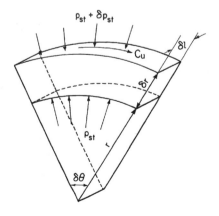

Figure 6.1. Stability of a fluid element

and

$$\frac{dh_{st}}{dr} = T_{st}\frac{ds_{st}}{dr} + v_{st}\frac{dp_{st}}{dr}$$

Therefore,

$$g_c\frac{dh_0}{dr} - g_c T_{st}\frac{ds_{st}}{dr} = g_c\frac{1}{\rho_{st}}\frac{dp_{st}}{dr} + \frac{1}{2}\frac{d}{dr}C_x^2 + \frac{1}{2}\frac{d}{dr}C_u^2 + \frac{1}{2}\frac{d}{dr}C_r^2 \qquad (6.3)$$

We can substitute for the radial pressure gradient from equation 6.2 and for the tangential-velocity variation we can use the following construction:

$$\frac{d}{dr}(r^2 C_u^2) = r^2\frac{dC_u^2}{dr} + 2r C_u^2$$

Therefore,

$$\frac{1}{2}\frac{dC_u^2}{dr} = \frac{1}{2r^2}\frac{d}{dr}(r^2 C_u^2) - \frac{C_u^2}{r}$$

Equation 6.3 becomes

$$g_c\frac{dh_0}{dr} - g_c T_{st}\frac{ds_{st}}{dr} = \frac{C_u^2}{r} + \frac{1}{2}\frac{d}{dr}C_x^2 + \frac{1}{2r^2}\frac{d}{dr}(r^2 C_u^2) - \frac{C_u^2}{r} + \frac{1}{2}\frac{dC_r^2}{dr}$$

or

$$g_c\frac{dh_0}{dr} - g_c T_{st}\frac{ds_{st}}{dr} = \frac{1}{2}\frac{d}{dr}C_x^2 + \frac{1}{2r^2}\frac{d}{dr}(r^2 C_u^2) + \frac{1}{2}\frac{dC_r^2}{dr} \qquad (6.4)$$

For

$$\frac{dC_r^2}{dr} = 0 \quad \text{and} \quad \frac{dh_0}{dr} = 0 \quad \text{and} \quad \frac{ds_{st}}{dr} = 0$$

we obtain the equation of simple radial equilibrium (SRE):

$$\frac{1}{r^2}\frac{d}{dr}(r^2 C_u^2) + \frac{d}{dr}C_x^2 = 0 \qquad (6.5)$$

6.3   Use of the SRE equation for velocity distributions

The designer has considerable freedom to choose either a variation of axial velocity with radius or a variation of tangential velocity with radius, at any one inter-row plane in an axial-flow turbomachine. This choice, and the choice of a velocity diagram at any one radius (for instance, the mean or the hub radius) then determines the axial and tangential velocities in that plane and the neighboring plane in the following way.

In the first plane the choice of either the axial- or the tangential-velocity variation will allow the other to be found from the simple-radial-equilibrium equation. The tangential-velocity variation in the neighboring plane will then be determined because the radial distribution of work (usually constant) is given from the Euler equation as $u_1 C_{u,1} - u_2 C_{u,2}$. From this tangential-velocity variation in the second plane, together with the specified axial velocity at the reference radius, the distribution of axial velocity can be found from the SRE equation.

A particular case of interest is when

$$rC_u = r_m C_{u,m} = \text{constant} \tag{6.6}$$

where the subscript m signifies the mean-diameter condition. This is the relation for a free vortex. The SRE equation (equation 6.5) and this relation lead directly to the axial-velocity distribution, which is $C_x = \text{constant}$.

Free-vortex designs were formerly almost universal in the turbines of gas-turbine engines, and some older designs still use them, although others are supplanting them. Figure 6.2 shows velocity diagrams for the hub, mean-diameter, and shroud sections of a turbine with zero-reaction (impulse) conditions at the hub ($r' \equiv r/r_m$). Even though the hub-shroud ratio is not extreme (0.6) the reaction at the shroud is 64 percent. This large radial variation in reaction is normal. Figure 6.3 shows some variations in reaction for turbines having different designs. It will be seen that the principal determinant of, say, shroud reaction is the choice of reaction at the hub, rather than other aspects, discussed next, of the radial variation of flow.

A general form studied by Carmichael and Lewis (Horlock, 1958) is the following:

$$C_{u,1} = ar^n - b/r$$
$$C_{u,2} = ar^n + b/r \tag{6.7}$$

where a, b and n are constants for any one type of swirl distribution.

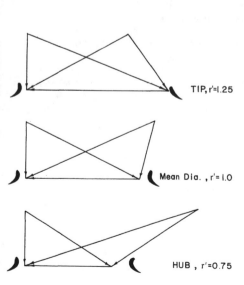

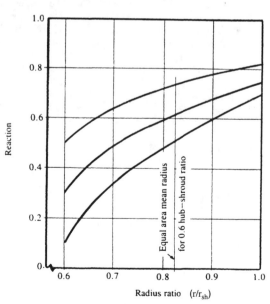

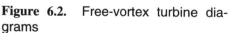

Figure 6.2. Free-vortex turbine diagrams

Figure 6.3. Typical reaction distributions for different values of hub reaction

The great length of the last low-pressure blades of this steam turbine is the result of the huge increase in specific volume of steam as it expands through a turbine. The high-pressure turbine will be a single rotor having blades of perhaps only 25-mm length. This low-pressure unit is double-flow (the steam comes in the middle of the two low-pressure turbines that can be seen, and flows outwards in two directions) and there might well be two double-flow low-pressure units. The next two rotor stages can be seen between the large disks, and it is obvious that the major part of the volume expansion occurs at low pressure. This turbine can be compared with wide-chord fans in fan-jet engines: the low-pressure blades have wide chords and no part-span "shrouds" or "snubbers" that are required to inhibit blade vibration when small-chord blades are used. It can be seen that the low-pressure blades are twisted to accommodate radial equilibrium. It is remarkable that steam-turbine designers use untwisted "strip" blading everywhere else. We have not been able to find why it would not be worth the extra cost of using twisted blading in view of the improved expansion efficiency that would result.

Illustration. Rotor of double-flow low-pressure steam turbine Courtesy Brown Boveri Co., now ABB

6.4   Prescribed reaction variation

There are too many criteria for judging a "good" type of radial-equilibrium solution for us to arrive easily at a simple rule. In axial-compressor design the most important criterion is to have the relative-velocity ratio W_2/W_1 (the de Haller ratio), stay above, say, 0.7 for both rotor and stator from hub to shroud. It is often also important to keep the relative Mach number below either, say, 0.9 or (if a transonic design is acceptable) 1.4, say, for the rotor-shroud conditions and below 0.9 for the stator-hub radius.

In compressors and turbines we like to keep the reaction at the hub above zero and the reaction at the shroud below unity. Since we know that the distortion of the velocity diagram resulting from extreme reactions produces high relative Mach numbers, and also gives diffusing conditions in turbines, a reasonable approach to the choice of an acceptable radial-equilibrium solution is to start with the requirement that the variation in reaction along the blade length shall be such that the hub shall have some positive value and the shroud shall be below 1.0. Then the results can be examined for variations in relative Mach number, in velocity ratio, and in blade angle for the particular mean-diameter velocity diagram chosen for the design being examined.

Some typical variations in reactions across the blade annulus for different value of the reaction at the hub are shown in figure 6.3.

Let us define new variables $C'_u \equiv C_u/u_m$, $C'_x \equiv C_x/u_m$, and $r' \equiv r/r_m$, so that the equation of simple radial equilibrium (equation 6.5) becomes

$$\frac{1}{(r')^2}\frac{d}{dr'}[(r')^2(C'_u)^2] + \frac{d}{dr'}[(C'_x)^2] = 0 \tag{6.8}$$

The reaction variable R'_n is defined as

$$R_n' \equiv 1 - \frac{1}{2r'}(C'_{u,1} + C'_{u,2}) \tag{6.9}$$

This is equal to the actual reaction R_n only where the velocity diagram is "simple" (having constant C_x and u from rotor inlet to rotor outlet for any one streamline).

The class of variations of tangential velocity suggested by Carmichael and Lewis (equation 6.7) is defined by

$$\begin{aligned} C'_{u,1} &= a'(r')^n - b'/r' \\ C'_{u,2} &= a'(r')^n + b'/r' \end{aligned} \tag{6.10}$$

For this class of velocity distributions

$$R_n' = 1 - a'(r')^{n-1} \tag{6.11}$$

$$\psi' \equiv \frac{C_{u,1} - C_{u,2}}{u} = \frac{C'_{u,1} - C'_{u,2}}{r'} = -2b'/(r')^2, \qquad \text{for} \quad u_1 \approx u_2 \tag{6.12}$$

Therefore,

$$\begin{aligned} a' &= 1 - R'_{n,m} \\ b' &= -\psi_m/2 \end{aligned}$$

These can be inserted into the SRE equation 6.8 and integrated to give

$$\begin{aligned} \left[\frac{C'_{x,1}}{\phi_m}\right]^2 &= \left(\frac{C_{x,1}}{C_{x,m}}\right)^2 = 1 + \left(\frac{n+1}{n}\right)\left[\frac{1 - R'_{n,m}}{\phi_m}\right]^2 \\ &\quad \times \left\{[1 - (r')^{2n}] + \left(\frac{n}{n-1}\right)\left[\frac{\psi_m}{(1 - R'_{n,m})}\right][1 - (r')^{n-1}]\right\} \end{aligned} \tag{6.13}$$

$$\left[\frac{C'_{x,2}}{\phi_m}\right]^2 = \left(\frac{C_{x,2}}{C_{x,m}}\right)^2 = 1 + \left(\frac{n+1}{n}\right)\left[\frac{1 - R'_{n,m}}{\phi_m}\right]^2$$
$$\times \left\{[1 - (r')^{2n}] - \left(\frac{n}{n-1}\right)\left[\frac{\psi_m}{(1 - R'_{n,m})}\right][1 - (r')^{n-1}]\right\}$$

This form for the prescribed radial variations in axial velocity is more useful than the simpler versions because the constants are velocity-diagram parameters ($R'_{n,m}$, ϕ_m, and ψ_m) that must be chosen in the preliminary-design stage, not arbitrary constants. However, equations 6.13 cannot be used for $n = 1.0$, for which value equation 6.8 should be integrated directly.

The actual reaction R_n, is found from its definition, equation 5.7:

$$R_n \equiv \frac{\Delta h_{st}}{\Delta h_0} = 1 - \frac{(C_2^2 - C_1^2)/2}{u_2 C_{u,2} - u_1 C_{u,1}} = 1 - \frac{(C_2^2 - C_1^2)/2}{-\psi_m u_m^2}$$

$$= 1 + \frac{1}{2\psi_m}\left[\left(\frac{C_{x,2}}{u_m}\right)^2 - \left(\frac{C_{x,1}}{u_m}\right)^2 + \left(\frac{C_{u,2}}{u_m}\right)^2 - \left(\frac{C_{u,1}}{u_m}\right)^2\right]$$

$$= 1 + \frac{1}{2\psi_m}\left[-2\left(\frac{n+1}{n}\right)\frac{(1 - R'_{n,m})^2}{\phi_m^2}\frac{n}{(n-1)}\frac{\psi_m}{(1 - R_{n,m}')}\phi_m^2[1 - (r')^{(n-1)}]\right.$$

$$\left. -2(1 - R_{n,m}')\psi_m(r')^{n-1}\right]$$

$$= 1 + (1 - R_{n,m}')\left\{-(r')^{n-1} - \frac{n+1}{n-1}[1 - (r')^{n-1}]\right\}$$

$$R_n = 1 + (1 - R_{n,m}')\left[\frac{2(r')^{n-1} - n - 1}{n-1}\right] \tag{6.14}$$

There are too many variables for recommendations to be made that would be useful in all cases. Rather, the use of equations 6.13 is best illustrated by an example.

Example 1   Calculation of flow variation across an annulus

An axial-flow-compressor stage has the following mean-diameter velocity-diagram parameters: $\phi_m = 0.4$, $\psi_m = -0.25$, $R_{n,m}' = R_{n,m} = 0.50$.

It is desired to use the Carmichael and Lewis tangential-velocity distributions, and to set $R_n' = 0.25$ at $r' = 0.75$. Find the required variations of axial velocity, the reaction variable R_n', and the true reaction R_n at radial stations where $r' = 0.75, 0.80, 0.90, 1.0, 1.10, 1.20$, and 1.25.

Solution.

With the substitution of the given values into equations 6.11 and 6.12, we can find a', b', and n. At mean diameter $r'_m = 1.0$

$$R'_{n,m} = R_n = 1 - a' = 0.5$$

Therefore, $a' = 0.5$ and

$$b' = -\frac{\psi_m}{2} = 0.125$$

At the hub, $r'_{hb} = 0.75$

$$R'_{n,hb} = 0.25 = 1 - 0.5(0.75)^{n-1}$$

Therefore,

$$(n-1) = \frac{\ln 1.5}{\ln 0.75} = -1.40942 \quad \text{and} \quad n = -0.40942$$

Equations 6.13 become

$$\left(\frac{C'_{x,2}}{\phi_m}\right)^2 = 1 - 2.25387\{[1 - (r')^{-0.8188}] \pm 0.14524[1 - (r')^{-1.40942}]\}$$

and equation 6.14 becomes

$$R_n = 1 - 0.5\left[\frac{2(r')^{-1.40942} - 0.59058}{1.40942}\right]$$

The results can be tabulated and plotted as in table 6.1 and figure 6.4.

Table 6.1. Results of calculation of flow variation across an annulus

	Hub			Mean			Shroud
$r' \equiv \dfrac{r}{r_m}$	0.75	0.80	0.90	1.0	1.1	1.2	1.25
$C'_{x,1} \equiv \dfrac{C_{x,1}}{u_m}$	0.4792	0.4614	0.4291	0.400	0.3735	0.3491	0.3375
$C'_{x,2} \equiv \dfrac{C_{x,2}}{u_m}$	0.5310	0.5016	0.4482	0.400	0.3555	0.3132	0.2927
$C'_{u,1} \equiv \dfrac{C_{u,1}}{u_m}$	0.3958	0.3916	0.3832	0.3750	0.3672	0.3599	0.3563
$C'_{u,2} \equiv \dfrac{C_{u,2}}{u_m}$	0.7292	0.7041	0.6609	0.6250	0.5945	0.5682	0.5563
R_n'	0.250	0.315	0.420	0.500	0.563	0.613	0.635
R_n	0.145	0.238	0.386	0.500	0.589	0.661	0.691

Comment

The actual reaction variation exceeds that of the reaction variable. The reaction at the hub is still positive, however. The table gives all the information necessary to draw the velocity diagrams and to measure or calculate the relative-velocity ratio, (W_{ex}/W_{in}). If a blade speed or a required pressure rise and estimated efficiency were given, the relative Mach numbers could also be calculated.

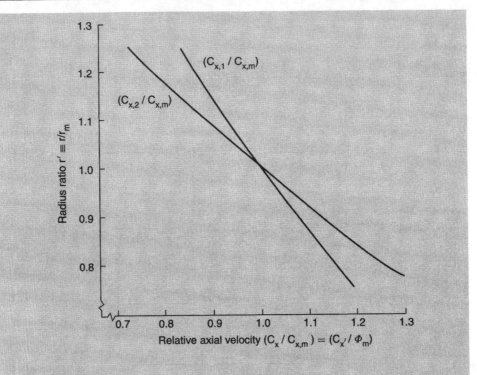

Figure 6.4. Results of calculation of flow variation across an axial-compressor annulus

It would be simple to produce an optimizing computer program if explicit trade-offs for relative Mach numbers at hub and shroud, for minimum relative-velocity ratio, and for minimum reaction, could be given. All the trade-off curves would be highly nonlinear, and at the end of the process the designer would not know what other designs the computer has eliminated in the process of finding its optimum. For that reason we feel that the integration of the optimization of such trade-offs is best done in the designer's mind.

6.5   Advantageous values of the index *n*

In a study of preliminary design methods for axial-flow turbines (Wilson, 1987), the effect of choosing different values of the index *n* was studied. The first conclusion was that the variation of reaction was little affected by the choice of index (figure 6.5).

The principal determinant of the variation of reaction across the annulus was simply the starting value (figure 6.3). An old rule for turbine design was that the hub diagram should be "above impulse" (it should have just-positive reaction). However, while the variation of efficiency with reaction shows that the peak is reached for reactions near 0.5 (chapter 7), the efficiency drops off far more sharply towards lower reactions than towards higher. Therefore a guideline for good design would be to set the reaction of the hub diagram of axial-flow turbines at 0.3, or at that value given by setting the mean-diameter reaction at 0.5, whichever gives the higher value.

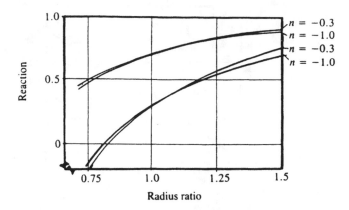

Figure 6.5. Variation of reaction with radius ratio in axial-flow turbines for different values of hub reaction, showing small effect of index n. From Wilson (1987)

This guideline would remove a source of considerable loss in many existing axial-flow turbines having overall diffusion, sometimes called "recompression", at the hub section.

Variations in the index n were found to have a strong influence on the radial variation of the nozzle-exit flow angle, and a lesser effect on the variation of flow coefficient. It is very desirable to use a swirl pattern that produces an increasing axial velocity towards the hub, because, for constant-enthalpy-drop blading, the loading coefficient is proportional to the inverse power of the square of the radius, and therefore is highest at the hub. For a given stage loading coefficient, the flow deflection decreases as the axial velocity (the flow coefficient) is increased. Therefore a flow pattern that produces a high hub axial velocity will reduce the amount of flow turning required in the blading, and should result in lower losses.

Some values of the index n produce swirl distributions that result in very little change in the nozzle-outlet flow angle. A nozzle blade having little twist is desirable because it is easier to manufacture and measure a nozzle with a specified throat area if it has little twist than it is when the nozzle blades are highly twisted. The throat area is the major factor governing the matching of the compressor and turbine (i.e., the assurance that both components will simultaneously be operating at near their optimum efficiencies). It was found that the swirl distributions that gave nearly constant nozzle flow-outlet angles also gave increased axial velocities at the hub. Therefore it seems desirable to start a preliminary-design study with this type of swirl. The appropriate values of the index n were found to be a function of the vector-diagram reaction at mean diameter, as shown in figure 6.6.

6.6 Practical considerations governing blade twist

Until at least the 1970s, most compressors, certainly most axial compressors for industrial applications, were designed for "full" or "simple" radial equilibrium but were manufactured to be at least partially mismatched. After the first low-pressure stages,

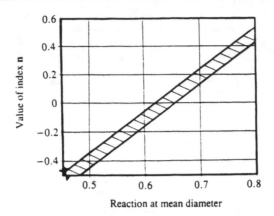

Figure 6.6. Values of the index n giving nearly constant nozzle-exit flow angles for axial-flow turbines. From Wilson (1987)

which might be transonic, stage design usually "settled down" to a repeating pattern. If a constant outer diameter was employed, therefore, all the stator blades in the repeating-stage part of the compressor could be identical except that they would be "cropped" or "trimmed" to the appropriate blade length for whichever stage they were to be used in. However, because the rotor-blade roots would be at different diameters, for full matching each rotor-blade row would have to be machined specifically for use in that particular stage. To save considerable machining costs, most manufacturers used identical rotor blades, appropriately cropped, in perhaps three successive rows, fully matched only in the middle of the three rows. We do not know what penalty was incurred by this practice. The highest-efficiency compressors we know of, with peak stagnation-to-static polytropic efficiencies measured on test from 0.92 to 0.94, were manufactured in this way, so that the penalty could not have been large.

The advent of numerically controlled machining and precision casting has reduced the cost of producing specially tailored blades for each row, so that modern compressors should be fully matched on all blade rows.

The loading coefficient of axial-flow turbines is so much higher than it is for compressors that the high density ratio so produced results in a much greater angle of "flare" (i.e., equivalent cone angle) in turbines than compressors. The mismatch in using a common blade for neighboring rows having differing root diameters would be, in general, too large, and in gas turbines it is usual to manufacture each stator, or nozzle, row specifically for each stage. If the turbine has a constant hub diameter, and if the blades are uncooled, it is possible to use one blade form for several neighboring rotor rows, cropped appropriately.

In steam turbines there remains the seemingly strange practice of using untwisted ("strip") blades for all except the one or two lowest-pressure stages. This seems to be a holdover from early design practice. Whittle introduced radial equilibrium into axial-turbomachinery design. (There was an old joke that when Whittle and many of his

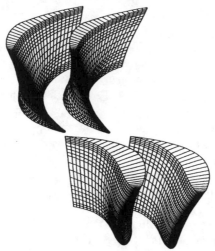

Advanced computer programs have been used to arrive at the three-dimensional, highly curved shapes of the axial-turbine nozzle blades and axial-compressor rotor blades. Loss reductions of ten percent have been reported compared with "stacked two-dimensional" designs.

Illustration. Three-dimensional blade shapes. Courtesy Rolls-Royce plc

associates left government service at the end of World War II to form Power Jets Ltd. he first patented free-vortex flow, and then non-free-vortex flow, thus requiring everyone involved in turbomachinery development to pay the firm royalties.) Those companies that remained with the government were brought into the National Gas-Turbine Establishment (NGTE), and an early project compared the performances of a turbine fitted first with untwisted blading and then with free-vortex blading. The NGTE measured a significant difference in peak efficiency, increasing from about 0.86 with strip blading to perhaps 0.88 with free-vortex blading. This gain in efficiency, which would be considerably greater with modern vortex patterns and blade profiles, almost proportionally affects steam-plant performance and capital cost. We have asked several people employed by two major steam-turbine manufacturers in the USA and Europe why strip blading is still used, and we have not heard an answer that makes sense to us. Therefore we would advocate the use of full, or at least simple, radial equilibrium in at least the medium-pressure and low-pressure stages of steam turbines.

6.7 Streamline-curvature calculation methods

In a typical multi-stage axial compressor or turbine the streamlines change radius along the stages, or axial length of the component. This lengthwise change in radius creates an additional pressure gradient acting on the flow. The simple-radial-equilibrium equation can be used for preliminary considerations on an individual stage, but it does not take into effect this additional pressure gradient from the streamwise change of streamline curvature.

Contemporary preliminary-design systems usually run on computers, and start from a through-flow analysis. This is based on assumptions of steady quasi-three-dimensional axisymmetric flow in a series of meridional planes (through-flow analysis in the axial-radial direction), based on the approach initially described by Wu (1952). The flow is analyzed as a sequence of two-dimensional calculations on meridional planes through the components (Denton, 1978; Dring and Joslyn, 1986). Flow solutions are obtained along streamsurfaces of revolution by writing the flow equations along lines orthogonal to the streamsurfaces. The axial-radial variation of properties can be described by the streamfunction approach, which leads to the matrix through-flow method (Marsh, 1968). This method is not easily used for locally supersonic designs because the solution is not unique. Alternatively, the axial-radial variation of properties is governed by the radial-equilibrium equations, which leads to the streamline formulation (Wu and Wolfenstein, 1950; Smith, 1966; Novak, 1967; Denton, 1978). Some of the streamline formulations lead to a singularity around Mach number $M = 1$ (Novak, 1967), and others do not (Denton, 1978). Some streamline methods require two nested iteration loops to provide a converged solution: an outside iteration loop for the mass-flow balance; and an inside iteration loop to solve the radial momentum equation at each flow station.

Here we present a through-flow streamline method proposed by Korakianitis and Zou (1992). This method solves the three-dimensional radial-momentum equation while manipulating the annulus area at each flow station until a desirable velocity diagram is obtained. The method still requires the iteration loop for the mass-flow balance, but the radial-momentum equation at each flow station is solved using a one-pass numerical predictor-corrector technique, thus reducing the computational effort substantially. The method can be used for subsonic or supersonic designs. The flow is considered at a series of axial stations located between blade rows. The energy, momentum, and continuity equations are coupled with the radial-equilibrium conditions to provide a differential equation for the axial component of velocity. The method is used to obtain (design) the velocity diagrams and the location of streamlines (as well as the three-dimensional description of the annulus) at each axial station from hub to tip using typical guidelines for the tangential component of velocity. This method can be coupled with any empirical or analytic technique to predict the stagnation-pressure drop due to friction of each streamline in the stators, and for the stagnation-to-stagnation polytropic efficiency of each streamline of each stage. (For example such models have been described by Jansen and Moffatt (1967), and additional ones are included in chapters 7 and 8.) Therefore this method for the solution of the radial-momentum equations can also be used for analyses to predict the performance of a multi-stage component.

Derivation of radial equilibrium with streamline curvature

In the following the radial-equilibrium equation is derived assuming that the radial components of velocity do not contribute to friction. The resultant equation is coupled numerically with the enthalpy rise per streamline in each stage. The performance of each stage is analyzed with suitable models for the stagnation-pressure drops in stators, and stagnation-to-stagnation polytropic efficiencies for each stage. These losses can be applied in each individual streamline, accounting for losses as a function of geometry, using various models published in the open literature.

The inviscid momentum equation in the r direction is given by Bird et al. (1960):

$$-\frac{1}{\rho}\frac{\partial p}{\partial r} = C_x\frac{\partial C_r}{\partial x} + C_r\frac{\partial C_r}{\partial r} - \frac{C_u^2}{r} \tag{6.15}$$

By definition, the stagnation enthalpy can be written as

$$h_0 \equiv h + \frac{C^2}{2} = h + \frac{C_x^2 + C_u^2 + C_r^2}{2} \tag{6.16}$$

Taking the partial derivative with respect to r gives

$$\frac{\partial h_0}{\partial r} = \frac{\partial h}{\partial r} + C_u\frac{\partial C_u}{\partial r} + C_x\frac{\partial C_x}{\partial r} + C_r\frac{\partial C_r}{\partial r} \tag{6.17}$$

Gibbs' equation for a simple one-component system ($T ds = du + p dv$) is combined with the definition of enthalpy ($h \equiv u + pv$) and with property variations along the radial direction

$$dh = \frac{\partial h}{\partial r}dr, \qquad ds = \frac{\partial s}{\partial r}dr, \qquad \text{and} \qquad dp = \frac{\partial p}{\partial r}dr$$

to give

$$-\frac{1}{\rho}\frac{\partial p}{\partial r} = T\frac{\partial s}{\partial r} - \frac{\partial h}{\partial r} \tag{6.18}$$

Combining equations 6.17 and 6.18 gives

$$-\frac{1}{\rho}\frac{\partial p}{\partial r} = T\frac{\partial s}{\partial r} - \frac{\partial h_0}{\partial r} + C_u\frac{\partial C_u}{\partial r} + C_x\frac{\partial C_x}{\partial r} + C_r\frac{\partial C_r}{\partial r} \tag{6.19}$$

Substituting the last equation into equation 6.15 gives

$$T\frac{\partial s}{\partial r} - \frac{\partial h_0}{\partial r} = C_x\frac{\partial C_r}{\partial x} - C_u\frac{\partial C_u}{\partial r} - C_x\frac{\partial C_x}{\partial r} - \frac{C_u^2}{r} \tag{6.20}$$

Since $C_r = C_x\tan\alpha_r$ and $C_u = C_x\tan\alpha_u$, equation 6.20 can be written as

$$\frac{\partial h_0}{\partial r} - T\frac{\partial s}{\partial r} = -C_x\frac{\partial(C_x\tan\alpha_r)}{\partial x} + C_x\tan\alpha_u\frac{\partial(C_x\tan\alpha_u)}{\partial r}$$
$$+ C_x\frac{\partial C_x}{\partial r} + \frac{C_x^2\tan^2\alpha_u}{r} \tag{6.21}$$

or

$$\frac{\partial h_0}{\partial r} - T\frac{\partial s}{\partial r} = -C_x\tan\alpha_r\frac{\partial C_x}{\partial x} + C_x(1 + \tan^2\alpha_u)\frac{\partial C_x}{\partial r}$$
$$+ C_x^2\left[\frac{\sin\alpha_u}{\cos^2\alpha_u}\frac{\partial\alpha_u}{\partial r} - \frac{1}{\cos^2\alpha_r}\frac{\partial\alpha_r}{\partial r}\right] + \frac{C_x^2\tan^2\alpha_u}{r} \tag{6.22}$$

Specifying that $\partial s/\partial r = 0$, and substituting $C_x \equiv C_x(r, u)$, equation 6.22 becomes

$$\frac{\partial C_x(r, u)}{\partial r} - \frac{\tan \alpha_r}{1 + \tan^2 \alpha_u} \frac{\partial C_x(r, u)}{\partial x} + \frac{C_x(r, u)}{1 + \tan^2 \alpha_u} \left[\frac{\sin \alpha_u}{\cos^3 \alpha_u} \frac{\partial \alpha_u}{\partial r} \right.$$

$$\left. - \frac{1}{\cos^2 \alpha_r} \frac{\partial \alpha_r}{\partial x} + \frac{\tan^2 \alpha_u}{r} \right] = \left[\left[\frac{\partial h_0}{\partial r} \right] \right] = 0 \qquad (6.23)$$

where $C_x(r, x)$ is the unknown distribution of axial velocity through the radii (r) and across the axial stages (x). Sometimes radial variations of stagnation enthalpy (total temperature) are eliminated in order to design constant-work stages (the terms corresponding to $\partial h_0/\partial r$ shown in double square brackets are set to zero), but this is not a requirement to solve equation 6.23.

Equation 6.23 must be solved numerically via an iterative process. One of several numerical techniques can be used, coupled with a double-loop iteration to match the mass flow rate while the hub, or mean, or tip diameters at each station are manipulated to match the mass flow. One iteration loop is along the radii; the second is along the length of the component. Korakianitis and Zou (1992) suggested an alternative iterative solution (figure 6.7) with one single iteration loop, incorporating the use of equations 6.7 and

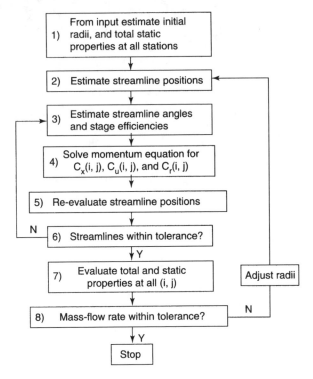

Figure 6.7. Flow diagram for the solution of equation 6.23 with one iteration loop on $C_x(r, u)$. From Korakianitis and Zou (1992)

MacCormack's explicit predictor-corrector algorithm (Anderson et al., 1984). Besides the obvious advantage of reduced computation time, this technique provides additional insight via the use of equations 6.7, so that all comments of section 6.5 can be applied for the choice of three-dimensional velocity diagrams.

References

Anderson, D.A., Tannehill, J.C., and Pletcher, R.H. (1984). Computational fluid mechanics and heat transfer. Hemisphere Publishing Corporation.

Bird, R.B., Stewart, W.E., and Lightfoot, E.N. (1960). Transport phenomena. John Wiley & Sons.

Denton, J.D. (1978). Throughflow calculations for transonic axial flow turbines. Trans. ASME, J. Engineering for Power, vol. 100, pp. 212–218.

Dring, R.P., and Joslyn, H.D. (1986). Through-flow analysis of a multistage compressor: part 1—aerodynamic input. Trans. ASME, J. of Turbomachinery, vol. 108, pp. 17–22.

Horlock, J.H. (1958). Axial-flow Compressors. Butterworths, London.

Jansen, W., and Moffatt, W.C. (1967). The off-design analysis of axial-flow compressors. Trans. ASME, J. of Engineering for Power, pp. 453–462.

Korakıanıtis, T., and Zou, D. (1992). Through-flow analysis for axial-stage design including streamline-slope effects. ASME paper 92-GT-461, New York, NY.

Lakshminarayana, Budugur (1996). Fluid dynamics and heat transfer of turbomachinery. John Wiley & Sons, New York, NY.

Marsh, H. (1968). A digital computer programme for the through-flow fluid mechanics in an arbitrary turbomachine using a matrix method. ARC, R&M 3509, U.K.

Novak, R.A. (1967). Streamline curvature computing procedures for fluid-flow problems. Trans. ASME, J. of Engineering for Power, pp. 478–490, New York, NY.

Smith, L.H., Jr. (1966). The radial-equilibrium equation of turbomachinery. Trans. ASME, J. of Engineering for Power, pp. 1–12, New York, NY.

Whitney, Warren J., and Warner Stewart (1972). Velocity diagrams, in Turbine Design and Application, ed. by Arthur J. Glassman. Special publication SP-290, vol. 1, NASA, Washington, DC, pp. 69–98.

Wilson, David Gordon (1987). New guidelines for the preliminary design and performance prediction of axial-flow turbines. Proc. I. Mech. E, part A, issue A4, November 1987, London UK.

Wu, C.H., and Wolfenstein, L. (1950). Application of radial equilibrium condition to axial flow compressor and turbine design. NACA report 955 (also NACA TN 1795, 1949).

Wu, C.H., (1952). A general theory of three-dimensional flow in subsonic and supersonic turbomachines of axial-radial and mixed-flow type. NACA TN 2604.

Problems

1. Using figure P6.1, calculate the flow incidence angles on to the rotor blades relative to design incidence at the hub ($r' = 0.75$) and the tip ($r' = 1.25$) of an axial-flow fan having untwisted constant-section inlet guide vanes and rotor blades. There are no downstream stator blades.

The mean-diameter ($r' = 1.0$) velocity diagram is 50-percent reaction, with flow coefficient 0.4 and work coefficient 0.3. The inlet guide vanes produce the same flow angle, α_0, at all radii. The solution of the simple-radial-equilibrium equation for this case is $C'_x = (r')^{-\sin^2 \alpha_0}$.

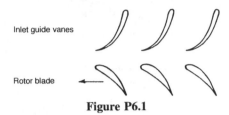

Inlet guide vanes

Rotor blade

Figure P6.1

2. Find the fluid angles, de Haller velocity ratios, and velocity vectors at hub, mean, and tip diameters of a compressor stage with a mean-diameter "simple" velocity diagram ($C_x =$ constant) characterized by $R_{n,m} = 0.5$, $\phi_m = 0.4$, and $\psi_m = -0.33$. The fluid compressed is carbon dioxide, and the maximum relative Mach number based on the inlet relative velocity at the mean diameter is to be 0.8. Find the relative Mach numbers at the hub and tip sections for three cases:

   free vortex (varying $R'_{n,m}$),
   constant R'_n,
   $R'_{n,hb} = 0.2$.

   The diameter ratio is 80 percent. The enthalpy rise is to be constant along the radius. The inlet total temperature is 145 °C. Comment on your findings for the constant R'_n case.

3. Find the velocity diagrams at hub and tip diameters for two alternative designs of water pump. One has a free-vortex type of blade "twist", and the other has $C_{u,1} = a - b/r$, $C_{u,2} = a + b/r$. The hub, mean, and tip diagrams are in the ratio 0.6, 0.8, and 0.1. The mean-diameter velocity diagram has axial inlet and outlet, rotor followed by stator, with a flow coefficient of 0.3 and a work coefficient of 0.208.

4. Comment on the apparent advantages and disadvantages of a forced-vortex type of flow distribution in which the tangential component of velocity at rotor entrance, $C_{u,1}$, is proportional to the square root of the radius. The mean-diameter diagram ($r' = 1.0$) is shown in figure P6.4; estimate qualitatively the shapes of the diagram at hub ($r' = 0.75$) and tip ($r' = 1.25$).

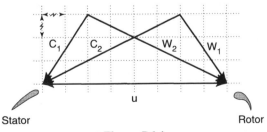

Stator   Rotor

Figure P6.4

5. Calculate the angles and velocities for the diagrams of an axial-compressor stage at mean diameter, and at diameters 20 percent greater and 25 percent smaller. The mean-diameter velocity diagram has 50 percent reaction, a flow coefficient of 0.362 and a work coefficient of 0.25. The relative Mach number of the flow entering the rotor blades is 0.85 at mean diameter. Calculate the relative Mach numbers and the degrees of reaction at the other diameters if the total enthalpy rise is constant across the rotor and if the variation of tangential velocity is given by $C_u = a \pm b/r$. The air inlet temperature is 300 K.

6. (a) Sketch the tip and hub velocity diagrams for an axial-flow compressor, the mean diameter velocity diagram of which is defined by $R_{n,m} = 0.5$, $\phi_m = 0.4$, $\psi_m = -0.3$, C_x = constant. The tangential component of velocity at entry to the stator is to be constant with radius, and the enthalpy rise must also be constant. The hub-tip diameter ratio is 0.6, and the mean diameter can be taken to be at 0.8 of the tip diameter. (b) By inspecting the diagrams, state where you would look for design limitations, such as high relative Mach number, high deflections, high static-pressure-rise coefficients, and so forth. (c) Comment on the desirability of this type of tangential-velocity distribution.

7. It is often desirable in axial-compressor design to have higher axial velocities at the hub than at the shroud, so that the diffusion ratios there will be lessened. By using the equation of simple radial equilibrium, show that a condition for the axial velocity to increase from shroud to hub at the rotor-entry plane is for the distribution of tangential velocity at the same plane to be

$$(C'_{u,1})^2 = ar' + \frac{(C'_{u,m,1})^2}{(r')^2} - \frac{a}{r'},$$

where a is a constant and is positive; r' is the ratio of the radius with the mean radius; C'_u is the ratio of the tangential velocity with the mean-radius blade speed; and $C'_{u,m}$ is the value of that quantity at mean radius.

8. Calculate, draw, and comment on the hub, mean, and shroud velocity diagrams for an axial-flow turbine of hub-shroud ratio 0.6. The mean-diameter (area mean) diagram is a rectangular diagram with a nozzle flow angle of 65 degrees. Use constant axial velocity, constant nozzle flow angle, and constant enthalpy drop along the radius. Comment in particular on the qualitative variation of reaction that you find.

9. Calculate the tangential velocity at rotor entrance, the axial velocity, the relative velocity on to the rotor, and the relative flow angle at the hub of an axial-flow stator-rotor compressor stage. The mean-diameter velocity diagram has a flow coefficient of 0.3, axial flow leaving the rotor, and a work coefficient of 0.446. The hub diameter is 0.75 of the mean diameter. We want to force the axial velocity to be inversely proportional to radius. Use the equation of simple radial equilibrium to comment (by estimating) on whether or not the hub diagram will be within the diffusion limits set for the mean diameter.

Chapter 7

The design and performance prediction of axial-flow turbines

We shall be concerned in this chapter first with methods for choosing the shape, size, and number of turbine blades to fulfill given initial specifications. Second, we shall look at ways of finding the efficiency and the enthalpy drop that will be given by turbine blade rows in different conditions of operation.

7.1 The sequence of preliminary design

This chapter fits in the overall progress of preliminary design in the following way.

1. The starting point will be given turbine specifications. These should include at least:

> the fluid to be used,
> the fluid stagnation temperature and pressure at inlet,
> the fluid stagnation or static pressure at outlet,
> either the fluid mass flow or the turbine power output required,
> and perhaps the shaft speed.

These specifications will (presumably) be for the so-called "design point". The specifications should also contain some information on the conditions, and to what extent the machine will operate at other conditions. The specifications must also give "trade-offs", or an objective function. This will enable the designer to aim for maximum efficiency, or minimum first cost, or minimum weight, or minimum rotating inertia, or maximum life, or some combination of these and other measures. (The designer is often left to guess at the trade-offs.)

2. The velocity diagrams at the mean diameter (or some other significant radial station such as at the hub) must be chosen by the approach described in chapter 5. This choice will include a selection of the number of stages to be used. This in turn implies some foreknowledge of permissible blade speeds, or an initial choice can be made and changed in later iterations.

3. The velocity diagrams at other radial locations are next chosen by the methods described in chapter 6.

4. The number of blades for each blade row, and their shape and size for each radial station for which velocity diagrams are available, are obtained by the methods described in this chapter.

5. The turbine performance at design-point and off-design conditions is predicted by methods described in this chapter.

6. When iterations involved in these steps have resulted in an apparently satisfactory blade design, it is drawn in detail. Initially, the various blade sections must be "stacked" in such a way that the centers of area are in a straight radial line to prevent the centrifugal forces from bending the blade. The leading and trailing edges must also be "faired" along continuous, regular, straight lines or curves. We do not deal with this stage of the design process in this book.

7. Centrifugal and fluid-bending stresses are calculated for the turbine blades so produced.

8. For high-temperature turbines, heat-transfer calculations are made to give the transient and steady-state temperature and stress distributions. Some fundamentals needed for heat-transfer analysis are given in chapter 10.

9. The mechanical and thermal stresses are combined. The acceptability of the combined stresses is checked against various criteria of failure. For high-temperature turbines, the most important criteria are the maximum permissible creep rate, the maximum rate of crack growth, or fatigue failure from low-cycle (i.e., low-frequency) thermal stresses. These aspects of materials selection are discussed in chapter 13. Full mechanical-plus-thermal stress analysis is beyond the scope of this book, however.

10. The natural frequencies of the individual blades are calculated for rotational speeds up to overspeed. The aerodynamic excitations, principally the wakes of upstream blade rows and struts and the potential-flow interaction from upstream and downstream blade rows and struts, or asymmetric inlets or exhausts, are also found. Coincidence of excitations with first- or second-order natural frequencies must be avoided by redesign or by changing the natural frequencies (for example, through the use of connected tip shrouds, or part-span shrouds, "snubbers", or lacing wires). The design process to avoid vibration failures is briefly described in section 7.6. A full vibrational analysis is beyond the scope of this book.

Once initial blade shapes have been obtained, computer-based calculation methods can predict and optimize the unsteady forcing functions (forces and moments) at one engine operating condition by refining the blade shapes. The blade shapes, ranges of numbers of stators and rotors of blade rows, and the axial gap between them can be selected from the preliminary-design stages in ways to ensure that unsteady forcing functions can later be minimized by the more-advanced methods. Guidelines for these choices have been developed, and they should be used from the preliminary-design stages. These guidelines are described in sections 7.4 to 7.6.

7.2 Blade shape, spacing, and number

This is the fourth phase of axial-turbine design, as detailed in section 7.1. It can be broken down into these steps.

1. Calculation of approximate optimum solidity for design-point operation.

2. Interpolation of "induced incidence" from correlations.

3. Interpolation of outlet-flow deviation from correlations.

4. Choice of leading- and trailing-edge radii.

5. Selection of near-optimum "stagger" or setting angle, from a correlation or by trial and error, and choice of a blade shape giving a passage shape of desirable characteristics.

6. Selection of number of blades from consideration of performance, vibration, and heat-transfer analyses.

Optimum solidity

Suppose that we have a row of blades whose shape and number we can change at will. For instance, they may be made of sheet metal and may therefore be bent. The inlet flow approaches the blade row at a predetermined angle. We have to bend the blades so that a desired outlet angle is produced. We try first with a small number of widely spaced blades and find that we have to bend the blades excessively, and that this leads to high stagnation-pressure losses from inevitable separation of the flow from the suction surface.

When we use many more blades at close spacing, the blades do not have to be as strongly curved, and there will be no separation. However, it is obvious that there will be unnecessary friction losses, again leading to high stagnation-pressure losses.

It is clear that there must be a range of spacings for which the losses are near a minimum (see figure 7.1). This optimum solidity must be related somehow to the aerodynamic loading of the blade row. Zweifel (1945) found that minimum-loss solidities could be well correlated by setting the tangential lift coefficient, C_L, at a constant value of 0.8, where

$$C_L \equiv \frac{\text{tangential aerodynamic force}}{\text{tangential blade area} \times \text{outlet dynamic head}}$$

Improved blade designs have been increasing the optimum value of C_L since 1945. For example, it was between 0.9 and 1.0 in the early 1970s (Stewart and Glassman, 1973). Current design practice uses values between 0.9 and 1.2, and occasionally (supersonic

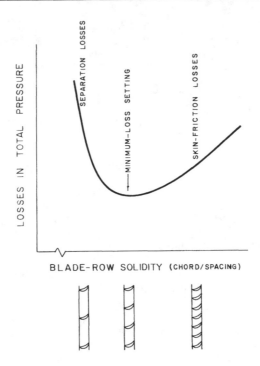

Figure 7.1. Diagrammatic variation of loss with blade solidity

stages) uses marginally higher values. The higher the value of C_L the harder it is to control the surface Mach-number or pressure distribution (and therefore the efficiency) of the blade (see section 7.5).

Angle convention

We use the convention given in chapter 5: absolute tangential components of velocity, and their associated flow angles with the axial direction, are positive in the blade-speed direction, and relative components are negative, and vice versa.

Subscript convention

It is not possible to have unique definitions for all symbols. Some must serve double or triple duty. When we are considering blade rows by themselves, the subscript *in* pertains to the blade or blade-row inlet and the subscript *ex* to the blade outlet. When we are dealing with the complete stage, we try to restrict the subscript 1 to "rotor inlet" and the subscript 2 to "rotor outlet". For the thermodynamic conditions in the engine cycle the subscript 1 means "compressor-inlet conditions", 2 means compressor-outlet conditions, and 4 and 5 are normally the turbine-inlet and -outlet conditions. We shall try to make it clear which of these definitions is being used.

Tangential aerodynamic force, F_u

The tangential loading is calculated by applying Newton's law to the fluid deflection through a blade row, figure 7.2.

$$F_u = \frac{\dot{m}(W_{in}\sin\alpha_{in} - W_{ex}\sin\alpha_{ex})}{g_c}$$

$$= \frac{sh\cos\alpha_{ex}\rho_{st,ex}W_{ex}(W_{in}\sin\alpha_{in} - W_{ex}\sin\alpha_{ex})}{g_c} \qquad (7.1)$$

where subscript $in \equiv$ blade-row inlet, subscript $ex \equiv$ blade-row outlet (exit) and h is the blade length. The mean pressure on the blade to give this force is

$$p_{st} = \frac{sh\rho_{st,ex}W_{ex}\cos\alpha_{ex}}{g_c b_x h}(W_{in}\sin\alpha_{in} - W_{ex}\sin\alpha_{ex}) \qquad (7.2)$$

where b is the blade axial chord (figure 7.2). This pressure can be converted to a lift coefficient by dividing by the quantity $\rho_{0,ex}W_{ex}^2/2g_c$, which in low-speed or incompressible flow is the relative outlet dynamic head.

Tangential lift coefficient

$$C_L = \frac{\rho_{0,ex}s\,W_{ex}^2\cos\alpha_{ex}[(W_{in}/W_{ex})\sin\alpha_{in} - \sin\alpha_{ex}]}{g_c b_x \rho_{0,ex}W_{ex}^2/2g_c}$$

$$= 2\left(\frac{s}{b_x}\right)\cos^2\alpha_{ex}\left[\left(\frac{W_{in}\sin\alpha_{in}}{W_{ex}\cos\alpha_{ex}}\right) - \tan\alpha_{ex}\right] \qquad (7.3)$$

If the axial velocity, C_x, is constant through the blade row (a usual design condition),

$$C_x = W_{in}\cos\alpha_{in} = W_{ex}\cos\alpha_{ex}$$

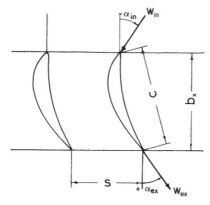

The blade height or length, h, is into the paper

Figure 7.2. Diagram for calculating the tangential lift on a cascade of blades

Therefore,

$$C_L = 2 \left(\frac{s}{b_x} \right) \cos^2 \alpha_{ex} \, (\tan \alpha_{in} - \tan \alpha_{ex}) \qquad (7.4)$$

[The use of the "incompressible" dynamic head in the derivation of this lift coefficient does not preclude it from being used for compressible flow. The correlation based on this defined lift coefficient has been found to be a useful guide for compressible and incompressible flow. See chapter 4, equations 4.4 through 4.8, for a similar argument for an incompressible diffusion coefficient].

If the optimum lift coefficient is $C_{L,op}$, then the optimum "axial solidity", $(b_x/s)_{op}$, is

$$\left(\frac{b_x}{s} \right)_{op} = \left| \frac{2}{C_{L,op}} \cos^2 \alpha_{ex} \, (\tan \alpha_{in} - \tan \alpha_{ex}) \right| \qquad (7.5)$$

The vertical bounding lines indicate that the magnitude of this quantity should be used. Axial solidity (b_x/s) is smaller than the solidity σ, defined later by $\sigma \equiv c/s$, where c is the chord length.

Number of blades

Up to this point the number of blades is open to the designer to choose. Axial turbines have been built with:

1 to 24 blades for wind turbines,
3 to 30 blades for water turbines, and
11 to 110 blades for gas (including steam) turbines.

Gas-turbine rotors would normally have an even number of blades for balancing reasons, and stators would usually have an odd (preferably prime) number of blades. These numbers normally change from stage to stage to provide randomness in the excitation between blade rows. The number of blades in any row does not have to be chosen at this phase, although the designer usually has a desirable range in mind. Lightweight aero engines employ a large number of rotor blades to keep the engine length short (axial chord, b, decreases as the number of blades is increased, also decreasing the spacing, s). The number of stator vanes (nozzles) in aircraft engines is often considerably smaller than the number of rotor blades — the added length does not incur much weight penalty, because the vanes are usually hollow, and do not require a massive disk to carry them. Stationary industrial machines often have a small number of large blades, because manufacturing costs are reduced, relative profile accuracies will be increased, blade boundary layers will be relatively smaller (higher Reynolds numbers), and, if used, internal blade cooling may be easier to incorporate. The normal procedure is to make an initial choice of number of blades at this point and to change it later, if desired, from consideration of calculated aspect-ratio losses or of vibration frequencies.

Induced incidence

The circulation around the blades that enables them to give lift to the flow also causes the incoming streamlines to turn as they approach the blade row, figure 7.3. Dunavant and Erwin (1956) correlated the induced incidence, $\Delta \theta_{id}$ over a range of conditions, as

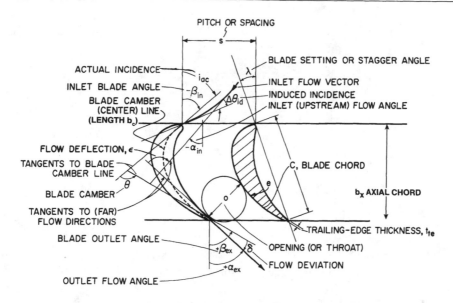

Figure 7.3. Cascade notation for turbine-blade rows

shown in figure 7.4a. A reasonable correlation of these data is given by

$$\Delta\theta_{id} = 14°\left(1 - \frac{\alpha_{in}}{70°}\right) + 9°\left(1.8 - \frac{c}{s}\right) \tag{7.6}$$

The validity of this expression outside the range of α_{in} between 0° and 70° is uncertain. This uncertainty is not of great significance. Turbine blades are often designed without the induced incidence being incorporated. This results in the actual incidence, i_{ac}, being somewhat greater than that calculated, and in the blade camber, θ (figure 7.3), being lower than it would be were the induced incidence taken into account. This neglect of the induced incidence is sometimes not so much optional as necessary, particularly for low-reaction rotor blading with little overall acceleration through the blade row. In this case the addition of camber in the leading-edge region can produce an upstream throat, which is obviously undesirable and is certainly worse than some positive incidence.

The specification of the camber line is more critical to (relatively thinner) compressor blades than to (relatively thicker) turbine blades. As will be seen, initial turbine-blade shapes, which may be obtained using the material in sections 7.2 and 7.3, should be further refined by methods such as those presented in section 7.5, which typically pay more attention to blade-surface shapes (and resultant surface Mach-number and pressure distributions) than to camber line and induced incidence.

The nominal blade inlet angle, β_{in}, from which the actual incidence, i_{ac}, is measured in figure 7.3 is the tangent to the blade centerline 5 percent of the chordal distance from the leading edge. Similarly, the blade outlet angle, β_{ex}, is the tangent to the blade

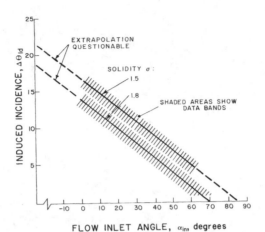

(a) Induced incidence angle. From Dunavant and Erwin (1956)

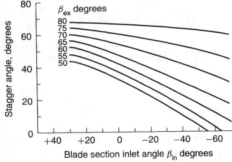

(b) Stagger angle versus inlet and outlet angles. From Kacker and Okapuu (1982)

Figure 7.4. Data for low-speed turbine-blade rows

centerline at 95 percent of the chordal distance from the leading edge. These definitions are used for all NACA/NASA correlations.

The upstream and actual fluid and blade inlet angles, α_{in}, $\alpha_{in,ac}$, and β_{in} and the upstream, actual and induced incidences, i, i_{ac}, and $\Delta\theta_{id}$ are related by

$$i_{ac} = i + \Delta\theta_{id}$$

$$\alpha_{in,ac} = \alpha_{in} + \Delta\theta_{id}$$

$$\beta_{in} + i = \alpha_{in}$$

Therefore,

$$\beta_{in} + i_{ac} = \alpha_{in} + \Delta\theta_{id} \tag{7.7}$$

Deviation

The same circulation around an aerodynamically loaded blade causes the leaving-flow direction to deviate from that of the blade trailing edge, as shown in figure 7.3. However, while the induced incidence can be considered or ignored without significant effects on blade performance, the deviation, designated by δ, is of critical importance to the turbine designer. If ignored, the turbine blade will produce a lower change of tangential velocity, and therefore a lower torque, work output, and enthalpy drop, than the designer would predict based on velocity diagrams using the blade-outlet angle to indicate flow direction. Conversely, an assumption of a large deviation would result in a turbine that, in practice, would give more specific work output (work per unit mass) than it is designed for, but would pass a reduced mass flow if the throat area were thereby reduced.

In computer-based blade-design methods the inputs can be blade shapes and inlet and outlet flow angles, but then the resultant exit Mach number (and thus mass flow rate) is an output. Alternatively one can input inlet flow angle and outlet Mach number, but then the outlet flow angle becomes an output. In either case the lack of foreknowledge of deviation affects turbine specific work.

In view of the importance of being able to predict the deviation accurately, it is disappointing to have to report that there is a strong lack of agreement among the several correlations that have been published for turbine-blade deviation. This uncertainty is probably as responsible as any other factor for the mismatching between compressor and turbine that has been a disappointing feature of many gas-turbine engines. The cost of re-blading one component when mismatching has been found has often seemed too large, particularly for industrial engines. The engine efficiency has suffered, often to a considerable extent, because one or both of the major components, even though intrinsically capable of high-efficiency operation, has been forced to work in a low-efficiency area of its characteristic.

Among the alternative correlations for deviation that have been published, we have chosen that due to Ainley and Mathieson (1951) for two reasons. First, in the necessarily limited experience of the authors and of those of their associates with turbine design and test information, it has seemed to provide the most accurate and reasonable of the alternatives. Second, the correlation is based on the blade opening or throat, whereas some of the alternatives use other measures of the cascade configuration. The blade throat (especially the throat of the first-stage nozzle vanes) controls the mass flow passed by a turbine, and is a very important specification in the manufacturing drawings. It is therefore advantageous to bring out this quantity directly in the design process, rather than to infer it indirectly subsequently.

The Ainley and Mathieson correlation is normally given graphically, with corrections for high subsonic Mach numbers and for the degree of curvature on the turbine blade downstream of the throat (the radius e in figure 7.3). We have translated these into three analytic expressions, for different Mach-number conditions. The correlations are for flow outlet angle rather than deviation.

1. Predicted outlet angle for incompressible-fluid turbines, and for compressible-fluid turbines with throat Mach numbers up to 0.5:

$$|\alpha_{ex}|_{0<M_t<0.5} = \left[\frac{7}{6}\left\{\left|\cos^{-1}\left(\frac{o}{s}\right)\right| - 10°\right\} + 4°\left(\frac{s}{e}\right)\right] \qquad (7.8)$$

where

$$o \equiv \text{diameter of throat opening (figure 7.3)},$$
$$s \equiv \text{blade pitch},$$
$$e \equiv \text{radius of curvature of blade convex surface downstream of throat},$$
$$M_t \equiv \text{Mach number at throat}.$$

2. Predicted outlet flow angle for turbines with sonic flow at the throat ($M_t = 1.0$):

$$|\alpha_{ex}|_{M_t=1.0} = \left|\cos^{-1}\left(\frac{o}{s}\right)\right| - \sin^{-1}\left(\frac{o}{s}\right)\left|\left(\frac{s}{e}\right)^{(1.786 + 4.128(s/e))}\right| \qquad (7.9)$$

3. Predicted outlet flow angle for throat Mach number between 0.5 and 1.0:

$$|\alpha_{ex}|_{0.5<M_t<1.0} = |\alpha_{ex}|_{0<M_t<0.5} - (2M_t - 1)\left(|\alpha_{ex}|_{0<M_t<0.5} - |\alpha_{ex}|_{M_t=1.0}\right) \qquad (7.10)$$

Selection of surface curvature downstream of the throat

Early turbine blades were usually designed to be plane downstream of the throat, partly because of the limitations of the milling methods normally used for blade manufacture, and partly because designers felt that the flow would not be turned after the throat. Now, with improved casting methods, numerically controlled machining, and improved blade-design methods, these limitations do not exist, and blades are curved downstream of the throat.

The velocity of the core of the flow (mid-passage) will accelerate monotonically from inlet to outlet from W_{in} to W_{ex} (figures 7.2 and 7.3, and figures 7.11 to 7.20). The velocity of the flow at the edge of the boundary layer along the blade convex (suction) surface reaches a maximum somewhere near the throat, and from there it must decelerate to the velocity at the edge of the boundary layer near the trailing edge (since the flow is viscous, there will be lower-velocity wakes near the trailing edge). The region of the suction surface between the throat and the trailing edge is called the "region of unguided diffusion". In subsonic turbines this diffusion is accomplished by continuing to curve the suction blade surface from the throat to the trailing edge (for instance, with average curvature e). This curvature is reduced (blade surface more straight) for blades with higher exit Mach numbers, and for some supersonic turbines the blade suction surface may slightly inflex. (The curvature will change sign from the throat to the trailing edge, the flow condition resembling that of supersonic diffusers contracting downstream. The blade-surface inflection is more noticeable in supersonic compressor cascades.)

For preliminary design, ratios of the blade spacing, s, to the surface curvature, e, of up to 0.75 are permissible. A range that appears to give favorably shaped blades is $0.25 < (s/e) < 0.625$.

Blade leading and trailing edges

For the purposes of obtaining an initial blade shape, the turbine-blade leading-edge radius is often specified to be a given proportion of either the blade chord, c, or the spacing, s. Typical values usually fall between $0.05s$ and $0.10s$. In addition, specifying the angle between the tangents made by the two blade surfaces (upper and lower, or convex and concave) to the leading-edge circle (say, 20 degrees) can help to define an acceptable blade or passage shape.

The trailing edge is sometimes specified as having a radius, but it is more usual to specify a trailing-edge thickness, t_{te}, along the blade-row-exit plane with the blades being cut off sharply. Typical values are between $0.015c$ and $0.05c$.

Choice of blade setting or stagger angle

One variable remains to define the "framework" of the blade-row configuration at any one radial station after the axial solidity, (b_x/s), the induced incidence, $\Delta\theta_{id}$, the blade curvature, e, and outlet angle β_{ex}, have been found. This variable is the setting or stagger angle λ (figure 7.3). It is related to the axial chord, b_x, and the blade chord, c, by

$$\cos\lambda = \frac{b_x}{c} \qquad (7.11)$$

The setting angle determines, to a large extent, the overall blade shape. More important, for turbines it determines the passage shape. One does not have the freedom to choose the setting angle arbitrarily. We are restricting our attention to subsonic turbines, and in these the throat (the narrowest part of the passage) must occur at the trailing edge. Nonetheless, it is possible to arrange that the throat occur at the trailing edge over only a small range ($10°$ to $30°$) of setting angles. Finding which part of the range will be best still appears to be more of an art than a science, even when sophisticated computer programs are used to judge the "quality" of a blade passage (suggestions suitable for preliminary-design efforts are included in the subsection "stagger-angle effects" in section 7.5, and figures 7.17 to 7.20). In other words, even computer programs work on a trial-and-error basis, with each new trial being checked for quality by approximation, rather than an optimum setting angle being found by precise computation. A guide for obtaining the setting angles of low-speed turbines has been published by Kacker and Okapuu (1982); their data are summarized in figure 7.4b.

In a computer program to design turbine blades, the criterion of quality would be some function of the calculated stability, and the stagnation losses, of the boundary layers. The type of preliminary design with which we are concerned here has to be an approximation of this criterion, in qualitative rather than quantitative terms. We look for at least a throat at the trailing edge, for a steady reduction in area along the passage, and for gradual curvatures of the passage walls. Finally there is the undefinable prerogative of the designer: to ask the question, "Does it look right?" An example, given after the following formal restatement of the quantitative component of turbine-blade design, should make the procedure clearer.

Blade-row design procedure

The necessary preliminary work for a blade-row design will include the selection of the velocity diagram for the radius in question, from which the fluid angles will be obtained. The throat Mach number, which for subsonic blade rows can be approximated to be the blade-exit Mach number, can also be found for compressible-fluid turbines. The tangential loading coefficient will also be a design input. The optimum axial solidity can be found from equation 7.5.

The procedure is, then, as follows.

1. Choose one or more values of the spacing/curvature ratio, (s/e), and of the leading-edge radius, r_{le} and trailing-edge thickness, t_{te}.

2. Calculate the throat/spacing ratio (o/s), from equations 7.8, 7.9, or 7.10, for the appropriate Mach number.

3. Draw the throat-to-trailing-edge regions for the values found. Choose one or more for further exploration.

4. Choose several values of the setting angle, λ, guided by the values suggested by figure 7.4b. Draw in the leading edges using the previously chosen leading-edge radius and a blade inlet angle chosen in the range between α_{in} and $(\alpha_{in} + \Delta\theta_{id})$, where α_{in} is the fluid inlet angle and $\Delta\theta_{id}$ is the induced incidence (equation 7.5 and figure 7.4a).

5. Connect the leading-edges to the throat sections with smooth curves. Draw circles in the passages so formed to touch the convex and concave blade surfaces. Acceptable blade passages are those giving continuous reductions of area (or inscribed-circle diameter) up to the throat.

   In preliminary design (which for most turbines built in the past was the final design) the selection of the "best" blade-row design from those produced by this procedure will be largely aesthetic. There may, however, be some constraints favoring one configuration rather than another when the blade-row designs chosen for other radial sections are assembled into a complete blade. For turbines where the need to attain the ultimate in performance is critical, these procedures can produce the input for a computer program that will adjust the blade shape until a favorable, or even a prescribed, velocity distribution around the blade surface is produced.

Example 1 **Blade-section preliminary design**

Design and draw at least two neighboring blade sections for a turbine rotor-blade row at a radial station where the velocity diagram is 50-percent reaction, with a work coefficient of 2.0 and a flow coefficient of 0.55. The throat Mach number is 0.75.

Use a leading-edge diameter of one-tenth of the spacing. The trailing-edge thickness should be 0.02c. The ratio of the spacing to the surface curvature downstream of the throat may be chosen at 0.333.

Solution.

We solve the velocity diagram to find the flow angles (figure 7.5). These will allow the optimum solidity to be obtained. Then we make a table in which we try various values of the setting angle. We draw blade and passage shapes for some of the setting angles and choose that set giving the best passage shape.

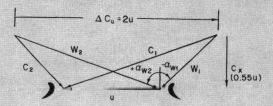

Figure 7.5. Velocity diagram in turbine-blade preliminary design

Velocity diagram

$$\tan\alpha_{W1} = \frac{-0.5u}{0.55u} \qquad \alpha_{in} \equiv \alpha_{W1} = -42.27°$$

$$\tan\alpha_{W2} = \frac{1.5u}{0.55u} \qquad \alpha_{ex} \equiv \alpha_{W2} = 69.86°$$

Optimum solidity

We will use equation 7.5 with the optimum lift coefficient set at $C_L = 1.0$:

$$\left(\frac{b_x}{s}\right)_{op} = |2.0\cos^2\alpha_{ex}(\tan\alpha_{in} - \tan\alpha_{ex})| = 0.862$$

Throat opening

With the throat Mach number of 0.75 the outlet angle will be the mean of the values given by equations 7.8 and 7.9. We interpolate to find the value of o/s that will give an outlet angle of 69.86°:

Interpolating, we find we need $\cos^{-1}(o/s) = 69.71°$ to obtain the outlet angle desired, 69.86°. Therefore

$$\frac{o}{s} = 0.3468 \quad \text{or} \quad o = 0.3468s$$

Predicted outlet flow angle, α_{ex}		
$\cos^{-1}(o/s)$	69°	70°
$\alpha_{ex,M_t=0.5}$	70.167°	71.333°	(equation 7.8)
$\alpha_{ex,M_t=1.0}$	68.971°	69.971°	(equation 7.9)
$\alpha_{ex,M_t=0.75}$	69.569°	70.652°	(by interpolation)

It is convenient to give all the geometrical parameters in terms of the spacing, s, which will be known from the specifications of the turbine (mass flow, fluid properties, velocity diagram, hub-shroud diameter ratio, and the number of blades). If we use the value of $(b_x/s)_{op}$ derived earlier,

$$b_x = 0.862s$$

$$c = \left(\frac{0.862}{\cos\lambda}\right)s$$

Before the trailing-edge section downstream of the throat can be drawn, an estimate must be made of the intercept, Δs_{te}, of the trailing edge on the outlet plane. We can start by assuming that the trailing-edge slope is $\cos^{-1}(o/s)$ so that in the present case,

$$\frac{\Delta s_{te}}{s} = \frac{0.02(b_x/s)}{(o/s)\cos\lambda} = \frac{0.05}{\cos\lambda}$$

where λ is the setting angle. A first guess at the setting or stagger angle can be made from figure 7.4b to be 40°.

We can then begin to lay out the blade section on a drawing board, perhaps ten times full size for accuracy. Along the trailing-edge plane we mark a blade pitch, s, and the trailing-edge intercept. Then, as shown in figure 7.6, we find the center of curvature of the convex surface downstream of the throat by striking radii of e and $(e + o)$ appropriately. We can draw in this part of the blade surface, and it is useful also to draw in the throat circle.

This first iteration of the blade surface downstream of the throat will be common to all subsequent trials of various setting angles. We have chosen a setting angle of 40 degrees to illustrate the remainder of the procedure. We therefore draw the chord line from the trailing-edge point on the convex surface (the point that defines one side of the throat), and the leading-edge line using the axial chord, b_x. The leading-edge circle is drawn tangent to the chord and leading-edge lines. The initial directions of the concave and convex surfaces were specified to have an included angle of 30 degrees, and these will be equally disposed around the blade angle. For zero incidence this will be given by the flow angle plus the induced incidence, which is found using equation 7.6. A list can be made up such as the following.

1. Setting angle, deg.   40°
2. $b_x/c = \cos\lambda$ 0.766
3. Solidity $c/s = 0.862/(b/c)$ 1.125
4. $\Delta\theta_{id} = [5.546 + 9(1.8 - c/s)]$ deg.   11.62°
5. Flow inlet angle α_{in} deg.   −42.27°
6. Flow outlet angle α_{ex} deg.   +69.86°

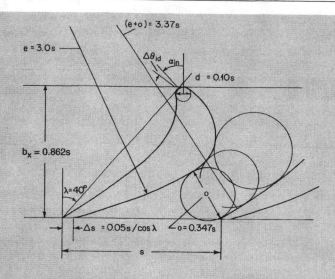

Figure 7.6. Construction of turbine-blade profiles

The remainder of the design procedure is non-quantitative and iterative. The leading-edge tangents are joined to the throat surfaces by smooth curves, perhaps using French curves. Initially, we draw just the surfaces bounding a passage, rather than the two surfaces of a single blade. Then a series of circles of increasing size touching both sides of the flow passage is drawn, starting at the throat circle and proceeding upstream, to check that the passage does indeed converge steadily to the throat. Vectors showing the range of probable design-point flow angles at inlet α_{in} to $(\alpha_{in} + \Delta\theta_{id})$, are drawn to check, by eye, the leading-edge shape. Finally, one blade surface is redrawn to produce a complete blade profile. The trailing-edge thickness is checked and the blade and passages are redrawn if necessary.

The whole procedure may be repeated for other blade setting angles, as was done in the present case, for $\lambda = 30°$, $45°$, and $50°$. The choice of the best passage and blade shapes from among these alternatives is made purely by the designer's judgment, in this preliminary-design stage. A later stage of refinement would be to examine the velocity distributions in a flow-analysis computer program and, if necessary, to modify the passage shape to produce minimum-loss flow.

7.3   More-detailed design sequence emphasizing aircraft engines

Selection of number of stages

For high-pressure stages in aircraft engines, the choice is most often between one stage and two, although there are exceptions. With present construction methods involving metal blades cooled by compressor air it is desirable to use one stage if possible, to reduce engine cost and size and the diversion of costly compressor air. If uncooled non-metallic blading becomes viable, the efficiency advantages of using two or more stages may override the size advantages of a single stage, particularly for long-range aircraft and for land and sea applications.

In order to force the choice towards a single stage, a high mean-diameter loading coefficient (e.g., 1.8) may be used in the high-pressure turbine. It is permissible to design on the mean-diameter rather than the hub loading in the high-pressure turbine because the hub-shroud diameter ratio is high and therefore the hub loading will not be extreme. The stagnation-enthalpy drop should be given in the specifications, or can be calculated from the pressure ratio with the assumption of a stage or turbine polytropic efficiency. A first guess could be 0.9, so that the approximate turbine-outlet stagnation temperature is found from equation 2.79.

$$\frac{T_{0,ex}}{T_{0,in}} = \left(\frac{p_{0,ex}}{p_{0,in}}\right)^{\left[\left(\frac{R}{C_p}\right)\eta_{p,c}\right]}$$

The enthalpy drop from inlet to outlet is then calculated from:

$$\Delta_{ex}^{in} h_0 = \Delta h_0 = C_p(T_{0,in} - T_{0,ex})$$

and the required mean-diameter blade speed from:

$$u_m = \sqrt{\Delta h_0 / \psi_m}$$

A guideline may therefore be summarized as follows: use a single-stage high-pressure turbine if possible, by specifying a mean-diameter loading coefficient of up to 1.8. If the turbine overall pressure ratio is above, say, 8:1, the limiting loading coefficient in the low-pressure stages should be that at the hub, where very high values, eg 4.0 and even 5.0, must occasionally be used. The required reduction in mean-diameter loading for a given value of hub loading is given in figure 7.7 as a function of hub-shroud diameter ratio. The relation between hub-shroud ratio and pressure ratio will be shown later.

Choice of velocity diagrams

The mean-diameter velocity diagram is normally chosen on the assumption that the stream surface has a constant radius through the stage, so that the blade peripheral speed, u, is the same at rotor inlet as at rotor outlet, and so that the axial component of velocity, C_x, is constant through the stage. This special form was referred to in chapter 5 as a "simple diagram", figure 7.8a, whereas a general diagram having varying u and C_x is shown in figure 7.8b.

Guidance for the initial choice of optimum flow coefficient, or stator-outlet flow angle, for chosen values of loading coefficient and reaction is given in figure 7.9a-f. Each plot is for a specified reaction, from zero (impulse) to 0.9. Three sets of curves of isentropic efficiencies are given on each plot. One set (full lines) is of stagnation-to-stagnation efficiency across the blading of a single stage. It is appropriate to use these curves for all except the last stage of multistage turbines. A second set of curves (shown dotted) is for stagnation-to-static efficiencies across the blading. This set of curves is appropriate

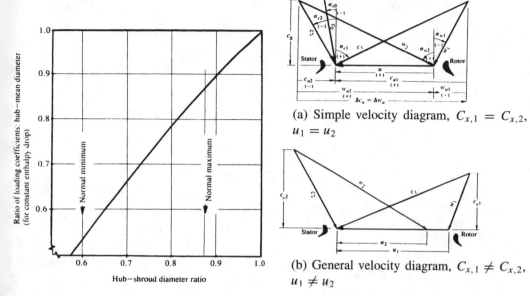

(a) Simple velocity diagram, $C_{x,1} = C_{x,2}$, $u_1 = u_2$

(b) General velocity diagram, $C_{x,1} \neq C_{x,2}$, $u_1 \neq u_2$

Figure 7.7. Reduction of mean-diameter loading for long blades. From Wilson (1987)

Figure 7.8. Simple and general velocity diagrams for axial-flow turbine stage

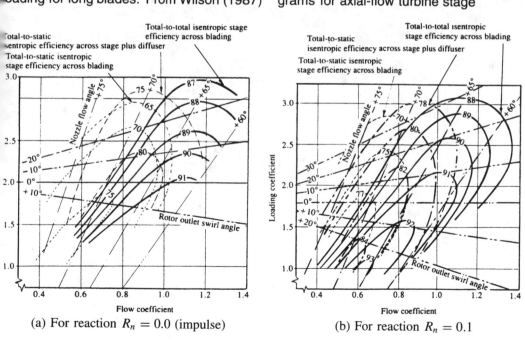

(a) For reaction $R_n = 0.0$ (impulse)

(b) For reaction $R_n = 0.1$

Figure 7.9. Axial-flow-turbine stage efficiency vs. loading and flow coefficients, for various reactions. From Wilson (1987). Figs. 7.7–7.10 Courtesy I. Mech. E.

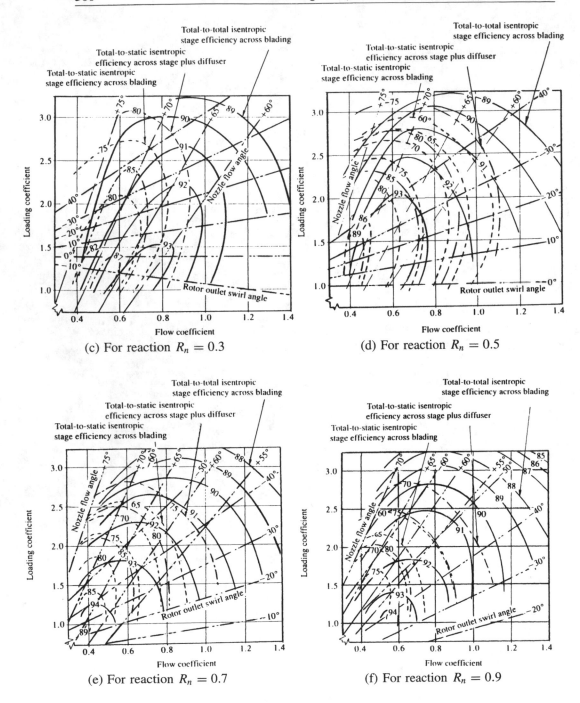

(c) For reaction $R_n = 0.3$

(d) For reaction $R_n = 0.5$

(e) For reaction $R_n = 0.7$

(f) For reaction $R_n = 0.9$

Figure 7.9. Continued

for single-stage turbines having no diffuser. A third set of curves (chain-dotted lines) is approximately the stagnation-to-static efficiency that would be obtained if a single-stage turbine were followed by a diffuser producing fifty-percent pressure recovery.

These plots were calculated by the methods discussed later. In all cases the rotor and stator aspect ratio (blade span/mean-diameter chord) was 4, the rotor tip clearance was one percent of the passage height, the trailing-edge thickness was 2.5 percent of the blade chord, and the Reynolds number was taken as 100,000.

Cross-correlations of the plots are shown in figure 7.10a-c. In figure 7.10a the peak isentropic stage efficiency is plotted against stage loading, with reaction as a parameter, for the different definitions of efficiency. In all cases the efficiency drops as loading is increased, and in all cases the fifty-percent-reaction diagrams have the highest indicated efficiency. The drop in stagnation-to-static efficiency with loading is particular strong for higher reactions.

The effects of loading on rotor-outlet flow swirl and optimum nozzle angle are shown in figures 7.10b and 7.10c.

The curves of figures 7.9 and 7.10 should be used in the initial-design phase when the number of stages and therefore the stage loading has been chosen. With these, initial choices can be made of mean-diameter reaction and flow coefficients. The predicted efficiencies will be changed at a later phase in the design process as more realistic values of Reynolds number, aspect ratio, tip clearance and other parameters are substituted for the typical values used for the plots. However, these predictions are probably close

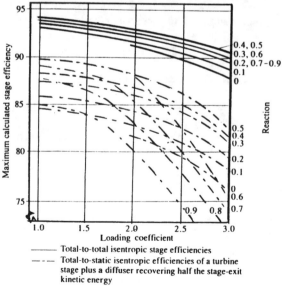

(a) Maximum stage isentropic efficiencies vs loading coefficient and reaction

Figure 7.10. Optimum characteristics of axial-flow-turbine velocity diagrams. From Wilson (1987), by permission Inst. Mech. Engrs.

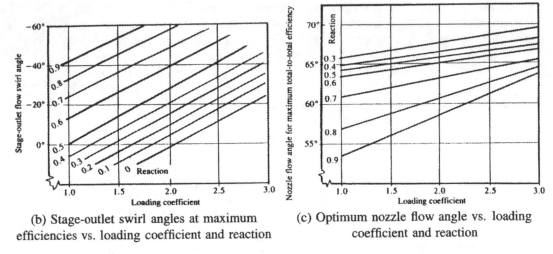

(b) Stage-outlet swirl angles at maximum efficiencies vs. loading coefficient and reaction

(c) Optimum nozzle flow angle vs. loading coefficient and reaction

Figure 7.10. Continued

enough in most cases for an iteration in the initial choices of stage loading and reaction to be made, if the designer judges that the optimum has been missed.

The preliminary-design guideline that emerges from these charts is, then: Choose the number of blades between 12 for small turbines and 120 for large units. The radial clearance between the rotor-blade tips and the shroud must also be specified. A value that is claimed by manufacturers of turbines of a wide range of sizes is 0.25 mm, or 0.010 inches, and this value can be used in the absence of better information.

7.4 Blade-surface curvature-distribution effects

In the earlier days of steam and gas turbines some turbine blades were designed by arranging a thickness distribution around a camber line. As tangential-loading coefficient increased and solidity decreased it gradually became harder to ensure a continuous decrease in area from inlet to outlet of subsonic turbine blades designed with thickness distributions. From the 1950s on most turbine blades have been designed by specifying the shape of the suction (more curved) and pressure (less curved) surfaces of the blades, thus specifying the passage defined by the blades rather than the blades themselves.

Figure 7.11 illustrates the computed performance of a typical two-dimensional section designated B4-0 (Korakianitis, 1993b, 1993d). Figures 7.11a and 7.11b show the velocity distribution in terms of Mach-number surface distribution and passage contours (some investigators prefer to show pressure distributions, which look the inverse of velocity distributions). The surface Mach number corresponds to the flow velocity at the edge of the boundary layer. As expected, the Mach numbers are higher on the more-curved suction surface, and lower on the less-curved pressure surface. The area prescribed by the surface-Mach-number vs. axial-distance figure is a measure of the tangential loading of the blade, i.e., blades with higher loadings will have higher velocity differences between

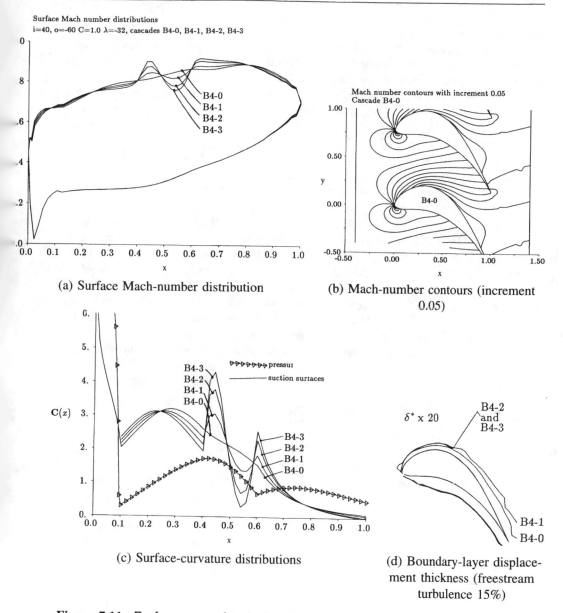

(a) Surface Mach-number distribution

(b) Mach-number contours (increment 0.05)

(c) Surface-curvature distributions

(d) Boundary-layer displacement thickness (freestream turbulence 15%)

Figure 7.11. Performance of typical turbine cascade with $\alpha_{in} = 40°$, $\alpha_{ex} = -60°$, $\lambda = -32°$, and $C_L = 1.0$. From Korakianitis (1993b, 1993d)

the suction and pressure surfaces. The stagnation point near the leading edge is typically a little distance downstream on the pressure surface from the geometric leading edge because of induced incidence (see section 7.2). The Mach-number contours indicate that in this case the maximum velocity on the suction surface occurs a little downstream of the throat. High-efficiency turbine geometries would exhibit continuously accelerating flow on the suction surface up to the maximum velocity near the throat, and then minimum deceleration (minimum diffusion) from there to the trailing edge. This is a measure of diffusion in the "region of unguided diffusion", and determines the thickness of the boundary layer at the trailing edge, and consequently the blade efficiency. On the pressure surface one would prefer a continously accelerating flow from the leading to the trailing edge. This is normally controlled by the shape of the pressure surface near the leading edge. This condition is easier to achieve at design point and in blades with lower tangential loading, and much harder to achieve off-design in highly-loaded blades. (In high-temperature cooled blades, which also tend to be of higher loading for reasons described above, locally high velocities should be avoided near the leading edge for heat-transfer considerations.)

Figures 7.11c and 7.11d show the surface-curvature distribution of four blades designated B4-0 to B4-3, and the corresponding computed boundary-layer displacement thickness. The Mach-number performance of the four blades is also shown in figure 7.11a. All four blades are plotted in figure 7.11d, but the differences in blade surfaces caused by the curvature differences shown in figure 7.11c are too small to be seen in figure 7.11d. Nevertheless, the curvature differences are "mirrored" in the surface-Mach-number plots of figure 7.11a. Figure 7.11d indicates that the boundary layers of the two blades with the worst slope-of-curvature discontinuities separate, a condition expected to lead to losses even if the boundary layer re-attaches. These surface-curvature effects are explained below.

Direct and inverse blade-design methods

Various investigators use different definitions for the terms direct method and inverse, semi-inverse, full-inverse or full-optimization methods (Meauze, 1989), analysis and design modes (Stow, 1989), optimization and design methods (Bry, 1989), and others, for the approaches and methods used in turbine design sequences. For the purposes of this book we *define* as direct the method in which the designer inputs the geometry of the cascade and the output is the performance (from an analysis code) in terms of surface Mach-number and pressure distributions. The performance provides guidelines for where to increase or decrease the loading, and how to modify the surface geometry in successive iterations. This requires iterative use of a geometric description package and of an analysis code until a desirable performance is obtained from the analysis code. Also for the purposes of this book we *define* as inverse the various methods in which the designer specifies the performance of the cascade to obtain the geometry, or specifies modifications to a portion of the surface velocity or pressure distribution to obtain modifications to the geometry. This latter definition includes what other investigators define as fully-inverse, semi-inverse, or simply design methods.

Both direct and inverse methods are based on the assumption of steady flow conditions, and require numerous iterations until a desirable blade shape and performance are obtained. The direct method is laborious, and it requires considerable insight. On the other hand, the designer has direct control of the various geometric parameters and structurally- or dynamically-infeasible blade shapes are excluded before they are analyzed. One would like to start the inverse methods with a blade shape close to an acceptable design, since this will significantly reduce the number of inverse-design iterations. Therefore the designer needs to start inverse-design iterations from a good blade shape obtained by a direct method. The inverse method is less laborious, but poses mathematical problems for handling the trailing-edge thickness, and many (although not all) inverse-design programs must start from a blade geometry generated by a direct method or from the preliminary-design method of section 7.2. The designer has less control on the blade shape (by the very purpose of inverse design). It can result in blades that are structurally undesirable due to stress considerations, location of cooling passages, or impossible to manufacture. If one decides that some geometric parameters (such as stagger angle) must be changed, one must start from the beginning (direct method) again. The final design is usually obtained by a judicious combination of both methods.

Surface-curvature and surface-velocity distributions

Both the curvature and the slope of curvature of blade surfaces affect boundary-layer development and blade performance. The slope-of-curvature effects are less obvious than the curvature effects. Theoretical and experimental evidence of their importance is presented below.

Theoretical evidence

The flow in turbomachinery *curves* around the blades. The dependence of flow on curvature is seen if one writes the compressible-flow Navier-Stokes equations in the limiting case of cylindrical coordinates (r, θ, z). In particular since we are interested in "stacking" two-dimensional solutions and geometries we present here the equations of conservation of mass and momenta in polar coordinates (r, θ), where r is the local radius and θ is the local angle of the point we are considering the flow from an origin, taken from the text by Bird et al. (1960).[1]

$$\frac{\partial \rho}{\partial t} + \frac{1}{r}\frac{\partial}{\partial r}(\rho r C_r) + \frac{1}{r}\frac{\partial}{\partial \theta}(\rho C_\theta) = 0 \qquad (7.12)$$

$$\rho\left(\frac{\partial C_r}{\partial t} + C_r\frac{\partial C_r}{\partial r} + \frac{C_\theta}{r}\frac{\partial C_r}{\partial \theta} - \frac{C_\theta^2}{r}\right) =$$
$$-\frac{\partial p}{\partial r} - \left(\frac{1}{r}\frac{\partial}{\partial r}(r\tau_{rr}) + \frac{1}{r}\frac{\partial \tau_{r\theta}}{\partial \theta} - \frac{\tau_{\theta\theta}}{r}\right) + \rho g_r \qquad (7.13)$$

$$\rho\left(\frac{\partial C_\theta}{\partial t} + C_r\frac{\partial C_\theta}{\partial r} + \frac{C_\theta}{r}\frac{\partial C_\theta}{\partial \theta} + \frac{C_r C_\theta}{r}\right) =$$
$$-\frac{1}{r}\frac{\partial p}{\partial \theta} - \left(\frac{1}{r^2}\frac{\partial}{\partial r}(r^2\tau_{r\theta}) + \frac{1}{r}\frac{\partial \tau_{\theta\theta}}{\partial \theta}\right) + \rho g_\theta \qquad (7.14)$$

---

[1]The nomenclature of this and the following two sections (7.5 and 7.6) contains special symbols defined here rather than in the general nomenclature.

where in the above equations p is the static pressure, and for Newtonian fluids the stress components τ_{ij} are related to the dynamic viscosity μ by the equations:

$$\tau_{rr} = -\mu\left[2\frac{\partial C_r}{\partial r} - \frac{2}{3}(\nabla \cdot \vec{C})\right]$$

$$\tau_{\theta\theta} = -\mu\left[2\left(\frac{1}{r}\frac{\partial C_u}{\partial \theta} + \frac{C_r}{r}\right) - \frac{2}{3}(\nabla \cdot \vec{C})\right] \qquad (7.15)$$

$$\tau_{r\theta} = \tau_{\theta r} = -\mu\left[r\frac{\partial}{\partial r}\left(\frac{C_u}{r}\right) + \frac{1}{r}\frac{\partial C_r}{\partial \theta}\right]$$

$$\left(\nabla \cdot \vec{C}\right) = \frac{1}{r}\frac{\partial}{\partial r}(rC_r) + \frac{1}{r}\frac{\partial C_u}{\partial \theta}$$

A similar dependence of incompressible flow on curvature is also shown in most fluid dynamics texts in cylindrical coordinates, and Schlichting (1960, page 112) attributes a set of similar equations for incompressible flow in curvilinear coordinates (x, r, y) to W. Tollmien. The $1/r$ and $1/r^2$ terms in these equations suggest a strong dependence of local pressure and velocity on curvature. Turbine-blade performance depends on the behavior of the boundary layer. While one expects that the boundary layer shields the core of the flow from surface discontinuities, the boundary-layer behavior depends on the slope of the velocity distribution as it approaches the solid wall, and equations 7.15 indicate that the boundary layer and its behavior are affected by the local radius. Smooth velocity distributions along the blade surfaces require smooth surface-curvature distributions (continuous slopes of velocity and curvature along the blade surface). Continuous slope of curvature requires continuous third derivatives at the knots of the parametric splines or polynomials used to describe the blade surfaces (i.e., fourth-order splines). This is illustrated in the following two equations for curvature and slope of curvature for the function prescribing the (x, y) locations of the points on the blade surface.

$$\frac{1}{r} = \frac{(dy/dx)}{\left[1 + (dy/dx)^2\right]^{(3/2)}} \qquad (7.16)$$

$$\frac{d}{dx}\left(\frac{1}{r}\right) = \frac{(d^3y/dx^3)\left[1 + (dy/dx)^2\right] - 3(dy/dx)(d^2y/dx^2)^2}{\left[1 + (dy/dx)^2\right]^{(5/2)}} \qquad (7.17)$$

Experimental evidence

Most parametric splines currently in use for direct blade design (e.g., cubic, B-splines, Bezier splines, etc.) have continuous first and second derivatives (third-order splines) and they result in smooth-looking surfaces with continuous curvatures, but discontinuous slopes of curvature at the spline knots (junctions). One must distinguish here between the small local slope-of-curvature discontinuities from surface roughness and fouling, with which turbines must operate, and the slope-of-curvature discontinuities in the as-designed shape at the junctions of the splines. As shown in figure 7.11 the latter are invisible to the eye, but they may produce unusually-loaded cascades and thicker wakes. This has resulted in many test and production airfoils (isolated and in turbomachine cascades) that exhibit spikes or dips of various magnitudes in Mach-number and pressure-coefficient distribution, which occasionally result in unexpected loading distributions along the blade length and in local separation bubbles. These effects are visible as small local "kinks" in surface pressure or Mach-number distributions in some of the computational and experimental data published, for example, by: Okapuu (1974), Gostelow (1976), Wagner et al. (1984), Hourmouziadis et al. (1987, with separation bubble), Sharma et al. (1990) and others. One case using Kiock's (1986) published experimental results and computations (Korakianitis, 1993c) is shown in figure 7.12.

These "kinks" affect boundary-layer performance and blade efficiency, and they could be avoided by the use of fourth-order splines. However, fourth-order splines have other "stability" problems (wiggles) of their own. An alternative method to prescribe blade surfaces with continuous curvature and continuous slope of curvature has been developed (Korakianitis, 1993c) and is described in section 7.5.

Blending a leading-edge circle or ellipse with the blade surfaces usually results in a local curvature or slope-of-curvature discontinuity that may cause separation and substantial leading-edge surface-Mach-number spikes that are detected only if the probes are located at the correct location. Such a local leading-edge laminar-separation bubble due to blending of a leading-edge circle with the blade suction surface occurs in the turbine geometry published by Hodson and Dominy (1987), seen in the data published

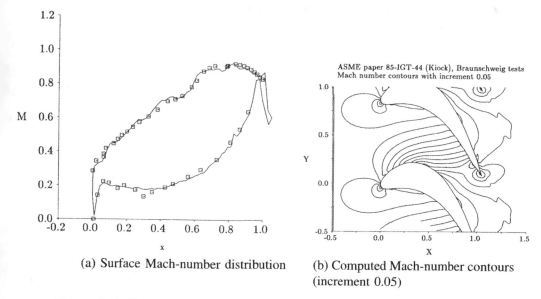

(a) Surface Mach-number distribution

(b) Computed Mach-number contours (increment 0.05)

Figure 7.12. Tested and computed Mach-number distributions for Kiock's (1986) turbine cascade. From Korakianitis (1993c)

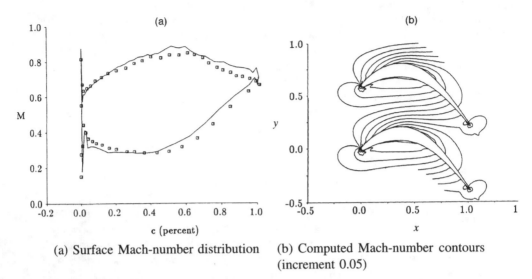

(a) Surface Mach-number distribution   (b) Computed Mach-number contours
                                            (increment 0.05)

Figure 7.13. Effect of curvature discontinuity caused by the leading-edge circle. Tested and computed Mach-number distributions for Hodson and Dominy's (1987) turbine cascade. From Korakianitis and Papagiannidis (1993d)

in figure 11 of Hodson (1985). Test results (from Hodson, 1985) and computations (from Korakianitis and Papagiannidis, 1993d) are shown in figure 7.13. (This blade also exhibits local deceleration effects in two places on the pressure surface. The first near the leading edge is due to the leading-edge slope-of-curvature discontinuity. The second around mid chord can be removed by lowering the (x, y) location of the pressure surface in that region, which will raise the suction-surface numbers accordingly to accommodate the tangential blade loading.) The leading-edge spike and separation region were removed by modifying the geometry of the blade in the vicinity of the slope-of-curvature discontinuity with an inverse design technique as explained in figures 11, 12, and 13 of Stow (1989).

7.5   Prescribed-curvature turbine-blade design

Despite continual advances in inverse-design methods, they can be used only for the design of blades with pointed trailing edges (requiring corrections for boundary-layer thickness), and with them one loses the control of blade section and the intuition gained by the use of direct methods. As will be shown, the local increases and decreases in surface Mach number are a direct consequence of the surface-curvature distribution. We therefore advocate the use of a direct blade-design method (Korakianitis, 1993c) that is based on prescribing the blade surface-curvature distribution. The method, illustrated in figure 7.14, can start from the velocity diagram, or from an initial blade shape developed by the method described in section 7.2.

The trailing-edge region

Typically the Mach number at the outlet-flow plane, M_2, will be a design specification. The values of α_1, α_2, and C_L also specify s/b; the throat diameter $o/s \approx \cos\alpha_2$, and can be approximated by equations 7.8, 7.9, and 7.10. Choosing the value of λ and the trailing-edge thickness specifies the trailing-edge points P_{s2} and P_{p2} (these are shown at the same point in figure 7.14a only because the designs are analyzed with an inviscid-flow program).

The trailing-edge region of the suction surface is specified by a third-order polynomial $y = f(x)$ evaluated using four boundary conditions: the location of P_{s2}; the blade angle β_{s2} at the trailing edge; and the angle ϕ of the throat-circle diameter with the x axis. The latter specifies two boundary conditions: the location of point P_{sm}; and the tangent of the third-order polynomial at that point, equal to $(\pi/2 + \phi)$. The analytic polynomial has continuous first, second, and third derivatives. With these we compute the curvature of the resulting blade surface $\mathbf{C}_s(x) = 1/r(x)$ (using equation 7.16) between points P_{sm} and P_{s2}, and plot it between points $C4_s$ and $C5_s$ in figure 7.14c.

The trailing-edge region of the pressure surface is specified by a third-order polynomial $y = f(x)$ evaluated using four boundary conditions: the location of P_{p2}; the blade angle β_{p2} at the trailing edge; the (x, y) coordinates of a point P_{pm} on the surface; and the angle β_{pm} of the blade surface at point P_{pm} with the x axis. With the first and second derivatives of this polynomial we compute the curvature of the resulting blade surface $\mathbf{C}_p(x) = 1/r(x)$ (using equation 7.16) between points P_{pm} and P_{p2}, and plot it between points $C4_p$ and $C5_p$ in figure 7.14c.

The slopes of the curvatures $\mathbf{C}_s(x)$ and $\mathbf{C}_p(x)$ at points $C4_s$ and $C4_p$ respectively are computed and become inputs for the rest of the design. A gradually-increasing curvature distribution in the trailing-edge regions results in mechanically- and aerodynamically-feasible blade shapes. Different types of curvature distributions would be required for transonic and supersonic blades, but their slope of curvature would also need to be continuous.

The main part of the blades

The design is converted into an effort of specifying a curvature distribution for the shape of the blade surfaces $\mathbf{C}_s(x)$ and $\mathbf{C}_p(x)$. In order to ensure slope-of-curvature continuity, the curvature from points P_f to points P_m is specified using a Bezier spline in curvatures using the following conditions (see figure 7.14c). On the curvature of the suction surface we specify points $C1_s$, $C2_s$, and $C3_s$. Point $C4_s$ is already specified from the corresponding value of curvature evaluated from the trailing-edge third-order polynomial. The x location of point $C3_s$ is variable, but the value of curvature there is evaluated such that line $C3_s C4_s$ is tangent to the surface-curvature line at point $C4_s$. Point $C2_s$ is user specified. Point $C1_s$ is specified at an x location corresponding to P_f, or at an x location just ahead of P_f. Since the slope of the Bezier curve is tangent to the line of knots at its ends, the tangency condition at point $C4_s$ ensures slope-of-curvature continuity from $C1_s$ to $C5_s$ (from P_{sf} to P_{s2}). Similarly points $C1_p$, $C2_p$, and $C3_p$ are specified on $\mathbf{C}_p(x)$.

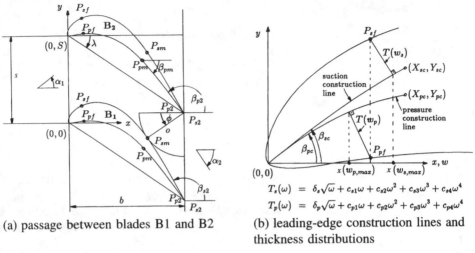

(a) passage between blades B1 and B2

(b) leading-edge construction lines and thickness distributions

$$T_s(\omega) = \delta_s\sqrt{\omega} + c_{s1}\omega + c_{s2}\omega^2 + c_{s3}\omega^3 + c_{s4}\omega^4$$
$$T_p(\omega) = \delta_p\sqrt{\omega} + c_{p1}\omega + c_{p2}\omega^2 + c_{p3}\omega^3 + c_{p4}\omega^4$$

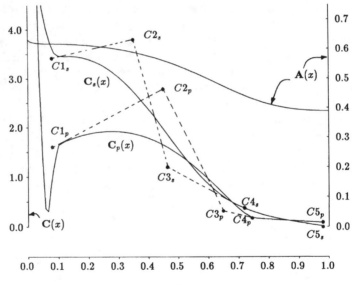

(c) Non-dimensional ($b = 1.0$) curvature and area distributions

Points P on the blade surfaces shown in figure 7.14a divide the blade-surfaces in three parts: the leading-edge region with details shown in figure 7.14b; the middle region specified by the prescribed-curvature-distribution method; and the trailing-edge region prescribed by an analytic polynomial. In figure 7.14c solid line $A(x)$ shows the monotonically-decreasing passage width (area) between the two blades from passage inlet to outlet. The other two solid lines are the curvature distributions of the two blade surfaces from leading to trailing edge; and the dotted lines show the control points used to prescribe the curvature distributions in the middle region of the blade.

Figure 7.14. Parameters of prescribed-curvature-distribution blade-design method. From Korakianitis (1993c)

Using central differences for three consecutive blade points $i-1, i, i+1$, equation 7.16 gives the curvature $1/r_i$ at point i as

$$
\frac{1}{r_i} = \frac{2\left(\frac{y_{i+1} - y_i}{x_{i+1} - x_i} - \frac{y_i - y_{i-1}}{x_i - x_{i-1}}\right) / (x_{i+1} - x_{i-1})}{\left\{\left[\frac{1}{2}\left(\frac{y_{i+1} - y_i}{x_{i+1} - x_i} + \frac{y_i - y_{i-1}}{x_i - x_{i-1}}\right)\right]^2 + 1\right\}^{3/2}}
\tag{7.18}
$$

Given (x_{i-1}, y_{i-1}), (x_i, y_i), x_{i+1} and $1/r_i = C_i$ from figure 7.14c we can now compute y_{i+1} using equation 7.18. This can be done by manipulating the equation into a sixth-order linear algebraic equation in y_{i+1}, or by a numerical solution. We start from points P_m and progress explicitly to points P_f. The Bezier spline is iteratively manipulated until the slope and the y location of the blade surfaces at points P_f, the shape of the surface-curvature distributions, and the passage-width distribution are acceptable.

Leading-edge geometry

The leading-edge geometry is designed by specifying thickness distributions added perpendicularly to parabolic construction lines that pass through the leading edge, as shown in figure 7.14c. The parabolic construction lines are specified by the angle of the construction line at the origin β_κ and the (x, y) location of the origin $(0, 0)$ and of one additional point (X_κ, Y_κ) on the construction line. For the suction line we add a thickness distribution above the suction-side parabolic construction line and for the pressure line we add a thickness distribution below the pressure-side parabolic construction line. The thickness distribution is of the form:

$$
\begin{align}
t_s(x) &= t_{s,sq}\sqrt{x} + t_{s,1}x + t_{s,2}x^2 + t_{s,3}x^3 + t_{s,4}x^4 \\
t_p(x) &= t_{p,sq}\sqrt{x} + t_{p,1}x + t_{p,2}x^2 t_{p,3}x^3 + t_{p,4}x^4
\end{align}
\tag{7.19}
$$

where $t_{s,sq}$ and $t_{p,sq}$ are specified by the designer and the coefficients $t_{s,1}$, $t_{s,2}$, $t_{s,3}$, $t_{s,4}$ $t_{p,1}$, $t_{p,2}$, $t_{p,3}$, and $t_{p,4}$ are derived with the conditions that the absolute values, first, second, and third derivatives of the thickness distribution and of the downstream blade segment at point P_f with respect to the parabolic construction line are equal. This ensures curvature and slope-of-curvature continuity at point P_f, and avoids the need to blend a circle, ellipse or other shape with the blade surfaces near the leading edge. The continuous slope of curvature at point P_f also ensures that the blade will not exhibit unexpected leading-edge separation bubbles. The values of parameters $t_{s,sq}$ and $t_{p,sq}$ are used to specify thicker or thinner leading edges, which affect the design-point and off-design-point operating performance of the cascade. The origin $(0, 0)$ itself is a singular point. In order to achieve first-derivative continuity at that point the angle of the parabolic construction lines should be equal for the suction and pressure surfaces.

Tangential-loading effects

This blade-design method has been used to design numerous cascades of various loading distributions. It was found that the designer has a lot of freedom in specifying the curvature (and location of the blade surface) in cascades of medium and low loading (C_L) (Korakianitis and Papagiannidis, 1993d). Blades with high turning and high values of C_L require very fine adjustments in blade curvature (and location) in order to attain good performance. For some of these cases it may be necessary to modify the above method to use more Bezier splines to specify the surface curvatures in figure 7.14c.

Figure 7.15 shows the computed performance of a blade with loading coefficient $C_L = 1.2$ and exit Mach number about 0.80. This figure illustrates the difficulty of increasing C_L much above 1.2 in subsonic turbines while maintaining accelerating flow on both the suction and pressure surfaces. The suction-surface Mach number is just subsonic (just below M=1.0) for most of the length of the suction surface, and it is very sensitive to extremely small changes in blade geometry and the surface-curvature distribution of both suction and pressure surfaces. One way to increase loading further is by a lossier "transonic" suction surface. (Transonic means that even though the velocity diagram is subsonic, the Mach number locally exceeds unity somewhere along the blade surfaces.) Another way to increase loading is by lowering the pressure-surface velocities, probably by introducing a region of (undesirable) deceleration along the pressure surface. Therefore further increases in loading introduce undesirable effects.

Leading- and trailing-edge wedge angles

We have found it useful to keep the x location of point P_{pm} near the x location of point P_{sm} (see figure 7.14a) to control the surface angles and the wedge angle between the surfaces (inputs $phi + \pi/2$ and β_{pm}) at that region of the airfoil. Similarly β_{s2}

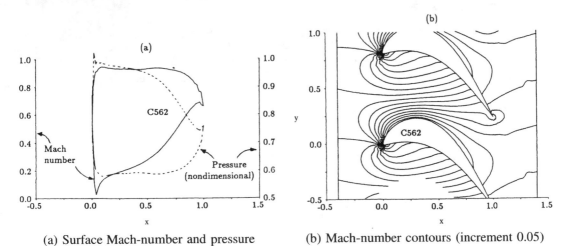

(a) Surface Mach-number and pressure distributions

(b) Mach-number contours (increment 0.05)

Figure 7.15. Performance of highly-loaded blade with $\alpha_1 = 50°$, $\alpha_2 = -60°$, $C_L = 1.2$ and $\lambda = -30°$. From Korakianitis (1993b)

and β_{p2} are used to specify the trailing-edge wedge angle (it becomes an input). The curvature distribution results in (x, y) points near P_{sf} and P_{pf} that give as output the wedge angle near the leading edge. The surface-curvature distributions are manipulated until desirable wedge angles are obtained. The surface-angle distributions are part of the graphical output during use of the method.

Surface-curvature effects

The exact locations of points $C1_s$, $C2_s$, $C3_s$, and $C4_s$ for the suction surface and $C1_p$, $C2_p$, $C3_p$, and $C4_p$ for the pressure surface (figure 7.14c) are not as critical as the resulting shape of the curvature distribution. In using the method we compare desired changes in Mach-number distribution with changes in curvature distribution and location of blade surface. After the first iteration (first geometric design and analysis) the user examines the resulting loading (surface-Mach-number or surface-pressure) distribution and appreciates where to increase and decrease local curvature (and local loading). After the second iteration the user gains an appreciation of the magnitude of the required changes in curvature to cause the desired changes in Mach-number distribution. The procedure is repeated until a desirable geometry and performance is obtained, or until the designer decides that the shape is good enough to proceed to the next design stage (three-dimensional integration of blade shapes).

Figure 7.16 shows five variations of a blade designed with $\alpha_1 = 40°$ $\alpha_2 = -60°$ $\lambda = -25°$ and $C_L = 1.0$. Relatively large changes near the trailing edge of the suction-surface curvature cause relatively small changes in blade-surface locations. However, closer to the leading edge similar curvature changes cause large blade changes. More important is the correspondence between surface curvature and surface Mach-number distributions, throughout the blade surfaces, which can be used to "front-load" or "aft-load" the turbine (front-loading means pushing the maximum of the suction-surface Mach number towards the leading edge, and aft-loading means pushing it towards the trailing edge). Thus the initial choice of (s/e) in very early stages of design (see subsection "selection of surface curvature downstream of the throat" in section 7.2) affects the type of loading distribution of the blade. Design philosophies have developed, with some engine manufacturers preferring front-loaded designs, and others mid-loaded designs. Our investigations indicate that a mid-loaded design minimizes boundary-layer losses (for a detailed discussion, beyond the scope of this book, refer to Korakianitis and Papagiannidis, 1993d).

Stagger-angle effects

Figures 7.17 to 7.20 show the effect of increasing stagger angle from $\lambda = -25°$ to $\lambda = -40°$. In these figures the main cascade-design variables except the stagger angle are kept constant (for example $\alpha_1 = 40°$, $\alpha_2 = -60°$, $M_2 = 0.800$, $C_L = 1.00$, and $o/s = 0.50$). Increasing the stagger angle results in thinner and more-front-loaded cascades, with performance that is more sensitive to small surface-curvature and blade-geometry changes. This is because as λ is increased the throat circle pushes both blade

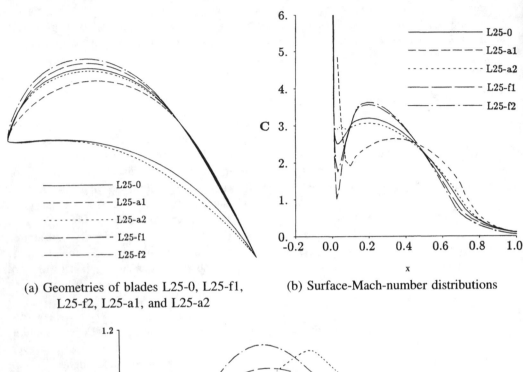

(a) Geometries of blades L25-0, L25-f1,
L25-f2, L25-a1, and L25-a2

(b) Surface-Mach-number distributions

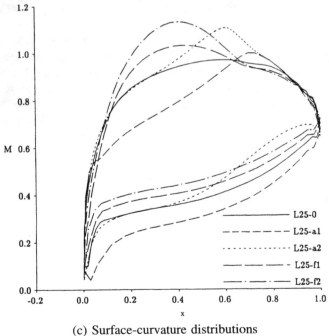

(c) Surface-curvature distributions

Figure 7.16. Surface-curvature-distribution effectss on blade loading and boundary layer. From Korakianitis and Papagiannidis (1993d)

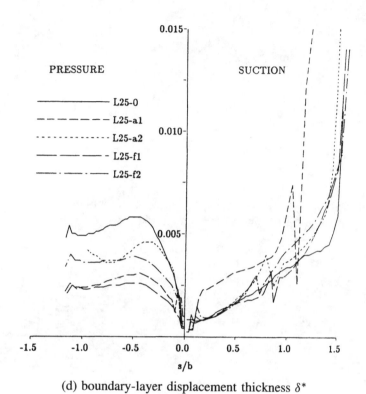

(d) boundary-layer displacement thickness δ^*

Figure 7.16. Concluded

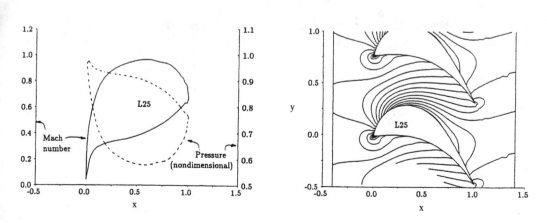

Figure 7.17. Stagger-angle effects: performance of blade L25 with $\alpha_1 = 40°$, $\alpha_2 = -60°$, $C_L = 1.0$, and $\lambda = -25°$. From Korakianitis (1993c)

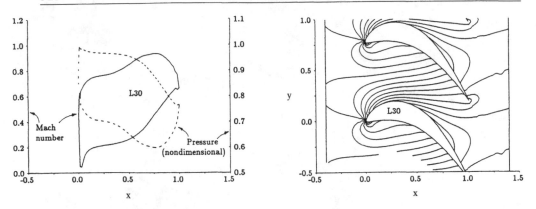

Figure 7.18. Stagger-angle effects: performance of blade L30 with $\alpha_1 = 40°$, $\alpha_2 = -60°$, $C_L = 1.0$, and $\lambda = -30°$. From Korakianitis (1993c)

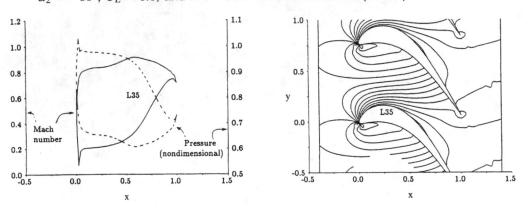

Figure 7.19. Stagger-angle effects: performance of blade L35 with $\alpha_1 = 40°$, $\alpha_2 = -60°$, $C_L = 1.0$, and $\lambda = -35°$. From Korakianitis (1993c)

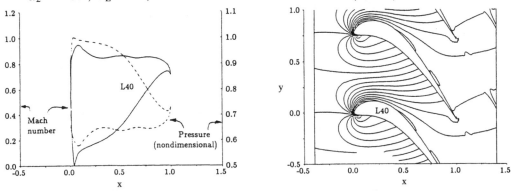

Figure 7.20. Stagger-angle effects: performance of blade L40 with $\alpha_1 = 40°$, $\alpha_2 = -60°$, $C_L = 1.0$, and $\lambda = -40°$. From Korakianitis (1993c)

surfaces further down, thus locating the throat further upstream on the suction surface and thinning the blade. In order to fit the rest of the blade into the passage the designer is forced to increase the curvature of the suction surface of the blade in the vicinity of point P_{sf} (figure 7.14a). The increased curvature in turn increases the Mach number in that vicinity, thus loading the front part of the suction surface of the blade accordingly. While it is relatively easy to change the type of loading distribution of cascade L25 (figure 7.16), it is considerably harder to change the front-loaded cascade L40.

The relatively-aft-loaded cascade of figure 7.17 ($\lambda = -25°$) has a relatively-short length of relatively-more-adverse pressure gradient. As the stagger angle is increased this changes until the relatively-front-loaded cascade of figure 7.20 ($\lambda = -40°$) has a relatively-long length of relatively-not-as-adverse pressure gradient. This clearly affects the development of the pressure- and suction-surface boundary layers. Our calculations indicate that a mid value of stagger angle provides the thinnest boundary layer with minimum losses. This agrees with a previous investigation by Hoheisel et al. (1987), who found that a mid-loaded cascade had a relatively-thinner boundary layer (although their blades were designed with inverse-design methods and did not have the rest of the design parameters identical).

The role of steady-flow computer programs

Clearly the type of analysis presented in this section requires computer programs for steady-flow analyses. Such programs first appeared in the early 1960s, and became increasingly available starting in the mid 1980s. Their availability will continue to increase. Even though the use of these programs is beyond the scope of this book, there is an inescapable inter-dependence between the simpler graphical blade-design method presented in section 7.2, the next design steps, such as those presented in this section, and the performance-prediction methods described in section 7.7.

One will reduce the initial effort to find a first blade with the prescribed-curvature blade design by drawing to scale initial blade shapes as described in section 7.2. On the other hand, using the analytic results presented in this section one has some foreknowledge of the later effects of initial design choices made while using the graphical blade-design method of section 7.2. The analytic results of this section have provided information on the effects of higher or lower stagger angle on front versus aft loading, thicker versus thinner blades (affecting in turn structural considerations, cooling-passage geometry, etc.), and how to change the surface geometry to front-load or aft-load the blade.

7.6   Stator-rotor interactions

Blades are designed assuming (average) steady inflow and outflow conditions, but the flow between blade rows is inherently unsteady. Sources of flow non-uniformity and blade-row excitation include: one or more excitations per revolution from circumferential flow non-uniformities at the inlet, or from elsewhere (for instance from diffuser separation or from rotating stall in compressors); viscous wakes propagating from the trailing edges of upstream blade rows or struts to downstream ones; potential-flow interaction (an inviscid effect) from each individual lifting surface (blade) propagating to both upstream and downstream blade rows; small-amplitude high-frequency vortices in the wakes; and larger-amplitude vortices from three-dimensional flow phenomena (for instance the passage vortex generated by secondary flow perpendicular to boundary-layer flow, see figure 10.24).

Figure 7.21 illustrates the wake disturbance and the potential-flow interaction from one blade. The wake is a velocity defect generated by the boundary layers of the blade surfaces, and if undisturbed by other blades it would move downstream along the direction of outlet-flow angle while decaying slowly over three or four chord lengths. The velocity perturbation from the wake generates a vorticity perturbation. Some investigators model the wake as a velocity defect, and others model it as a sheet of vortices. The potential-flow interaction is a static-pressure disturbance radiating out from the leading and trailing edges that would propagate upstream and downstream respectively approximately in the directions shown, while decaying faster than the wake over one or two chord lengths.

All flow excitations contribute to high-frequency flow, force, and momentum un-steadiness, leading to what is known as "high-frequency" or "high-cycle" fatigue, which can lead to rapid failures. Campbell diagrams (figure 7.22) are used to select the numbers

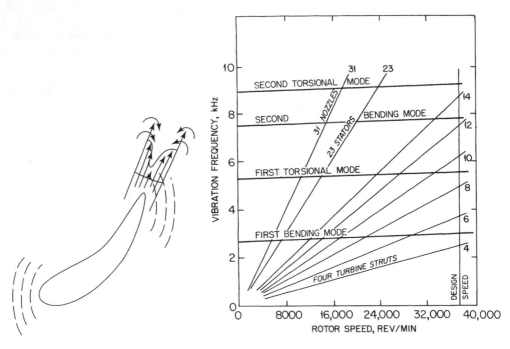

Figure 7.21. Wake disturbance and potential-flow interaction from a two-dimensional blade section

Figure 7.22. Campbell diagram for turbine rotor blades

of rotor and stator blades so as to remove exciting frequencies from blade natural frequencies at operating speeds. The Campbell diagram of figure 7.22 is for a typical turbine rotor blade, and it shows which blade natural frequencies will be excited at different rotating velocities by different numbers of upstream (or downstream) struts (or stators). As the rotor speed increases the centripetal force acting on the blade makes it stiffer, so the horizontal lines of rotor-blade excitation frequencies shift slightly at higher rotor speeds, a phenomenon known as "centrifugal stiffening". (Stator natural frequencies would not change with rotational velocity.) The Campbell diagrams of compressor blades, which are typically thinner than turbine blades, look similar to those for turbines, except the natural frequency for the second bending mode is usually lower than that for the first torsional mode.

With the aid of Campbell diagrams one can consider stator and rotor blade rows excited by disturbances induced by other upstream and downstream blade rows, and refine the number of blades in each blade row to avoid obvious resonant conditions at operating speeds.

However, this does not mean that the excitation or the blade response is eliminated. It means it should not occur at a natural frequency. Several spectacular fan, compressor and turbine failures due to stator-rotor interactions have occurred over the years, but details are unfortunately proprietary.

Excitation prediction and minimization

The problem of unsteady flows and stator-rotor interactions received significant analytic attention from the late 1950s. In the mid 1980s computer programs for stator-rotor or disturbance/blade-row interactions were developed, and engine manufacturers started utilizing them in stages that exhibited problems. By the mid 1990s engine manufacturers were using some of these computer programs for their new design cases. With faster computers these programs will be used earlier in the design process.

Korakianitis (1987, 1993a, 1993e) has used Giles' (1988) computer program to derive design rules to predict, and stage-geometry changes to minimize, unsteady forcing functions from wake and potential-flow interaction. Principal parameters are the ratio $R_{sr,rr}$, the number of rotor blades to the number of stator blades, which is equal to the stator pitch divided by the rotor pitch,

$$R_{sr,rr} \equiv \frac{N_{rr}}{N_{sr}} = \frac{s_{sr}}{s_{rr}} \tag{7.20}$$

and the axial gap between rotor and stator non-dimensionalized by rotor axial chord

$$d_x = \frac{\text{stator-rotor axial gap}}{b_{rr}} \tag{7.21}$$

Figure 7.23 shows the same stage-velocity diagram and the same rotor affected by the size of the stator (all that changes between the left and right sides of the figure is the non-dimensional size, not the shape, of the upstream stator). Figure 7.24 shows the same stage-velocity diagram and the same stator and rotor affected by the axial gap

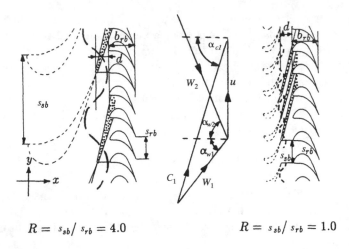

$$R = \, s_{sb}/ \, s_{rb} = 4.0 \qquad\qquad R = \, s_{sb}/ \, s_{rb} = 1.0$$

Figure 7.23. Rotor excitation influenced by number of stators and stator size. From Korakianitis (1993a)

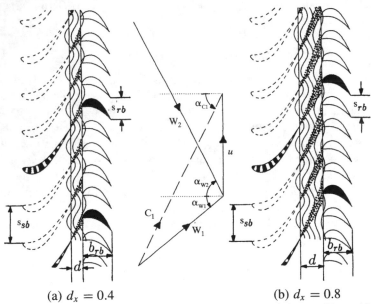

(a) $d_x = 0.4$ (b) $d_x = 0.8$

Figure 7.24. Rotor excitation influenced by stator-rotor axial gap. From Korakianitis (1993e)

between the rotor and the stator. The main conclusions of these investigations, illustrated in figures 7.23 and 7.24, and resultant preliminary-design guidelines, are the following.

1. The flow is unsteady and cannot be accurately predicted by quasi-steady-flow considerations.

2. The wake interaction dominates the unsteadiness for (low) values of $R_{sr,rr} \approx 1$ (the potential-flow effect is negligible by the time it reaches the rotor).

3. The potential-flow interaction dominates the unsteadiness for (high) values of $R_{sr,rr} > 2.5$ (the wake disturbance is a small portion of the potential disturbance engulfing one or more rotors).

4. The wake disturbance is chopped and sheared by the downstream blade. It propagates in segments in the downstream blade-row passage, where circulation effects create a region of increased pressure upstream of the wake-segment centerline and a region of decreased pressure downstream of the wake-segment centerline.

5. Regions of increased and decreased pressure from the potential-flow interaction are cut by the downstream blade row, where they propagate as static-pressure disturbances.

6. The phases of propagation of the two disturbances in downstream blade rows are different. This leads to the next conclusion.

7. For intermediate values of $R_{sr,rr}$, specifically $1.5 < R_{sr,rr} < 2.5$ where the two effects are of comparable magnitude, there exist opportunities to counteract the effects of the two disturbances, thereby reducing stator-rotor interaction excitation. This procedure is beyond the scope of this textbook, but is accomplished by varying d_x (figure 7.24) and the surface Mach-number distribution of the blades (section 7.5).

8. The potential-flow interaction from the stator downstream of the rotor under investigation can be used to further minimize the rotor disturbance, creating an opportunity for coupling the excitation-minimization effort along several stages of the machine, row by row, while specifying the relative azimuthal displacement of the stators.

9. The most significant conclusion for preliminary-design purposes is that for excitation minimization stator-blade rows should have approximately twice the number of downstream rotor-blade rows. For further details of how one can minimize the excitation refer to Korakianitis (1993a, 1993e).

The shock structure and wake propagation in a typical stator-rotor interaction case is illustrated in figures 10.21 and 10.22 in chapter 10.

The role of unsteady-flow computer programs

Just as in section 7.5, the type of analysis presented in this section requires computer programs for unsteady-flow analyses. The availability of these programs will continue to increase. Even though the use of these programs is beyond the scope of this book, there is an inescapable inter-dependence between the simpler graphical blade-design method

presented in section 7.2, and the next design steps. These steps include, but are not limited to: selecting the number of blades in each blade row with the purpose of minimizing
stator-rotor interaction excitation; and selecting blade stagger angles, surface-curvature
distributions and loading distributions that enhance excitation-minimization efforts. Indeed, we have found cases in which these initial choices were not considered from the
preliminary-design stages, and the initial choices impede excitation-minimization efforts,
leading to undesirable solutions.

7.7   Performance (efficiency) prediction of axial turbine stages

For the purposes of calculating the efficiency, a turbine can be thought of as a perfect
machine to which losses have been added. We can categorize the losses into three groups
(although some losses refuse to behave in so orderly a fashion and straddle two or more
groups).

1. Losses that appear as pressure losses in the fluid: for instance, boundary-layer
   friction.

2. Losses that appear as enthalpy increases in the fluid: for instance, disk friction,
   which extracts power from the turbine output and in effect adds this energy as heat
   to the turbine outlet flow.

3. Losses that reduce the shaft-power output of the turbine through friction, the energy
   of which is dissipated elsewhere and not to the working fluid; for instance, friction
   losses in external bearings. Group 3 losses were referred to as "external energy
   losses" in section 2.6.

The "perfect machine" of the first paragraph is the turbine expanding from inlet stagnation conditions 1 in figure 7.25 isentropically through the design turbine enthalpy drop, Δh_0 to 2_{tl}. The Euler equation, (5.1), and the first law, equation (2.10),
show that a turbine designed to produce the flow angles called for in a velocity diagram will in fact produce the required work and enthalpy drop regardless of losses,
so long as a sufficient pressure ratio is provided for the flow to attain the designed
velocities.

In the treatment of losses outlined here, the various losses are compartmentalized as
if all the effects of a given loss, for instance, the tip-clearance loss, could be treated as a
pressure drop. In practice, losses have a wider influence. In particular, losses introduce
non-uniformities into the flow. These non-uniformities may then induce secondary losses
in downstream components, and these might be higher than the original primary losses.
For instance, a large turbine tip-clearance loss could produce so large a boundary layer
at entrance to the exit diffuser that the flow could stall, producing a loss much greater
than the original tip-clearance loss. The diffuser designer may insist that the loss be
allocated to the turbine-designer's ledger. The analyst simply calculates the losses in
each component, without assigning blame. The overall engine or system designer (and
of course all three could be the same person) tries to ensure that upstream components

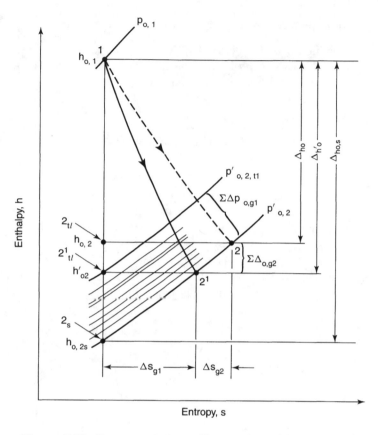

Figure 7.25. Representation of losses in turbine expanders

either have lower losses or produce more-uniform velocity distributions, or uses instead downstream components that are less sensitive to poor velocity profiles.

Turbine performance estimation

In this section we offer what we believe to be improvements on the method of Craig & Cox (1971), which is the most comprehensive of the published approaches to the performance estimation of axial-flow turbines. The changes we advocate are in two principal areas: the form of the loss factors, and the manner in which the loss factors are added. We also want to be precise in the definition of turbine efficiency. Lastly, we point out here that the loss predictions of the Craig & Cox method have been found by manufacturers of aircraft gas turbines to be slightly higher than is found in tests. We believe that the higher efficiencies actually achieved in advanced aircraft engines are probably due to a combination of the effects of improved blade profiles, careful attention to the smoothness of the annular walls of the blading flow path, and to how leakage and cooling air are discharged into the stream (e.g., never being discharged normal to the main flow).

Thermodynamic principles of performance estimation

The general definition of turbine efficiency is *the actual power delivered as a proportion of the power delivered by an ideal turbine.* Exact definition is required in four respects, discussed in chapter 2 (section 2.6), and not repeated here.

Formal thermodynamic definition of efficiency

The isentropic efficiency of an adiabatic turbine without mass addition or leakage is:

$$\eta_s = \frac{\dot{m}\,\overline{C}_p(T_{0,in} - T_{0,ex})}{\dot{m}\,\overline{C}_p(T_{0,in} - T_{0,ex,s})} = \frac{1 - T_{0,ex}/T_{0,in}}{1 - T_{0,ex,s}/T_{0,in}} \qquad (7.22)$$

where the temperature that would be reached in an isentropic expansion, $T_{0,ex,s}$, is shown in figure 7.26.

The values of the mean specific heats in the numerator and the denominator are those over the temperature ranges of the appropriate expansion. There will be a small difference in the two mean specific heats, but this is usually taken as negligible.

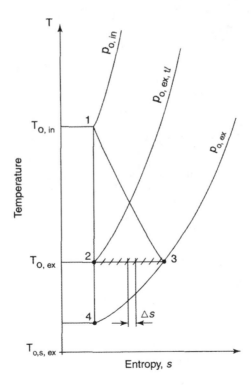

Figure 7.26. Incremental entropy losses of a turbine shown on a temperature-entropy diagram

The thermodynamically correct modelling of losses, insofar as they can be considered to be separable and capable of being added, is shown in figure 7.26 as increments of entropy on the outlet of a turbine giving the designed work. (In the Craig & Cox approach, losses are modelled as increases of enthalpy, or losses of potential work output, at the outlet of a turbine producing the isentropic work, figure 7.25. This model has two drawbacks. First, the isentropic work is more than the turbine vector diagrams are capable of delivering. Therefore the model of a high-output machine from which losses dissipate work is incorrect. Second, because of the "reheat" effect of the enthalpy increments at the actual outlet temperature, the magnitude of each individual loss will depend on the local temperature at which it is assumed to occur, and this temperature is not modelled precisely. Both drawbacks are removed by considering the losses as entropy increases added to the outlet state of an ideal turbine producing the actual work output; i.e., having the real enthalpy drop.)

For semi-perfect gases, which model the compressible fluids in turbines sufficiently closely, entropy increments can be represented as pressure losses divided by the local stagnation pressure, as follows from the Gibbs equation:

$$T\, ds = C_p\, dT - v\, dp$$

Therefore an increment of entropy along a constant-temperature line is:

$$ds_0 = -v\, dp_0/T = -R\, dp_0/p_0$$

or, for finite increments,

$$\Delta s_0/R = -\Delta p_0/p_0$$

In this form, losses can be added regardless of where in the turbine expansion they occur. We will later introduce greater specificity in the conversion of a loss to the relative-pressure-drop form.

The isentropic and polytropic efficiencies can be reformulated to incorporate the losses as relative pressure drops.

Isentropic

$$
\begin{aligned}
\eta_s &= \frac{1 - (T_{0,ex}/T_{0,in})}{1 - (T_{0,ex}/T_{0,in})(T_{0,ex,s}/T_{0,ex})} \\[2mm]
&= \frac{1 - T'}{1 - T'[p_{0,ex}/p_{0,ex,tl}]^{R/C_p}}
\end{aligned}
$$

where

$$
\begin{aligned}
T' &= T_{0,ex}/T_{0,in} \\[2mm]
\eta_s &= \frac{1 - T'}{1 - T'[1 + \Sigma(\Delta p_0/p_0)]^{R/C_p}}
\end{aligned}
\qquad (7.23)
$$

Polytropic

$$\frac{T_{0,ex}}{T_{0,in}} = \left(\frac{p_{0,ex}}{p_{0,in}}\right)^{\left[\left(\frac{R}{C_p}\right)\eta_{p,e}\right]}$$

$$\ln T' = \frac{R}{C_p}\eta_p \ln\left[\frac{p_{0,ex}}{p_{0,ex,tl}}\right]\left[\frac{p_{0,ex,tl}}{p_{0,in}}\right]$$

$$\eta_p = \left\{1 + \frac{(R/C_p)\ln\left[1 + \Sigma\,(\Delta p_0/p_0)\right]}{\ln T'}\right\}^{-1} \tag{7.24}$$

7.8 Treatment of air-cooled turbines

In cooled turbines, compressor air is ducted separately to the nozzle and to the rotor blades, in which it passes along internal passages, picking up heat through convection, and then is discharged through holes in the tip or trailing edge and/or through holes around the blade profile. The usual approach is to calculate a mixed-mean temperature at the nozzle-exit plane from the mass flows and inlet enthalpies of the gas flow from the combustor and the cooling flow from the compressor. A loss factor associated with the discharge of the cooling air will also result in a revised stagnation pressure at the nozzle-exit plane.

This combined nozzle-exit flow will then produce work in the turbine through its enthalpy drop. The cooling air discharged from the rotor blade is not considered to contribute to the work output. However, in an ideal turbine the cooling flow would add work in its expansion. The formulation of this treatment is beyond the scope of this text; refer to Wilson (1987) for the appropriate equations.

7.9 Loss correlations

There are several alternative forms of loss correlation, all having the aim of enabling loss factors measured in or calculated for one set of conditions to be used over a range of conditions. Here we will use one recommended by Brown (1972).

With this method, loss factors, χ_i, correlated by Craig & Cox (1971) or by other investigators, are converted to pressure-loss ratios by means of the following relations for the stators and the rotors. The expressions differ because the relative stagnation temperature is required as a reference in the moving blade row(s).

Stators

$$\left(\frac{\Delta p_0}{p_0}\right)_i = -\chi_i\,\frac{(C_p/R)}{\left[\dfrac{2(T_{0,ex}/T_{0,ne})}{(C'_1/u'_1)^2 u_{fn}} - 1\right]} \tag{7.25}$$

where

$$C'_1 \equiv C_1/u_m$$
$$C_1 \equiv \text{nozzle-exit velocity}$$
$$u'_1 \equiv u_1/u_m$$
$$u_1 \equiv \text{mean-diameter blade speed of first stage}$$
$$u_m \equiv \text{mean-diameter blade speed of local stage}$$
$$u_{fn} \equiv u_1^2/(C_p T_{0,ne})$$

Rotors

$$\left(\frac{\Delta p_0}{p_0}\right)_i = -\chi_i \frac{(C_p/R)}{\left[\dfrac{2\left(\dfrac{T_{0,ex}}{T_{0,ne}}\right)\left\{1 + \left[(W'_2)^2 - (C'_2)^2\right]\left(\dfrac{u_{fn}}{2}\right)\left(\dfrac{T_{0,ne}}{T_{0,ex}}\right)(u'_1)^{-2}\right\}}{\left(\dfrac{W'_2}{u'_1}\right)^2 u_{fn}} - 1\right]} \tag{7.26}$$

where

$$C_2 \equiv \text{blade-row-outlet absolute velocity}$$
$$C'_2 \equiv C_2/u_m$$
$$W_2 \equiv \text{blade-row-outlet relative velocity}$$
$$W'_2 \equiv W_2/u_m$$

Efficiency prediction

The steps required to predict the efficiency of a single- or multi-stage turbine are summarized as follows.

1. Tabulate all relevant loss factors, χ_i, for each blade row, using the Craig & Cox or other preferred correlations.

2. Convert each loss factor in each blade row to a pressure-loss ratio, $(\Delta p_0/p_0)$, using the relations for a stator or rotor row as appropriate.

3. Add the values of $(\Delta p_0/p_0)$ producing $\Sigma(\Delta p_0/p_0)$, and insert this into the equations for isentropic or polytropic efficiency, equations (7.23) and (7.24).

Caution

The methods of loss correlation and addition recommended above are thermodynamically more justifiable than are some alternatives. However, correlations of loss factors, such as those in Craig and Cox, were adjusted to agree with the mass of data available when the losses were added according to the system used in their paper. While greater consistency over a range of turbines of different designs would be expected from the method advocated here, there might need to be some adjustment of the loss correlations to conform with this method.

However, initial trials on about twenty multistage turbines whose test data were known to a high degree of accuracy (and which unfortunately are proprietary), indicated that these methods gave better accuracy of prediction than other methods in the majority of cases.

7.10   Loss-coefficient data for axial-flow turbomachinery

In practice, loss data, which are generally obtained from correlations of systematic wind-tunnel or turbine tests, do not fall as neatly into categories as the theoretical breakdown in section 7.3 seems to indicate. Some losses appear to be best correlated by adding increments to the principal factors, while in other cases the effects of some design variables can best be included by a multiplier to the principal factor.

Whichever method is used, it is obviously necessary to be consistent. Accordingly, we shall give the correlations published by Craig and Cox (1971). These methods were developed for the design of steam turbines but were acclaimed by workers in the gas-turbine as well as the steam-turbine industry. We shall give in addition some of the data confirming and extending the method, published by the Institution of Mechanical Engineers in extensive discussions of the Craig-Cox approach. The data appear to agree well with a later method by Kacker and Okapuu (1982).

Profile losses

The basic profile-loss parameter, χ_{pl}, is given in figure 7.27 as a function of the modified lift coefficient (or parameter) for the blade row, obtained from figure 7.28, the blade-row contraction ratio, figure 7.29, and the length of the blade camber line, b_c, figure 7.3. For circular camber lines,

$$\left(\frac{s}{b_c}\right) = \left(\frac{\cos\lambda\,\sin(\theta/2)}{(b/s)_{op}\pi(\theta,\,deg/360°)}\right) \tag{7.27}$$

In this equation the blade camber, θ, can be estimated from the flow deflection, ε, by adding about five degrees to the flow angles at inlet and outlet: thus the camber will be about ten degrees greater than the deflection. The stagger angle, λ, can be taken from figure 7.4b. The axial solidity, $(b/s)_{op}$, is calculated from equations 7.4 and 7.5.

The profile-loss parameter so obtained would be valid for a throat Reynolds number, Re_0, of 10^5, for a low Mach number (below 0.75), for zero trailing-edge thickness, and for optimum blade-row incidence.

Trailing-edge-thickness correction

Figure 7.30 shows two corrections, a loss-factor increment, $\Delta\chi_{pl}$, and a profile-loss ratio, given as functions of the ratio of the trailing-edge thickness to the blade pitch, (t_{te}/s), and the outlet flow angle α_2.

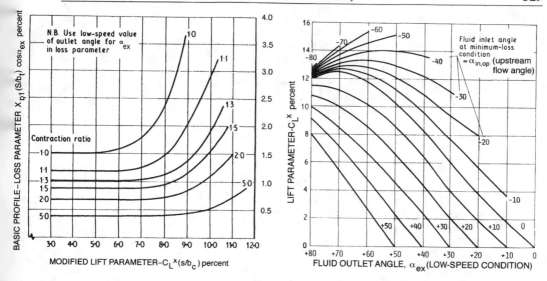

Figure 7.27. Basic profile loss. From Craig and Cox (1971)

Figure 7.28. Lift parameter versus flow angles. From Craig and Cox (1971)

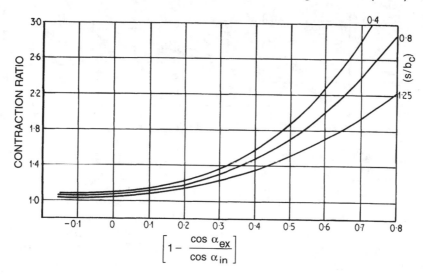

Figure 7.29. Contraction ratio for average profiles. From Craig and Cox (1971)

Reynolds-number correction

For throat Reynolds numbers, based on blade-row opening, o, outlet relative flow velocity, and local (static) flow properties at the blade outlet, other than 10^5, figure 7.31 can be used to find a loss-factor ratio. This curve is an approximate mean of several curves given in the discussion of the Craig and Cox paper.

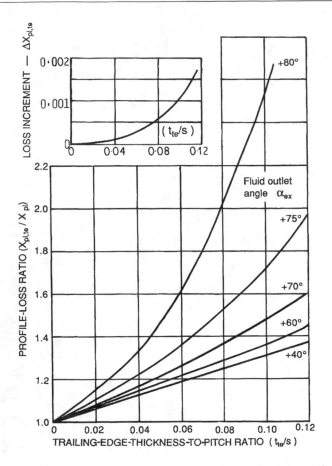

Figure 7.30. Trailing-edge-thickness loss. From Craig and Cox (1971)

Secondary-flow and annulus losses

Losses resulting from "secondary" (transverse boundary-layer) flows and from friction on the annulus walls are largely a function of the blade-row aspect ratio, (h_b/b). Figure 7.32 is a combination of mean data presented by V. T. Forster in his discussion of the Craig-Cox paper. One should select the blade row closest to that being studied, or interpolate or extrapolate, to obtain a secondary-loss estimate.

Tip-clearance losses

The simplest of the many loss correlations available is to multiply the "zero-clearance" efficiency by the ratio of the blading annulus area to the casing annulus area for each blade row.

This approach is appropriate for unshrouded blade rows in smooth-wall casings. Blades with rotating shrouds will incur losses from leakage flows through the shroud

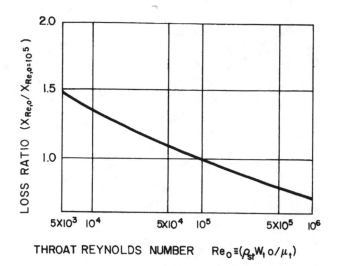

Figure 7.31. Effect of Reynolds number on profile losses. From Craig and Cox (1971)

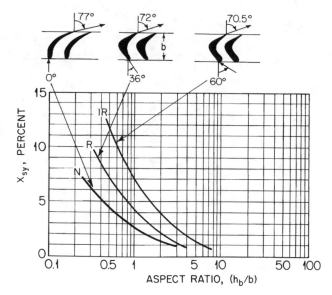

Figure 7.32. Secondary-flow and annulus losses. From Forster's discussion of Craig and Cox (1971)

labyrinths and disturbance of the wall boundary layers. Craig and Cox (1971) give some correlations for the clearance losses around blade shrouds and for the losses found in non-smooth annulus walls. Cordes (1963) gives data reproduced as figure 7.33 showing the effect of reaction on tip-clearance losses.

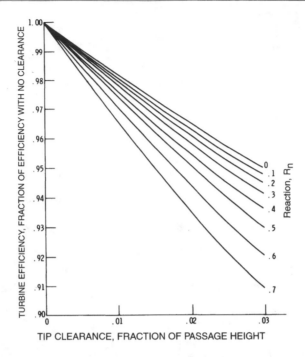

Figure 7.33. Tip-clearance correlation for unshrouded blades. From Cordes (1963)

Example 2   Estimation of axial-turbine efficiencies

Estimate the design-point polytropic and isentropic stagnation-to-stagnation efficiencies of a single-stage axial turbine of the following specifications. (As is appropriate for preliminary design, only the mean-diameter configuration will be given and calculated.)

Fluid: combustion gas, fuel/air ratio 0.014
Mass flow: 1.54 kg/s
Inlet pressure: 2.0 bar
Inlet temperature: 1,200 K
Exhaust stagnation pressure: 1.25 bar
Mean-diameter diagram: work coefficient = 1.0
                                      flow coefficient = 0.6
                                      reaction = 0.5
Number of nozzle blades: 16
Number of rotor blades: 17
Tip clearance (rotor, on radius): 0.50 mm
Nozzle and rotor pitch-to-back-radius ratio: $s/e = 0.50$
Hub-shroud ratio, rotor entrance: 0.6
Optimum tangential lift coefficient: 1.0
Solidity, nozzle and rotor: 1.6
Trailing-edge thickness ratio, nozzle and rotor: 0.025

Solution.

The blading design is carried out by the methods suggested earlier in this chapter. Only the principal results will be given here.

An initial estimate of the efficiency must be made to enable the velocity diagram and outlet conditions to be calculated. A polytropic stagnation-to-stagnation efficiency of 90 percent across the blading was chosen. This turned out to be close to the final calculated value. If it had not been, a second iteration would have been desirable.

Some of the results of the calculations follow.

> Outer (shroud) diameter, rotor entrance: 173.8 mm
>
> Inner (hub) diameter, rotor entrance: 104.3 mm
>
> Circular pitch (nozzle) at mean diameter (139.1 mm): $s = 27.3$ mm
>
> Flow angles (symmetrical diagram): 0 degree and 59.04 degrees
>
> Optimum axial solidity (equation 7.5): 0.882
>
> Rotor-blade axial chord: $b_x = 22.7$ mm
>
> Rotor blade opening/spacing: $o/s = 0.505$
>
> Relative outlet velocity: 434.0 m/s
>
> Mach number, rotor-exit relative: $M = 0.685$
>
> Reynolds number, rotor exit. 49,100
>
> Reynolds-number loss ratio (figure 7.31): 1.17
>
> Lift parameter for figure 7.28: 10.1% ("modified lift parameter")
>
> Contraction ratio for input 0.486 (figure 7.29) = 1.66
>
> Basic profile-loss parameter (figure 7.27) (nozzle and rotor): 2.07%
>
> Profile-loss ratio (figure 7.30): 1.1
>
> Profile-loss increment (figure 7.30): 0.0
>
> Secondary and annulus losses (figure 7.32): 2.9% (nozzle) and 2.35% (rotor) for aspect ratio = 1.44 and 1.53 (interpolating between curves N and R)

The sum of the nozzle and rotor corrected profile and secondary loss factors = 0.0545 and 0.0501. These were inserted in equations 7.25 and 7.26 to give the stagnation-pressure-loss ratios. By adding these ($\Sigma(\Delta p_0/p_0) = -0.0314$) and using the sum in equation 7.23, the predicted no-clearance isentropic efficiency was found to be 0.933. The tip-clearance as a fraction of passage height was calculated to be 0.0144, and entering this and fifty-percent reaction in figure 7.33 gave an efficiency multiplier of 0.974. Thus the corrected isentropic efficiency became 0.909.

The polytropic efficiency was given by equation 2.88:

$$\eta_{p,e} = \frac{\ln\left\{1 - \eta_{s,e}\left(1 - \left(\frac{p'_{0,5}}{p_{0,4}}\right)^{R/\overline{C_{p,e}}}\right)\right\}}{\ln\left(\frac{p'_{0,5}}{p_{0,4}}\right)^{R/\overline{C_{p,e}}}} = 0.904$$

(Alternatively, equation 7.24 could have been used directly.) This is sufficiently close, for preliminary design, to the initial estimate of 0.90 for no further iteration to be necessary.

7.11   Turbine performance characteristics

The data and methods given in the previous section make it possible to estimate the design-point performance of single-stage or multistage axial-flow turbine. These methods can be extended to cover off-design performance (see Craig and Cox, 1971).

Another approach, often adequate in preliminary design, is simply to estimate the off-design performance from the characteristics of other, similar, machines. Turbine flow is less complex than is compressor flow, and it leads to less-complex characteristics. Here are some comparisons.

1. In a compressor the rotor blades slice into the fluid, so that the rotor speed largely determines the shape of the mass-flow vs. pressure-ratio performance characteristic. In a turbine the flow is first "squirted" through nozzles, and the resistance, or area, of these nozzles largely determines the flow characteristics. The speed of the downstream turbine rotor has a rather minor effect on the relationship of flow to pressure drop.

2. Because the fluid in a compressor is flowing generally from low to high pressure, and because turbo-compressors have no "positive-displacement" action, sometimes the high-pressure fluid can leak backward through the compressor, and can lead to a violent flow breakdown. Therefore there is a large region in all turbo-compressor mass-flow vs. pressure-ratio characteristics where steady-state efficient operation is impossible. In turbines the flow is generally from high pressure to low, and there are no regions of unstable flow in a turbine characteristic "map."

Let us first examine simplified cases to understand turbine characteristics and how they affect the operation of a circuit of which a turbine may be a component.

For the lossless flow of an incompressible fluid through a nozzle, the steady-flow energy equation gives the restricted form of equation (2.12):

$$p_0 - p_{st} = \frac{\rho C^2}{2g_c}$$

yielding

$$\frac{\dot{m}}{A} = \sqrt{2g_c \rho (p_0 - p_{st})} \tag{7.28}$$

The flow of real fluids through frictional nozzles can be given by multiplying the right-hand side of equation 7.28 by a coefficient of discharge and a compressibility factor. Thus nozzle flow has a flow-pressure-drop relationship generally known as a "square-law" form, figure 7.34a.

The effect of having a rotor downstream of the nozzles is that of adding a small resistance in series with the controlling resistance. The added resistance of the rotor varies with rotor speed, but because the effect is small in any case, the effect of rotor speed on the turbine flow:pressure-drop characteristics is small, figure 7.34b.

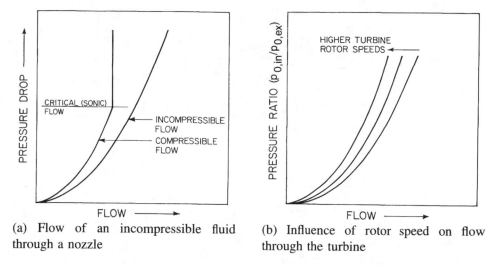

(a) Flow of an incompressible fluid through a nozzle

(b) Influence of rotor speed on flow through the turbine

Figure 7.34. Illustration of typical turbine characteristics

The torque, power output, and efficiency of a turbine are of course greatly influenced by rotor speed. Consider a single-stage turbine connected to a test-stand brake and supplied with fluid at constant pressure. To a first approximation, we can deduce from figure 7.34b that the mass flow will be constant regardless of rotor speed (there will in fact be a small variation as rotor speed changes). To predict the general shape of the output characteristics, we will therefore take the axial velocity, C_x, in the velocity diagrams shown below the abscissa of figure 7.35 as constant. The diagram at the center is taken to be the design-point velocity diagram. It is a "simple" diagram, as defined in chapter 5: constant axial velocity and constant peripheral velocity. As a further simplification we consider the off-design velocity diagrams also as if they were "simple", although we know that in compressible-fluid turbines at Mach numbers above, say, 0.5 there would be significant variations in axial velocity at best-efficiency point.

We will make one further simplifying assumption: that the fluid outlet angles from rotor and stator blade rows remain at their design-point values even at extreme off-design conditions. For turbines with their intrinsically-favorable pressure distribution and generally accelerating relative flow, this assumption is reasonably accurate. The outlet flow vectors are shown as bold lines.

Now suppose that the turbine, running at its design point, is gradually brought to a standstill by the application of the brake. We move left along the abscissa to the origin, where $u = 0$. The effect on the velocity diagram is that the peripheral speed, u, is reduced to zero while the axial (flow) velocity, C_x and the stator and rotor flow vectors remain constant. Therefore the change of tangential velocity, ΔC_u, increases to $(\Delta C_{u,dp} + u_{dp})$. The torque on the rotor is proportional to the product of mass flow (which is constant) and ΔC_u and it rises linearly to a maximum value as the rotor is brought to rest. The power output is proportional to the product of torque and shaft speed, and it falls to

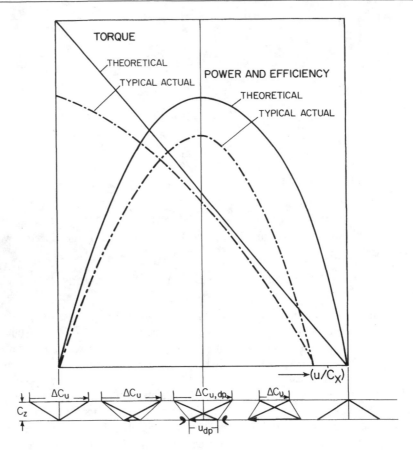

Figure 7.35. Torque, power, and efficiency characteristics of a turbine

zero. If, as we suppose, the turbine is supplied with a constant flow of fluid at constant pressure, the input energy is constant, and the turbine efficiency is directly proportional to shaft power output.

The reverse occurs if we gradually reduce the braking torque to zero, the free-running or so-called "runaway" condition. The peripheral-velocity vector, u, on the velocity diagrams increases in length, and ΔC_u shrinks, passing through the design-point value and eventually going to zero. At this point the rotor is delivering zero torque, and the work output and efficiency are also zero. The "theoretical" shapes of the characteristics are plotted in figure 7.35 from these arguments, together with typical actual curves.

Test results on actual turbines are shown in figures 7.36 to 7.38.

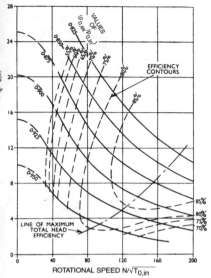

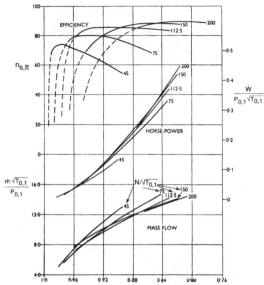

(a) Torque-speed curves of low-pressure-ratio four-stage reaction turbine

(b) Pressure-ratio vs. speed characteristics of four-stage low-pressure-ratio reaction turbine

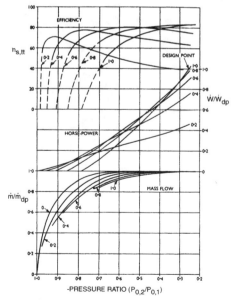

(c) Pressure-ratio vs. speed characteristics of two-stage low-pressure-ratio impulse turbine

Figure 7.36. Axial-turbine characteristics. From Ainley (1948)

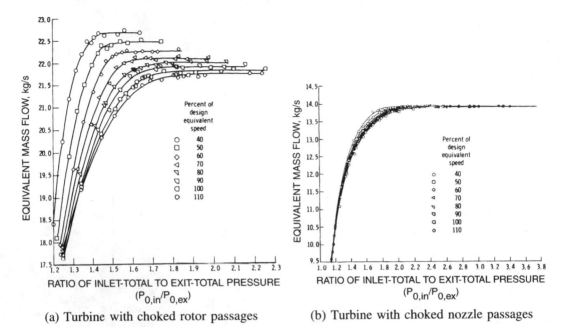

(a) Turbine with choked rotor passages   (b) Turbine with choked nozzle passages

Figure 7.37. Axial-turbine mass-flow vs. pressure-ratio performance "map". From Szanca and Schum (1975)

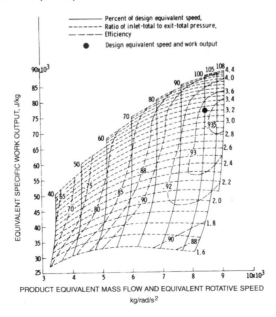

Figure 7.38. Axial-turbine enthalpy-drop vs. mass-flow performance "map". From Szanca and Schum (1975)

References

Ainley, D.G. (1948). Performance of axial-flow turbines. In Proc. Inst. Mech. Engrs., vol. 159, War Emergency Proc. 41, pp. 230–244. London, UK.

Ainley, D.G., and Mathieson, G.C.R. (1951). An examination of the flow and pressure losses in blade rows of axial-flow turbines. R&M no. 2892 (March). Aeron. Research Comm., U.K.

Bird, R.B., Stewart, W.E., and Lightfoot, E.N. (1960). Transport Phenomena. John Wiley & Sons, New York, NY.

Brown, L.E. (1972). Axial flow compressor and turbine loss coefficients: a comparison of several parameters. ASME paper 72-GT-18, New York, NY.

Bry, P.F. (1989). Blading design for cooled high-pressure turbines. In "Blading design for axial turbomachines", AGARD Lecture Series 167, AGARD-LS-167.

Cordes, Gerhard (1963). Stromungstechnik der gasbeaufschlagten Axialturbine. Springer-Verlag, Berlin, W. Germany.

Craig, H.R.M., and Cox, H.J.A. (1971). Performance estimation of axial-flow turbines. Proc. Inst. Mech. Engrs., vol. 185 32/71, paper and discussion. London, UK.

Dunavant, J.C., and Erwin, J.R. (1956). Investigation of a related series of turbine-blade profiles in cascade. Technical note TN 3802. NACA, Washington, DC.

Giles, M.B. (1988). Calculation of unsteady wake/rotor interactions. AIAA Journal of Propulsion and Power, vol. 4, no. 4, July/August. Washington, DC.

Gostelow, J.P. (1976). A new approach to the experimental study of turbomachinery flow phenomena. ASME paper 76-GT-47, New York, NY.

Hodson, H.P. (1985). Boundary-layer transition and separation near the leading edge of a high-speed turbine blade. Journal of Engineering for Gas Turbines and Power, vol. 127, pp. 127–134, January. ASME, New York, NY.

Hodson, H.P., and Dominy, R.G. (1987). Three-dimensional flow in a low pressure turbine cascade at its design condition. Journal of Turbomachinery, vol. 109, no. 2, pp. 177–185, April. ASME, New York, NY.

Hoheisel, H., Kiock, R., Lichtfuss, H.J., and Fottner L. (1987). Influence of free-stream turbulence and blade pressure gradient on boundary layer and loss behavior of turbine cascades. Journal of Turbomachinery, vol. 109, pp. 210–219, April. ASME, New York, NY.

Horlock, J.H. (1966). Axial-flow turbines. Butterworths, London, UK.

Hourmouziadis, J., Buckl, F., and Bergmann, P. (1987). The development of the profile boundary layer in a turbine environment. Journal of Turbomachinery, vol. 109, no. 2, pp. 286–295, April. ASME, New York, NY.

Kacker, S.C., and Okapuu, U. (1982). A mean-line prediction method for axial-flow-turbine efficiency. Journal of Engineering for Power, vol. 104, no. 2, pp 111–119, ASME, New York, NY.

Kiock, R., Lehthaus, F., Baines, N.C., and Sieverding C.H. (1986). The transonic flow through a turbine cascade as measured in four European wind tunnels. Journal of Engineering for Gas Turbines and Power, vol. 108, pp. 277–284, April. ASME, New York, NY.

Korakianitis, T. (1987). A design method for the prediction of unsteady forces on subsonic, axial gas-turbine blades. Doctoral dissertation (Sc.D.), Massachusetts Institute of Technology, Cambridge, MA.

Korakianitis, T. (1993a). On the propagation of viscous wakes and potential flow in axial-turbine cascades. Journal of Turbomachinery, vol. 115, no. 1, January, pp. 118–127, ASME, New York, NY.

Korakianitis, T. (1993b). Hierarchical development of three direct-design methods for two-dimensional axial-turbomachinery cascades. Journal of Turbomachinery, vol. 115, no. 2, April, pp. 315–324, ASME, New York, NY.

Korakianitis, T. (1993c). Prescribed-curvature-distribution airfoils for the preliminary geometric design of axial-turbomachinery cascades. Journal of Turbomachinery, vol. 115, no. 2, April, pp. 325–333, ASME, New York, NY.

Korakianitis, T. and Papagiannidis, P. (1993d). Surface-curvature distribution effects on turbine-cascade performance. Journal of Turbomachinery, vol. 115, no. 2, April, pp. 334–342, ASME, New York, NY.

Korakianitis, T. (1993e). Influence of stator-rotor gap on axial-turbine unsteady forcing functions. AIAA Journal, vol. 31, no. 7, July, pp. 1256–1264, AIAA, Washington, DC.

Meauze, G. (1989). Overview on blading design methods, in "Blading design for axial turbomachines", AGARD Lecture Series 167, AGARD-LS-167.

Okapuu, U. (1974). Some results from tests on a high work axial gas generator turbine. ASME paper 74-GT-81, New York, NY.

Roelke, Richard J. (1973). Miscellaneous losses, in Turbine Design and Application, vol. 2, ed. by Arthur Glassman. Special publication SP-290, pp. 127–128, NASA, Washington, DC.

Schlichting, H. (1960). Boundary-Layer Theory. 4th edition, McGraw-Hill, New York, NY.

Sharma, O.P., Pickett, G.F., and Ni, R.H. (1990). Assessment of unsteady flows in turbines. ASME paper 90-GT-150, New York, NY.

Stewart, Warner L., and Glassman, Arthur J. (1973). Blade design, in Turbine Design and Application, vol. 2, ed. by Arthur J. Glassman. Special publication SP-290, pp. 1–25, NASA, Washington, DC.

Stow, P. (1989). Blading design for multi-stage hp compressors, in "Blading design for axial turbomachines", AGARD Lecture Series 167, AGARD-LS-167.

Szanca, Edward M., and Schum, Harold J. (1975). Experimental determination of aerodynamic performance, in Turbine Design and Application. vol. 3, ed. by Arthur J. Glassman. Special publication SP-290, pp. 103–139, NASA, Washington, DC.

Wagner, J.H., Dring, R.P., and Joslyn, H.D. (1984). Inlet boundary layer effects in an axial compressor rotor: part 1—blade-to-blade effects. ASME paper 84-GT-84, New York, NY.

Wilson, David Gordon. (1987). New guidelines for the preliminary design and performance prediction of axial-flow turbines. Proc. Inst. Mech. Engrs., part A, issue A4, London, UK.

Zweifel, O. (1945). The spacing of turbomachine blading, especially with large angular deflection. Brown Boveri Review, vol. 32, no. 12, Baden, Switzerland.

Problems

1. Find the apparent optimum solidity and the blade inlet and outlet angles for an axial-flow turbine having an inlet flow angle of $+30°$ and an outlet flow angle of $-60°$. Make the actual

incidence zero. Use two stagger angles, $-20°$ and $-40°$, and comment on their apparent relative advantages. Do your preferences agree with the recommendations of figure 7.4b?

2. Explain why a multi-stage turbomachine would give a higher total-to-static flange-to-flange isentropic efficiency than a single-stage machine having the same velocity diagram and working between the same pressures.

3. Calculate the camber angle, and sketch the camber line of an axial-flow turbine blade having inlet and outlet flow angles of $+20°$ and $-50°$. The induced incidence angle is $10°$, the actual incidence angle is $15°$, and the trailing-edge deviation angle is $5°$.

4. Choose blade shapes for the mean diameter of a single-stage axial-flow nitrogen turbine. Use a velocity diagram with axial-flow entry and exit; a flow angle out of the nozzle blades of $70°$; and a loading coefficient of 1.5. The nitrogen total conditions are 172 kN/m^2 and 583 K at inlet and 103 kN/m^2 at outlet. The flow is 5 kg/s. The hub-tip ratio at nozzle exit is 0.75. Use an estimated total-to-total polytropic efficiency across the blading of 0.89 to find the blade speed, diameters, and relative nozzle-throat Mach number. Choose other parameters to aim at high efficiency. Specify 19 nozzle vanes, 23 rotor blades, and 0.25 mm tip clearance. Calculate $\eta_{p,tt}$ across the blading based on group 1 losses.

5. Find the number of stages required, the mean-diameter blade speed, and the optimum solidity for the first stage of a steam turbine for a deep-submergence submarine rescue vessel.

>   Inlet total conditions: 750 lbf/in^2, 100 °R superheat.
>   Outlet total pressure: 2 lbf/in^2.
>   Total-to-total isentropic efficiency: 83 percent across blading.
>   Maximum mean-diameter blade speed: 1, 500 ft/s.
>   Mean-diameter velocity diagrams: $\psi = 1.0$, $\phi = 0.5$, $R_n = 0.5$.

6. Design and draw two neighboring blades of the rotor of a turbine having velocity-diagram parameters of $\psi = 2.0$, $\phi = 0.55$, and $R_n = 0.5$. Calculate the spacing using Zweifel's criterion ($C_L = 0.8$). Use $t_{te}/s = 0.05$, $r_{le}/s = 0.1$, and $s/e = 0.4$. Use the low-Mach-number correlation for outlet angle.

7. (a) Choose a suitable blade shape for an axial-flow water turbine of the following specifications.

>   Mean diameter, d_m: 3.00 in.
>   Blade speed at d_m: 13.09 ft/s
>   Flow axial velocity, C_x: 20.73 ft/s.
>   Change in tangential velocity, $\Delta C_u =$: 7.40 ft/s.
>   Reaction at mean diameter, R_n: 0.5.
>   Axial chord, b: 0.76 in.
>   Number of rotor blades: 10.
>   Number of stator blades: 9.

(b) For your final blade shapes estimate the total-to-total efficiency across the stage blading. The hub diameter is 2.05 in, and the rotor-tip diameter is 3.75 in. There is a tip clearance radially of 0.02 in between the rotor tip and the casing. This turbine was operated with drilling mud in an oil-well drilling string (pipe).

8. Calculate the nozzle-exit outside and inside diameters and static pressure and the turbine-rotor rpm of a single-stage partial-admission axial-flow turbine used for expanding the gases given off during the "blowing" parts of the cycles of six basic-oxygen steel furnaces operating under high pressure. Each furnace supplies one-sixth of the 360-degrees of the nozzle annulus. Each furnace has a design-point mass flow of 54 kg/s of gas of molecular weight 22 at $1,000,000$ Pa and 1700 K stagnation conditions. The flow direction at nozzle exit is to be $68°$ to the axial direction; the rotor-exit actual flow direction is axial. The mean blade speed is 600 m/s. The nozzle-exit hub-shroud diameter ratio is 0.75. There is a one-percent drop in stagnation pressure in the nozzles. The mean $C_{p,e}$ is 1457 J/(kg·K). Draw the enthalpy-entropy and stage velocity diagrams.

9. Suppose that you had completed the design of an axial-flow turbine, including the choice of the shape and number of blades, but your felt concerned that the small chords of the nozzle blades might put them in danger of increased losses because of low Reynolds numbers. Accordingly, you decide to use about half the number of nozzle blades, keeping exactly the same shape, but increasing the chord. What penalties associated with the use of a smaller number of nozzle blades do you think you should keep in mind? What are the benefits in addition to those coming from the use of higher Reynolds numbers?

10. Why, in the turbine characteristics shown, is there no area of instability as there is in typical compressor characteristics?

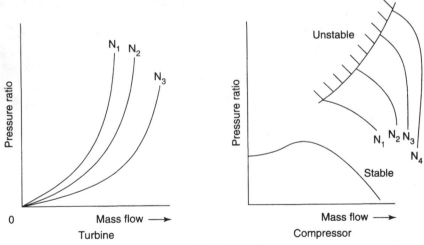

Fig. P7.10. Turbine and compressor characteristics

Chapter 8

The design and performance prediction of axial-flow compressors

This chapter is the counterpart for axial-flow compressors of the blade-design methods of chapter 7 for axial-flow turbines. There are, however, major differences in the approaches to blade design. Turbine blades are drawn from first principles for the passages they form between them. Compressor (and pump) blades are chosen for their own shapes, and at this stage of design are generally selected from a set of precisely shaped blade sections that have been tested in cascade (a straight row of similar blades) in a wind tunnel. Increasingly, however, blade shapes are being produced by "inverse" design methods (see section 7.4) from computer analyses aimed at optimum velocity distributions around the profiles.

In other respects the phases of design outlined for axial-flow turbines at the beginning of chapter 7 apply also to axial-flow compressors, except that the heat-transfer calculations of phase 8 are not normally of interest.

8.1 Introduction

It is easy to design a turbine that will work: all that is needed is to squirt fluid out of a nozzle at the blades of a turbine and the rotor will experience a torque.

To design an axial compressor or pump, on the other hand, requires considerable skill. Pressure-increasing machines are a collection of sets of parallel diffusers (the blade rows) in series, with alternate rows moving with respect to the others. All manner of influences can prevent diffusers from operating to increase static pressure, and many early axial compressors worked more as stirring devices (see the history section for examples). Despite all of these problems many, if not most, axial compressors designed during the decade 1955–1965 and beyond had peak polytropic efficiencies higher than those of axial turbines of that period.

This situation is not as illogical as it sounds. It came about simply because there had been so many failures in axial-compressor design. In several countries and in many private (mainly aircraft-engine) companies an extraordinary effort was made to develop rational compressor-design methods. This many-pronged attack was so successful that the rational design of turbines was left behind.

The reasons for high axial-flow-compressor efficiencies can be summarized as these.

1. Because compressors involve successive diffusing blade rows while, in general, turbines use successive nozzles, compressors are much more difficult to design. Much more attention has therefore been given to compressor design than to turbine design, and overall a higher standard has been reached.

2. Again because axial-flow compressors use successive diffusers, the permissible stage work is less than in axial-flow turbines, and the stage-outlet Mach number (which governs the outlet dynamic head) is much lower. "Leaving losses" are therefore naturally lower in axial compressors than in axial turbines.

3. "Flow-straightening" diffuser stator blades are used at the outlet of axial-compressor blading (sometimes up to three rows being employed) to produce axial flow into the diffuser. Annular diffusers may suffer flow breakdown if the inlet swirl angle is high (see chapter 4). Therefore the annular diffusers downstream from axial-flow compressors usually have unseparated flow and achieve a rise in static pressure. Axial turbines seldom have straightening vanes. (Why?) Less diffusion in their downstream annular diffusers is thereby achieved.

The principal published methods of axial-compressor design were developed at the National Gas-Turbine Establishment in Britain and at the National Advisory Committee for Aeronautics (NACA, a predecessor of NASA) in the United States. We shall give here the NACA method, as modified by Mellor (1956) with other inputs from several sources, such as Horlock (1958).

Cumpsty (1989) gives a fuller discussion of the topics covered by this chapter (and by chapter 9), including in particular the choice of blade profiles, performance and loss data, the effect of casing treatment on surge performance, blade-row matching, and high-Mach-number and transonic design, and is recommended for readers who want to "dig deeper."

8.2 Cascade tests

Compressor and pump blades are normally constructed from a chosen "base profile", such as those listed in tables 8.1 and 8.2 and shown in figures 8.1 and 8.2. The reference, mean, or camber lines can then be given a circular, parabolic, or elliptic shape. However, the camber line of the NACA 65-series is usually defined as in table 8.2. The y coordinates defining the upper and lower (suction and pressure) surfaces can all be multiplied by a factor to increase or decrease the maximum thickness of the blade shape. The base profile is usually specified to give a maximum thickness of 10 percent of the camber-line length.

Experimental data on compressor-blade performance are usually taken from a single row of blades arranged in a linear row, called a linear "cascade" (figure 8.3) rather than in an annulus, to simplify the test conditions. Cascade tests are usually made with blades

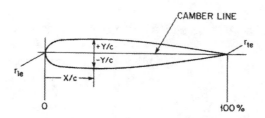

Figure 8.1. Construction of C4 and C7 compressor-blade base profiles (see table 8.1)

Table 8.1. Coordinates of British C4 and C7 compressor-blade base profiles

| x/c (percent) | Airfoil form $(\pm y/c)^*$ | |
	C4	C7
0	0	0
1.25	1.65	1.51
2.5	2.27	2.04
5.0	3.08	2.72
7.5	3.62	3.18
10	4.02	3.54
15	4.55	4.05
20	4.83	4.42
30	5.00	4.86
40	4.89	5.00
50	4.57	4.86
60	4.05	4.43
70	3.37	3.73
80	2.54	2.78
90	1.60	1.65
95	1.06	1.09
100	0	0

$(t_{mx}/c) = 0.10$; $(r_{le}/t) = 0.12$; $(r_{te}/t) = 0.06$.

*These airfoils are symmetric about their camber lines. For a cambered airfoil, the camber line is usually given a parabolic shape, but it may also be circular.

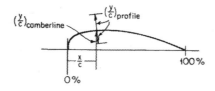

Figure 8.2. Coordinates of NACA 65-series airfoils (see table 8.2)

Table 8.2. NACA 65-series compressor-blade base profile

Station, x/c (percent)	Base profile, $\pm(y/c)_{pl}$ (percent)	Ordinates and slope of mean line (camber line) for $C_{L,tl} = 1.0^*$	
		Ordinate, y/c (percent)	Slope, dy/dx
0	0.000	0.000	
0.5	0.772	0.250	0.42120
0.75	0.932	0.350	0.38875
1.25	1.169	0.535	0.34770
2.5	1.574	0.930	0.29155
5.0	2.177	1.580	0.23430
7.5	2.647	2.120	0.19995
10.0	3.040	2.585	0.17485
15.0	3.666	3.365	0.13805
20.0	4.143	3.980	0.11030
25.0	4.503	4.475	0.08745
30.0	4.760	4.860	0.06745
35.0	4.924	5.150	0.04925
40.0	4.996	5.355	0.03225
45.0	4.963	5.475	0.01595
50.0	4.812	5.515	0
55.0	4.530	5.475	−0.01595
60.0	4.146	5.355	−0.03225
65.0	3.682	5.150	−0.04925
70.0	3.156	4.860	−0.06745
75.0	2.584	4.475	−0.08745
80.0	1.987	3.980	−0.11030
85.0	1.385	3.365	−0.13805
90.0	0.810	2.585	−0.17485
95.0	0.306	1.580	−0.23430
100.0	0.000	0.000	

Leading-edge radius 0.687

* The amount of camber is expressed as design lift coefficient, $C_{L,tl}$, for the isolated airfoil, and that system has been retained. Ordinates and slopes are given in the third and fourth columns for the camber corresponding to $C_{L,tl} = 1.0$. Both ordinates and slopes are scaled directly to obtain other cambers. Cambered blade sections are obtained by applying the thickness perpendicular to the mean line at stations laid out along the chord line (see figure 8.2).

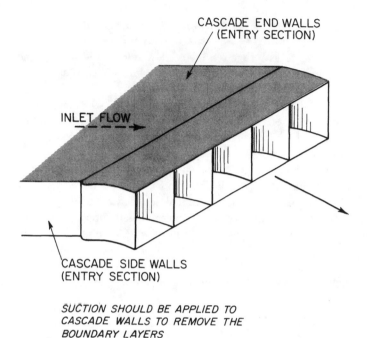

Figure 8.3. Axial-flow compressor-blade cascade

of maximum thickness equal to 10 percent of the chord length, which is a little shorter than a curved camber line.

If we ignore for the moment the possibility of varying the base profile, the camber-line shape, and the maximum thickness, we still have three variables (figure 8.4) to specify a linear cascade:

the blade camber, θ

the setting or stagger angle, λ, and

the spacing of the blades, s or the solidity, $\sigma \equiv (c/s)$. (This is different from the axial solidity b_x/s used in chapter 7.)

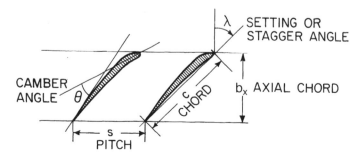

Figure 8.4. Blade-setting variables

To choose a set of values, the resulting linear cascade should be tested over a range of fluid inlet angles from negative to positive stall to obtain the stagnation-pressure loss and the change of outlet angle as a function of the flow inlet angle or incidence, figure 8.5. It is desirable also to take some readings over a range of Reynolds and Mach numbers of interest.

The magnitude of undertaking a testing program with so many variables at a time (1938-1957) when there were no automatic readout and data-processing systems makes it understandable that organizations tended to remain attached to the base profiles and camber-line shapes that they chose initially for testing.

In Britain test results such as those shown in figure 8.5 were correlated for a range of setting angles and cambers into a single curve in which the deflection $\epsilon \equiv |\alpha_{ex} - \alpha_{in}|$, for instance, was given as a ratio with the "nominal" deflection, arbitrarily set at 80 percent of the stalling deflection. Such correlations were not attempted in the NACA method. Separate curves for each blade setting were grouped by Mellor (1956) to produce plots for each blade camber over a range of useful setting angles and useful incidences (figure 8.6). This useful range was defined as that from negative stall to positive stall, these being the incidences, or inlet flow angles, where the stagnation-pressure loss rose to 1.5 times its

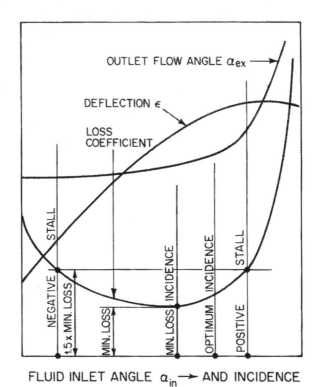

Figure 8.5. Cascade-test results

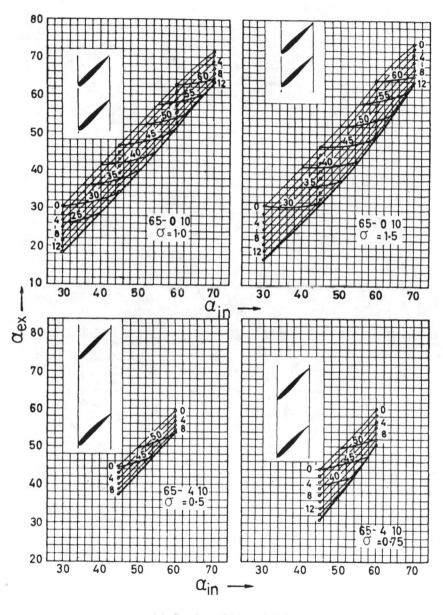

(a) Cambers 010 and 410

Figure 8.6. NACA 65-series cascade data, Mellor charts. From Horlock (1958) and Mellor (1956)

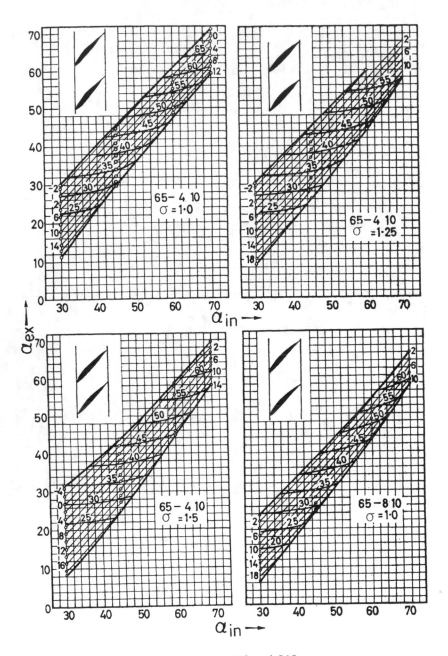

(b) Cambers 410 and 810

Figure 8.6. Continued

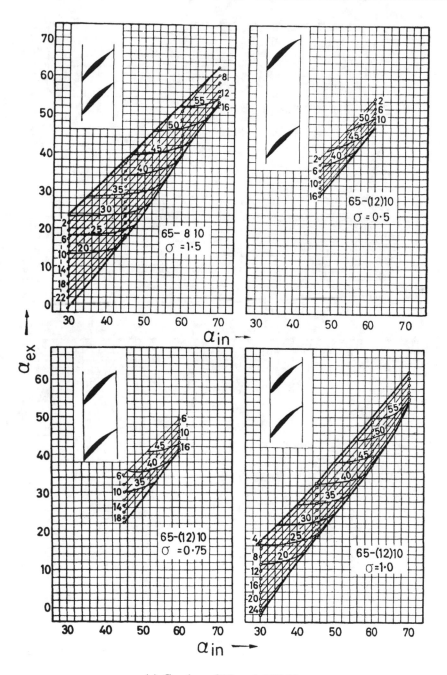

(c) Cambers 810 and (12)10

Figure 8.6. Continued

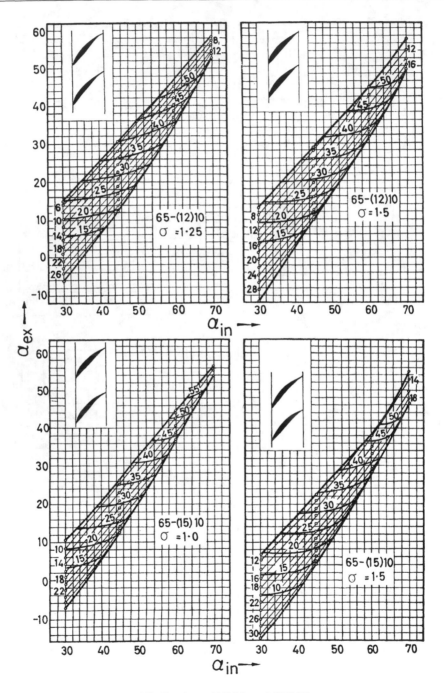

(d) Cambers (12)10 and (15)10

Figure 8.6. Continued

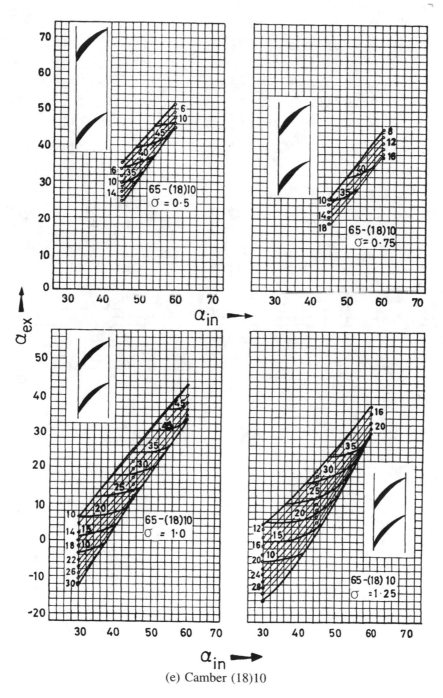

(e) Camber (18)10

Figure 8.6. Continued

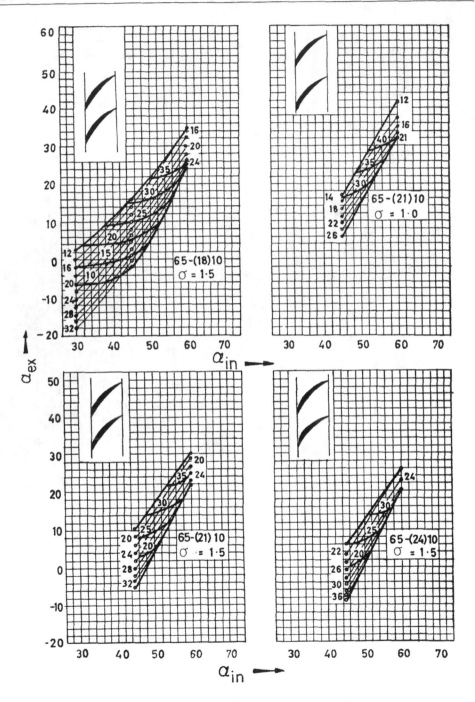

(f) Cambers (18)10, (21)10, and (24)10

Figure 8.6. Concluded

minimum value (figure 8.5). The definition of incidence (figure 8.7) is that for aircraft wings: the angle between the inlet flow direction and the chord line i^* (whereas the more usual definition of incidence in turbomachinery is the angle between the inlet flow direction and the blade inlet angle i). Some geometrical relations between the NACA airfoils and blades with circular-arc camber lines of the same mean height are shown in figure 8.8.

8.3 The preliminary design of single-stage fans and compressors

This section is the counterpart to phase 4 of the design procedure for axial turbines. We suppose that the stage velocity diagram has been chosen for the diameter of concern. If the diffusion ratios, W_2/W_1 and C_1/C_2, have been held at above the de Haller limit of about 0.71 (chapter 5), we are assured that viable blade cascades can be found.

Axial-compressor cascade selection for subsonic conditions

Fluid inlet and outlet angles for each blade row will be specified in the velocity diagram. For blade sections and settings of axial-flow compressors and fans these angles can be selected easily and rapidly by means of the Mellor-NACA charts, figures 8.6 (Mellor 1956), as follows.

1. Mark on tracing paper the ordinates and scales of the Mellor charts, as shown in figure 8.9. Show the desired fluid inlet and outlet angles.

2. Place the sheet, correctly aligned, over the 65-series cascade-data charts shown in figure 8.6. Note the cascade designations and setting angles that include the desired fluid angles between positive and negative stall.

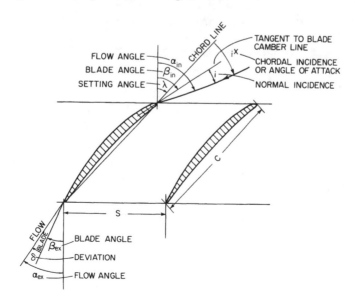

Figure 8.7. Incidence definitions

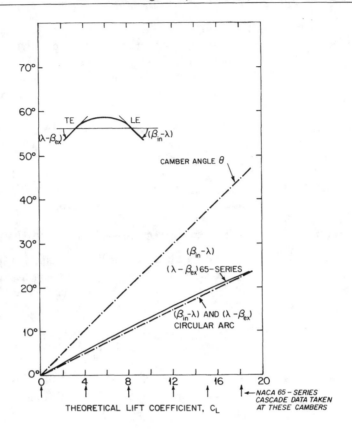

Figure 8.8. Airfoil relationships

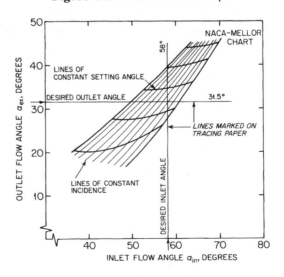

Figure 8.9. Use of NACA-Mellor charts

3. Choose the most suitable blade and setting. Because the curves give relative rather than absolute losses, it is not possible to select the most efficient blade and setting from these data alone. However, the following is a general design rule. Minimum loss for axial-compressor blade rows having moderate loading (typical of mean-diameter conditions) and of moderate hub-shroud ratio (above 0.6, say) is usually given by blade settings at a solidity of about unity and with approximately zero actual incidence. Hub (inner diameter) sections will then have higher solidity and shroud sections will have lower solidity. The actual incidence, i, is obtained from the chordal incidence, i^*, and $(\beta_1 - \lambda)$ given by figure 8.8, through $i = i^* - (\beta_1 - \lambda)$ for the theoretical lift coefficient $C_{L,tl}$.

The meaning of the NACA cascade designation is as follows. The first two numbers indicate the basic airfoil section (in this case the 65-series). The middle number, one or two digits, is ten times the theoretical lift coefficient. The camber angle corresponding to this lift coefficient may be found from figure 8.8. The third number, usually ten, is the maximum profile thickness as a percentage of the chord length. Then the solidity, c/s, the setting angle, λ, and the chordal incidence, i^*, are given as parameters.

The NACA-Mellor charts are for nearly constant axial velocity across the blade rows Where the axial velocity changes by more than 10 percent, correction factors for deviation and loss must be applied. A recent method is that by Starke (1980).

For high-pressure-ratio multistage axial-flow compressors, and for fans that must face varying circuit resistances, it is not always desirable that the design-point condition be selected at the peak-efficiency point (which will normally be close to positive stall). Considerations of off-design operation may dictate design-point operation closer to negative stall. Such considerations are discussed below with regard to stage design of multistage axial compressors.

8.4 Prescribed-curvature compressor-blade design

Similarly to turbine blades the surface-curvature distribution of compressor blades affects the surface Mach-number and pressure distributions (see sections 7.4 and 7.5). The prescribed-curvature-distribution blade-design method described in section 7.5 can be applied to compressor as well as turbine blades. Most modern compressor-blade design techniques (AGARD, 1989) involve inverse, semi-inverse, or optimization methods. Their advantages and disadvantages have been discussed in subsection "direct and inverse blade-design methods" in section 7.4. Computer-aided-design methods for controlled-diffusion axial-compressor blades, cascades, stages and complete machines have been presented by Sanger (1983) and Massardo (1990). Stator-rotor-interaction, forced-response, and flutter problems in compressor and fan blades require consideration of unsteady-flow phenomena from the preliminary-design stages (see section 8.12). Some computational-fluid-dynamic (CFD) methods for blade-optimization in unsteady flow are emerging, but these involve much smaller changes in geometry than those needed in preliminary design. These preliminary-design global-optimization problems cannot be handled by inverse methods. It is therefore possible that future compressor-blade and turbine-blade preliminary-design methods will use techniques such as those described in sections 7.4 to 7.6.

8.5   Performance prediction of axial-flow compressors

A real compressor can be thought of as an ideal machine taking in a gas at $p_{0,1}, h_{0,1}$ and delivering it at $p'_{0,2,tl}, h'_{0,2}$ (figure 8.10) with added losses which make the actual delivery conditions $p'_{0,2}, h_{0,2}$. To make them manageable, we treat these losses as if they were separable. In fact they are to some extent interactive and connected.

Group-1 losses

Group-1 losses are pressure losses resulting from fluid friction. Examples are boundary-layer friction on blade and casing walls and flow-separation losses in areas of stall and blade trailing edges.

The sum of group-1 pressure losses is designated $\Sigma \Delta p_{0,g,1}$ in figure 8.10. These pressure losses reduce the outlet pressure from the theoretical $p'_{0,2,tl}$ to $p'_{0,2}$ which is identical to $p_{0,2}$.

Group-2 losses

Group-2 losses take shaft power and degrade it to an enthalpy increase in the gas-discharge condition. For instance, the "windage" gas friction on the compressor disks

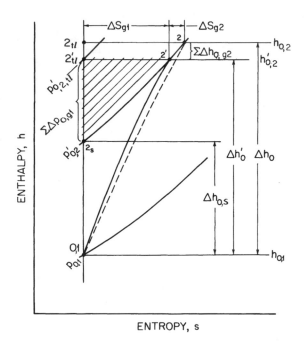

Figure 8.10. Loss representation in adiabatic compression

usually appears in the discharge gas state either by heat transfer through the annulus (hub) wall or, more likely, by a mass exchange of the main flow with the gas partly trapped in the cavity around the disks. In either case the result of all such degradations of shaft power is an increase in discharge enthalpy from $h'_{0,2}$ to $h_{0,2}$. The sum of group-2 losses is designated $\Sigma \Delta h_{0,g,2}$ in figure 8.10.

Group-3 losses

Group-3 losses are losses of shaft power through friction, the energy from which is dissipated away from the working fluid. Examples are the friction losses in external bearings and seals. The power losses appear as increases in compressor power requirements, (whereas in turbines group-3 losses decrease power output). In chapter 2 group-3 losses are designated "external energy losses" (section 2.6). Their treatment is obvious, and we do not need to be concerned further with group-3 losses here.

Isentropic efficiency

By definition,

$$\eta_{s,c} \equiv \frac{\Delta h_{0,s}}{\Delta h_0} = \frac{\Delta h_{0,s}}{\Delta h'_0 + \Sigma \Delta h_{0,g2}} \tag{8.1}$$

$$\eta_{s,c} = \frac{\overline{C}_{p,1-2s}(T_{0,2s} - T_{0,1})}{\overline{C}_{p,1-2'}(T'_{0,2} - T_{0,1}) + \overline{C}_{p,2'-2} \Sigma \Delta T_{0,g2}}$$

$$= \frac{[(T_{0,2s}/T_{0,1}) - 1]}{\left(\dfrac{\overline{C}_{p,1-2'}}{\overline{C}_{p,1-2s}}\right)[(T'_{0,2}/T_{0,1}) - 1] + \left(\dfrac{\overline{C}_{p,2'-2}}{\overline{C}_{p,1-2s}}\right) \Sigma(\Delta T_{0,g2}/T_{0,1})}$$

$$\eta_{sc} = \frac{r^{(R/\overline{C}_{p,1-2s})} - 1}{\left(\dfrac{\overline{C}_{p,1-2'}}{\overline{C}_{p,1-2s}}\right)\left\{\left[r + (\Sigma \Delta p_{0,g1})/p_{0,1}\right]^{(R/\overline{C}_{p,1-2'})} - 1\right\} + \dfrac{\Sigma \Delta h_{0,g2}}{(\overline{C}_{p,1-2s} T_{0,1})}} \tag{8.2}$$

where $r \equiv p_{0,2'}/p_{0,1}$.

In preliminary design the specific-heat ratio $(\overline{C}_{p,1-2'}/\overline{C}_{p,1-2s})$ is often taken to be unity, and the group-2 losses are often merely guessed at by deducting one or two points from the value of isentropic efficiency calculated without accounting otherwise for group-2 losses.

The final state after compression is $p_{0,2}, h_{0,2}$. This point can be considered to be arrived at directly by the process line shown dashed in figure 8.10. Or two processes could be conceptually involved: points 01 to 2', with just group-1 (pressure) losses; and 2' to 2, with just shaft-power (energy) losses (group 2). Or, we could think of an ideal (isentropic) process over the actual enthalpy rise from 01 to 2_{tl} followed by a pure increase of entropy from 2_{tl} to 2, a throttling process, being the sum of the entropy increases due to group-1 and group-2 losses.

Calculation of group-1 losses

We give here simplified, approximate methods for estimating losses in the following categories:

1. profile diffusion,
2. trailing-edge thickness,
3. high-Mach-number losses,
4. end-wall boundary-layer friction, and
5. blade-end clearance.

Of the published loss correlations we have chosen the method of Koch and Smith (1976) as a basis, with that of Lieblein (1959) providing simplifications. We have also made our own approximations, as seems appropriate for preliminary design.

Blade-profile losses

Lieblein showed that the losses around the blade profile appeared as a boundary-layer "momentum" thickness, θ_{ie}, at the trailing edge or, in the wake, θ_{ex} (θ_{ex} will be larger than θ_{ie} for highly loaded blades because there will be a mixing loss as the suction-surface and pressure-surface boundary layers join to form the wake). Lieblein also showed that as the aerodynamic loading on a compressor blade increased, the diffusion on the suction surface increased, but that on the pressure surface stayed approximately constant.

He therefore defined an "equivalent diffusion ratio" (D_{eq}) as the ratio of the suction-surface peak velocity to the outlet velocity:

$$D_{eq} \equiv \left[\frac{W_{mx}}{W_{ex}} \right] \tag{8.3}$$

and he showed that this could be correlated by

$$D_{eq} \equiv \frac{\cos \alpha_{ex}}{\cos \alpha_{in}} \left[1.12 + 0.61 \frac{\cos^2 \alpha_{in}}{\sigma} (\tan \alpha_{in} - \tan \alpha_{ex}) \right] \tag{8.4}$$

where α_{in} and α_{ex} are the blade-row inlet and outlet mean flow angles, and σ is the solidity (chord/spacing, c/s). This correlation could be extended to cover operation of the blade row at incidence, i, defined here as the difference between the inlet flow angle with that for minimum loss,

$$i \equiv (\alpha_{in} - \alpha_{in,be}):$$

$$D_{eq} = \frac{\cos \alpha_{ex}}{\cos \alpha_{in}} \left[1.12 + 0.0117 i^{1.43} + 0.61 \frac{\cos^2 \alpha_{in}}{\sigma} (\tan \alpha_{in} - \tan \alpha_{ex}) \right] \tag{8.5}$$

Koch and Smith (1976) introduced factors correlating the airfoil maximum-thickness ratio, t_{mx}/c and the streamtube contraction ratio, A_{ex}/A_{in}. They also used cascade data for which the boundary layers were turbulent to a greater extent than were those used by Lieblein, and which therefore were somewhat more representative of the conditions in a compressor:

$$
D_{eq} \equiv \frac{W_{in}}{W_{ex}} \left[1 + 0.7688 \left(\frac{t_{mx}}{c} \right) + 0.6024\Gamma \right] \left\{ (\sin \alpha_{in} - 0.2445\sigma\Gamma)^2 \right.
$$

$$
\left. + \left(\frac{\cos \alpha_{in}}{1 - [0.4458\sigma(t_{mx}/c)/\cos((\alpha_{in} + \alpha_{ex})/2)][1 - [1 - (A_{ex}/A_{in})]/3]} \right)^2 \right\}^{1/2} \tag{8.6}
$$

where the circulation, Γ, is given by

$$
\Gamma = \frac{r'_{in} C'_{u,in} - r'_{ex} C'_{u,ex}}{\sigma W'_{in}} \tag{8.7}
$$

and

$$
\begin{aligned}
r' &\equiv r/r_m \\
r_m &\equiv \text{mean radius} \equiv (r_{hb} + r_{sh})/2 \\
W' &\equiv W/u_m \\
u_m &\equiv \text{blade peripheral speed at } r_m \\
\sigma &\equiv \text{solidity, } c/s \\
C' &\equiv C/u_m
\end{aligned}
$$

The subscripts "*in*" and "*ex*" are for inlet to and outlet from the blade row. The symbol W is used for relative velocity rather than C in equation 8.6 to indicate that the correlation can be used for moving as well as for fixed blade rows.

The Lieblein and the Koch-Smith correlations, equations 8.5 and 8.6, can be regarded as alternatives, with equation 8.6 being used, presumably, where greater precision is desired.

Lieblein's data can be correlated with blade-outlet momentum thickness, θ_{ex}, by

$$
\left(\frac{\theta_{ex}}{c} \right) = 0.00258 e^{0.886 D_{eq}} \tag{8.8}
$$

The momentum thickness can be regarded as an intermediate parameter that need not be defined.

Koch and Smith's data are correlated by

$$
\left(\frac{\theta_{ex}}{c} \right) = 0.00138 e^{1.1127 D_{eq}} + 0.0025 \tag{8.9}
$$

and by a trailing-edge boundary-layer shape factor, H_{te}:

$$
H_{te} = 1.26 + 0.795(D_{eq} - 1)^{1.681} \tag{8.10}
$$

The shape factor is the ratio of the boundary-layer displacement thickness δ^* to the momentum thickness θ, but again may be treated as an intermediate parameter.

In the Koch-Smith correlations the values of θ_{ex}/c and H_{te} given by equations 8.9 and 8.10 are for nominal conditions of:

1. an inlet Mach number below 0.05,

2. no contraction of the streamtube (annulus or blade) height h_b,

3. a Reynolds number, Re (based on chord and blade-row inlet velocity) of 10^6, and

4. hydraulically smooth blades.

Multipliers are given for conditions other than nominal. These are reproduced in figure 8.11. The corrected values of θ_{ex}/c and H_{te} for each blade row can then be used in the following relation, due to Lieblein, to find the blade-profile stagnation-pressure loss:

$$
\frac{\Delta p_{0,pl}}{[\rho_{0,in} W_{in}^2/2g_c]} = +2\left(\frac{\Theta_{ex}}{c}\right) \frac{\sigma}{\cos\alpha_{ex}} \left(\frac{\cos\alpha_{in}}{\cos\alpha_{ex}}\right)^2 \left[\frac{2}{3 - 1/H_{te}}\right]
$$

$$
\times \left[1 - \left(\frac{\Theta_{ex}}{c}\right)\frac{\sigma H_{te}}{\cos\alpha_{ex}}\right]^{-3} \tag{8.11}
$$

If the more-approximate Lieblein approach is used (equation 8.5) a constant value may be taken for H_{te} of 1.08.

High-Mach-number leading-edge losses

Koch and Smith give the following correlation, ascribed to D. C. Prince:

$$
\Delta p_{0,le} = -p_{0,in} \ln\left\{1 - \frac{t_{le}}{b\cos\alpha_{in}}[1.28(M_{in} - 1) + 0.96(M_{in} - 1)^2]\right\} \tag{8.12}
$$

Here t_{le} is the blade thickness at the leading edge and b the axial chord. In all these expressions the subscripts "*in*" and "*ex*" denote "relative conditions at inlet to (and outlet from) the blade row". Also, to conform with figure 8.10 and equation 8.2, the stagnation-pressure losses are treated as positive, although for consistency pressure losses are by definition negative.

Stage free-stream efficiency

By adding the profile and leading-edge stagnation-pressure losses for a rotor and a succeeding (or preceding) stator-blade row (depending upon whether the stage is defined as a rotor-plus-stator or a stator-plus-rotor), and inserting these in equation 8.2, the stage isentropic compressor free-stream efficiency is obtained, $\eta_{s,c,fs}$. (Group-2 losses are not accounted for at this point and are set to zero in equation 8.2.)

For greater accuracy, Koch and Smith recommend two further steps and an iteration. The combination of θ_{ex}/c and H_{te} will give a boundary-layer displacement thickness,

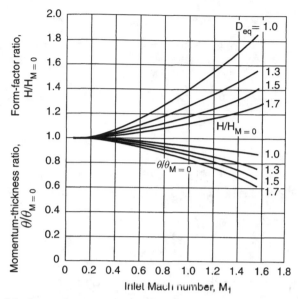

(a) Effect of inlet Mach number on calculated trailing-edge thickness and form factor

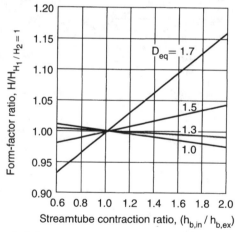

(b) Effect of streamtube height variation on calculated trailing-edge form factor

Figure 8.11. Multipliers for blade momentum thickness and form factor. From Koch and Smith (1976)

δ^*, which would in effect thicken the blade profile and change the velocity distribution. For a more-exact calculation, beyond the scope of this book, the calculation should be iterated until the calculated value of δ^* conforms with the trailing-edge relative velocity. The second step is to calculate the mixing losses that will occur in the wake downstream

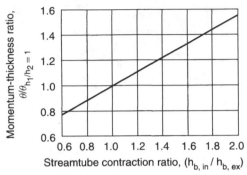

(c) Effect of stream-tube height variation on calculated trailing-edge momentum thickness

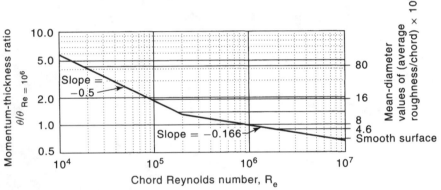

(d) Effect of Reynolds number and surface finish on calculated trailing-edge momentum thickness

Figure 8.11. Concluded

of the trailing edge. However, it is pointed out that these are small except at blade loadings higher than are recommended here, and therefore this step also is considered to be beyond the scope of this book.

End-wall losses

The effects on stage efficiency of hub-shroud ratio, blade-shroud clearance, axial gap between blade rows and, to a lesser extent, blade aspect ratio (blade height over chord, h_b/c) can be estimated using the Koch and Smith formulation of end-wall losses. We give here a simplified, approximate version.

The test data were correlated by Koch and Smith with the sum of the relative displacement thicknesses of the two end-wall boundary layers plotted against the stage pressure-rise coefficient relative to the maximum pressure-rise coefficient of which the stage is capable (figure 8.12a). Rotor-blade clearance is a parameter. (There is no guid-

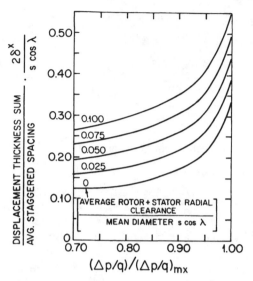

(a) Sum of end-wall displacement thicknesses

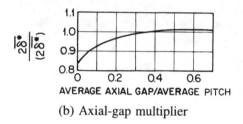

(b) Axial-gap multiplier

Figure 8.12. Correlations of end-wall losses. From Koch and Smith (1976)

ance for the type of disk-and-drum construction where the stator blades are unshrouded and also have an end clearance.) A multiplier for mean blade-row axial gap (relative to mean blade pitch) different from 0.35 is shown in figure 8.12b.

The stage isentropic stagnation-to-stagnation efficiency is then given by a modification of the stage free-stream efficiency, already obtained:

$$\eta_{s,c,se} = \eta_{s,c,fs} \left[\frac{1 - (2\delta_{12a}^*/h_b)(2\delta^*/2\delta_{12a}^*)}{1 - \frac{1}{2}(2\delta_{12a}^*/h_b)(2\delta^*/2\delta_{12a}^*)} \right] \qquad (8.13)$$

where $2\delta_{12a}^*$ is the value of end-wall boundary-layer displacement thickness (for the two end walls) given by figure 8.12a and h_b is the mean blade height.

For preliminary-design purposes it is probably sufficiently accurate to estimate the relative static-pressure-rise coefficient with which to enter figure 8.12a. For somewhat greater accuracy one can use the velocity diagram and the blade-row fluid-turning curves (figure 8.6) to obtain α_{in}, α_{ex}, $\alpha_{in,mx}$ and $\alpha_{ex.mx}$ for each blade row, the maximum values being those at positive stall.

$$\frac{(\Delta p/q)_{tl}}{(\Delta p/q)_{tl,mx}} = \frac{1 - \dfrac{\cos^2 \alpha_{in}}{\cos^2 \alpha_{ex}}}{1 - \dfrac{\cos^2 \alpha_{in,mx}}{\cos^2 \alpha_{ex,mx}}} \qquad (8.14)$$

Aspect ratio as an independent variable

As in the case of axial-flow turbines the designer has considerable freedom to choose the number of blades in a blade row. A large number of blades leads to a shorter compressor, and vice versa. As the number of blades in rotors and stators is changed, there will be proportional changes in the excitation frequencies from blade wakes, and nonlinear changes in blade natural frequencies, thus allowing the designer to use the choice of number of blades as a principal method of avoidance of critical excitations.

In aircraft and some ground-transportation engines the requirement for the smallest possible engine often outweighs other considerations, and high-aspect-ratio compressors result from the need to reduce compressor lengths. When there is design freedom to do otherwise, the designer should be aware that the use of low-aspect-ratio compressor blading brings many benefits. In experimental study of such blading Reid and Moore (1980) confirmed earlier studies, not all published, that low aspect ratios produce:

1. higher peak pressure ratio, higher stage efficiency, greater stall margin;

2. improved performance over the whole blade span;

3. good operation at higher diffusion factors and higher incidences; and

4. improved high-Mach-number performance.

In addition, substantial cost savings result from the use of fewer, larger, blades and vanes.

Example 1   Calculation of compressor-stage efficiency

This example is of the calculation of the efficiency of the hub (inner diameter) section of a low-speed fan. The flow is specified to be almost incompressible, and several of the factors that would normally be accounted for in the calculation of a compressor stage can be ignored. The example then demonstrates the method but not the potential complexity.

The problem is to calculate the stagnation-to-stagnation isentropic efficiency, from stage inlet to stage outlet, of the hub section of an axial-flow fan (figure 8.13) of the following specifications:

$p_{0,in} = 101 \text{ kN/m}^2$

$T_{0,in} = 310 \text{ K}$

Fluid = air

$\overline{C}_{p,c} = 1.005 \text{ kJ/kg-}^\circ\text{K}$

$R = 286.96 \text{ J/kg-}^\circ\text{K}$

$\Lambda \equiv \text{(hub-shroud diameter ratio)} = 0.5$

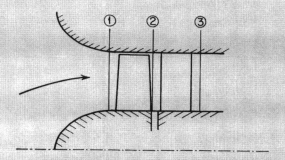

Figure 8.13. Calculation planes for axial-flow-fan example

$\phi_{hb} \equiv$ (flow coefficient at hub) $= 0.6$
$u_{hb} \equiv$ (blade speed at hub) $= 14.30$ m/s
$(\Delta p/q)_{tl,hb} \equiv$ (pressure-rise coefficient at (rotor) hub) $= 0.5$
Axial-flow inlet and outlet flow in a rotor-stator stage.
$d_{sh} \equiv$ (shroud diameter) $= 0.4$ m
Rotor-blade radial clearance $= 1$ mm
Number of rotor blades $= 16$
Number of stator blades $= 15$

Solution.

The hub velocity diagram was calculated and compressor-blade sections were chosen from figure 8.6 as given in table 8.3 and figure 8.14.

The axial gap between rotor and stator is 0.35 of the blade axial chord (thus no boundary-layer correction will be needed from figure 8.12b).

Table 8.3. Choice of hub-section blade profiles

		Rotor	Stator
Flow inlet angle:	design point	$-59.04°$	$35.90°$
	+stall point	$-67.0°$	$46.08°$
Flow outlet angle:	design point	$-43.31°$	$0°$
	+stall point	$-48.0°$	$4.8°$
Blade sections		65(12)10	65(18)10
Maximum thickness, t_{mx}/c		0.125	0.075
Solidity, σ		1.25	1.25
Stagger, λ		$49°$	$15°$
Annulus-area ratio		1.0	1.0
Relative inlet velocity, m/s		16.68 (W_1)	10.59 (C_2)
Work coefficient, ψ_{hb}		0.4343	

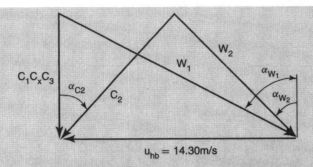

$u_{hb} = 14.30$m/s

Figure 8.14. Hub velocity diagram of axial-flow fan

We first calculate the circulation for rotor and stator from equation 8.7. The ratio of the (area) mean diameter to the shroud diameter,

$$\frac{d_m}{d_{sh}} - \sqrt{\frac{1}{2}(1 + \Lambda^2)} = 0.791$$

Therefore $r'_{hb} \equiv r_{hb}/r_m = 0.632$.

We have given the principal results of calculations in table 8.4. The number of significant figures carried along in the calculation is much greater than the accuracy warrants.

Table 8.4. Loss calculation for compressor stage

	Rotor	Stator
Circulation at hub (equation 8.7), Γ_{hb}	0.1883	0.2965
Assumed (specified), $(t_{mx}/c)_{hb}$	0.125	0.075
(recognizing that figure 8.4 is for t_{mx}/c of 0.10)		
$(W_{in}/W_{ex})_{hb}$	1.4142	1.2345
D_{eq} (equation 8.6)	2.9337	1.5263
θ_{ex}/c (equation 8.9)	0.03715	0.00981
H_{te} (equation 8.10)	3.592	1.5065
Circular pitch, s_{hb} m	0.0491	0.0524
Mean density, ρ_{st} kg/m^3	1.412	1.412
Mean viscosity, μ_{st} $Ns/m^2 \times 10^5$	1.845	1.845
Reynolds number, $\rho_{st} W_{in} c_{hb}/\mu_{st}$	78,300	53,060
$\theta/\theta_{Re=10^6}$ (figure 8.9d)	2.06	2.45
θ_{ex}/c corrected for Re	0.0765	0.02403
$\Delta p_0/q_{in}$ (equation 8.11)	0.0442	0.0294
Inlet dynamic head, $\rho_{0,in} W_{in}^2/2$ N/m^2	196	79.2
Stagnation-pressure loss, profile, Δp_0 N/m^2	8.665	2.326
$\Sigma(\Delta p_0)$ N/m^2		10.991

To use this stagnation-pressure drop to calculate the free-stream efficiency, we calculate $p_{2',tl}$ from $\Delta h'_0$ (which, accounting only for group-1 losses, is the actual work, found from the work coefficient, table 8.3, and the blade speed) (figure 8.10). The stagnation-pressure losses are deducted to give $p'_{0,2}$, from which $\Delta h_{0,s}$ can be calculated. The free-stream isentropic efficiency for the stage is then found to be, for this example, 0.891.

We now have merely the end-wall losses to account for, using figure 8.12a and equations 8.13 and 8.14. The curves of figure 8.12a are for the average rotor and stator clearance (we assume that the stator blades have zero clearance, and we therefore use half the rotor-blade radial clearance) normalized by the mean-diameter value of $s \cos \lambda$. We know this last quantity for the hub radius, but we do not know precisely how it varies with radius. The spacing will increase proportional to radius, and the stagger or setting angle, λ, will increase slightly with radius for the rotor blades, which is the appropriate blade row because it is the one having clearance. The value of (radial clearance$/s_{hb} \cos \lambda_{hb}$) for the rotor is 0.031; the average rotor-stator value is half this, 0.0155, and an estimate of the mean-diameter value is 0.011.

For the rotor, the stall flow angle is estimated from figure 8.6d and listed in table 8.3. (Very conservative airfoil cambers were chosen to allow for blade-surface fouling in the projected application.) Equation 8.14 can therefore be solved (giving for this case 0.759 for the rotor) and used as input to figure 8.12a. The value of $2\delta^*/s_m \cos \lambda_m$ in this figure is then read at approximately 0.15. The hub values in tables 8.3 and 8.4 enable this to be converted to $(2\delta^*/h) = 0.0483$; the axial-gap multiplier in figure 8.12b is 1.0; the multiplier from equation 8.13 is 0.975; and the stage efficiency becomes $0.975 \times 0.891 = 0.869$.

This example, and the loss correlations that preceded it, were for single compressor stages and for stagnation-to-stagnation efficiencies. For a full calculation of the efficiency across a multi-stage axial-flow compressor and diffuser, similar calculations to those given here would be made for the hub, mean, and shroud streamlines for all stages (with the stage efficiencies being the average of the three values) to arrive at a stagnation-to-stagnation isentropic efficiency across the blading. (This may be too complex. The major aircraft-engine companies currently spend about $1 billion to develop a new engine, and, according to one senior compressor designer, "momentous decisions are made on the basis of mean-line analyses".) Then the stagnation-pressure losses in the diffuser (see chapter 4) would be added, and, if a stagnation-to-static efficiency were desired, the dynamic pressure at diffuser outlet would also be treated as a loss.

Some other considerations of multi-stage-compressor design follow.

8.6 The design and analysis of multi-stage axial compressors

At first sight it may seem that one can simply add compressor stages together (with, of course, progressive reductions in annulus area or blade height to accommodate the increase in density) to reach any desired pressure ratio. In fact high-pressure-ratio "fixed-geometry" compressors have severe starting and off-design losses, for the reasons explained in this section, so that the practical limit of pressure ratio for such machines working on air is about ten to one. This range can be considerably extended, perhaps up to thirty to one, by one or more of the approaches described in section 8.8.

However, let us suppose that we are given design specifications acceptable for a fixed-geometry compressor (say, a pressure ratio of six to one) to compress air from normal ambient temperature and pressure. We can calculate the isentropic enthalpy rise and, making an estimate of the appropriate efficiency, we can estimate the actual enthalpy rise, figure 8.15.

The first two questions to be answered are related: what type of first-stage velocity diagram and blades should be used? And how many stages should be employed?

The choice of the first-stage diagram will entail the determination of the mean-diameter and shroud-diameter peripheral speed. We may then decide to keep the shroud diameter constant for later stages (or the mean or the hub diameters could have been held constant with different advantages and disadvantages). While the first-stage velocities will be chosen largely on the question of the relative Mach number allowable, subsequent stages will have lower Mach numbers, and there will be some freedom to modify somewhat the velocity diagram used for the first stage to yield higher-work downstream stages. The number of stages to give the overall (estimated) enthalpy rise will result from these choices of first-stage and subsequent velocity diagrams.

Normally, the vector-diagram mean-diameter axial velocity at design point is kept approximately constant (at the reference diameter) throughout the machine. In practice, this is accomplished by changing the annulus area between each blade row and "fairing" the annulus shape between these locations.

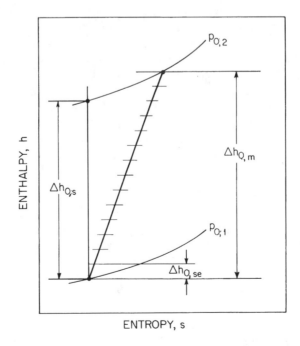

Figure 8.15. Stage enthalpy rises in multi-stage compressor

Off-design performance of a multi-stage axial compressor

The most important off-design situation for a compressor is when it is started from rest, usually against a fixed circuit resistance. This resistance will usually be of the "square-law" type: the pressure ratio will increase approximately as the square of the volume flow. In a gas-turbine engine without interstage air bleed-off to atmosphere, the circuit resistance will in addition increase sharply when heat is added to the compressed air, especially by combustor ignition. Then the gas passing through the high-pressure-turbine nozzles will undergo a sharp increase in temperature and, therefore, specific volume, bringing about an increase in the pressure ratio required to pass the flow through the nozzle passages.

The following argument is made in rather specific terms so that the reader may have a distinct picture in mind. However, the argument may obviously be generalized.

Let us take the case of an air compressor designed to produce a pressure ratio of 6:1 at design point from atmospheric air at inlet. The first stage is designed to a maximum relative Mach number of 0.9 with respect to the rotor-blade tips. Let us suppose that this stage design is then repeated fourteen times with successively shorter blades to give the overall pressure ratio desired. (The relative Mach number will of course be greatly reduced as the temperature and the speed of sound increase through the machine.) If the polytropic efficiency $\eta_{p,c,tt}$ across the blading at design point is 0.9, the density ratio (static) from inlet to outlet will be about 3.5. Therefore, since the design-point axial velocity is kept constant (by the use of similar stages), and because the mass flow is constant (no air bleed), the annulus area at outlet is smaller than that at inlet by a factor of 3.5 (figure 8.16a).

Now let us consider what will happen at off-design conditions. By "off-design" we mean that we impose on the compressor a different combination of shaft speed and mass flow (or, more strictly, inlet Mach number). Let us take the following conditions in turn.

1. The compressor operates at low shaft speed and low mass flow to give the design-point value of C_x/u ($\equiv \phi$) at the first stage (line 1, figure 8.16b). At low speed there will be little or no increase of density. So in succeeding stages, u is constant, but C_x increases because the cross-sectional annulus area decreases and ϕ goes up beyond ϕ_{dp} into the region of negative stall (line 1). With so many of the later stages in negative stall, the small pressure rise produced by the first stages will be more than dissipated by losses, and the result will be a negative pressure ratio (that is, there would have to be external pressurization to force this amount of flow through the compressor). This is shown as point 1 in the compressor characteristics of figure 8.16c.

2. The compressor operates at low speed and low mass flow such that the first two or three stages are in positive stall (line 2). In this case there are more stages working, and a small pressure ratio should be produced across the compressor. The point is plotted as 2 in the compressor-characteristics chart (figure 8.16c).

3. Low speed, even lower mass flow (see line 3 and point 3 in figures 8.16b and c). The first stages are deeper into positive stall, but the later stages should be

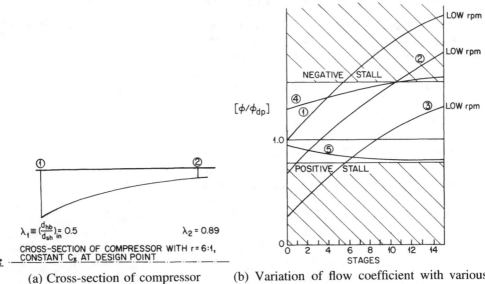

(a) Cross-section of compressor

(b) Variation of flow coefficient with various combinations of rotational speed and mass flow

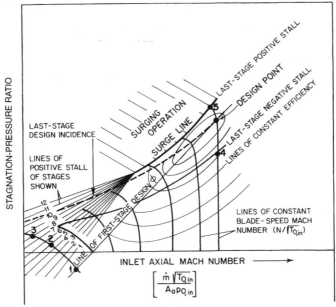

(c) Typical multi-stage axial-flow-compressor characteristics

Figure 8.16. Off-design operation of a high-pressure-ratio axial-flow compressor

working in the high-efficiency region of their characteristics and should produce a pressure rise. (However, the early stages could be deeply into a strong rotating stall which could propagate downstream, forcing later stages to operate along the low-efficiency line shown in the accompanying diagram of stage characteristics.)

4. High (design-point) speed, higher-than-design-point mass flow, line 4. There will some increase in C_x because the stages will be producing a lower work coefficient at higher flow coefficients, but the increase in flow coefficients along the compressor will not be as strong as at low speeds because there will be some increase in density. But when the later stages run into negative stall at full speed, the high local velocities around the blades will sooner or later cause local sonic velocities to be reached, with consequent shock losses. The mass flow will not be susceptible to further increase, and a vertical constant-speed line will result on the compressor characteristics.

5. High (design-point) speed, lower-than-design-point mass flow. Each stage will give more than design-point work because of the reduced mass flow. The density will be increased to above that of design. This increase of density will further reduce C_x and ϕ. The line 5 has been shown as one where the last stage is on the verge of stalling. If it were to be pushed into stall, by a momentary reduction in mass flow, for instance, the next-to-the-last stage would suddenly experience the full compressor outlet pressure. Being so close to positive stall, the stage could not deliver flow, but would go into stall itself. This domino-effect stall sequence would generate what is called a "surge" flow breakdown. The transient forces imposed during surge are so large that no known high-pressure-ratio axial compressor could withstand repeated full-speed, full-density surging for long without losing blades. In addition, with little throughflow the large energy dissipation produces rapidly increasing air temperatures that can degrade the properties of the blade materials.

Intermediate speeds and mass flows would complete the characteristics to show not only lines of positive and negative stall and lines of design-point incidence for various stages but also lines of constant isentropic or polytropic efficiency, which are also drawn in figure 8.16c.

8.7 Compressor surge

Compressor surge is therefore essentially a compressible-flow high-Mach-number phenomenon. The boundary between steady-flow and surging operation is called the "surge line" (figure 8.16c), and it exists strictly only where the blade-speed Mach number and axial Mach number are in the compressible region, say above 0.3. (However, in most representations of compressor characteristics, the surge line is incorrectly continued down to the zero-mass-flow origin.) At lower speeds the compressor can operate in a stable manner with one or more low-pressure stages stalled. Whereas low-speed stalled operation depends very little on downstream conditions, surging is a function of the capacity and loss characteristics of the system into which the compressor discharges.

Let us suppose that, when operating the previously discussed compressor in the condition represented by point 5 in figure 8.16c, the circuit resistance is increased slightly so that the last stage stalls. The next-to-last stage is exposed to the full outlet pressure when it is working near positive stall and will in turn stall. In this manner the stall propagates backward through the compressor, with reverse flow becoming fully established when the stall reaches the first stage. The downstream circuit will then discharge itself through the compressor, which is still rotating forward at full speed. After a period of time, the length of which will depend on the volume and loss characteristics of the circuit, the pressure will have fallen to the point where most if not all of the compressor stages will again be able to operate unstalled. Forward flow becomes re-established, the discharge pressure gradually rises, and, unless the circuit resistance has been lessened by some control change, the compressor will surge again.

The period of this surge cycle may be several seconds for a process compressor discharging into a large chemical reaction vessel, or may be a fraction of a second for an aircraft gas-turbine compressor. A surge sounds like a succession of pops or explosions, depending on the machine Mach number and on how well the ducts are muffled.

Control systems are designed to avoid high-speed surge, although compressors can usually withstand a few such surges. Multiple or long-duration surging usually results in the destruction of the compressor.

8.8 Axial-compressor stage stacking

As was intimated in section 8.6, most early compressors were designed with all stages at their optimum-efficiency condition, which is near positive stall at the design point. Such compressors frequently had very poor "sagging" surge lines even though the design-point efficiency of the whole machine could be high. A so-called "knee" in the surge line means that the compressor may run into severe stalling or mild surging during the normal run up to full speed, necessitating the installation of devices such as interstage blow-off valves to increase the flow coefficients of the low-pressure stages. Such measures may be necessary for two reasons. First, even half-speed surges can impose severe fluctuating loads on the blading and possibly also on the thrust bearing. Second, there tends to be a hysteresis connected with both stalling and surging (chapter 5, section 5.3). Once a blade row has stalled, it often has to be brought to well below the stalling incidence before it becomes unstalled. Therefore, although the starting line of a compressor may only just graze the surge line at perhaps 60-percent speed and then pass through a normally unsurged region, the machine might in fact remain surged all the way up to full speed.

A compressor for which one of the authors (DGW) was given the responsibility to test exhibited this behavior to a high degree. Although of only four-to-one pressure ratio and thirteen stages (which were therefore, on the average, lightly loaded), it would prevent the gas turbine of which it was a part from being started if there was any unsteadiness of flow (shown by a tuft in a window in the outlet duct) even at 5-percent speed. This compressor had low-stagger blades, which were found later to have a large hysteresis loop when stalled (Wilson, 1960) (figure 5.13) and a very poor inlet, which probably helped to force many stages into stall. Koch (1981) has quantified the stalling characteristics of a wide range of compressor stages.

The starting and surging conditions influence stage "stacking". This term is used to mean the choice of types of stages, and their points of design operation, to be used in different parts of a compressor. The earlier description of surging and of the flow coefficients imposed on different stages showed that the low-pressure stages operate in the region from design-point incidence into full positive stall. Therefore it would seem sensible to choose a blade-row design that has little hysteresis in positive stall (for instance, one having a high stagger angle) and a design-point flow coefficient such that the blade row operates toward negative stall. Likewise, the high-pressure stages normally operate in the range between design point and negative stall and should therefore be chosen to have good negative-stall and high-Mach-number characteristics (for instance, low-stagger blade rows).

This type of compromise involves a trade-off of a slight reduction in peak efficiency for a broader range of high efficiency and a "pushed-out" surge line. This philosophy was applied to a small degree by one of the authors (DGW) in the aerodynamic design of a Ruston & Hornsby compressor of 5:1 design-point pressure ratio and a mass flow of 110 lbm/s. The first build and test, with no adjustments or stage tests whatsoever, gave the characteristics shown in figure 8.17. Despite a missing region in the characteristic (in which data could not be taken because of instabilities in the naval-destroyer steam-turbine drive, and not because of any deficiencies in the compressor), it can be seen that there is a good surge line, a broad plateau of high efficiency, and a peak polytropic stagnation-to-stagnation blading efficiency of over 0.91.

A second aspect of stage stacking is the need to adjust the annulus area to allow for boundary-layer growth. Larger boundary layers also mean lower stage efficiencies. To some extent these can be estimated by the Koch-and-Smith method introduced in the previous section. However, there has to be some guesswork in choosing values of wall-boundary-layer displacement thicknesses that will give the incremental amounts by which the casing radius should be increased, or will give annulus "blockage" factors, and the values of reduced efficiency. The following observations are made on the basis of personal and vicarious experiences.

1. In compressors that have blade rows designed to operate away from stall, wall boundary layers grow through the first three or four blade rows and then remain at constant thickness. In compressors in which the blade rows operate near stall, boundary layers have been found to continue growth through the machine. See DeRuyck and Hirsch (1980).

2. If the designer assumes large boundary layers or low efficiencies, these assumptions tend to become self-fulfilling prophecies. The reason is that they produce increased blade-row incidences and greater likelihood of local stalling. Therefore it is actually more conservative to assume high efficiencies, even if they are on the high side of what is actually attained.

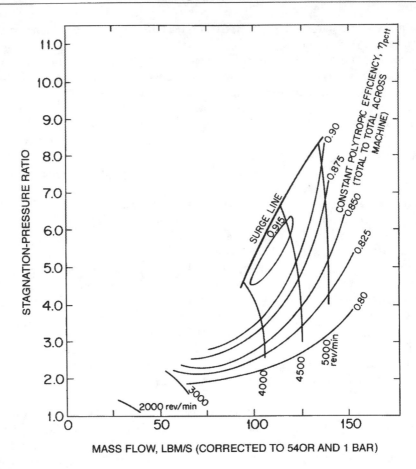

Figure 8.17. Compressor characteristics, Ruston and Hornsby TG

3. The use of the "work-done factor" should be avoided. It will not be defined here for fear of contaminating any remaining pure minds reading these words. Suffice it to say that it is a thermodynamically unacceptable "fudge" factor invented to bring cascade data and compressor data into line. The early British cascade data were taken without boundary-layer bleed, and hence with boundary-layer growth through the cascade, leading to accelerating flow. The NACA and NASA data reproduced here were taken with boundary-layer bleed and do not require correcting factors other than those described.

8.9 Alternative starting arrangements to reduce low-speed stalling

The foregoing discussion shows that it is inevitable that a compressor designed for a medium-to-high pressure ratio will force its low-pressure stages into a low-flow-coefficient condition at low speeds unless some special steps are taken. The alternatives are these.

1. For medium pressure ratios the designer can simply choose the design-point operating point of the low-pressure stages towards negative stall, so that at low speeds they will not be deeply into stall, as suggested above. (These stages should not have hysteresis-stall characteristics; however, we know of no way other than tests to determine such characteristics. High stagger is better than low stagger in this respect.)

2. The flow through the low-pressure stages can be increased during low-speed operation by opening up "mid-stage" blow-off valves. Between a third and a half-way along an axial compressor a "bleed" plenum belt is fitted around the compressor, with circumferential slits or spaced holes to the shroud surface between two stages. A valve, opened during starting, allows a considerable proportion of the flow to go to atmosphere.

3. The low-pressure stators can be fitted to rotatable platforms that can be moved towards the "closed" or high-stagger position at low speeds (figure 8.18). The simplest such approach is to pivot just the inlet guide vanes. For higher-pressure-ratio machines the "IGVs" and up to six rows of stator vanes can be linked together to be moved by a single actuator. This approach is known, somewhat ungrammatically, as "variable geometry".

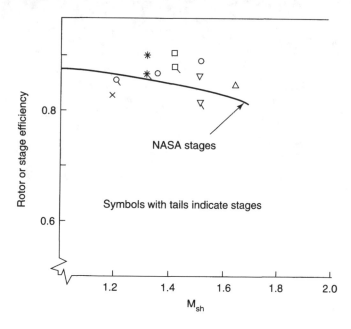

Figure 8.18. Reduction of isentropic stage efficiency with tip Mach number for some typical transonic stages. From Kerrebrock (1992)

4. A compressor of high pressure ratio can be divided into two or more compressors of low pressure ratio, driven by separate turbines. Each "spool" will then operate as if it were an independent low-pressure-ratio unit.

The pressure ratio of some modern aircraft engines is about 40, requiring both multi-spool compressors and variable-setting stators.

8.10 Axial-radial compressors

In relatively low-flow and/or high-pressure-ratio compressors the high-pressure blades are small and short. Clearance losses are likely to be high, and the thin blades are highly subject to erosion by sand and dirt. A frequently used alternative is to substitute a single radial stage for the last few axial rotors. Early attempts to use axial-radial compressors

The Allison 250 engine dates from the 1950s. We believe that it incorporated the smallest multistage axial compressor up to that point. The pressure ratio of the axial-centrifugal combination was originally 7.2 at 52,000 RPM, with an air mass flow of 1.6 kg/s. The six rotors in this version are "blisks": the disk and blades are integral, and appear to have been machined from solid blanks or from forgings. The centrifugal wheel appears to have been cast, including the inducer shroud, the purpose of which may have been to reduce leakage or inhibit vibration. Many versions of this successful engine have been produced, the compressor in some having four axial stages plus one centrifugal stage, and in others having just a single centrifugal stage.

Illustration. Axial-centrifugal compressor rotor for Allison T63/250 engine. Courtesy Allison Engine Co.

were largely unsuccessful because of the difficulty in obtaining good flow matching, but this problem has been overcome. Increasingly the radial rotor will use backward-swept blades and will have very high tip speeds, up to 700 m/s. Because of the extremely low blade height that results, the radial rotors are often fitted with a thin (and highly stressed) rotating shroud to reduce clearance losses.

8.11 Transonic compressors and fans

All aircraft engines designed since about 1960 have, at full speed, supersonic relative inlet flow at the first-stage rotor-blade and fan tips. The hub regions are subsonic. These blade rows are thus designated "transonic", and, in the case of the compressor, the whole compressor is known as transonic, even though the supersonic region is confined to a relatively small annular region at the entrance to the first rotor row.

The design of transonic stages is a topic beyond the scope of this text. The following discussion is intended merely to indicate the advantages and disadvantages of transonic stages, and the technology involved in their design. Much of it is taken from Kerrebrock (1977 and 1992). A more detailed treatment is given in Bölcs and Suter (1986).

It was recognized in the early days of gas turbines that an alternative method to the use of diffusion to convert dynamic pressure to static pressure was to pass the flow through a normal or oblique shock. Data for the loss of stagnation pressure in a normal shock at low supersonic conditions indicated that high polytropic efficiencies could be expected. However, all attempts at producing a fully supersonic compressor stage led to devices with many times the predicted losses (not fully understood) and a very limited range of operation. Fully supersonic compressors, by which is meant compressors with high hub-tip diameter ratios so that the flow is supersonic along the whole span of the blade, have not, therefore, found a useful role in general propulsion gas turbines.

The breakthrough allowing transonic compressors to be used occurred in several countries independently in 1950 (Serovy, 1985), when it was found that a supersonic section on long blades having subsonic flow over most of their length not only had higher efficiencies (though still less than would be predicted from shock-loss considerations alone) but a wide range of operation. The combination of supersonic flow with its capability of high pressure ratio at high relative velocities and subsonic flow with its capability of flow adjustment was apparently the key to the success.

Relative Mach numbers at the tip of the first-stage rotor blades in axial-flow compressors and fans rose rapidly from 0.85 to 1.2, and is now over 1.6 in advanced units, with Mach numbers up to 2.0 in experimental compressors. The tip speed is over 550 m/s, similar to that in axial-flow turbines. The pressure ratio of the inlet stage rose from 1.15 for subsonic stages to over 2.0. The axial Mach number is now typically 0.7 in modern transonic fans and compressors. Therefore the diameter required to pass a given flow has been substantially reduced, and the number of stages to achieve a specified pressure ratio has been even more drastically reduced.

The cost of this substantial increase in performance is some reduction in efficiency, as shown in figure 8.18, from Kerrebrock (1992). The isentropic efficiency is plotted; the polytropic would be more appropriate, and would show less drop-off with Mach number because the higher Mach number is associated with a higher pressure ratio.

The blading configuration used for the supersonic sections of these transonic rotor blades is shown in figure 8.19. Only the rotor-inlet flow is supersonic, leading to blades of the "shock-in-rotor" type. The passage shock is formed from the pressure-surface side of the bow shock. The suction-side bow shock propagates upstream as

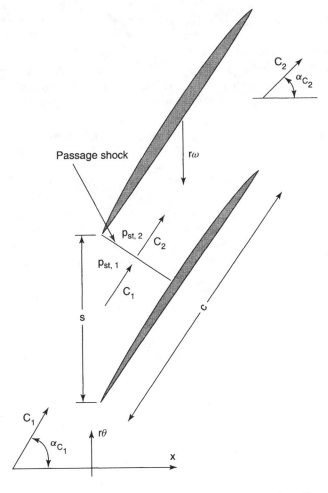

Figure 8.19. Configuration of a typical supersonic cascade. From Kerrebrock (1992)

weak shocks alternating with expansion waves. The inlet flow is unturned, so that the suction side of the blade leading edges must be aligned with the incoming flow at design point.

The relative flow leaving the rotor, and the flow entering the following stator, is everywhere subsonic. The relative Mach number on to the tips of the second-stage rotor blades is greatly reduced by the large temperature increase in the first stage, and sometimes additionally by the reduction of annulus diameter, but in some cases there is a small region of supersonic flow at the tips of the second-stage rotor blades. Subsequently there is no essential difference between the design of the subsonic stages of a transonic compressor and those of a completely subsonic compressor.

8.12 Improved compressor-blade geometries and flutter

The surface-curvature effects described for turbine blades in section 7.4 are also valid for compressors. By specification the compressor-blade thickness distributions described in section 8.2, and other similar compressor-blade thickness distributions, have smooth surface-curvature distributions, with the exception of where the leading-edge circle meets the blade surfaces. Even though in compressor blades this is very near the leading-edge stagnation point and velocities are low, it may lead to local separation bubbles near the leading edge.

Many production engines use compressor blades constructed by arranging thickness distributions around camber lines. However, the performance of some modern compressor blades, which approximate a pointed trailing edge better than the thicker turbine blades, has benefited enormously from "inverse" blade-design methods. Whether compressor blades have been improved by direct or inverse blade-design methods, the surface-curvature effects discussed in section 7.4 are still valid: discontinuities in the slope of curvature should be avoided; highly-loaded blades (and regions of high surface velocity, particularly transonic and supersonic blades) are extremely sensitive to small changes in curvature and surface geometry; local increases in curvature increase local loading; and small suction-surface curvature variations can be used in some transonic blades to remove shocks.

Stator-rotor interactions such as those discussed in section 7.6 are also present in compressor blade rows, and excitation-minimization considerations similar to those in section 7.6 apply to compressor and fan blade rows. However, compressor and fan blades are thinner than turbine blades, and the major problem with them is flutter. The condition of flutter arises when the energy absorbed by an airfoil due to negative aerodynamic damping equals or exceeds the energy dissipated by structural damping at the equilibrium vibratory-stress level. Since in most cases the structural damping is small, the design criterion to avoid flutter reduces to designing for positive aerodynamic damping. A full discussion of flutter couples blade vibration with flow disturbances, and is beyond the scope of this book, but it is discussed in an AGARD manual (Platzer and Carta, 1988). Regions of several types of flutter in a typical compressor performance map are shown in figure 8.20 (from Fottner, 1989).

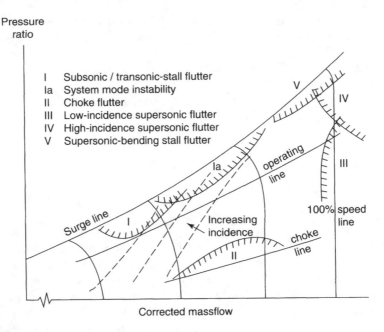

Figure 8.20. Regions and types of fan and compressor flutter. From Fottner (1989)

8.13 Axial-flow pump design

Compressor-design methods such as those given in the earlier part of this chapter can be used for other fluids, including Newtonian liquids (water, for instance). (The design of machines to work on non-Newtonian liquids, such as free-molecule flows or fluids having viscosities that vary with shear rate, is well beyond the scope of this book.)

This section on pump design is included because the cascade data given earlier may not wholly cover the axial-pump region for two reasons. First, an overriding design consideration is often the avoidance of cavitation. This consideration puts a premium on the use of a low axial velocity and low relative velocities in the choice of velocity diagrams. Cavitation avoidance is reviewed in chapter 9.

Second, axial-flow pumps are most usually single-stage designs, or even single rotors without stators. This factor again puts a premium on low axial velocities to give low leaving losses.

These two considerations combine to lead to velocity diagrams of low flow coefficient and to blades of high stagger or setting angles. The cascade data given in figure 8.6 do not adequately cover the area of interest for pump designs. Accordingly, we give in figure 8.21 cascade data for double-circular-arc (DCA) hydrofoils (Taylor et al., 1969)[1].

These data were taken for NASA by United Aircraft Research Laboratories in a series that also included multiple-circular-arc hydrofoils and slotted hydrofoils. We have chosen

---

[1]Taylor, W.E., Murrin, T.A., and Colombo, R.M. (1969). Systematic two-dimensional cascade tests, double-circular-arc hydrofoils. Report CR 72498, vol. 1, NASA, Cleveland, OH.

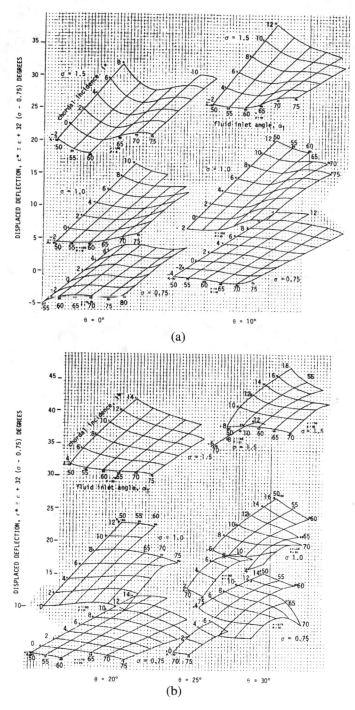

Figure 8.21. Turning data for double-circular-arc hydrofoils. From Taylor, et al. 1969

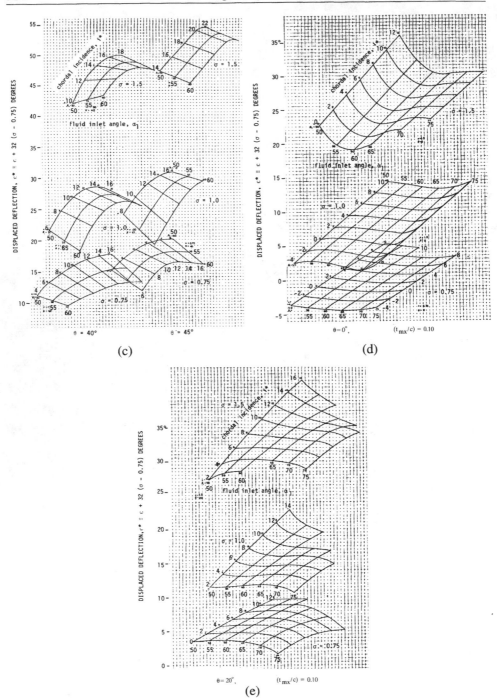

(c)

(d)

(e)

Figure 8.21. Continued

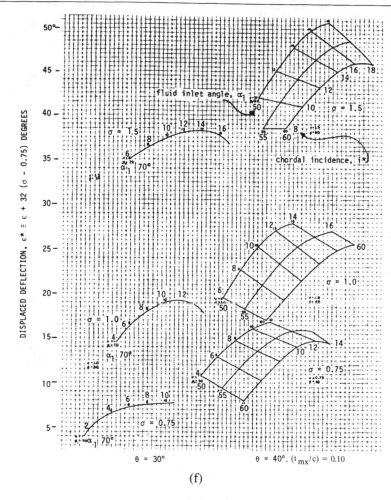

(f)

Figure 8.21. Concluded

to give only the DCA data because this configuration appears to be the best general choice for pumps, and because it is popular also for axial-flow compressors. (For compressor blades that are not computer-generated, the NACA 65-series are recommended for relative Mach numbers up to 0.78, and DCA blades for all relative Mach numbers up to 1.2.) The coordinates of the tested shapes are given in table 8.5 and the cascade nomenclature in figure 8.22.

The information given in Taylor et al. (1969) includes loss coefficients, diffusion coefficients, trailing-edge boundary-layer thicknesses, and cavitation indexes. We are reproducing just the turning-angle data in the form of a carpet plot, the use of which we illustrate by means of an example. Readers are recommended to go to the original reports for the profile-loss and other performance data.

Table 8.5. Coordinates for double-circular-arc profiles

Camber angle θ (deg)	0°		10°	20°	
Thickness (t_{mx}/c) ratio (%)	6%	10%	6%	6%	10%
Chordal station (%) (x/c)	y/c upper surface (%)				
0	0.10	0.10	0.10	0.10	0.10
8.33	0.93	1.53	1.62	2.29	2.94
16.67	1.67	2.80	2.92	4.14	5.29
25.00	2.23	3.77	3.94	5.56	7.09
33.33	2.67	4.43	4.65	6.57	8.37
41.67	2.90	4.87	5.09	7.18	9.12
50.00	3.00	5.00	5.23	7.38	9.38
58.33	2.90	4.87	5.09	7.18	9.12
66.67	2.67	4.43	4.65	6.57	8.37
75.00	2.23	3.77	3.94	5.56	7.09
83.33	1.67	2.80	2.92	4.14	5.29
91.67	0.93	1.53	1.62	2.29	2.94
100.00	0.10	0.10	0.10	0.10	0.10
	y/c lower surface (%)				
0	−0.10	−0.10	−0.10	−0.10	−0.10
8.33	−0.93	−1.53	−0.23	0.42	−0.19
16.67	−1.67	−2.80	−0.42	0.76	−0.35
25.00	−2.23	−3.77	−0.58	1.03	−0.47
33.33	−2.67	−4.43	−0.68	1.22	−0.55
41.67	−2.90	−4.87	−0.76	1.33	−0.61
50.00	−3.00	−5.00	−0.77	1.38	−0.62
58.33	−2.90	−4.87	−0.76	1.33	−0.61
66.67	−2.67	−4.43	−0.68	1.22	−0.55
75.00	−2.23	−3.77	−0.58	1.03	−0.47
83.33	−1.67	−2.80	−0.42	0.76	−0.35
91.67	−0.93	−1.53	−0.23	0.42	−0.19
100.00	−0.10	−0.10	−0.10	−0.10	−0.10

Table 8.5. Continued

Camber angle θ (deg)	25°	30°		40°		45°
Thickness (t_{mx}/c) ratio (%)	6%	6%	10%	6%	10%	6%
Chordal station (%) (x/c)	y/c upper surface (%)					
0	0.10	0.10	0.10	0.10	0.10	0.10
8.33	2.64	3.00	3.67	3.77	4.48	4.15
16.67	4.76	5.41	6.58	6.74	7.95	7.41
25.00	6.40	7.25	8.80	9.01	10.58	9.87
33.33	7.55	8.55	10.35	10.58	12.40	11.59
41.67	8.24	9.33	11.27	11.54	13.49	12.61
50.00	8.47	9.58	11.58	11.84	13.84	12.95
58.33	8.24	9.33	11.27	11.54	13.49	12.61
66.67	7.55	8.55	10.35	10.58	12.40	11.59
75.00	6.40	7.25	8.80	9.01	10.58	9.87
83.33	4.76	5.41	6.58	6.74	7.95	7.41
91.67	2.64	3.00	3.67	3.77	4.48	4.15
100.00	0.10	0.10	0.10	0.10	0.10	0.10
	y/c lower surface (%)					
0	−0.10	−0.10	−0.10	−0.10	−0.10	−0.10
8.33	0.79	1.10	0.48	1.80	1.17	2.15
16.67	1.41	2.00	0.88	3.26	2.12	3.89
25.00	1.89	2.69	1.18	4.40	2.87	5.23
33.33	2.20	3.19	1.40	5.19	3.40	6.19
41.67	2.40	3.48	1.54	5.68	3.72	6.76
50.00	2.47	3.58	1.58	5.84	3.82	6.95
58.33	2.40	3.48	1.54	5.68	3.72	6.76
66.67	2.20	3.19	1.40	5.19	3.40	6.19
75.00	1.89	2.69	1.18	4.40	2.87	5.23
83.33	1.41	2.00	0.88	3.26	2.12	3.89
91.67	0.79	1.10	0.48	1.80	1.17	2.15
100.00	−0.10	−0.10	−0.10	−0.10	−0.10	−0.10

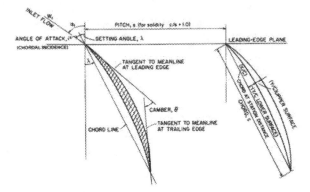

Figure 8.22. Hydrofoil cascade nomenclature

Example 2   Use of hydrofoil carpet plots

The problem is to recommend blade shapes for an axial-flow, rotor-only pump. The design specifications are the following.

> Flow = 8,000 gpm, 0.5047 M^3/s.
> Static-head rise = 15 ft, 4.57 m.

With an assumed efficiency and outlet dynamic-head allowance and an approximate boundary-layer allowance, the specifications used for the velocity-triangle calculations were

> Flow = 8,034 gpm, 0.5069 m^3/s.
> Theoretical total-head rise = 19.45 ft, 5.928 m.

After trials of various flow coefficients and hub-tip diameter ratios, it was decided to use a hub-tip diameter ratio of 0.8, a peripheral speed of at the hub of 40.0 ft/s (12.19 m/s), and an axial velocity of 25.0 ft/s (7.62 m/s). From these values the velocity-diagram values in the first part of table 8.6 were arrived at. The problem is to recommend blade shapes for those

Table 8.6. Preliminary design of axial-flow pumps

	Hub	Mean	Tip
Fluid inlet angle to rotor, deg.	57.99	60.94	63.43
Fluid outlet from rotor, deg.	44.26	51.20	56.30
Radius ratio	0.8	0.9	1.0
Deflection, $\varepsilon \equiv (\alpha_{in} - \alpha_{out})$, deg.	13.73	9.74	7.13
Solidity, $\sigma \equiv c/s$ (first choice)	0.75	0.75	0.75
Ordinate $\varepsilon^x \equiv \varepsilon + 32(\sigma - 0.75) = \varepsilon$	13.73	9.74	7.13
Possible blades: camber, θ deg.	45	30	15
Angle of attack (chordal incidence) i^* deg.	7.3	6.15	8.25
Solidity, σ (second choice)	1.0	1.0	1.0
Ordinate $\varepsilon^x \equiv \varepsilon + 32(\sigma - 0.75)$	21.73	17.74	15.13
Possible blades: camber, θ deg.	30	20	10
Angle of attack (chordal incidence) i^* deg.	10.6	7.2	7.5

flow angles. We do this by continuing the table. We will select blade forms for two values of solidity: 0.75 and 1.0.

The blades were chosen by calculating the ordinate ε^x for figure 8.21, drawing a horizontal line corresponding to the value of ε^x, and choosing an operating point near positive or negative stall, or in the middle of the range, for plots of the appropriate solidity. Each separate "carpet plot" is for a cascade of blades of fixed camber and set at a given solidity, with setting or stagger angle and angle of attack (incidence) varied (figure 8.22). Usually the designer has a choice of blades of two or more camber angles and therefore can interpolate between them.

References

AGARD (1989). "Blading design for axial turbomachines", AGARD Lecture Series 167, AGARD-LS-167.

Bölcs, Albin and Suter, Peter (1986). Transsonische Turbomaschinen. G. Braun, Karlsruhe, W. Germany.

Cumpsty, N.A. (1989). "Compressor aerodynamics." Longman and John Wiley & Sons, Harlow, UK and New York, NY.

DeRuyck, J., and Hirsch, C. (1980). Investigations of an axial-compressor end-wall boundary-layer prediction method. Paper 80-GT-53. ASME, New York, NY.

Emery, James C., Herrig, L. Joseph, Erwin, John R., and Felix, Richard (1959). Systematic two-dimensional cascade tests of NACA 65-series compressor blades at low speeds. Report 1368, NACA, Langley, VA.

Fottner, L. (1989). Review on turbomachinery blading design problems. In "Blading design for axial turbomachines", AGARD Lecture Series 167, AGARD-LS-167.

Horlock, J.H. (1958). Axial-flow compressors. Butterworths, London, UK.

Kerrebrock, Jack L. (1992). Aircraft engines and gas turbines, second edition. The MIT Press, Cambridge, MA.

Kerrebrock, Jack L. (1981). Flow in transonic compressors. AIAA Jl. Jan. 1981, New York, NY.

Koch, C.C. (1981). Stalling pressure-rise capability of axial-flow compressor stages. Paper 81-GT-3. ASME, New York, NY.

Koch, C.C., and Smith, J.H., Jr. (1976). Loss sources and magnitudes in axial compressors. Trans. ASME J.Eng. Power, A 98 (July): 411–424. ASME, New York, NY.

Lieblein, Seymour (1959). Loss and stall analysis of compressor cascades. Trans. ASME J. Basic Eng. (September): 387–400. ASME, New York, NY.

Massardo, A.F, Satta, A., and Marini, M. (1990). Axial-flow-compressor design optimization: part I. Pitching analysis and multivariable objective-function influence; part II. Throughflow analysis. Trans. ASME J. of Turbomachinery, vol. 119, pp. 399–410, July. ASME, New York, NY.

Mellor, George L. (1956). The NACA 65-Series cascade data. Gas-Turbine Laboratory charts. MIT, Cambridge, MA.

Platzer, M.F., and Carta, F.O., eds. (1988). "Aeroelasticity in axial-flow turbomachines." vol. 1 and 2, AGARD manual AGARD-AG-298.

Reid, L., and Moore, R.D. (1980). Experimental study of low-aspect-ratio compressor blading. Paper 80-GT-6. ASME, NY.

Sanger, N.L. (1983). The use of optimization techniques to design controlled-diffusion compressor blading. Trans. ASME J. of Engineering for Power, vol. 105, pp. 256–264, April. ASME, New York, NY.

Serovy, G.K. (1985). Axial-flow-compressor aerodynamics. In "Aerothermodynamics of aircraft-engine components", ed. Gordon C. Oates, AIAA, New York, NY.

Starke, J. (1980). The effect of the axial-velocity-density ratio on the aerodynamic coefficients of compressor cascades. Paper 80-GT-134. ASME, New York, NY.

Taylor, W.E., Murrin, T.A. and Colombo, R.M. (1969). Systemic two-dimensional cascade tests. Double-circular-arc hydrofoils, vol. 1. Report CR 72498. NASA, Cleveland, OH.

Wilson, David Gordon. (1960). Patterning stage characteristics for wide-range axial compressors. Paper 60-WA-113. ASME, New York, NY.

Problems

1. Choose the mean-diameter stage arrangement (rotor-stator or stator-rotor), the rotational speed, the hub and tip diameters velocity diagrams at hub and tip diameters, and blades for a human-powered axial-flow irrigation pump of the flowing specifications.

> Vertical lift = 2 m
> Steady power to blading = 150 W
> Delivery-pipe length = 8 m
> Delivery-pipe diameter = 0.1 m
> Pump hub-tip ratio = 0.7.

The delivery pipe has three 45-degree bends, each contributing an estimated pressure drop equal to a length of pipe of 15 diameters. The friction factor C_f of the pipe is 0.0007 in the relation

$$\Delta p = 4C_f \left(\frac{L}{d}\right) \left(\frac{\rho C^2}{2g_c}\right)$$

Assume that one pipe dynamic head is lost at outlet (the use of a diffuser would be desirable). Assume a pump efficiency of 0.88, stagnation-to-stagnation across the blading.

2. If you have chosen the mean-diameter velocity diagram for the single-stage cooling fan of problem 5.1, choose suitable blades from figure 8.6. Then calculate the stagnation-to-stagnation isentropic efficiency, using the methods of chapter 8. Convert your value to a polytropic efficiency, and compare it with the value given in problem 5.1. You will have to assume that the loss values you obtain for the mean diameter apply over the whole annulus. Make your own design choices.

3. (a) On the axial-compressor characteristics provided in figure P8.3, draw in lines of constant design-point incidence for the first stage and for the last stage. The design point is shown. You will have to make various simplifications because you are not given many data that would be required for a more precise calculation. State your assumptions.

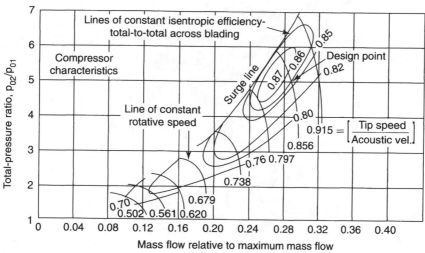

Figure P8.3

(b) Then select suitable mean-diameter blade cascades from figure 8.6. Find the flow angles for positive stall of the first stage and negative stall of the last stage, and draw lines on the characteristics when these apply. Assume that the velocity diagrams are 50 percent reaction, that the work coefficient is 0.25 for the first stage and 0.4 for the last stage, that the flow coefficients are 0.3 for each stage, and that the solidities are unity.

(c) From your results discuss the flow conditions in the blading as the compressor is started from rest in a flow circuit with a "square-law" resistance curve going through the design point.

4. (a) Choose NACA 65-series compressor blades for both rotor and stator of a helium compressor having the following specifications. Choose blades that give a deflection about 20 percent less than that for positive stall. For this preliminary design choose to have all mean-diameter velocity diagrams similar.

$p_{0,1} = 981 \text{ kN/m}^2$
$T_{0,1} = 40°\text{C}$
$p_{0,2} = 2 \times p_{0,1}$ (at last-stage exit).
Maximum blade speed = 500 m/s.
Reaction = 0.5.
Work coefficient = 0.5 (mean-diameter, first stage).
$W_2/W_1 = 0.7$.
$\eta_{p,tt} = 0.90$.

(b) Find the mean-diameter blade speed necessary to give an integral number of stages.

5. Find the flow coefficients relative to the design-point coefficients for the first-stage inlet and last-stage outlet for the two conditions marked A and B on the axial-flow compressor characteristics shown in figure P8.5. At these two locations the flow is axial at the design point. The axial Mach number before the first stage at design point is 0.4.

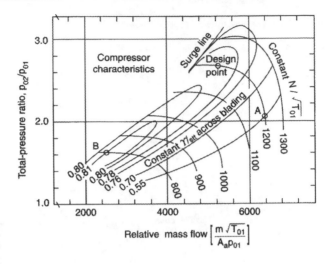

Figure P8.5

6. Use the charts of figure 8.6 and figure 8.21, as appropriate, to list all the alternative stator and rotor blades (cambers, solidities, setting and incidence angles) for a jet-propulsion pump for a high-speed vessel.

   The pump is a rotor-stator combination having axial inlet and outlet flow at design speed and flow. The flow coefficient is equal to $\tan 20°$, and the minimum value of the velocity ratio for either stator or rotor, W_2/W_1 or C_x/C_2, is 0.75. The axial velocity should be held constant through the stage.

7. If high-speed surge in an axial-flow compressor is triggered by last-stage stall, it seems to be sensible to use, in the high-pressure stages, a type of blade row that does not produce a drop in pressure rise at stall, as shown in figure 5.13b. Assuming that high-stagger blading gives a rising characteristic of this type, discuss the probable implications in the following aspects.
   (a) What is the likely effect on the high-speed surge line?
   (b) What is the likely effect on the number of stages required?
   (c) What is the probable effect on the area controlling last-stage choke?
   (d) What could be the effect on the clearance losses?
   (e) What is the likely effect on the peak total-to-total efficiency of the compressor?

8. Calculate the mass flow and the annulus area at rotor outlet for an axial-compressor stage. The flow enters the rotor without swirl, at a Mach number of 0.4, at a stagnation pressure of 200,000 Pa, and stagnation temperature of 370 K, in an annulus of area 0.17 m². The axial velocity is constant across the rotor, but the flow leaves at 30.6° to the axial direction. The stagnation pressure at rotor outlet is 230,000 Pa, and the stagnation-to-stagnation polytropic efficiency to this point is 0.94. You may use a single mean value of specific heat (for air).

9. (a) Give two reasons why it might be desirable to design axial-compressor blade rows to operate as near to positive stall as possible.

   (b) Give five reasons why in practice the design-point operation of axial-compressor blade rows (calculated on mean conditions) is not set very close to positive stall.

10. Choose blades from the NACA 65-series for the mean diameter of a cooling fan for a computer power supply. Also calculate the shroud diameter and the rotational speed. Choose the hub-shroud diameter ratio at 0.8; the blade solidity at 1.0; the mean-diameter work coefficient at -0.25; and the flow coefficient at 0.4. There is no stator. The flow enters axially without swirl and leaves with swirl. The stagnation-to-static polytropic efficiency across the rotor is estimated to be 0.75. The fan must remove 600 watts of heat energy, and the maximum permissible temperature rise of the cooling air is 25 K. The flow resistance of the power-supply circuit is 5 kPa for a flow of 0.1 m^3/s; the pressure drop varies with the square of the flow rate.

Treat the flow as incompressible. Its density is 1.2 kg/m^3; its specific heat is 1010 J/(kg · K); and the gas constant is 286.96 J/(kg · K);

11. Suppose that you have completed the preliminary design of an axial-flow compressor and have chosen to have a large number of rotor blades, say 58, in the first stage. You find, in going through the loss calculations, that the small chords result in a low Reynolds number, and you decide to look at the effects of reducing the number of rotor blades to, say, 14. This gives a much higher Reynolds number, and manufacturing costs would decrease while the relative profile accuracy would increase. These are at least some of the benefits. What are some costs associated with using a smaller number of blades? If 14 is a good choice, would 7 be better still? Mention three costs if you can.

12. Write out the sequence of steps you would use to calculate the design-speed surge-line mass flow as a ratio with the design-point mass flow for an axial-flow air compressor. The surge line at design speed may be taken to be defined by last-stage stall. The last-stage vector diagram is fifty-percent reaction with a value of $W_2/W_1 = 0.71$ and the loading coefficient $\psi = -0.3$ at design point. The stage stalls when the value of (W_2/W_1) drops to 0.65. As an approximation, the vector diagram can be taken to remain "simple" with constant axial velocity through the stage and unchanged blade-row-outlet flow angles. The static density at the blading outlet varies as the nth power of the last-stage enthalpy rise.

Draw the vector diagrams and write down the initial equations, but make no calculations.

13. Plot the work coefficient versus the flow coefficient for an axial-compressor stage of fifty-percent reaction, 0.5 flow coefficient, and -0.3 work coefficient, for three incidence levels: -10 degrees, zero (design point) and $+10$ degrees. (Incidence is the inlet flow angle minus the design-point inlet flow angle.) Approximate by keeping the axial velocity and blade velocity constant even at off-design conditions.

Plot two lines. One is for the case where the outlet flow angle of each blade row is constant. The other is for the case where the deviation [(outlet flow angle) $-$ (design-point outlet flow angle)] is 0.2 × (incidence angle).

14. An axial-flow compressor of design pressure ratio 16 : 1 and polytropic efficiency 0.92 has been designed with several blow-off points between the first and last stages. In low-speed operation, what approximate percentage of the inlet flow would have to be blown off (the sum of all the blow-off streams) to keep the first and last stages operating at design incidence?

To solve this problem several of the following approximations are necessary. Ignore the differences between stagnation and static properties. Assume that the compressor was designed for a constant axial velocity at design point. The last-stage outlet flow conditions can be used as a guide to the last-stage inlet flow conditions. The compressor inlet conditions may be taken as 300 K and 1 bar, and the mean specific heat is 1027 J/(kg · K).

Chapter 9

Design methods for radial-flow turbomachines

Some preliminary design methods for radial-flow turbomachines have been given in chapter 5. In this chapter we will discuss some more-detailed aspects of design of radial-flow machines.

The advantages and disadvantages of using radial-flow machines in preference to those having predominantly axial flow are first reviewed. It is shown that a radial-flow compressor can be designed with many times the head or enthalpy rise per stage than is possible with axial-flow machines. This gives radial-flow compressors an advantage where it is desirable to reduce to a minimum the number of stages in an application. In small sizes, in which case the flow Reynolds numbers will be low, and the relative blade-shroud clearances will be high, radial-flow machines also have very large cost and even efficiency advantages over multistage axial machines. Radial-flow machinery in almost every size will cost less to manufacture than the equivalent multistage axial machinery, but in the size range in which both types can be used the efficiencies will be, in general, lower than for axial machines.

In contrast, radial-flow turbines produce generally lower head or enthalpy drops per stage at efficiencies which, in the smaller sizes, are higher than is possible with axial-flow machines running at the same peripheral speeds. The areas of application of radial-flow turbines are limited predominantly to those in which the lower manufacturing cost of small, single-stage radial turbines is of overriding importance and to special cases where the configuration is advantageous, such as large Francis water turbines.

Simple design methods are given, and illustrated with examples, for most types of radial-flow machines.

9.1   The difficulties of precise design

Radial-flow machines present a challenge and an enigma. The preliminary design of a radial-flow machine can be quite simple. However, the flow is extremely complex. Attempts over many years to solve the three-dimensional flow analytically and numerically, even with the most advanced computer programs, have not yielded procedures that guarantee good performance in machines designed with their aid. (However, these programs are being rapidly improved.)

The two most challenging categories of radial-flow turbomachines are radial-inflow water turbines and high-pressure-ratio radial-outflow compressors. The authors know of no instance in either of these categories where a new machine, when run in the as-designed configuration, achieved outstanding performance. Normally a fairly extensive program of development, of models in the case of water turbines, has to be undertaken, in which "cut-and-try" methods, perhaps disguised, play a prominent part. What is meant by a "new" machine has to be defined for this statement to have validity. Some new machines in fact are merely small extrapolations from, or interpolations between, previous designs. Most new machines are designed by a person or by a team with years of experience in designing and developing radial-flow machines. Features of those machines that have given good performance will obviously tend to be carried forward as part of design practice. This is the art that, for radial-flow machines particularly, complements the science.

It should not be inferred that flow analysis is of no utility. It seems at least justifiable to claim that application of advanced methods of flow analysis to, particularly, the flow in the rotor of a radial-flow turbomachine will probably greatly reduce, but equally probably not eliminate, the period of subsequent development testing that will be required.

For the less demanding types of radial-flow turbomachines, for instance, low-head-rise centrifugal pumps and compressors, the preliminary design methods given in chapter 5, supplemented by the guidance in this chapter, may be wholly sufficient. Before these methods are developed, the areas of potential application of radial-flow machines should be discussed.

9.2   Advantages and disadvantages and areas of application

Radial-flow machines have advantages that enable them to rule unchallenged in some applications. In others, they are used preferentially just over the smaller end of the size range.

Air and gas compressors

More work per stage (a larger head rise, or larger pressure ratio) can be given by a radial-flow machine than by an axial-flow stage. Often this fact is attributed to the "centrifugal effect": the radial pressure gradient that will be set up when a mass of fluid is rotated by an impeller. But the enthalpy rise is still the result of the torque given by the impeller to accelerate the fluid, and is evaluated by Euler's equation (equation 5.1). Therefore we can look to the velocity diagrams of radial-flow and axial-flow machines to find out why the former should give more work per stage (figure 9.1).

The work coefficient, ψ, which is the specific work for a given blade speed, is limited in an axial stage by the requirement that the relative-velocity ratios stay above some limit, such as the de Haller limit of 0.71 (figure 4.9). In a radial-flow machine the relative velocity of the flow entering the rotor is low simply because it comes in at

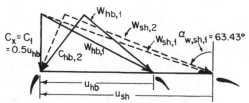

(a) Axial-flow compressor diagram: axial entry and exit; free vortex; hub-tip ratio 0.6; relative-velocity ratio 0.71

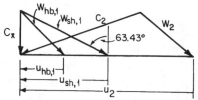

(b) Radial-flow diagram: axial entry; inducer-periphery diameter ratio 0.5; inducer hub-tip ratio 0.5; relative-velocity ratio 0.71

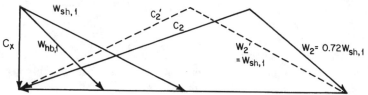

(c) Radial-flow diagram similar to (b) but with identical shroud inlet relative velocity to (a)

Figure 9.1. Comparison of axial-flow and radial-flow vector diagrams

a smaller radius than that of the outlet flow. If one uses the same de Haller limit for radial-flow rotors as for axial, the permissible increase in outlet tangential velocity is similar to the increase in rotor peripheral velocity between the entering and leaving radii.

In figure 9.1 velocity diagrams for axial-entry stages are compared. The diagrams for the radial-flow machines (figure 9.1b and c) are for a configuration in which the diameter of the inducer-blade tips, or of the inner surface of the shroud, is one-half that of the rotor periphery. The radius of the hub at impeller (or inducer) entry is one-half that of the shroud at the same plane. The axial-machine diagram (figure 9.1a) is for a hub-shroud diameter ratio of 0.6, which has been regarded (chapter 6) as the lower limit at which the simple design methods outlined in this book can be used with success. For simplicity, both diagrams have constant axial velocity with radius at the rotor-entry plane.

For the axial stage, the limiting diagram insofar as relative-velocity ratio is concerned is at the hub. The diagram has been drawn with a hub flow coefficient of 0.5 (so that the shroud value is 0.33) and with a relative-velocity ratio of 0.71. The angle of flow into the rotor at the hub is then 63.43 degrees.

The velocity diagram for the shroud streamline to have the same enthalpy rise as that at the hub is shown in broken lines.

In the radial machine the limiting diagram for rotor relative-velocity ratio is that involving the shroud streamlines at the impeller inlet. The reason for this is that both the hub and the shroud streamlines (or stream surfaces) emerge from the rotor at the same radius (the rotor periphery) and the same relative flow angle. (That of course is a simplifying assumption. In fact the flow is extremely nonuniform within a centrifugal-impeller channel. Even in those few cases where streamlines have been found experimentally, the patterns do not hold for other machines. Therefore we are really referring here to the mixed-flow mean streamline.) Then the inlet streamline having the larger relative velocity, which is obviously the shroud streamline, will undergo the larger flow deceleration.

Therefore we have chosen $W_{sh,1}$ in the radial diagram to have the same relative flow angle, $63.43°$ as $W_{hb,1}$ in the axial diagram and W_2 to be $0.71\,W_{sh,1}$, just as W_{h2} was $0.71\,W_{hb,1}$ in the axial case. (It will be shown later that this flow angle happens to be close to the optimum for the relative flow at the inlet shroud for most radial-flow machines.) When we complete the diagram, we find that the work coefficient for the radial machine is 0.69, whereas for the axial machine at the same peripheral speed, which is for the shroud radius of the axial diagram, the work coefficient is 0.23. The radial machine has exactly three times the work coefficient of the axial machine using the same relative-velocity ratio.

But is it fair to use the same peripheral speed, and the same relative-velocity ratio, in the two cases? The peripheral speed might be set by stress limits. However, this will be the case only for compressors of low-molecular-weight gases such as hydrogen and helium and for experimental air compressors of very high pressure ratio, such as 10:1. We regard these as special cases. In more usual compressors and pumps the highest relative velocity at inlet will normally set the limiting speed because of Mach-number limits in compressors and cavitation limits in pumps.

In both the axial and radial diagrams, figure 9.1a and b, the highest relative velocity at inlet is $W_{sh,1}$, the relative velocity of the shroud streamlines. But as these two diagrams are drawn for the same value of peripheral speed, the relative velocity as drawn for the radial diagram is much lower than that for the axial diagram. In figure 9.1c the radial diagram has been enlarged so that $W_{sh,1}$ is the same as for the axial diagram, figure 9.1a. The work that can be given by the radial stage is now 10.56 times that for the axial stage.

Now we return to the second question: is it fair to use the same relative-velocity ratio for the radial machine as for the axial? A study by Rodgers (1978) concludes that, perhaps coincidentally, it is conservative to use the same ratio, 0.71, for the limiting value of W_{ex}/W_{in} in the channel of a radial-flow compressor if stalling is to be avoided. This seems surprising, because the flow entering a radial-flow impeller has to go through such sharp turns, and has to withstand such continued lateral acceleration, that it would be natural to expect that, for the flow to withstand the tendency to separate from the walls of the channel, the deceleration rate should be lessened. In figure 9.1c a broken line shows an extreme case in which W_2 has the same value as $W_{sh,1}$. That is, there is no deceleration of the mean flow. The work coefficient falls from 0.69 to 0.50, and the ratio of radial-machine work to axial work falls to 7.74, which is still large.

To produce a relative-velocity vector in either the higher-work direction, W_2, or the lower-work direction, W'_2, the rotor blades would have to be slanted backward with respect to the direction of rotation. They are termed "backward-swept" blades. The blade-root bending stress increases with higher degrees of "sweepback" or backslope, and with blades that are relatively longer in the axial direction.

However, in radial-flow compressors it is merely desirable, not necessary, that the flow be unseparated in the rotor channels. Rotors designed for very high pressure ratios must use a high peripheral speed, and radial, or near-radial, blades. The relative-velocity ratio in the rotor becomes too low for efficient flow, and separation within the rotor must be expected. When the relative-velocity-ratio limitation is relaxed, the impeller flow will be highly separated, and it may not be possible to obtain much diffusion in a following diffuser fed with highly nonuniform flow. Despite this, a radial-flow compressor can give fifteen or more times the enthalpy rise of an axial stage with the same inlet relative Mach number. The efficiency will, however, be much lower: for instance, 83 percent instead of 90 percent.

To sum up, radial-flow compressors can produce much higher enthalpy rises per stage than can axial machines at the same relative inlet velocity principally because there is no direct coupling between the magnitude of the relative inlet velocity and that of the peripheral speed and of the tangential component of the absolute velocity at outlet. Furthermore, flow separation in the radial-machine rotor may actually increase the enthalpy rise, whereas in the axial rotor passage separation acts to reduce the deflection of the flow, and with it the enthalpy rise.

Relative efficiencies

The efficiency of radial pumps and compressors is usually lower than that of alternative axial machines, for four reasons.

1. The flow in the rotor is much more complex in the radial-machine rotor than in the more streamlined passages of an axial compressor or pump. In addition to the sharp turn from the axial to the radial direction, which is required and will introduce losses, the severe aerodynamic loading and the long flattened passage shape encourages the growth of secondary flows in the boundary layers (figure 9.2). These secondary flows can combine to facilitate flow separation within the channels and to impose highly nonuniform flow at the entrance to the diffuser. (The "relative eddy" shown in figure 9.2 is in fact hypothetically produced by irrotational potential flow passing through a centrifugal impeller.   It can explain the deviation ("slip") of the flow from the blade direction.)

2. The absolute outlet velocity, C_2, is high in a radial-flow compressor or pump, so that much of the enthalpy rise is in the form of velocity. Even an ideal conical diffuser fed with almost uniform inlet flow cannot convert much more than 80 percent of velocity head into pressure head (figure 4.14).

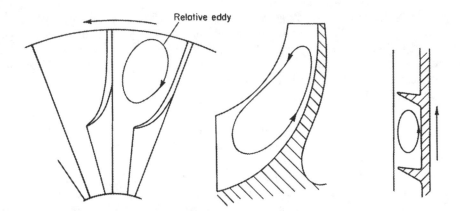

Figure 9.2. Secondary flows in radial-flow pumps and compressors

3. Radial diffusers of any type (see figure 4.27) have lower actual pressure-rise co-efficients than do axial-flow diffusers, particularly when the radial flow at inlet is highly nonuniform.

4. Radial-flow machines have relatively large wetted areas, both on the surface of the flow channels and on the back faces of the disks, and on the rotating shrouds, if used. Friction must therefore be relatively large.

Effect of backslope

The first two of the above-listed causes of reduced efficiency can be made much less penalizing by utilizing large angles of backslope, when stress limits allow (as in hydraulic pumps and low-pressure-ratio fans and blowers). As the blade angle is increased to, perhaps, 50 degrees from the radial direction, the flow angle will be greater than this and the magnitude of the absolute outlet velocity (the diffuser-entry velocity) will be greatly reduced. While the work per stage is also reduced, the efficiency increases both because of the reduced loading on the diffuser and because of the reduced diffusion in the rotor. (The mean relative flow may even accelerate in the rotor.) The uniformity of the flow entering the diffuser will also be improved. Kluge (1953) showed the results of tests on similar impellers with only the blade angle at outlet varying and found a peak efficiency at an angle of about 50 degrees (figure 5.19). The efficiency levels of Kluge's results are low. For high Reynolds numbers and near-optimum specific speeds it should be possible to reach ten points or more higher than Kluge's curve, as indicated by the shaded area. Large centrifugal pumps can give efficiencies along this upper line.

Effect of size

When the shroud diameter of the first stage of an axial-compressor has to be much below say, 500 mm, the blades in the high-pressure stages (assuming that the overall pressure ratio is four or above) become small, fragile, expensive to produce, and severely affected by tip-clearance losses. (Tip clearances have been greatly reduced following the introduction of high-speed preloaded ball bearings, in place of the lightly loaded plain bearings previously required, and of abradable shroud materials.) Therefore for small, high-pressure-ratio compressors the only choice is a radial machine. For intermediate-size high-pressure-ratio compressors the axial-radial compressor is often used (for instance, for helicopter engines of 400 to 800 kW). Often two to six axial stages are followed by a single radial stage. Envelope curves showing the effects of size and pressure ratio on radial-outflow compressors are given in figure 3.13.

Axial-flow pumps, because they operate on incompressible fluids, retain the same blade sizes in the high-pressure stages and could therefore be made in smaller sizes than would be economic for compressors. However, the high-backslope multistage centrifugal pump is preferred in practice even for the larger sizes, giving very high efficiencies (often well over 90 percent). Axial-flow pumps presently seem to be used only for high-flow applications, when the head rise is small enough for a single stage to be sufficient (high-specific-speed applications).

Specific-speed effects

Radial-flow machines are also be preferred when rotational speeds are so limited that the blade lengths of axial-flow machines would be very small. (In other words, the hub-shroud ratio at machine entry might be above 0.85. As developed in chapter 5, a high hub-shroud ratio signifies a low specific speed.) The efficiency of a radial machine may exceed that of an axial-flow machine if the blade lengths are small compared with the shroud diameter (which will in turn influence the blade clearance); see figure 5.20. Rodgers (1991) recommends using a (non-dimensional) specific speed of 0.7 for centrifugal compressors of maximum efficiency, and 1.0 for compressors of minimum size and weight (figure 5.20).

Cost considerations

Radial-flow compressors and pumps are normally much less expensive to produce, for the same flow and head rise, than are their axial-flow equivalents, because fewer stages are required, and because one-piece rotors, and possibly one-piece stators, can be produced, often by highly accurate but inexpensive investment casting.

Space considerations

Radial-flow compressors and pumps may be shorter than their axial-flow equivalents, but they have a much larger overall diameter (frontal area). Whether or not there is a weight penalty depends upon many factors.

Radial-inflow turbines

Some of the considerations made with regard to the areas of application of radial-flow pumps and compressors apply equally to radial-flow turbines. In particular, the remarks about the effects of size, specific speed, cost and space requirements all apply almost equally to radial turbines.

Relative work output

There is one very large difference that affects the relative applications of radial-flow turbines, on the one hand, and compressors and pumps, on the other. Whereas radial-flow compressors and pumps can be designed to give ten or twenty times the enthalpy rise per stage of the equivalent axial machines, radial-flow turbines produce, in some cases, less than the enthalpy drop possible with an axial turbine of the same rotor diameter and speed.

Velocity diagrams

The reasons for this rather extraordinary lack of symmetry lie in the velocity diagrams and in the very different nature of accelerating flow in a favorable pressure gradient compared with decelerating flow in an unfavorable pressure gradient.

The velocity diagram of a low-specific-speed radial-inflow turbine (one having the tip diameter of the rotor blades at the outlet plane considerably smaller, say under 60 percent, of the rotor peripheral diameter) will be similar to the equivalent radial-outflow compressor, figure 9.3a.

The blades of hot-gas expansion turbines are virtually always radial to minimize blade bending stress. Because there is no velocity-ratio limit in accelerating flow, there is no fluid-mechanic compromise involved in using radial blades. The peak efficiency appears to occur when the relative flow angle is similar to that which would be found to give peak efficiency if same rotor were used as a compressor. This means that the relative flow should be aimed slightly against the blade rotation, so as to provide a gradual increase of aerodynamic loading. The work coefficient would then be about 0.9.

The diagram for an axial turbine of near peak efficiency (see chapter 7) is shown in figure 9.3b. This is a 50-percent-reaction turbine with a work coefficient of unity.

The diagram of a high-work turbine is shown in figure 9.3c. It is possible to obtain a work coefficient of 3.0 without a major fall in the efficiency. However, a single-stage machine would have not have axial outflow, and the diffuser performance would suffer unless straightening vanes were used at diffuser entrance.

Cryogenic and hydraulic turbines

Cryogenic turbines are speed limited because of the low speed of sound at low temperatures. Accordingly, higher-work nonradial blades can be used. Hydraulic turbines are also not stress limited, and there is considerable freedom to choose the best blade shape. The very-high-specific-speed Francis turbines, which are nominally radial inflow but may have some stream surfaces increasing in radius through the blading, have blade shapes that are the result of decades of patient experiments aimed at eliminating flow reversals and cavitation (figure 1.3e). The blade shapes that have evolved cannot be fully designed by the methods of this book and must be regarded as a special case.

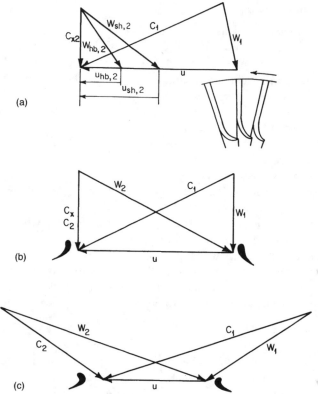

(a) Radial-inflow turbine (b) High-efficiency axial-flow turbine (c) High-work axial-flow turbine

Figure 9.3. Radial-inflow compared with axial-flow turbine vector diagrams

Relative efficiencies

Small radial-inflow gas turbines have efficiencies comparable to axial turbines that could be substituted. In some respects one would expect higher losses because of the higher turning in the rotor, higher aerodynamic loading, larger wetted area, and in addition the significant one that the conical diffusers that are almost always combined with radial turbines work very poorly with nonuniform or swirling inlet flow. It is very difficult to design the outlet section of the blades of a radial-inflow turbine (the exducer) so that the flow emerges without swirl. Even if the designer is successful in doing so, this condition will occur at only one flow coefficient. At slightly off-design conditions the flow will inevitably have swirl, which will cause the type of flow breakdown illustrated in figure 4.21. However, counterbalancing these unfavorable aspects is a significant favorable factor: the Mach number of the leaving flow is lower than for an equivalent single-stage axial turbine. Therefore high diffuser losses are relatively less important.

Disincentives to multistaging

Radial-inflow turbines are almost never used in multistage configurations. (The GM-Allison AGT-100 100-kW automotive engine developed in the 1980s under funding from the U.S. Department of Energy and NASA is an exception.) Any radial-flow machine makes severe demands on ducting. Centrifugal compressors and pumps have sufficiently large advantages in work per stage over axial-flow compressors and pumps that the heavy additional costs of the ducts and casings are compensated for. Radial-inflow turbines do not have these advantages. The configuration of axial-flow turbines makes the design and manufacture of multistage turbines relatively easy. For the highest-efficiency turbine, therefore, the designer can choose a multistage axial-flow turbine with an optimum diagram (figure 9.3b) and a low blade speed. This will probably have several percentage points advantage in efficiency over a single-stage radial-inflow turbine of high blade speed because the multistage axial turbine will have a low exit velocity. This velocity will then be diffused more efficiently because the flow will be more uniform (and because the annular diffusers used with axial turbines can incorporate rudimentary straightening vanes, so that flows with higher degrees of swirl can be efficiently diffused). These factors are additional to the intrinsically higher efficiency, at least in the larger sizes, of the more direct flow in axial turbines. The use of two or more axial stages reduces the leaving (kinetic-energy) loss to below that of a single-stage radial turbine.

Rotating inertia

Radial turbines have higher rotating inertia than do axial turbines designed to meet the same specifications with minimum inertia. This is the result of the necessarily long blades and consequently long hub (figure 9.4, Kronogard (1975)).

Thermal inertia

The long hub, with a continuous surface scrubbed by hot gases and extending over a large radial range, results in high thermal stresses, particularly in gas-turbine engines

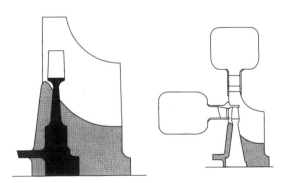

Figure 9.4. Comparison between axial and radial rotors and volutes plus nozzles designed for the same duty. From Kronogard (1975)

during start-up and shutdown. The absolute magnitude of thermal stress is a function of size. Thermal stresses are managed adequately in small hot-gas turbines (up to, say, 250-mm rotor diameter) by incorporating deep cuts ("scallops") between the blades at the periphery (figure 9.5).

Susceptibility to particle damage

Solid or liquid particles in the flow to a radial-inflow turbine will be accelerated in the nozzles only by the viscous drag of the surrounding fluid. Particles will emerge from the nozzles at velocities lower than that of the gas flow by amounts that increase with particle size. Larger particles (e.g. 0.1 mm) will therefore be struck by the leading side of the tips of the rotor blades, bounce off, and hit the nozzle vanes with very high velocity

This early USAF auxiliary turbine engine had a radial-inflow turbine that was repeatedly damaged as a result of hard carbon pieces shed from the combustor and coming through the turbine nozzles but not able to pass through the turbine until reduced in size by impact against the rotor blades and nozzle vanes. The scalloping of the backplate between the turbine blades is visible. This, and the large diameter of the "exducer" tips, distinguishes radial-inflow turbines from radial-outflow compressors. The centrifugal-compressor diffuser vanes can be seen between the turbine-inlet wall and the outer casing.

Figure 9.5. Hot-gas radial-inflow-turbine rotor

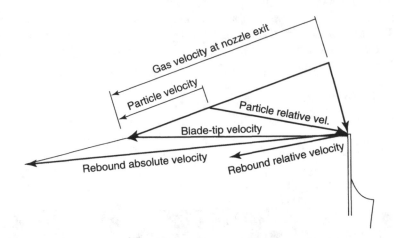

Figure 9.6. Typical path of particle in radial-inflow turbine

(figure 9.6). The particle, plus possible debris from the collisions, will then again hit a rotor blade. Only when a particle has been reduced in size to the point at which it will rapidly reach close to local gas speed will it be able to escape from the rotor with the exhaust flow. This process has destroyed radial-inflow turbines in less than an hour.

Reversible operation

Radial-inflow turbines have the potential advantage that the direction of rotation could be reversed by the incorporation of means to swivel the nozzle vanes (figure 9.7). The efficiency in the reverse direction would be lower than in the ahead direction because the exducer (the outlet axial-flow path of the rotor blading) would now add swirl to, instead of removing it from, the leaving flow. However, for transportation vehicles such as ships and tanks that operate for only short periods in reverse, but require rapid response, this concept could offer advantages. It has been incorporated into several proposed propulsion arrangements.

Summary of applications

The radial-inflow turbine emerges from the foregoing as being used in some special applications, such as for Francis hydraulic turbines and for cryogenic turbines, where low speeds permit optimal blade shapes to be used, and for small single-stage hot-gas expanders where low manufacturing cost is of major importance. Two examples in the latter category are small exhaust-gas turbosuperchargers for internal-combustion engines and small short-mission gas-turbine engines. In small sizes and for single-stage units, the efficiency of radial-inflow turbines is comparable to, and in certain cases higher than, that of axial turbines. Experimental ceramic turbines have achieved greater success in the radial configuration than in the axial, but particle erosion has been a serious problem.

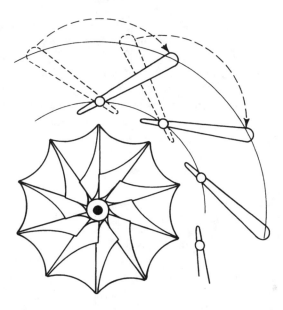

Figure 9.7. Reversible-rotation radial-inflow turbine

9.3 Design process for compressors, fans and pumps

The design process is expanded from the brief version in section 5.6.

1. The specifications give the pressure ratio, $p^{\otimes}_{0,2}/p_{0,1}$, or the total head rise, $H^{\otimes}_{T,2} - H_{T,1}$, to be obtained, as well as the mass flow rate, the inlet conditions, and possibly the shaft speed.

2. The efficiency corresponding to the definition of the useful outlet pressure, $p^{\otimes}_{0,2}$, or total head, $H^{\otimes}_{T,2}$, must be estimated. The experienced designer can usually make a fairly accurate guess based on an interpolation or extrapolation of previous known results. Envelope curves correlating the best published test efficiencies of radial-outflow compressors as a function of design-point pressure ratio are shown in figure 3.13. These best results will be at or close to the optimum specific speed and blade sweepback angle. For machines of non-optimum specific speed use the curves of figure 5.20. See below for estimates of efficiency decrements for high tip clearance, low Reynolds number, high Mach number, rough surface finishes, and so forth.

3. Initially at least the designer usually specifies that there be no rotation in the inlet flow (by requiring that, if necessary, there be straightening or "de-swirl" vanes) so that $C_{u,1} = 0$. Then $C_{u,2}$ is obtained from equations 5.20 or 5.23. The outlet velocity diagram can now be selected.

4. The outlet velocity diagram, figure 9.8a, is specified by two angles, α_{C2} and α_{W2}. The first of these, α_{C2}, is the absolute angle of flow leaving the rotor. (Under the conventions by which velocity diagrams are made, it would be the mean flow direction if the rotor-blade wakes, the other boundary layers, and the flow nonuniformities within the rotor-blade passages were to be fully mixed to give a uniform flow at the rotor-outlet diameter.) If the rotor flow passes into a diffuser, at least the first part of this will be a vaneless diffuser. The angle α_{C2} can then be chosen to give the desired margin from diffuser stall, using figure 4.22. In preliminary design a reasonable initial value to use for α_{C2}, if the diffuser configuration has not been selected, is 60 degrees. If the rotor is to exhaust direct into a plenum-collector, α_{C2} could be set somewhat higher, say 75 degrees, to reduce the losses in the outlet kinetic energy, so long as the rotor relative-velocity ratio, $W_2/W_{sh,1}$, is held above, say, 0.8 (see figure 5.26).

5. The rotor-outlet relative flow angle, α_{W2}, at the design point can be specified by the designer. The normal range is from -10 degrees, for a radial-bladed impeller with a large number of blades, to -60 degrees, for an impeller with a few highly

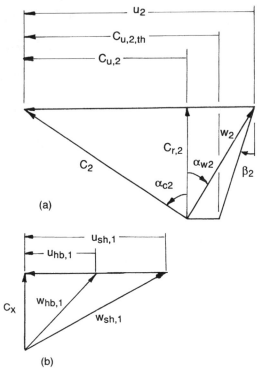

(a)

(b)

(a) Rotor-outlet vector diagram   (b) Rotor-inlet vector diagram

Figure 9.8. Radial-outflow-compressor vector diagrams

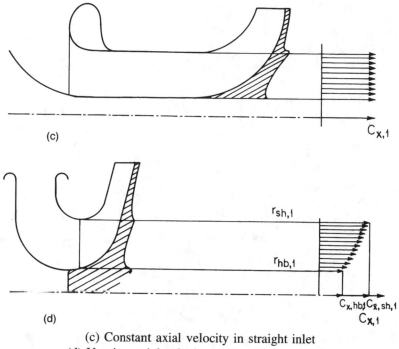

(c) Constant axial velocity in straight inlet
(d) Varying axial velocity in sharply curved inlet

Figure 9.8. Continued

backswept blades. A correlation due to Wiesner (1966) between the impeller-blade angle at the periphery, β_2, and the number of blades, Z, is given in equation 5.22.

$$\sigma_w \equiv \frac{C_{u,2,ac}}{C_{u,2,tl}} = 1 - \frac{\sqrt{\cos \beta_2}}{Z^{0.7}} \tag{9.1}$$

6. The outlet velocity triangle is now tentatively fixed. The work coefficient, $\psi = C_{u,2}/u_2$, can be calculated for impellers with no inlet swirl by the method of section 5.6.

7. It may be found that the blade speed for a single-stage machine is too high for stress reasons (see chapter 13). Therefore several stages must be used. The number of stages may be tentatively chosen at this point. Usually not all the swirl coming from the upstream diffusers is removed for the second and later stages, and the inlet velocity diagram may therefore need to be calculated to account for $C_{u,1}$.

There are reasons other than high stresses for choosing to use several stages rather than one. The overall stagnation-to-static efficiency will normally increase with the number of stages, because the outlet kinetic energy will decrease. The relative velocity at rotor inlet will decrease with increasing number of stages, alleviating

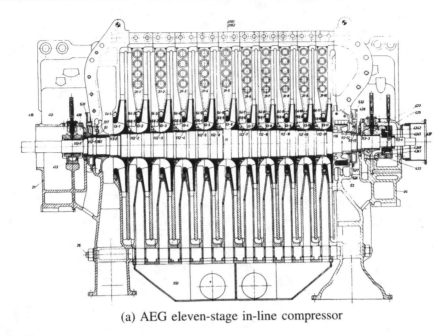

(a) AEG eleven-stage in-line compressor

Figure 9.9. Contrasting approaches to multi-stage intercooled centrifugal compressor

high-Mach-number problems. Last, using multiple stages will lower the required shaft speed and increase the size of the impellers, which may be desirable for small machines.

8. The rotor inlet can now be designed. Sometimes mechanical requirements dictate a minimum hub diameter, $d_{hb,1}$, at the inlet (figure 9.9a). Alternatively, "overhung" impellers used in many single-stage and geared multi-stage applications give the designer freedom to specify a desirable hub-shroud ratio (figure 9.9b). In this case the minimum hub diameter at inlet is set by blade blockage (i.e., there may be no room for a passage between the blades). In either case the shroud diameter at rotor inlet, $d_{sh,1}$, is now chosen to give minimum relative velocity at that point.

Optimization of rotor-inlet shroud diameter

If a small shroud diameter is used, the blade peripheral speed, $u_{sh,1}$, will be small, but the axial velocity, $C_{z,1}$, will be high because of the reduced area. Accordingly, the relative inlet velocity at the shroud, $W_{sh,1}$, will be high. If the area is increased to reduce the axial velocity, the peripheral speed is increased, and again $W_{sh,1}$ may be high, even though the mean axial velocity is reduced. It is obvious that there must be an inlet shroud diameter giving minimum relative velocity. This can be found analytically or iteratively. An example that follows will illustrate the method.

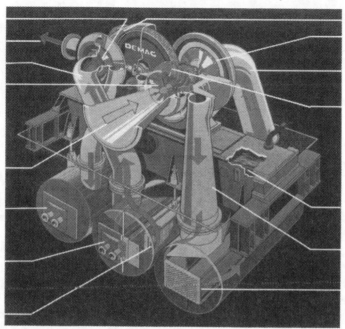

(b) DEMAG four-stage centrifugal compressor

Figure 9.9 shows different approaches to multi-stage intercooled centrifugal compressors. The AEG eleven-stage in-line compressor is compact, with the intercoolers in the return path from the radial diffusers. The specific speed of the impellers has to be smaller than optimum to keep the shaft length between the two bearings small enough to prevent severe shaft whirling. The shaft has its maximum diameter at its center, where the large hubs compromise the design of the stage inlets. The DEMAG compressor has four stages on two shafts driven by a "bull" gear. Each stage can be of near-optimum specific speed and can have excellent inlets to overhung impellers (ie there is no protruding shaft). The machine uses, consequently, considerably greater volume and floor area, and each stage must be given a separate pressure casing.

Figure 9.9. Concluded

A value of $C_{x,1}$ that is constant with radius can be used for each calculation if a gently curved inlet configuration can be incorporated (figure 9.8c). If the inlet flow is sharply curved, the axial flow velocity (if unseparated) will increase with radius, and a simple function, such as a power-law variation, of $C_{x,1}$ with radius can be assumed (figure 9.8d).

$$C_x = C_{x,hb} + (C_{x,sh} - C_{x,hb}) \left[\frac{r - r_{hb}}{r_{sh} - r_{hb}} \right]^f \tag{9.2}$$

The mean velocity $C_{x,m}$, for this velocity distribution is given by

$$\frac{C_{x,m}}{C_{x,sh}} = \frac{C_{x,hb}}{C_{x,sh}} + \frac{2[1 - (C_{x,hb}/C_{x,sh})]}{(1 + \Lambda)} \left[\frac{(1 - \Lambda)}{(f + 2)} + \frac{\Lambda}{(f + 1)} \right] \tag{9.3}$$

where $\Lambda \equiv r_{hb}/r_{sh}$, the hub-shroud ratio.

As an example of the use of this relation, suppose that an inlet of hub-shroud ratio 0.6 is being designed. Experimental results from a model of the inlet, or an approximation to a potential-flow solution, or a designer's intuition, are used to characterize the axial-velocity distribution by the following values of the two variables involved:

$$\frac{C_{x,hb}}{C_{x,sh}} = 0.5, \qquad f = 0.8$$

Then $C_{x,m}/C_{x,sh} = 0.798$.

The velocity diagram can be constructed from equation 9.2. The design process will be illustrated in a later example. Other design considerations for compressors, fans and pumps will be discussed subsequently.

9.4 Design process for radial-inflow turbines

The inlet velocity diagram is simple. The nozzle-exit flow angle, α_{C1}, is usually chosen at 70 degrees. (NASA data in figure 5.23b show that a slightly higher angle might be optimum.) The rotor blades are usually radial at the periphery. The velocity diagram is, therefore, qualitatively as shown in figure 9.10. The only uncertainty is the direction of the relative flow at inlet, α_{W1}. The angle will vary as the rotor peripheral speed, u_1, varies for a given nozzle-outlet velocity, C_1 (or as the flow coefficient, $\phi \equiv C_{r,1}/u_1$ varies). For design purposes, we want to be able to choose an angle that will give maximum efficiency. As has been stated in section 5.6, this angle has been found to be the same as would be calculated for the "slip factor" or slip angle were the same rotor to be operated as a compressor. For this we have recommended Wiesner's correlation (equation 9.1).

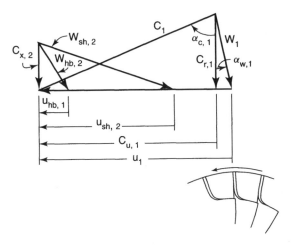

Figure 9.10. Inlet and outlet velocity diagram for a radial-inflow turbine

Another guideline is frequently used: to make the ratio of the peripheral speed, u_1, to the so-called "spouting velocity" equal to 0.7. We will show that this is approximately equivalent to the guidelines above. The spouting velocity is the hypothetical velocity reached if the fluid expanded in an isentropic nozzle through the whole turbine pressure ratio.

$$u_1 = 0.7\sqrt{2\Delta h_{0,s}} = 0.99\sqrt{\Delta h_{0,s}} \qquad (9.4)$$

Using the slip-factor correlation for a turbine with radial blades at inlet, the ratio $(C_{u,1}/u_1)$ is in the region of 0.82 (see table 5.1 in section 5.6).

The actual turbine enthalpy drop $\Delta h_0 = u_1 C_{u,1} = (C_{u,1}/u_1)u_1^2$ so that

$$u_1 = \sqrt{\frac{\Delta h_0}{(C_{u,1}/u_1)}} \approx 1.1\sqrt{\Delta h_0} \approx 1.1\sqrt{\eta_{s,e}\Delta h_{0,s}} \qquad (9.5)$$

If the turbine has an isentropic efficiency of 0.81 this value becomes almost identical to that of equation 9.4.

We prefer to use the slip-factor correlation because it better represents the fluid mechanics of the incident flow on to the turbine blades. Highly loaded blades (which in a compressor would have a relatively large angle of deviation of the leaving flow to allow for the blades to be gradually unloaded) would have a relatively large angle of negative incidence at best-efficiency point. The method allows for the choice of number of blades and even for non-radial blade angles to be accommodated.

Exducer design

Whereas the velocity diagram at the turbine-rotor inlet is similar to that at a compressor-rotor outlet, there is not the same parallel between the rotor outlet and the compressor inlet. The relative velocity at the outer diameter of the compressor inlet should be minimized to reduce Mach-number losses or, in pumps, the danger of cavitation. In radial-inflow turbines, rotor losses will be reduced if there is accelerating flow, so that the relative velocity at exducer outlet should be high. At the same time the axial-flow velocity at outlet should be low. Therefore the exducer outer diameter should be relatively large. The ratio between this diameter and the diameter of the rotor periphery can be chosen as follows.

The mass flow at rotor inlet and exducer outlet is given by:

$$\dot{m} = \rho_{st,1}\pi d_1 b_1 C_{r,1} = \rho_{st,2}\frac{\pi}{4}(d_{sh,2}^2 - d_{hb,2}^2)C_{x,2} \qquad (9.6)$$

$$\left(\frac{d_{sh,2}}{d_1}\right)^2 = \frac{4(b_1/d_1)(\rho_{st,1}/\rho_{st,2})}{(1-\Lambda^2)(C_{x,2}/C_{r,1})} \qquad (9.7)$$

$$\text{where} \quad \Lambda \equiv \frac{d_{hb,2}}{d_{sh,2}}$$

The ratio (b_1/d_1) is a function of the specific speed and other design parameters, as derived in equation 5.17.

$$N_s = 2\left[\pi\phi(b_1/d_1)\right]^{(1/2)}/\psi^{(3/4)})$$

$$\phi \equiv (C_{r,1}/u_1)$$

$$\psi = (C_{u,1}/u_1) \quad \text{for} \quad C_{u,2} = 0$$

$$\tan\alpha_{C1} = \psi u_1/C_{r,1}$$

$$\phi = \psi/\tan\alpha_{C1}$$

$$\left(\frac{b_1}{d_1}\right) = \frac{N_s^2 \tan\alpha_{C1}\psi^{(3/2)}}{4\pi\psi} = \frac{N_s^2 \tan\alpha_{C1}\psi^{(1/2)}}{4\pi} \tag{9.8}$$

At the optimum value of specific speed (about 0.65, figure 5.20), and for $\alpha_{C1} = 70°$ and $\psi = 0.82$, $(b_1/d_1) = 0.0836$, or approximately 1/12.

The value of the exducer-outlet hub-shroud ratio, Λ, is usually small and is usually fixed by the hub diameter. Although at first sight one might want the hub diameter to go to zero, the blades must be tapered in cross-section, thicker at the base, to withstand the stress, and reducing the hub diameter beyond some reasonable limit will increase the blade-root and hub stress while giving very little additional flow area. We shall also show below that a minimum value of the exducer hub-shroud ratio is required to ensure that the hub streamline through the rotor has an overall acceleration. Typical values for exducer hub-shroud ratios are 0.25 to 0.4.

With specified values of N_s, ψ, and with other known turbine parameters as inputs to equation 9.8, the exducer shroud diameter can be calculated for various values of $(C_{x,2}/C_{r,1})$ in equation 9.7. It is desirable that $(d_{sh,2}/d_1)$ be, say, less than 0.9. This equation can give the maximum value of the density ratio $(\rho_{st,1}/\rho_{st,2})$, which is a function of the turbine pressure ratio, for different values of $(C_{x,2}/C_{r,1})$. For a turbine of optimum specific speed, for which, as we have shown, (b_1/d_1) is about 1/12, and for an exducer hub-shroud ratio of about 0.32, the maximum density ratio that can be designed for efficiently is calculated to be 2.2 $(C_{x,2}/C_{r,1})$. Therefore the designer of a high-pressure-ratio radial-inflow turbine does not have the freedom to choose a low value of the outlet velocity.

For high efficiency it is also desirable that there be acceleration along the hub stream-line, i.e. that $(W_{hb,2}/W_1)$ be above, say, 1.1. This ratio can be calculated from the velocity diagram in figure 9.10 as follows.

$$\tan\alpha_{C1} = \frac{C_{u,1}}{C_{r,1}} = \left(\frac{C_{u,1}}{u_1}\right)\left(\frac{u_1}{C_{r,1}}\right) \Rightarrow \left(\frac{C_{r,1}}{u_1}\right) = \frac{\psi}{\tan\alpha_{C1}} \tag{9.9}$$

$$W_1^2 = C_{r,1}^2 + (u_1 - \psi u_1)^2 \Rightarrow \left(\frac{W_1}{u_1}\right)^2 = \left(\frac{C_{r,1}}{u_1}\right)^2 + (1-\psi)^2 \tag{9.10}$$

$$W_{hb,2}^2 = C_{x,2}^2 + u_1^2\left(\frac{d_{sh,2}}{d_1}\right)^2\Lambda^2 \Rightarrow$$

$$\left(\frac{W_{hb,2}}{u_1}\right)^2 = \left(\frac{C_{x,2}}{C_{r,1}}\right)^2\left(\frac{C_{r,1}}{u_1}\right)^2 + \left(\frac{d_{sh,2}}{d_1}\right)^2\Lambda^2 \tag{9.11}$$

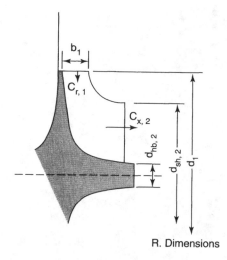

R. Dimensions

Figure 9.11. Exducer design

$$\left(\frac{W_{hb,2}}{W_1}\right)^2 = \frac{\left(\frac{C_{x,2}}{C_{r,1}}\right)^2 \left(\frac{C_{r,1}}{u_1}\right)^2 + \left(\frac{d_{sh,2}}{d_1}\right)^2 \Lambda^2}{\left(\frac{C_{r,1}}{u_1}\right)^2 + (1 - \psi)^2} \qquad (9.12)$$

The application of these design approaches will be illustrated in the turbocharger preliminary-design example below.

The outlet-side hub of a radial-inflow turbine is usually left "unstreamlined", and used as a balancing plane and/or as a point at which an extraction tool can grip the rotor. There must, therefore, be a pocket of "dead" flow downstream of this bluff hub that cannot help to promote efficient diffusion of the rotor-exit kinetic energy. There is no known case where a streamlined cone has been added to the hub, so that presumably the payoff is small.

Example 1   Radial turbosupercharger preliminary design

Calculate the principal dimensions of the centrifugal compressor, the rotor diameter of the radial-inflow turbine of a diesel-engine turbosupercharger, and the "back" pressure (stagnation) on the engine exhaust (figure 9.12). Some design information and suggestions are listed in table 9.1.

The vaneless-diffuser radius ratio should be designed to be 80 percent of that found for the stability limit. Both the compressor and turbine operating conditions at their design points should be for velocity diagrams at the rotor peripheries in which the tangential velocity is given by Wiesner's correlation (equation 9.1). Figure 9.13 is a version of figure 5.24 that can be applied to a compressor-rotor outlet or a turbine-rotor inlet.

Bearing and windage losses are 2 percent of the compressor fluid power. The fuel used can be regarded as "standard", giving no change in the molal mass of the products.

It is sufficient to make graphical or other approximate interpolations of the compressible-flow functions and to make only one iteration of, for instance, static conditions. Part of the

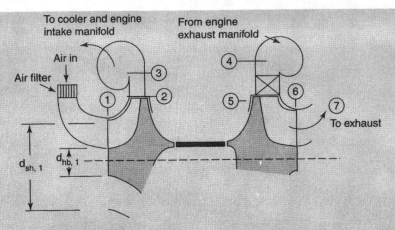

Figure 9.12. Diagram showing calculation planes of radial-flow turbosupercharger

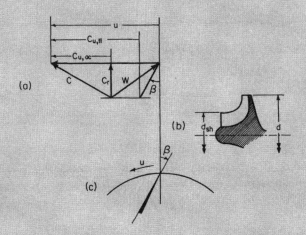

Figure 9.13. Diagram defining inputs to Wiesner's correlation of tangential velocities

design process will be to find the compressor-shroud diameter for minimum diffusion of the shroud streamline.

Calculations

The sequence of the calculations was to find the shaft speed from the specific speed and then to solve the velocity diagram at compressor-rotor outlet. A number of iterations on the inlet-shroud diameter was necessary to find the value that gives minimum relative velocity there. The vaneless-diffuser stability limit was found, and hence the diffuser diameter ratio for the specified safety factor. Last, the turbine rotor diameter was found, and the engine back pressure was calculated.

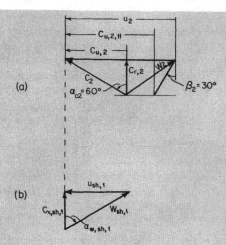

(a)

(b)

Figure 9.14. Turbocharger-compressor velocity diagrams. Top: rotor outlet; bottom: rotor-shroud inlet

Table 9.1. Turbocharger design data

	Compressor	Turbine
Mass flow, kg/s	1.0	1.04
Inlet stagnation temperature, K	300 ($T_{0,1}$)	800 ($T_{0,4}$)
Inlet stagnation pressure, N/m²	1×10^5 ($p_{0,1}$)	Find engine back pressure ($p_{0,4}$)
Outlet static pressure, N/m²	2×10^5 ($p_{st,3}$)	1.2×10^5 ($p_{st,7}$)
Fluid	Air	Combustion products, 100% theoretical air.
$\overline{C}_p$ J/(kg · K) & (C_p/R)	1,010 (3.52)	1,172 (4.084)
Blade angle at periphery (deg)	30 (β_2)	0 (β_5)
Specific speed, N_s (equation 5.10)	0.628	
$d_{hb,1}/d_{sh,1}$	0.60	
Polytropic efficiency	0.82 ($\eta_{p,ts,1-3}$)	0.82 ($\eta_{p,ts,(4-7)}$)
Flow angle leaving rotor, α_{C2} (deg)	60	0
Flow angle entering rotor (deg)	0	70
Number of rotor blades, Z	17	13
Polytropic efficiency	0.96 ($\eta_{p,tt,1-2}$)	0.96 ($\eta_{p,tt,5-6}$)

Enthalpy rise

$$\frac{T_{0,3}}{T_{0,1}} = \left(\frac{p_{0,3}^{\otimes}}{p_{0,1}}\right)^{\left[\left(\frac{R}{C_p}\right)\frac{1}{\eta_{p,c,ts}}\right]}$$

where $p_{0,3}^{\otimes} \equiv p_{st,3}$ when $\eta_{p,c,ts}$ is used.

$$
\begin{aligned}
T_{0,3}/T_{0,1} &= 2^{0.3484} = 1.2732, \\
\Delta T_{0,1-3} &= 0.2732 \times 300 = 81.95 \text{ K}, \\
\overline{C}_{pc} &= 1{,}010 \text{J/(kg} \cdot \text{K)}
\end{aligned}
$$

Therefore,

$$
\Delta h_{0,1-3} = 82{,}772 \text{ J/kg} \quad \text{and} \quad (g_c \Delta h_{0,1-3})^{3/4} = 4{,}879.9
$$

Inlet volume-flow rate

We start by guessing the value of the inlet axial velocity at plane 1 (figure 9.9) at 110 m/s (127.3 m/s). (Second-iteration values are given in parentheses.)

$$
C_x/\sqrt{g_c R T_{0,1}} = 110/293.41 = 0.375 \quad (0.434)
$$

From figure (2.10), for

$$
(C_p/R) = 3.52, \qquad M_1 = 0.32 \quad (0.371)
$$

From equation 2.59,

$$
\rho_0/\rho_{st} = \left[1 + \frac{M^2}{2\left(\dfrac{C_p}{R} - 1\right)} \right]^{[(C_p/R)-1]} = 1.0520 \quad (1.0703)
$$

$$
\rho_{0,1} = \frac{p_{0,1}}{R T_{0,1}} = \frac{10^5}{286.96 \times 300} = 1.1616 \text{ kg/m}^3
$$

$$
\rho_{st,1} = 1.1042 \text{ kg/m}^3 \quad (1.0853)
$$

$$
\dot{m} = \rho_{st,1} \dot{V}_1
$$

$$
\dot{V}_1 = 0.9056 \text{ m}^3/\text{s} \quad \sqrt{\dot{V}_1} = 0.9516 \quad (\dot{V}_1 = 0.9214 \text{ m}^3/\text{s}, \ \sqrt{\dot{V}_1} = 0.9599)
$$

$$
N = \frac{60 N_s (g_c \Delta h_0)^{3/4}}{2\pi \sqrt{\dot{V}_1}}
$$

$$
= 30{,}767 \text{ rev/min} \quad (30{,}500)
$$

Compressor-outlet velocity diagram

From Wiesner's correlation (equation 9.1):

$$
\frac{C_{u,ac}}{C_{u,tl}} \equiv \sigma_w = 1 - \frac{\sqrt{\cos \beta_2}}{Z^{0.7}} = 0.872 \qquad \text{for } \cos \beta_2 = \cos 30°, \qquad Z = 17
$$

From equation 5.23,

$$
\frac{C_{u,2}}{u_2} = \left[\frac{\tan \beta_2}{\tan \alpha_{C2}} + \frac{1}{\sigma_w} \right]^{-1} = 0.676 = \psi
$$

Therefore,

$$u_2 = \left\{ \frac{g_c \Delta h_0}{\psi} \right\}^{0.5} = \left\{ \frac{82,772}{0.676} \right\}^{0.5} = 350.0 \text{ m/s}$$

$$d_2 = 60 u_2 / (\pi N) = 219.2 \text{ mm}$$

At this point it seems likely that the compressor will fall well within the diameter-ratio limits for Wiesner's correlation to be valid. We will check this later.

Inlet-velocity diagram

The procedure to select minimum $W_{sh,1}$ is as follows.

1. Choose a value of $d_{sh,1}$.

2. Calculate $u_{sh,1} = (N\pi/60)d_{sh,1}$.

3. Calculate $A_a = (\pi d_{sh,1}^2/4)(1 - 0.6^2)$.

4. Calculate $\dot{m}\sqrt{RT_{0,1}}/A_a p_{0,1}$.

5. From figure 2.9 interpolate M_1, or solve eqn. 2.61.

6. Calculate $(p_{st,1}/p_{0,1}) \rightarrow p_{st,1} \rightarrow C_{x,1} \rightarrow W_{sh,1}$.

This procedure is given in table 9.2, in which second-iteration values are again given in parentheses. The optimum value of $d_{sh,1}$ is found to be ≈ 120 mm:

$$\alpha_{W,sh,1} = \cos^{-1}(C_{x,1}/W_{sh,1}) = 56.04°$$

Table 9.2. Calculation of optimum shroud diameter

$d_{sh,1}$ mm	100	125	120
$u_{sh,1}$ m/s $= 1610.97 d_{sh,1}$	161.10	201.37	193.32
$(=1575.22\, d_{sh,1})$	(157.22)	(199.64)	(191.65)
$A_a m^2 = 0.5027 d_{sh,1}^2 \times 10^3$	5.026	7.854	7.2389
$\dot{m}\sqrt{RT_{0,1}}/(Ap_{0,1})$	0.5838	0.3736	0.4053
M_1 from figure 2.9	0.61	0.34	0.37
$\rho_{st,1}/\rho_{0,1}$ (eqn 2.59)	0.836	0.944	0.934
$C_{x,1}$ m/s $= \dfrac{0.86088}{(\rho_{st,1}/\rho_{0,1})A_a}$	204.93	116.06	127.31
$W_{sh,1} = \sqrt{C_{x,1}^2 + u_{sh,1}^2}$	260.7	232.4	231.5
	(258.5)	(230.9)	(230.1)

The minimum value of $W_{sh,1}$ is found to be 230.1 m/s so that the de Haller velocity ratio (diffusion index) $W_2/W_{sh,1} = 0.83$. This should lead to no diffusion-induced separation.

Impeller-outlet and diffuser-inlet width, b_2

$$C_2 = \psi u_2 / \sin \alpha_{C_2} = 274.1 \text{ m/s}$$
$$T_{0,2} = 381.95 \text{ K}$$
$$C_2 / \sqrt{g_c R T_{0,2}} = 0.8281$$
$$M_2 \approx 0.740, \text{ from figure 2.10}$$
$$\text{From equation 2.59 } \rho_{st,2}/\rho_{0,2} \approx 0.771$$

We need to find the rotor-outlet stagnation pressure, using the rotor efficiency.

$$\frac{p_{0,2}}{p_{0,1}} = \left(\frac{T_{0,2}}{T_{0,1}}\right)^{\left[\left(\frac{\overline{c_p}}{R}\right)\eta_{p,c,tt,1-2}\right]}$$

Therefore,

$$p_{0,2} = 2.262 \times 10^5 \text{ N/m}^2$$
$$\rho_{0,2} = 2.0635 \text{ kg/m}^3$$
$$\rho_{st,2} = 1.5909 \text{ kg/m}^3$$
$$C_{r,2} = C_2 \cos 60° = 137.1 \text{ m/s}$$
$$b_2 = \dot{m}/(\pi d_2 \rho_{st,2} C_{r,2}) = 6.69 \text{ mm}$$

Radial-diffuser stability

We need the Reynolds number

$$R_{e,2} \equiv \frac{C_2 (d_2/2) \rho_{st,2}}{\mu_{st,2}}$$

From table A1 (appendix) for a mean $T_{st,2}$ of about 320 K,

$$\mu = 2.020 \times 10^{-5} \text{ Ns/m}^2,$$
$$R_{e,2} = 2.44 \times 10^6,$$
$$b_2/r_2 = 0.061.$$

From the stability limits in Jansen's curves (figure 4.22):

b_2/r_2	$(r_3/r_2)_{mx}$
0.125	4.0
0.08	2.9
0.061	2.0 (estimated by interpolation)

80 percent of 2.0 is 1.6. Therefore,

$$d_3 = 1.6 \times 219.2 = 350 \text{ mm}.$$

Turbine rotor diameter

The turbine has slightly increased mass flow, because of fuel addition, and must supply windage and bearing power in addition to compressor power.

Therefore,

$$\Delta h_{0,e} = \left[\frac{1.02}{1.04}\right] 82{,}772 \text{ J/kg} = \frac{u_5^2}{g_c} \frac{C_{u,5}}{u_5}$$

From equation 5.30:

$$\frac{C_{u,5}}{C_{u,5,tl}} = 1 - \frac{\sqrt{\cos\beta_5}}{Z_e^{0.7}}$$

$$C_{u,5,tl} = u_5 \quad \text{and} \quad \beta_5 = 0° \quad \text{and} \quad Z_e = 13$$

Therefore,

$$\frac{C_{u,5,ac}}{u_5} = 0.834$$

Therefore,

$$u_5^2 = \frac{1.02}{1.04} \times \frac{82{,}772.4}{0.834}$$

$$u_5 = 312.0 \text{ m/s} \quad \text{and} \quad d_5 = 198.1 \text{ mm}$$

Engine back pressure

We need to find the turbine pressure ratio:

$$\dot{m}_e \Delta h_{0,e} = 1.02 \dot{m}_c \Delta h_{0,c}$$

Therefore, using the specified specific heat,

$$\Delta T_{0,e} = \frac{1.02 \times 1.0}{1.04} \times \frac{1{,}010}{1{,}172} \times 81.953 = 69.27 \text{ K}$$

$$\frac{p_{0,in}}{p_{0,ex}^\otimes} = \left(\frac{T_{0,in}}{T_{0,ex}}\right)^{\left[\left(\frac{c_p}{R}\right)\frac{1}{\eta_{p,e,ts}}\right]} = \left(\frac{800}{730.73}\right)^{\left[\left(\frac{c_p}{R}\right)\frac{1}{\eta_{p,e,ts}}\right]} = 1.570$$

(where, as usual, $p_{0,ex}^\otimes$ and $\eta_{p,e,ts}$ are defined for the same locations so that $p_{0,ex}^\otimes \equiv p_{st,ex}$).
Therefore,

$$p_{0,in} = 1.570 \times 1.2 \times 10^5 = 1.884 \times 10^5 \text{ N/m}^3$$

Turbine exducer

We need to arrive at an acceptable exducer-shroud diameter, to check on the acceleration of the hub streamline, and to specify the flow outlet angles. We will use equations 9.6 to 9.12. These are based on a simple model of the turbine in which the inlet and outlet velocities are constant, but one would expect the flow at outlet to be affected by the considerable flow turning in the rotor. The flow is likely to become more constant the longer is the rotor. However, in most applications of radial-inflow turbines, the rotor inertia is kept as small as possible, and a long rotor is undesirable. Hence this method, along with many in the book, is good for preliminary design, probably adequate for many noncritical applications, but needs to be refined in a flow-analysis computer program for designs where a coherent nonswirling outlet flow is of considerable importance.

We need to find the nozzle-outlet velocity in order to calculate the static density and the volume flow at this point. This will enable us to calculate the specific speed and hence to use equation 9.8 to give the blade width/diameter ratio. Then equation 9.7 gives the exducer diameter ratio.

Previous calculations have given the blade speed (312.0 m/s) and the work coefficient (0.834). The nozzle angle is 70°, so that the nozzle outlet velocity is calculated from

$$\tan 70° = \frac{C_{u,1}}{C_{r,1}} = \frac{\psi u_1}{C_{r,1}}$$

$$\Rightarrow \quad \phi \equiv \frac{C_{r,1}}{u_1} = \frac{0.834}{\tan 70°} = 0.3035$$

$$\Rightarrow \quad C_{r,1} = 94.70 \text{ m/s}$$

$$\cos 70° = \frac{C_{r,5}}{C_5} \Rightarrow C_5 = 276.89 \text{ m/s}$$

$$\frac{C_5}{\sqrt{g_c R T_0}} = \frac{276.89}{\sqrt{286.96 \times 800}} = 0.5779$$

We enter 0.5779 and the specified value of $(C_p/R) = 4.084$ in figure 2.10 and read a Mach number of 0.513. (We specify that a "standard fuel" is used so that the gas constant does not change in combustion.) The Mach number can be entered into equation 2.59 to give the stagnation-to-static density ratio

$$\frac{\rho_0}{\rho_{st}} = \left[1 + \frac{M^2}{2[(C_p/R) - 1]} \right]^{[(C_p/R) - 1]} = 1.137$$

$$\rho_{0,5} = \frac{1.884 \times 10^5}{286.96 \times 800} = 0.8207 \text{ kg/m}^3 \Rightarrow \rho_{st,5} = 0.7215 \text{ kg/m}^3$$

The mass flow is 1.04 kg/s, so that the volume flow is 1.441 m³/s.

The enthalpy drop is 81,180 J/kg, so that the specific speed can be calculated:

$$N_s = \frac{2\pi N}{60} \frac{\sqrt{V_{in}}}{(g_c \Delta h_0)^{(3/4)}} = \frac{2\pi 30,500}{60} \frac{\sqrt{1.441}}{81,180^{(3/4)}} = 0.797$$

The optimum specific speed for radial-inflow turbines is shown in figure 5.23 to be about 0.65. The specific speed of this turbine is, therefore, high. We might expect that there might be a problem arriving at an exducer in which the outlet diameter reduces.

The turbine-inlet blade-width-to-diameter ratio is calculated from equation 9.8:

$$(b_5/d_5) = N_s^2 \tan \alpha_{C5} \psi^{(1/2)}/(4\pi) = 0.797^2 \tan 70° \sqrt{0.834}/(4\pi) = 0.1269 \approx 1/7.9$$

We need the outlet static density. The outlet static pressure is specified, together with the stagnation temperature. The Mach number here is low, and it will be sufficiently accurate to guess at the static temperature. (This will in fact vary very slightly with different choices for the velocity ratio.)

$$p_{st,6} = 1.2 \times 10^5 \text{ N/m}^2 \qquad T_{0,6} = 730.73 \text{ K}$$

$$T_{st,6} = 725 \text{ K} \quad \text{(guess)} \qquad \rho_{st,6} \approx \frac{1.2 \times 10^5}{286.96 \times 725} = 0.577 \text{ kg/m}^3$$

$$(\rho_{st,5}/\rho_{st,6} = 1.257)$$

We tabulate the results of the calculations below, made for an exducer hub-shroud ratio of 0.3.

$(C_{x,6}/C_{r,5})$	-	0.8	0.9	1
$(d_{sh,6}/d_5)$ (eqn. 9.7)	-	0.935	0.881	0.836
$(W_{hb,6}/W_5)$	-	1.15	1.208	1.314
$\alpha_{W,sh,6} = \tan^{(-1)}(d_{sh,6}/d_5)/(C_{x,6}/C_{r,5})\phi$	degrees	75.44	72.77	70.05
$\alpha_{W,hb,6} = \tan^{(-1)} \Lambda(d_{sh,6}/d_5)/[(C_{x,6}/C_5)\phi]$	degrees	49.12	44.06	39.57

The diameter ratio and the relative-flow acceleration are both marginal at 0.8 velocity ratio. Both are satisfactory at a velocity ratio of 0.9, and we select this value. The flow angles give a guide to the exducer blade angles, which should be set to somewhat higher values to allow for flow deviation. A necessarily approximate estimate of this deviation would be three degrees at the shroud and five degrees at the hub.

This completes the preliminary design of the turbocharger. Some of the following comments and guidelines may be used, however, to refine the design further.

9.5 Nozzles for radial-inflow turbines

Nozzle configurations vary widely. Small turbocharger turbines often have no nozzle vanes: the acceleration and flow direction is imparted by the "snail-shell" scroll or inlet volute. We know of no proven design methods for turbine or compressor snail-shell scrolls; many have been published, and some may be effective.

Medium-size radial-inflow gas turbines normally employ fixed nozzle vanes of the type shown in figure 9.7 in a variable-setting form. The shape of the nozzle vanes seems to be of minor importance because of the strong flow acceleration they impart. The design problem reduces to choosing the solidity (a value of 1.5 based on the spacing at the trailing edge is typical), the number of vanes (10 to 14 is again typical), and predicting the flow deviation. Fairbanks (1980) reviews previous work and recommends a streamline-curvature method, thus attaching more importance to nozzle-vane shape than was thought justifiable here. For a desired flow-outlet angle of about 60 degrees to the radial direction he found that twelve vanes appeared to give minimum loss, and the deviation with this design was 7.0 degrees.

9.6   Performance characteristics for radial-flow turbomachines

In chapters 7 and 8 the fundamentals of the changes in the velocity diagrams and the associated density changes which accompanied the changes in axial-flow turbines and compressors were reviewed. Examples of performance characteristics were given.

We are not doing the same for radial-flow machines because all the arguments presented for axial-flow machines apply equally well to radial-flow machines. The characteristics of the two types of machines are almost identical. If, for instance, one finds compressor characteristics for a machine of 4:1 design-point pressure ratio, there will be no strong indication that it was taken from a ten-stage axial-flow compressor or a single-stage centrifugal, except perhaps the peak efficiency should be higher for the axial machine if the mass flow is above, say, 5 kg/s (for atmospheric pressure at inlet). This similarity will occur of course only if similar design limitations were used, especially the relative-Mach-number limit at the tips of the inlet blades. In turbine comparisons radial-inflow turbines with radial blades have characteristics similar to axial-flow turbines of about 50-percent reaction having similar nozzle-outlet Mach numbers.

9.7   Alternative rotor configurations

Up to this point the type of compressor rotor that has been considered has been what Harada (1988) categorizes as "three-dimensional with inducer". He tested four types of impeller in a centrifugal compressor designed for a specified output. The four types are sketched in figure 9.15. Harada found that the full three-dimensional impeller gave a 12% greater head-rise coefficient, a 13.8% greater choke flow, and a 1.6% higher

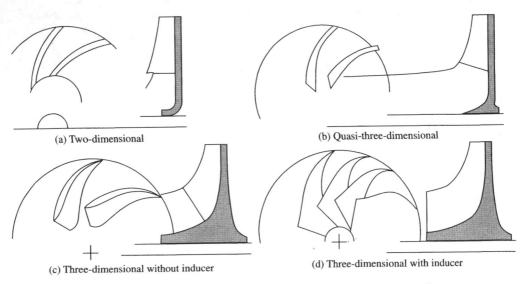

(a) Two-dimensional   (b) Quasi-three-dimensional

(c) Three-dimensional without inducer   (d) Three-dimensional with inducer

Figure 9.15. Alternative configurations of centrifugal-compressor impeller. Adapted from Harada (1988)

efficiency than the compressor with the two-dimensional impeller. The other types were intermediary.

Two-dimensional impellers with untwisted blades are frequently used in cases where manufacturing cost and short axial length are of greater importance than efficiency. Consumer products such as vacuum cleaners are an example.

The design of the rotor outlet follows the same procedure as for the three-dimensional examples above, and will produce a rotor-outlet diameter d_2, and a blade width b_2. The inlet conditions are, however, different. The velocity diagram and a sketch of the cross section of a two-dimensional fan are shown in figure 9.16.

For a good efficiency there should be some acceleration as the flow comes into the impeller axially and flows radially. We will specify an area ratio of the axial annulus to the radial cylindrical area, A', defined as:

$$A' \equiv \frac{\text{inlet annulus area}}{\text{inlet blading area}} = \frac{\pi(d_1^2 - d_{hb}^2)/4}{\pi d_1 b_1} \tag{9.13}$$

so that the continuity relation can be used to derive the inlet diameter, d_1, and the inlet width b_1:

$$d_1 = \sqrt{4A'b_2 d_2 \left(\frac{C_{r,2}}{C_{r,1}}\right)\left(\frac{\rho_{st,2}}{\rho_{st,1}}\right) + d_{hb}^2} \tag{9.14}$$

$$b_1 = \frac{b_2 d_2}{d_1}\left(\frac{C_{r,2}}{C_{r,1}}\right)\frac{\rho_{st,2}}{\rho_{st,1}} \tag{9.15}$$

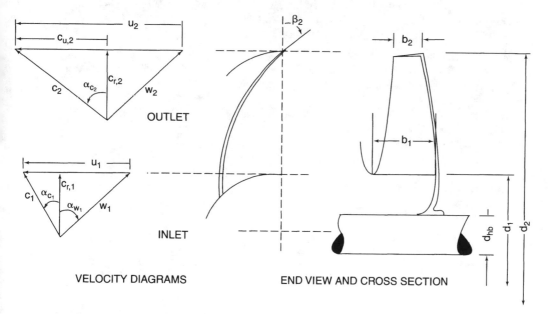

VELOCITY DIAGRAMS   END VIEW AND CROSS SECTION

Figure 9.16. Velocity diagram and cross section of two-dimensional fan

We want to minimize the inlet relative velocity and thereby to have an acceleration of the relative velocity through the rotor. We have used the radial-velocity ratio $(C_{r,2}/C_{r,1})$ as a parameter. For generality in the equations we have inserted the static density ratio, although most applications of two-dimensional impellers will be for very low Mach number conditions for which the density ratio will be unity within a close degree of possible error. We have also included the possibility of using flow swirl at inlet in the direction of impeller rotation, the angle with the radial direction at diameter d_1 being α_{C1}. The inverse of the inlet flow coefficient, the inlet flow angle α_{W1}, and the relative velocity ratio (W_2/W_1) are given by equations 9.16 to 9.18.

$$\left(\frac{u_1}{C_{r,1}}\right) = \left(\frac{d_1}{d_2}\right)\left(\frac{(C_{r,2}/C_{r,1})}{(C_{r,2}/u_2)}\right) \tag{9.16}$$

$$\alpha_{W1} = \tan^{-1}\left[\left(\frac{u_1}{C_{r,1}}\right) - \tan\alpha_{C1}\right] \tag{9.17}$$

$$\left(\frac{W_2}{W_1}\right) = \left(\frac{C_{r,2}}{C_{r,1}}\right)\sqrt{\frac{1 + (u_2/C_{r,2})^2(1-\psi^2)}{1 + [(u_1/C_{r,1}) - \tan\alpha_{C1}]^2}} \tag{9.18}$$

The optimization of a two-dimensional impeller is best illustrated by a design example.

Example 2   Design optimization of the inlet of a two-dimensional fan

A two-dimensional fan is required to produce a stagnation-to-static pressure rise (to the diffuser outlet) of 4,000 N/m² and a flow of 0.04 m³/s when running at 10,000 rpm. The inlet stagnation temperature is 288 K and the inlet stagnation pressure is 100,000 N/m². It has been decided to use eight rotor blades with an exit blade angle of 50° to the radial direction, and to specify the absolute flow direction at that point as 60°. The estimated polytropic stagnation-to-static efficiency to diffuser outlet is 0.65. The hub diameter is 20 mm.

Solution.

We have followed normal preliminary-design practice to arrive at a rotor outlet diameter d_2 of 185 mm and an axial width at outlet, b_2, of 2.3 mm. The slip factor, σ_w, is calculated to be 0.8501, leading to the work coefficient $\psi = 0.5364$ and the flow coefficient at outlet, $\phi_2 \equiv C_{r,2}/u_2 = 0.3097$. The results of calculations for four chosen radial-velocity ratios are given in the table below.

Radial-velocity ratio $C_{r,2}/C_{r,1}$	Eqn.		0.8	1.0	1.2	1.4
Inlet diameter d_1	9.14	mm	45.1	49.4	53.3	57
blade width at inlet b_1	9.15	mm	7.5	8.6	9.5	10.4
$(u_1/C_{r,1})$	9.16		0.629	0.862	1.117	1.394
Inlet relative flow angle α_{W1}	9.17	deg	32.2	40.8	48.2	54.3
relative velocity ratio (W_2/W_1)	9.18		1.52	1.36	1.2	1.05

Comments

Any of these designs should work well. The design represented by the last column provides sufficient acceleration of the relative velocity (work quoted earlier by Rodgers showed that radial-outflow machines can withstand relative-velocity ratios below 0.8) and the inlet relative flow angle is close to the blade angle at outlet so that the blade shape could be near to a spiral. However, the larger inlet would require more axial length for turning than the designs with lower radial-velocity ratios.

Mixed-flow compressors and fans

Mixed-flow compressors and fans take flow in axially, but the rotor discharge is between the radial and axial directions. Some guidance on design may be obtained from Musgrave and Plehn (1987), who write "The mixed flow compressor has a very modest history and it has rarely seen production". Deviation in mixed-flow fans is discussed by Matthews (1991).

9.8 Blade shape

In this text we concentrate on entry and exit conditions for rotors and stators. Some guidance on overall design, including blade shape, is given by Carre (1978), and in the published discussions of this paper by Moore and Whitfield. Moore quotes the Rossby number, defined as the mean relative velocity along the impeller meanline, divided by the angular speed of the rotor in radians per second and by the mean radius of curvature of the meanline streamline in the turn from axial to radial in a three-dimensional impeller (figure 9.17). When the Rossby number is less than or equal to one, the flow is dominated by the effects of impeller rotation. When it is higher than one, the flow is dominated by the effects of flow curvature. If space allows, therefore, the inducer and impeller inlet should be shaped so that the mean radius of curvature gives a Rossby number of one or lower.

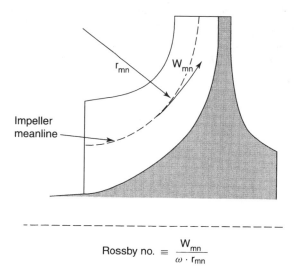

$$\text{Rossby no.} \equiv \frac{W_{mn}}{\omega \cdot r_{mn}}$$

Figure 9.17. Rossby number

Whitfield's discussion recommends impeller hub and shroud profiles that conform to the relation:

$$\left(\frac{r+a}{b}\right)^p + \left(\frac{z+c}{d}\right)^q = 1$$

where r and z are the radial and axial coordinates; a, b, c, and d are set by the end conditions; and p and q are varied to obtain sets of curves. (These variables are quoted from the paper and are not part of the standard symbols of this text.) Aungier (1995) gives recommendations and data on suggested shapes for blades, diffusers and return channels. Suga et al. (1996) investigated, among several other aspects, the placement of splitter blades and found that a 40-60 location closer to the suction side of the full blades gave improved performance over a 50-50 placement. The most comprehensive treatment of radial-flow compressors is given by Japikse (1996), including impeller shapes, loading patterns, and three-dimensional computer-aided design.

9.9 Surge range

The phenomenon of compressor surge has been discussed in connection with axial compressors in chapter 8. It is a compressible-flow effect occurring at high rotational speed when the flow is reduced by a restriction at compressor outlet, possibly triggered by other disturbances, for instance flow distortions at the compressor inlet. Eventually a component in the compressor suffers a stalling flow breakdown, and the sharp reduction in pressure rise that occurs in consequence allows a large proportion of the high-pressure air or gas in the upstream delivery pipe to reverse flow, more or less explosively, through the compressor. The pressure falls to a point at which the compressor can re-establish forward flow, the pressure rises again, and unless the downstream (or upstream) conditions have changed, the surge repeats. This process can be destructive to the compressor itself and/or to the process with which it is connected. In gas-turbine engines the combustion process is normally extinguished, and in aircraft engines it may be difficult to re-light the combustors at altitude. Therefore the "surge range" of a compressor, defined by Carre (1978) as

$$\left(\frac{r_{sl} - 1}{r_{dp} - 1} - 1\right) \times 100$$

where r_{sl} is the pressure ratio at the surge line, and r_{dp} is the pressure ratio at the design point, is a very important factor in the compressor characteristics. McCutcheon (1978) pointed out that turbocharger compressors must work over a much greater range than those for gas-turbine engines. The design-speed surge range for a turbocharger compressor was specified at 17%. Initial tests showed a surge range of 39% "because of an over-conservative diffuser design". The diffuser was re-matched, producing a surge range of 28%, still better than the specifications and accepted.

Rodgers (1991) stated that high-speed surge is often triggered by inducer stall. Regulation (variation) of the inlet guide vanes extended the surge range even though the diffuser (vaned in the case discussed) might be stalled. This interesting phenomenon is related to the topic of blade shape (section 9.7). The designer using a CFD (computer fluid dynamics) program or other guidance to vary the blade shape and loading can combine a highly loaded inducer with a more-lightly-loaded main impeller ("front-loaded") or can use a lightly loaded inducer and a more-highly-loaded main impeller ("rear-loaded"). The rotors studied by Rodgers are more likely to be front-loaded. He stated that at zero prewhirl, surge was precipitated by diffuser stall. Liquid-oxygen and liquid-hydrogen turbopumps for liquid-fueled rockets are examples where the inducers are extremely lightly loaded to avoid triggering cavitation.

Investigations by Greitzer and Epstein at MIT and others have shown that surge is preceded by a travelling wave. Inhibiting this wave passively or actively (by injecting timed opposing jets) has been shown to inhibit or delay surge.

9.10   Off-design performance prediction

Some published methods of performance prediction are the following. Dallenbach (1961) gives an empirical method based on tests of twelve different-design compressors, and reviewed different types of loading distributions. Musgrave (1980) develops a simple method based on mean-line specifications. Jansen (1967) predicts the distorted flow at the outlet of compressor impellers using the streamline-curvature method with the incorporation of distributed local loss factors, which are given. Korakianitis et al. (1994) use three-dimensional characteristics to compute steady and transient, design or off-design performance, with friction losses at the walls, using mean-line specifications. This model can also be used to predict the transient performance of centrifugal impellers (e.g., it can be used to predict "turbocharger lag") in true compressible unsteady flow (other transient-flow models use constant-density, or no-mass-flow storage in the component during the transients). Japikse (1996) includes off-design performance prediction in his comprehensive text.

9.11   Design of centrifugal (radial-flow) pumps

The methods used for the design of radial-flow compressors may be used for centrifugal pumps. However, because of the high density of liquids relative to the gases handled by compressors, pumps have much lower peripheral speeds. Consequently the rotors are seldom stress-limited. High "sweepback" rotor-exit angles can therefore be used: the range of 50° to 70° is identified with pumps in figure 5.19. The rotor-outlet kinetic energy is therefore a smaller proportion of the overall energy rise, and many pumps are designed to be operated without diffusers: the rotors exhaust directly into "scrolls" or "snail-shells" that have zero or very small in-built diffusion. In pumps in which high efficiency is important the flow will leave the rotor and pass to a vaneless and/or a vaned diffuser, as in compressors.

The optimum specific speed may be selected from figure 5.20 to 5.22 as for compressors. However, the Reynolds number of the flow through pumps is likely to be much higher than that in compressors because of the higher fluid density. Therefore the efficiencies should be higher for pumps than for compressors. As a general guide, one would expect the losses (1 - efficiency) to be about one-half that of compressors. The number of blades may be chosen from figure 5.25. Another guideline attributed to Stepanoff is that the number of blades should be (90° - blade angle at rotor periphery [degrees])/3. Thus a rotor with an outlet blade angle of 75° would have five blades while one with an angle of 30° would have 20 blades. (This guideline gives blade numbers that are high for "compressor" blade angles.)

9.12   Cavitation and two-phase flow in pumps

The principal differences between the design treatments of compressors and pumps is that cavitation must be considered for pumps in approximately the same location as high-Mach-number flow must be considered for compressors. This phenomenon is similar for axial-flow and radial-flow pumps because it occurs predominantly on the rotor blades close to the inlet. Cavitation manifests itself in two separate but related phenomena. The first is cavitation damage. When a liquid enters a pump, it may produce within the blade passages tiny vapor bubbles that grow rapidly and then collapse with violent decelerations and associated pressure waves able to cause catastrophic damage to the hardest of surfaces.

The second phenomenon is that of cavitation performance loss. Vapor bubbles may grow within the blade passages to relatively large sizes and often are present in the exit stream. The pumping performance suffers even though there may be no damage to the pump.

The third phenomenon is the loss of pump performance which occurs when the incoming stream is composed of a two-phase mixture of liquid and vapor of any proportion up to an all-vapor flow.

Korakianitis and Vlachopoulos (1997) used computational fluid dynamic programs to predict two-phase flow in turbine (drag) pumps for automotive-type fuel pumps. However, until the use of such programs becomes widespread, a combination of fundamentals with experimental data is needed.

Inception of cavitation

At normal atmospheric pressure we know that we can make water boil by heating it to 100 °C. If we contain the water in a pressure cooker so that the pressure is, say, 1.2 atmospheres, the water can be above 100 °C and still not boil. We can make it boil by releasing the vapor to lower the pressure in the cooker. We learn to do this slowly, as the boiling can be quite violent. We know, by experience or by extrapolation, that we can similarly cause water to boil, even at temperatures close to freezing, by reducing the pressure sufficiently.

In other words, we can cause boiling, or vapor formation, in a liquid, either by heating at constant pressure or by reducing the pressure at constant temperature, or by reducing the temperature at constant pressure, or both.

Cavitation is the formation of small bubbles from the reduction of stream pressure at constant temperature.

A pump can bring about a reduction in pressure in a liquid in the inlet flow in three ways.

1. The liquid may be lowered through a height in a gravitational field.

2. The liquid may experience friction in flowing past solid surfaces, and in frictional dissipation by shearing forces within the liquid, and so drop in pressure.

3. The liquid may be accelerated to higher velocities, so converting static pressure to velocity pressure.

Although the situation in which cavitation bubbles can grow may be set up by gravitational and frictional losses in pressure, the immediate precursor of bubble formation is normally the drop in pressure accompanying acceleration.

We will illustrate some of the mechanisms by which pressures are reduced to below the vapor pressure of the liquid. Before we do so, we will review the physics of bubble formation, in particular so that it can be appreciated that it is frequently necessary for the pressure to be reduced to considerably below vapor pressure before bubbles are formed.

Liquid tension and bubble formation

For a bubble to form in a liquid, molecules have to be separated from one another through the pressure of a vapor. Because of the small radius of curvature a microscopically small bubble must have, the pressure in the bubble must be higher than that in the surrounding liquid by the increment $2\sigma_s/r$, where r is the bubble radius and σ_s is the surface tension. Normally this over pressure is not large because there will be scratches, fissures, and cracks in bounding surfaces that will trap gas or vapor bubbles of relatively large radius (figure 9.18). These are responsible for the streams of vapor bubbles that can be seen in water coming to the boil in a glass beaker or a metal pan, for instance. Or the liquid may have dust particles or other contaminants able to form nuclei from which bubble can grow with only a small "superheat".

Experiments with ultrapure water heated on equally pure liquid mercury have shown that the water can exist in a metastable state without boiling even when far above the boiling point. The liquid can then be withstanding internal tensions of many atmospheres magnitude. When boiling eventually occurs in such an experiment, it does so explosively and somewhat dangerously.

In most circumstances in which pumps are used, there will be sufficient surface scratches and nuclei in the liquid for the formation of bubbles at only a little below vapor pressure. However, if the pressure reduction is very localized, which is often the case, nuclei may not be found at the precise location needed for bubble growth, and further reductions in pressure may be possible before cavitation occurs.

There are two other reasons why cavitation bubbles may not be evident, even though there is a local reduction of pressure to below vapor pressure. These reasons are based

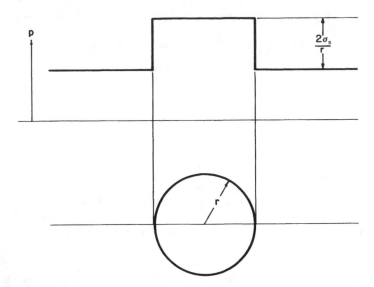

(a) Pressure rise induced by surface tension across a bubble boundary

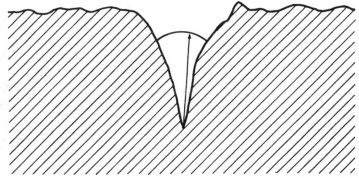

(b) Cavitation-bubble inception; potentially large radius of surface of gas bubble trapped in fissure allows bubble growth at higher stream pressure

Figure 9.18. Cavitation phenomena

on inertial and heat-transfer limitations. For a bubble to grow, the liquid surrounding it must acquire a radial outward velocity. In the early stages of growth the acceleration forces are small, because the incremental pressure is reduced by the surface-tension effect of $2\sigma_s/r$, and because the surface area is small. Bubbles may have grown to only a very small degree, therefore, when they pass out of the low-pressure region of the liquid; no further growth occurs, and they may even collapse. Bubble growth is thereby inertia limited. The collapse may occur before the bubbles are large enough to be perceptible, and/or the collapse may take place at a sufficient distance from the solid surface that the pressure waves experienced at the surface are very small.

For some liquids heat transfer from the liquid to the bubble surface may provide more of a limitation than inertia. The latent heat for vapor formation must be conducted to the surface during bubble growth and from the surface during collapse. For liquids with large evaporation enthalpies and/or small conductivities, the growth and collapse rates may be too slow for bubbles to be evident until comparatively large reductions of pressure to below the vapor pressure have taken place.

Vapor bubbles, therefore, do not necessarily appear immediately after the local pressure is reduced to below local vapor pressure. The incremental pressure reduction necessary before bubbles not only grow, but grow to a size large enough to be detected, and collapse near enough to a solid surface, and with enough intensity to cause damage, will be highly dependent on the properties of the liquid and on the configuration and operating conditions of the pump.

Modern theory attributes cavitation damage to the presence of a high-velocity microjet in the collapsing bubble (Gulich, 1989). When a bubble approaches a solid surface the relative flow becomes asymmetric, and bubble collapse also becomes asymmetric. The same reference makes these additional statements.

1. While boiling processes are dominated by wall-crevice nuclei, cavitation is governed by entrained nuclei (gas bubbles and particles).

2. There are always enough entrained nuclei available for bubble formation.

3. Increasing gas content in the pumped liquid actually lessens cavitation damage because of the cushioning effect of the noncondensible gas in an imploding bubble.

4. For minimum cavitation damage the rotor material should be resistant to corrosion from the liquid, and have high tensile strength, high fatigue strength, high hardness and high resilience. It should also be fine-grained at the surface.

5. Other conditions being equal, cavitation damage increases as the sixth power of the rotor peripheral speed, and the third power of the net positive suction head NPSH (see figure 9.19 and the discussion below). No consistent size effect has been found: small pumps and large suffer similar cavitation damage in similar conditions.

6. Cavitation is most likely to occur (in otherwise well-designed pumps) when there is flow recirculation at the inlet in low-flow conditions of operation (figure 9.20).

Pressure-velocity relation in a pump

The way in which the pressure or head in a liquid changes as it approaches and passes through a pump is illustrated in figure 9.21. Both the total head and the static head are shown. It is, of course, the static head that must fall to below the local vapor pressure or head before vapour bubbles can start growing.

If the axis of the pump is horizontal, we cannot treat the flow one-dimensionally, even if it is axisymmetric with respect to velocities, because the pressure will be lowest in the upper part of the inlet duct and pump. The reference height for the pump is therefore

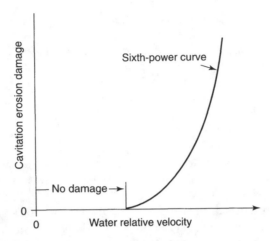

Figure 9.19. Cavitation damage versus relative velocity

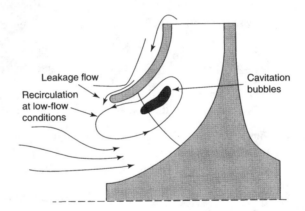

Figure 9.20. Cavitation from recirculation at low-flow operation. Adapted from Gulich (1989)

defined to be the top of the inlet blading circle. The use of this location rather than the centerline of the pump can be significant for large pumps with, for instance, inlets two meters in diameter, but has little significance for small pumps.

The pump in figure 9.21 can be either an axial-flow or radial-flow type, because only the inlet conditions have a deciding influence on cavitation. It is supposed that the inlet flow is led to the pump through a horizontal pipe of a diameter larger than that of the inlet blading. The head along a horizontal streamline through the top of the inlet blade circle is shown.

The total head is shown falling slowly along the inlet pipe under the effects of wall friction. The velocity will stay approximately constant until the pump inlet is reached, and the static head is shown falling at the same rate.

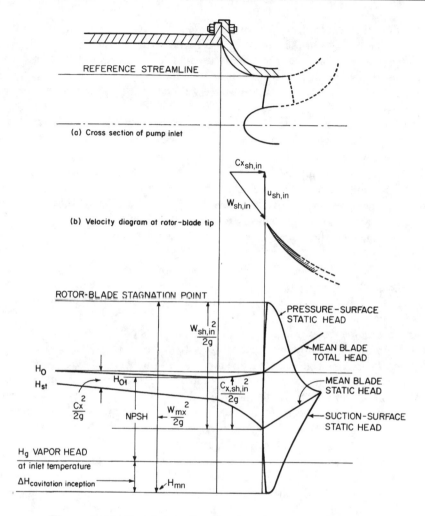

Figure 9.21. Head-velocity changes in a pump inlet

The flow is accelerated smoothly but sharply at the pump inlet. Friction is normally negligible under acceleration, so that the total head would remain almost constant, but the static head drops. It is possible, though not probable, that the cavitation in a pump is initiated by this acceleration-induced fall in static head.

If the pump is fitted with an initial stator-blade row, such as inlet guide vanes, there will be an additional acceleration and a lowering of the mean static head. Moreover the vanes will be loaded hydrodynamically, so that the head on the suction side of the vanes will be lower than the pressure-side head.

The diagram is drawn for the more usual situation where no inlet guide vanes are used. It would be easy, but complicating, to add the head distributions around vanes.

The flow then encounters the moving blades. The relative velocity of the blade tips at inlet, in the location of the streamline with which we are concerned, is $W_{sh,1}$. At the leading-edge stagnation point the flow is brought to rest relative to the moving blade, and the local static head at the blade surface is the stream static head plus the relative stagnation head. (At this point the relative static head is the same as the relative total head, because the relative flow velocity is zero.)

The inducer section of a radial-flow pump, or the first-stage rotor row of an axial-flow pump, is normally heavily loaded hydrodynamically. If there were zero loading, the relative flow would accelerate from the stagnation point to a value of $W_{sh,1}$ on both the pressure and the suction sides of the blade. The blades would have to be infinitely thin plates aligned exactly with the flow. With loading, the relative velocity on the pressure side is less than $W_{sh,1}$, and the relative velocity on the suction side increases to a local maximum, W_{mx}. (See chapter 8 and equation 8.3 for the significance of this local maximum in governing the diffusion in a pump or compressor.)

The ratio of W_{mx} to $W_{sh,1}$ may be large for airfoils of poor configuration, or where erosion or deposits have changed the basic shape of the airfoil, or where the flow incidence angle is large. It is this local maximum-velocity minimum-pressure point that usually initiates cavitation.

Any cavitation damage will occur downstream of the minimum-pressure point. The effect of the blade hydrodynamic loading is to increase the total head and the static head (see chapter 5 and Euler's equation, 5.1). Bubbles will grow while the local static head is below the vapor pressure, and inertia will cause continued growth for a short while after the local head is above the vapor pressure although the radial acceleration will become negative. At some point downstream the bubbles will collapse, causing local pressures calculated to be hundreds of atmospheres, and a resulting spherical elastic pressure wave radiates out from the collapse point, decreasing in intensity with the cube of the distance. Whether or not surface damage is caused, therefore, depends strongly on the maximum size of the bubble, the magnitude of the pressure gradient, and the distance of the solid surface from the point of bubble collapse, as well as on the fatigue-strength and the elastic properties of the surface material.

Pumps designed to handle liquids almost at their boiling points, such as the liquid-oxygen and the liquid-hydrogen pumps in some rockets, have very long spiral-blade inducers with near-zero incidence angles and very low loading, so that $W_{mx}/W_{sh,1}$ is close to unity.

Cavitation correlations

The cavitation performance of pumps of similar design can be correlated with a Thoma cavitation number, σ_c, defined as

$$\sigma_c \equiv \frac{NPSH}{\Delta H_T} \tag{9.19}$$

where

$NPSH \equiv$ net positive suction head (the head required to avoid cavitation)
$\equiv H_{T,in} - H_g$,
$H_g \equiv$ vapor head at inlet,
$\Delta H_T \equiv$ rise in total head through the pump,

and either the pump specific speed, N_s (section 5.5), or the suction specific speed, N_{sv},

$$\text{where} \quad N_{sv} \equiv N_s \left[\frac{\Delta H_T}{NPSH} \right]^{3/4} \equiv \left[\frac{N_s}{\sigma_c^{3/4}} \right] \tag{9.20}$$

The correlation arises as follows. From figure 9.21 we see by inspection that

$$H_{mn} = H_{T,in} + \frac{1}{2g}[(W_{sh,in}^2 - C_{x,sh,in}^2) - W_{mx}^2]$$

where W_{mx} is the maximum relative velocity on the blade profile, which is the maximum that will be found on the suction surface. Then

$$H_{mn} = H_{T,in} + \frac{1}{2g}[u_{sh,in}^2) - W_{mx}^2]$$

$$= H_{T,in} + \frac{u_{sh,in}^2}{2g} \left[1 - \left(\frac{W_{mx}}{W_{sh,in}} \right)^2 \left(\frac{W_{sh,in}}{u_{sh,in}} \right)^2 \right]$$

In the velocity diagram, figure 9.21, we see that

$$\left(\frac{W_{sh,in}}{u_{sh,in}} \right) = \frac{C_{x,sh,in}^2 + u_{sh,in}^2}{u_{sh,in}^2} = (\phi_{sh,in}^2 + 1)$$

where

$$\phi_{sh,in} \equiv \frac{C_{x,sh,in}}{u_{sh,in}}$$

$$\Rightarrow H_{T,in} - H_{mn} = (H_{T,in} - H_g) - (H_g - H_{mn})$$

$$= \frac{u_{sh,in}^2}{2g} \left[\left(\frac{W_{mx}}{W_{sh,in}} \right)^2 (\phi_{sh,in}^2 + 1) - 1 \right]$$

$$H_{T,in} - H_{mn} = NPSH + \Delta H_{ci} \tag{9.21}$$

where ΔH_{ci} is the incremental static head below the vapor pressure necessary to produce cavitation inception.

By dividing both sides of this equation by ΔH_T, the rise in total head through the pump, we will obtain Thoma's cavitation number, σ_c, on the left-hand side:

$$\sigma_c + \frac{\Delta H_{ci}}{\Delta H_T} = \frac{u_{sh,in}^2}{2g \Delta H_T} \left[\left(\frac{W_{mx}}{W_{sh,in}} \right)^2 (\phi_{sh,in}^2 + 1) - 1 \right]$$

When we substitute the pump head-rise coefficient, ψ_{pu},

$$\psi_{pu} \equiv \frac{g\,\Delta H_T}{u_2^2} \tag{9.22}$$

where u_2 is the rotor peripheral velocity. For an axial pump $u_2 = u_{sh,in}$; for a radial pump, $u_2 = u_{sh,in}(d_2/d_{sh,in})$. We will write the equation for the radial-pump case, noting that for an axial pump $d_2/d_{sh,in} = 1.0$:

$$\sigma_c + H' = \frac{1}{2\psi_{pu}} \left(\frac{d_{sh,in}}{d_2}\right)^2 \left[\left(\frac{W_{mx}}{W_{sh,in}}\right)^2 (\phi_{sh,in}^2 + 1) - 1\right] \tag{9.23}$$

where $H' \equiv (\Delta H_{ci}/\Delta H_T)$. This equation indicates that the value of Thoma's cavitation number for any pump is a function of H', which is partly dependent on fluid properties, on velocity-triangle values ψ_{pu} and $\phi_{sh,in}$, on the blade profile and incidence, which will govern $W_{mx}/W_{sh,in}$, and, for radial-flow pumps, on the diameter ratio $d_{sh,in}/d_2$.

We know from equations 5.11 through 5.14 that specific speed is also a function of ψ, ϕ, and the diameter ratio. It therefore seems reasonable to expect that for any one fluid at a given temperature, and any one type of pump, operating at the optimum efficiency condition, the Thoma number will be a function of the specific speed N_s.

This is indeed found to be the case. The Thoma cavitation number for traditional types of single-suction centrifugal pumps with backswept blades has been found to be well correlated with

$$\sigma_c = 2.789\, N_{s2}^{1.333} \tag{9.24}$$

where N_{s2} is the nondimensional specific speed for hydraulic machines given by equation 5.15:

$$N_{s2} \equiv \frac{2\pi N \sqrt{\dot{V}}}{60(g\,\Delta H_T)^{3/4}}$$

Correlations of Thoma's number with specific speed for many different types of pumps can be found in specialized texts such as the *Pump Handbook* (Karassik et al., 1986). Pump efficiency or suction specific speed are sometimes used as parameters. Means for applying water-cavitation data to other liquids are also given.

This handbook (and others) also offers guidance on material selection to reduce damage from cavitation when it cannot be avoided. In general very hard (such as Stellite) or very soft (polyurethane or neoprene) coatings are found to give the best protection.

All the previous discussion and analysis can be applied, with obvious changes, to hydraulic turbines, where cavitation tends to occur in the low-pressure (exit) part of the blading and in the downstream diffuser (draft tube).

9.13   Cavitation performance loss

When cavitation bubbles of the size that grow, collapse, and cause damage to solid surfaces in a pump, there is a negligible effect on pump performance. Suppose that one carries out a test in which a pump is run at constant speed and constant flow, and only the net positive suction head is slowly reduced. The head rise across the pump will remain constant up to and including the region of NPSH where cavitation bubbles first appear (figure 9.22).

When the NPSH is further reduced, more bubbles are formed, which grow larger and persist longer. Their collapse may occur downstream of the inlet blade row, or farther from the solid surface, or in a region of less-adverse pressure gradient than for the smaller bubbles, and cavitation damage lessens or disappears. The pump performance (head rise) begins to be affected, principally because of the influence of the vapor bubbles on the boundary layers. In some rare cases the initial appearance of persistent vapor cavities produces a slightly increased head. There is no agreed-upon reason for this improvement, but one possibility is that the vapor bubbles act to re-energize the boundary layer in the way accomplished by trapped toroidal vortices in ribbed diffusers (chapter 4).

Sooner or later, however, as the NPSH is reduced, the pump head falls. We define the region of cavitation performance deterioration as that in which the flow at inlet contains no vapor, although vapor forms within the pump and may persist at exit. We do not have a good correlation of cavitation performance deterioration, beyond knowing that it occurs at values of NPSH lower than that correlated by Thoma's cavitation number.

At still lower values of NPSH substantial vapor bubbles appear in the inlet flow. We refer to this condition as pump operation in two-phase flow. We are able to make approximate predictions of pump operation in certain circumstances.

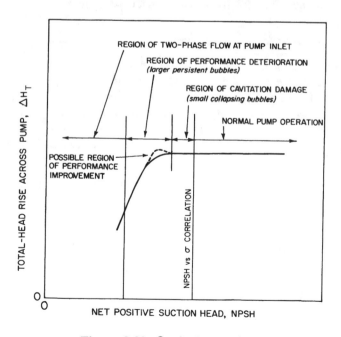

Figure 9.22. Cavitation regions

9.14   Pump operation in two-phase flow

Pumps are not normally operated with substantial vapor in the inlet. The significance of being able to predict the performance in this condition of operation is that the head produced might be of critical importance in certain emergency situations. The correlations outlined here were developed to enable the pump head to be predicted after the occurrence of a hypothetical break in the high-pressure-coolant pipe of a pressurized-water nuclear reactor. In these circumstances the flow might go forward or backward through the pump, depending on whether the break is in the inlet or the outlet pipe, and the pump rotation might be reversed during the rapid outflow which would occur if the break were in the pump-inlet pipe. The combination of flow direction and direction of pump rotation is referred to as a "quadrant". Forward flow and forward rotation is the first quadrant.

The following method of correlating the performance of centrifugal pumps with two-phase inlet flow, both vapor-liquid and gas-liquid, enables the performance of certain pumps to be predicted, at least in the first quadrant (forward flow and forward rotation).

The correlating parameter is the head-loss ratio, which is the ratio of losses in two-phase flow to the losses in single-phase flow. The losses are here defined as the difference between the theoretical and the actual change of head across the pump. The limited data available show a good correlation between peak head-loss ratio and peak single-phase efficiency for the first quadrant. Therefore the present correlations may be used to make predictions of two-phase performance of centrifugal pumps for which the single-phase performance, including the peak single-phase efficiency, is known.

Outline of the correlation method

The present correlation approach is based on writing the Euler equation (equation 5.1) for a flow consisting of two parallel streams, one of liquid and one of vapor, and on a correlation due to Lottes and Flinn (Thom, 1964) between losses in single-phase flow and those that will occur in a similar flow channel in two-phase flow. The consequence of this approach is to define a head-loss ratio, H_L and to show that this is a function of pump configuration, inlet void fraction, β_v, and flow coefficient ϕ:

$$H_L \equiv \frac{\psi^x_{tp,tl} - \psi^x_{tp}}{\psi^x_{sp,tl} - \psi^x_{sp}} = H_L(\beta_v, \phi, \text{pump configuration}) \qquad (9.25)$$

where ψ^x is the head-rise coefficient (equation 5.16), and the subscripts sp, single phase; tp, two phase; and tl theoretical. The void fraction β_v is the ratio of vapor or gas flow rate to total flow rate.

Correlation results

Very few two-phase data in the three quadrants of interest have been taken on reasonably large-scale models (one-fifth scale or larger) of typical circulating pumps used in pressurized-water nuclear reactors. The data taken at Babcock & Wilcox (B&W) by

Winks (1977) were found to correlate well by the head-loss ratio (figure 9.23a and b). These data were for air-water mixtures. Later data from steam-water experiments on a one-fifth-scale pump at Combustion Engineering (Kennedy, 1980) also correlated satisfactorily, as did experiments on much smaller pumps (Wilson et al., 1979). The lower flow coefficients are for operation in the pump region, (figure 9.24). At higher flow co-

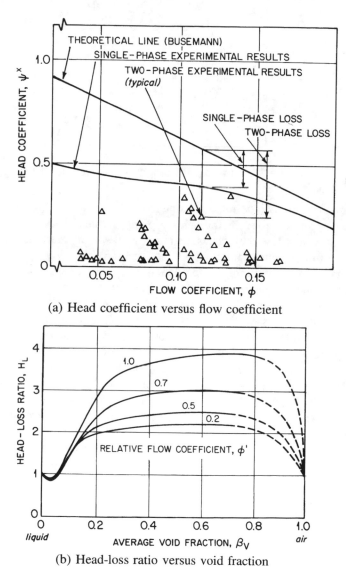

(a) Head coefficient versus flow coefficient

(b) Head-loss ratio versus void fraction

Figure 9.23. Correlation of two-phase flow in a centrifugal pump, first quadrant. From Winks (1977)

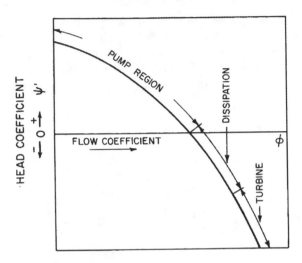

Figure 9.24. Operating regions of the first quadrant

efficients a pump will operate in the "dissipation" region, in which flow must be forced through the pump (the head falls through the pump) while the shaft still requires input power. At still higher flow coefficients the pump works in the "turbine" region, with the shaft capable of giving an output power.

The theoretical head used for the correlation of figure 9.23 was that given by assuming that the deviation angle between the direction of the mean relative flow leaving the pump impeller at the best-efficiency point and the tangent to the blade center line at the impeller periphery is constant throughout the "pump" region of figure 9.24. This is an unsatisfactory assumption, because it is obvious that, when the pump is operating in the "turbine" region, the deviation must be reversed. The change from positive to negative deviation will also occur gradually as a function of, at least, the relative flow coefficient. Although there are data on the variation of deviation with flow coefficient for particular pumps (see Keith, 1977; and Noorbakhsh, 1973), we have no satisfactory correlation of this variation over a range of pumps. The best-efficiency-point correlation of deviation due to Busemann, given in Dixon (1975), has been used for figure 9.23a. Its use is somewhat too complex to give fully here. For an approximate value the correlation due to Eck may be used:

$$\frac{C_{u,2,ac}}{C_{u,2,tl}} = \left\{ 1 + \frac{2\cos\beta_2}{Z[1 - (d_{sh,1}/d_2)]} \right\}^{-1} \tag{9.26}$$

where Z is the number of rotor blades. The Wiesner correlation (equation 9.1) may also be used.

The theoretical head coefficient, ψ_{tl}^x, is then given for any value of flow coefficient within the pumping region by

$$\psi_{tl}^{x} = \frac{C_{u,2,ac}}{C_{u,2,tl}}(1 - \phi_2 \tan \beta_2) \tag{9.27}$$

The values of theoretical head coefficient obtained from equation 9.27 may be used as a close approximation for both the single- and two-phase coefficient in equation 9.25. In certain circumstances there is a significant difference between the two values, and a method of calculating them is given by Wilson et al. (1979).

Extension to prediction of new-pump characteristics

The intended method of use of the correlation method is that several centrifugal pumps covering a range of configurations (for instance, pumps with varying specific speed, blade angle, and types of diffuser) should be tested throughout the two-phase-flow range of interest, and from these tests a library of head-loss-ratio correlations would be established. Then to predict the two-phase performance of a new pump, the single-phase performance of which would be known, one would simply select the head-loss ratio correlations of the pump with the nearest configuration to the new pump under investigation, and apply this together with the modified Euler equation to obtain a two-phase predicted characteristic.

At present, only data from tests by Winks (1977) and Kennedy et al. (1980) can be considered to be useful guides. These can be used for first-quadrant prediction with some modifications. The magnitudes of the head-loss ratios differ for different pumps. The first-quadrant data so far analyzed from several different pump tests show that the highest levels of head-loss ratio were obtained from pumps that were relatively efficient at design point in single-phase flow, while the inefficient pumps had low maximum head-loss ratios. This seemed easily explainable because for pumps to be efficient, they must use efficient diffusing channels in the rotor and diffuser, and diffuser performance is adversely affected by the presence of vapor bubbles.

This hypothesis is given credence by figure 9.25, in which the best efficiencies in single-phase flow are plotted as a ratio with the best efficiency of the Winks (B&W) data versus the peak head-loss ratios at best-efficiency flow coefficient for all first-quadrant test data available. These data were correlated so well by a straight line that a certain degree of good luck must be assumed.

Approximate prediction method

The full method of use of this prediction method is given in Wilson et al. (1979). The following approximate method should be sufficient for preliminary-design purposes for the first quadrant.

1. Read off the head-loss ratio from the Winks data of figure 9.23b for the desired values of void fraction and flow coefficient.

2. Multiply this head-loss ratio by the multiplier of figure 9.25 using the pump single-phase efficiency as input.

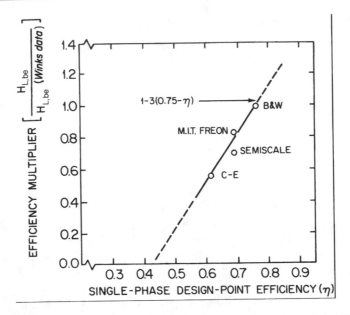

Figure 9.25. Efficiency multiplier versus single-phase design-point efficiency, first quadrant

3. Obtain the single-phase value of the head-rise coefficient for the desired flow co-efficient from the single-phase test curves (which have to be available).

4. Calculate the theoretical head-rise coefficient (equation 9.27) using a theoretical tangential-velocity ratio from the Busemann correlation (Dixon, 1975) or the Wiesner correlation (equation 9.1). Assume as an approximation that $\psi_{tp,tl}^{x} = \psi_{sp,tl}^{x}$.

5. Calculate the desired two-phase head-rise coefficient from

$$\psi_{tp}^{x} = \psi_{sp,tl}^{x} - H_L(\psi_{sp,tl}^{x} - \psi_{sp}^{x}) \qquad (9.28)$$

The method may be used for other quadrants, using the head-loss correlations given in Wilson et al. (1979).

9.15   Cryogenic pumps

When cryogenic fluids, such as liquid hydrogen and liquid oxygen, are pumped, the fluids are essentially boiling. The need to avoid a reduction in performance and cavitation damage, and the abrupt changes in output that accompanies large vapor cavities, is never more vital than in rocket turbopumps. Very-low-work screw-type booster inducers are used upstream of centrifugal pumps to produce a liquid flow with some positive suction head before the main impeller. The design of these inducers is too specialized for this text.

References

Aungier, R.H. (1995). Centrifugal compressor stage preliminary aerodynamic design and component sizing. Paper 95-GT-78, ASME, New York, NY.

Carre, P.M. (1978). The development, application and experimental evaluation of a design process for centrifugal compressors. Proc. Inst. Mech. Eng., vol. 192, London, UK.

Dallenbach, F. (1961). The aerodynamic design and performance of centrifugal and mixed-flow compressors. SAE paper 268A, SAE, Warrendale, PA.

Dean, Robert C., Jr. (1972). The fluid dynamic design of advanced centrifugal compressors. Report TN-153. Creare, Inc., Hanover, NH.

Dixon, S.L. (1975). Fluid Mechanics, Thermodynamics of Turbomachinery. 2nd ed. Pergamon Press, Oxford, pp. 196–203.

Fairbanks, F. (1980). The determination of deviation angles at exit from the nozzles of an inward-flow radial turbine. Paper 80-GT-147. ASME, New York, NY.

Gulich, J.F. (1989). Guidelines for the prevention of cavitation in centrifugal feed pumps. Final report no. GS-6398, by Sulzer Bros. Ltd. for EPRI, Palo Alto, CA.

Harada, H. (1988). "Performance characteristics of two- and three-dimensional impellers in centrifugal compressors." Trans. ASME, vol. 110, pp. 110–114, New York, NY.

Jansen, W. (1967). A method for calculating the flow in a centrifugal impeller when entropy gradients are present. Proc. of a Royal Society Wates Foundation conference at Cambridge, UK.

Japikse, D. (1996). Centrifugal compressor design and performance. Concepts, ETI, Wilder, VT.

Karassik, Igor J., William C. Krutzsch, Warren H. Frased, and Joseph P. Messina (1986). Pump Handbook, 2d ed. McGraw-Hill, New York, NY.

Keith, Stephen Wayne. (1977). Characteristics of first-quadrant one-phase flow in centrifugal pumps. BSME thesis. MIT, Cambridge, MA.

Keenan, Joseph H., and Kaye, Joseph (1948). Gas Tables. Wiley, New York, NY.

Kennedy, W.G., et al. (1980). Pump two-phase performance program. vols. 1–7. Report NP-1556. Electrical-Power Research Institute, Palo Alto, CA.

Korakianitis, T., Vlachopoulos, N.E., and Zou, D. (1994). Models for the prediction of transients in closed regenerative gas-turbine cycles with centrifugal impellers. Journal of Engineering for Gas Turbines and Power (in print). ASME paper 94-GT-342, New York, NY.

Korakianitis, T., and Vlachopoulos, N.E., (1997). Performance of turbine (drag) pumps in two-phase flow. Washington University report 97-ICE-2. St. Louis, MO.

Kronogard, S.O. (1975). Automotive turbine—advantages of three-shaft configurations. Gas Turbine International, Southport, CT.

Kluge, Friedrich. (1953). Kreiselgeblase und Kreisel verdichter radialer Bauert. Springer-Verlag, Berlin, W. Germany.

Lakshminarayana, B., W.R. Britsch, and W.S. Gearhart, eds. (1974). Fluid mechanics, acoustics and design of turbomachinery. Report SP-304. NASA, Washington, DC.

Matthews, R.D., Gasiorek, J.M., and Cavanagh, J. (1991). Deviation in a mixed-flow fan. Proc. Inst. Mech. Eng., vol. 205, pp. 241–249, London, UK.

McCutcheon, A.R.S. (1978). Aerodynamic design and development of a high-pressure ratio turbocharger compressor. Proc. Inst. Mech. Eng. 73/78, London, UK.

Musgrave, D.S. and Plehn, N.J. (1978). Mixed-flow compressor stage design and test results with a pressure ratio of 3:1. Journal of Turbomachinery, vol. 109, pp. 513–519, ASME, New York, NY.

Musgrave, D.S. (1980). The prediction of design and off-design efficiency for centrifugal-compressor impellers. In a symposium volume "Performance prediction of centrifugal pumps and compressors", ASME, New York, NY.

Noorbakhsh, A. (1973). Theoretical and real slip factor in centrifugal pumps. Technical note 93. Von Karman Institute for Fluid Dynamics, Rhode Saint Genese, Belgium.

Rodgers, C. (1978). A diffusion-factor correlation for centrifugal-impeller stalling. Journal of Turbomachinery, vol. 100, pp. 592–603, ASME, New York, NY.

Rodgers, C. (1991). The efficiencies of single-stage centrifugal compressors for aircraft applications. Paper 91-GT-77, ASME, New York, NY.

Rodgers, C. (1991). Centrifugal compressor inlet guide vanes for increased surge margin. Journal of Turbomachinery, vol. 113, p. 696, ASME, New York, NY.

Roelke, Richard J. (1973). Miscellaneous losses. In Turbine Design and Application, vol. 2, pp. 125–147, Arthur J. Glassman ed. Special publication SP-290. NASA, Washington, DC.

Rohlik, Harold E. (1975). Radial-inflow turbines. In Turbine Design and Application, vol. 3, pp. 31–58, Arthur J. Glassman ed. Special publication SP-290. NASA, Washington, DC.

Suga, S. et al. (1996). Aerodynamic design of centrifugal compressors with CFD. Bulletin of the Gas Turbine Society of Japan, 198, p. 44, Tokyo, Japan.

Thom, J.R.S. (1964). Prediction of pressure drop during forced-circulation boiling of water. J. Heat and Mass Transfer 7:709–724. Pergamon Press, Oxford, UK.

Wiesner, F.J. (1966). A review of slip factors for centrifugal impellers. ASME paper 66-WA/FE-18, New York, NY.

Wilson, David Gordon, Tak-Chee Chan, Juan Manzano-Ruiz. (1979). Analytical models and experimental studies of centrifugal-pump performance in two-phase flow. NP-677 final report (May). Electric Power Research Institute, Palo Alto, CA.

Winks, R.W. (1977). 1/3-scale air-water pump program, test program and pump-performance results. EPRI NP-135 (April).

Problems

1. Calculate the preliminary design dimensions (inlet hub and tip diameter, outlet diameter, and outlet blade width) of a pump suitable for a sewage-treatment plant. The specifications and design assumptions to be used follow. Sketch the meridional and transverse cross sections of the pump. Also sketch the performance characteristics, from zero flow to zero head, at constant design-point speed (for no cavitation).

   Specifications
   Design lift: 105 ft (flange-to-flange, total-to-static).
   Flow: 107 gal per day (U.S.).

Design choices suggested
Specific speed, N_s: 0.10 (nondimensional).
Outlet blade angle: $\beta_2 = 60°$.
Slip factor (deviation): $\tan \alpha_{W2} = \tan \beta_2 + (\pi \cos \beta_2 / Z \phi_{be,2})$.
Number of blades: Z: 7.
Hub-tip ratio at inlet: 0.6.
Shroud diameter, $d_{sh,in}$: giving minimum relative velocity.
Inlet axial-velocity profile: $C_{x,hb,in}/C_{x,sh,in} = 0.75$.
Velocity at outlet flange: 10 ft/s.
Efficiency, flange-to-flange, total-to-total: 0.88.
Flow angle (absolute) at rotor exit: 65°.

2. Find the optimum inlet shroud diameter of a radial-flow pump having 50-degree blade angle
at rotor exit, giving a relative flow angle of 55 degrees, and an absolute flow angle of 60
degrees. The nondimensional specific speed should be 0.754. The hub diameter at inlet is to
be 25 percent of the rotor peripheral diameter. The axial velocity at inlet varies linearly with
radius, being 50-percent greater at the shroud than at the hub. Also find the blade angle for
zero incidence at the hub and shroud at inlet, and the rotor diffusion (relative-velocity ratio)
along the shroud streamline. Also, $\dot{V} = 4.73$ m³/s; $\Delta p_0 - 1724$ kN/m², and the hydraulic
efficiency is 0.85.

3. Find the overall polytropic stagnation-to-stagnation efficiency, the stagnation-to-stagnation
pressure ratio, and the ratio of the rotor-blade heights at outlet (b_2/b_1) of the two-stage centrifugal air compressor sketched. Assume that each stage has the performance characteristic
shown in Figure P9.3. The design point for the first stage is shown. The working point for
the second stage should be chosen to give maximum efficiency for the non-dimensional speed
which is calculated. The second stage will be manufactured identical to the first stage except
that the blade height will be cut back further. For simplicity, assume that the fluid is a perfect
gas with $\gamma = 1.4$. The inlet conditions for the first stage are $300\ K$, $1 \times 10^5\ N/m^2$. In each
stage, $C_{x,1}/u_2 = 0.35$, $C_{u,2}/u_2 = 0.5$, $\alpha_2 = 65°$, where 1 refers to rotor inlet and 2 to rotor out-

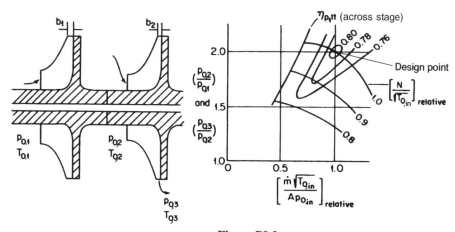

Figure P9.3

let. Also $\eta_{p,tt,1-2}$ for each stage is 5 percentage points above the stage efficiencies given on the performance characteristic. (This is the stagnation-to-stagnation efficiency across the rotor.)

4. Find the shroud diameter at inlet for minimum relative velocity in a radial-flow water pump designed to give a static lift of 100 feet at a specific speed $N_{s2} = 0.314$ (nondimensional). Also find the zero-incidence blade angle at the shroud, the inlet hub diameter, the peripheral diameter, the rotational speed, and the rotor diffusion ($W_{sh,1}/W_2$). The following are some specifications and assumptions:

> Blade angle at outlet: 60 degrees.
> Efficiency, total-to-static, flange-to-flange: 90 percent.
> Slip factor: $C_{u,2,ac}/C_{u,2,tl} = 0.88$.
> Hub diameter at inlet: $d_2/3$.
> Inlet swirl velocity: 0.
> Inlet axial velocity: proportional to $\sqrt{(radius)}$.
> Rotor-exit flow angle, absolute: 65 degrees.
> Mean flow velocity at outlet flange: 15 ft/s.
> Flow quantity: 500 gal/min (U.S.).

5. (a) Suppose that a solid particle enters the nozzle ring of a radial-inflow turbine (figure 9.5 with fixed nozzles in the position shown in broken lines in figure 9.7). The turbine is running in design conditions, with the inlet vector diagram shown at the top of figure 9.3. At nozzle exit the particle has, we suppose, been dragged along by fluid friction with the accelerating gas so that it has acquired half the gas velocity, and the same direction as the gas. Redraw the inlet velocity diagram lightly, and in heavier lines draw the absolute and the relative velocities of the particle at rotor entrance.

(b) It will obviously be probable that the particle will hit (or be hit by) a rotor blade. Assume that it bounces off the rotor blade as if it were a light ray reflected off a mirror, but with only half the prebounce relative velocity. Show dotted the relative and absolute velocity vectors. What will be the likely destination of the particle?

6. The following is extracted from a typical exercise given to students, one which is recommended in general. A manufacturer's paper, report, or brochure is copied, and the task is to check the design choices from the data given. This usually includes having to make imprecise measurements from the photographs or diagrams. Often there are some satisfying agreements between predictions and manufacturers' claimed performance values and some puzzling discrepancies that cannot be resolved except by assuming that the machine has been sensibly optimized for part-load operation, leading to compromise with the design-point or full-load performance.

The following data were extracted from a report on an experimental automotive gas-turbine engine. The compressor data will be given here, and those for the two radial-inflow turbines are in problem 9.7.

> *Centrifugal compressor*
> Number of blades: 32 (including splitter blades).
> Rotational speed: 86,256 rev/min.
> Stagnation pressure ratio: 4.5 (to diffuser outlet).
> Isentropic efficiency: 0.825 (to same outlet condition).

Inlet stagnation temperature: 288 K.
Inlet stagnation pressure: 100 kPa (assumed).
Rotor outlet blade angle: 45° (approximate from photograph).
Vaned diffuser inlet angle: 70° (approximate from photograph).
Hub diameter at inlet: 27.6 mm (approximate from photograph).
Shroud diameter at inlet: 61.4 mm (approximate from photograph).
Rotor outside diameter: 120.1 mm (approximate from photograph).
Rotor-outlet axial width: 6.2 mm (approximate from photograph).
Mass flow: 0.345 kg/s.

You will need to assume a polytropic efficiency for the rotor, stagnation-to-stagnation conditions. We found a large discrepancy in the rotor-outlet axial width and the reasonable agreement elsewhere. We therefore assumed that the designer had used a very large blockage factor to allow for flow separation, blade thickness, and boundary layers. (There is a point of view which maintains that using blockage factors to increase flow areas actually promotes separation and blockage.) The inlet shroud diameter should be checked for minimum relative velocity. We assumed no inlet swirl and obtained good agreement with the manufacturer's geometry as approximately measured. The machine had good inlet guide vanes which could be preset to give swirl in the inlet flow.

7. In table P9.7 are some data for the radial-inflow turbines driving the compressor (the "gasifier" turbine) and the output shaft (the "power" turbine) in problem 9.6. Make check calculations for the rotor outside diameters and axial widths, assuming that at design point the inlet velocity triangles are similar to the (compressor) correlations of Wiesner (1966), equation 9.1, of Eck (equation 9.26), and/ or one specifically for radial-inflow turbines given by Rohlik (1975): $C_{u,ac}/C_{u,tl} = 1 - (2/Z)$. Both turbines have 12 blades. (Assume efficiencies where needed. Not all the data are required for the calculations.)

Table P9.7

	Gasifier turbine	Power turbine
Inlet stagnation temperature, K	1,561	1,407
Outlet stagnation temperature, K.	1,407	1,206
Inlet stagnation pressure, kPa	400.5	236.9
Mass flow, kg/s	0.345	0.345
Fuel-air ratio	0.124	0.124
Outlet stagnation pressure, kPa	240	109
Power output, kW	65.77	82.5
Shaft speed, rev/min	86,256	68,156
Rotor outer diameter, mm	112.5	148.1 (a)
Rotor-inlet axial width, mm	8.0	9.4 (a)

(a) These are approximate values.

8. (a) Calculate the apparent polytropic turbine efficiency (flange-to-flange, total-to-total) of the hydrogen turbine driving the hydrogen booster pump of the space-shuttle main engine. Figures

taken from the space-shuttle Technical Manual are given in table P9.8. The liquid hydrogen flows from the main tank to the booster pump, goes on to the main turbopump, is circulated around the rocket-nozzle coolant passages, and returns as a gas to drive first the two-stage axial turbine driving the booster pump, and then, downstream, the turbine for the main pump. (The turbine drives only the pump. The molecular weight of hydrogen is 2.016; the liquid H_2 specific heat is 6,240 J/(kg · K), the liquid density is 74.62 kg/m³, and the mean specific heat of the gas, C_p, is 14,822 J/(kg · K).)

Table P9.8

	Centrifugal booster pump	Two-stage axial turbine
Mass flow kg/s	66.77	13.789
Inlet total temperature, K	20.56	315
Inlet total pressure, kPa	206.8	29,420
Outlet total temperature, K	21.83	308.9
Outlet total pressure, kPa	1,620.96	24,318

(b) Also calculate the power output from the turbine blading, the ideal work of the booster pump, the pump losses, and hence the pump efficiency. Comment on the inconsistencies you find. Suggest which one property value might be in error and which if corrected, would yield more believable results.

9. Your boss wants you to design a multistage centrifugal compressor for a new gas turbine. She wants it to have two compressor stages of maximum possible backslope that can be used, given the following specifications. Find this maximum backslope.

Compressor pressure ratio, stagnation-to-stagnation	8.5
Estimated overall polytropic efficiency, same points	0.86
Equal enthalpy rise per stage	
Inlet stagnation temperature	288 K
Inlet stagnation pressure	100, 000 Pa
Peripheral speed, both stages	500 m/s
Rotor-outlet absolute flow angle	65°

Wiesner's correlation for slip factor is given by equation 9.1.

The work coefficient can be calculated from equation 5.23.

Use the graphical estimate of number of blades, Z, and work coefficient versus backslope in figure P9.9.

10. Calculate the rpm of a radial-inflow turbine, and the nozzle-exit Mach number and static temperature and pressure. The rotor diameter is 250 mm, and the diameter of the nozzle exit is 254 mm. The height of the nozzle vane at exit is 10 percent of the nozzle-exit diameter. The mean flow direction produced by the nozzles at exit is 75° from the radial direction.

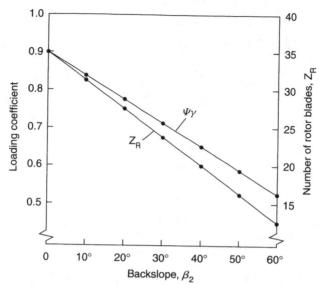

Figure P9.9 Blade number and work coefficient vs. backslope

The turbine is supplied with 1.5 kg/s of air at 2×10^5 Pa and 400 K stagnation conditions. You may approximate the flow through the nozzles as having no drop in stagnation pressure. The rotor peripheral velocity is 90 percent of the tangential component of the nozzle-exit velocity.

The turbine rotor is followed by a diffuser giving partial recovery of static pressure. Sketch the turbine expansion on an enthalpy-entropy diagram.

11. Make a quick analysis to determine if your boss's concept of a radial-inflow turbine expanding the gas from a high-pressure natural-gas (methane) pipeline is limited by stress or Mach-number considerations. The line pressure is 35 kPa above atmospheric $(100,000$ Pa$)$ and she suggests using the pressure drop from this level to that of an atmospheric burner to run the turbine. (The power from the turbine would then run the forced-air fan, but this is not your concern.) Methane has a mean specific heat in this region of 2153.9 J/(kg $\cdot$ K) (the gas is at 320 K) and the gas constant for methane is 518.22 J/(kg $\cdot$ K). Choose an appropriate velocity diagram and an efficiency for a small radial-inflow turbine, and make other choices as necessary.

Chapter 10

Convective heat transfer in blade cooling and heat-exchanger design

A knowledge of the physics of convective heat transfer is valuable in the design of most types of turbomachinery. It is vital in the design of advanced, high-temperature gas turbines. High turbine-inlet temperatures permit high thermal efficiency and high specific power, as explained in chapter 3. The turbine-inlet temperature of advanced gas turbines has been increasing at an average rate of 20 K per year, at least half of the rise being through improvements in blade cooling (figure 1.10).

The performance of simple-cycle gas turbines can be improved by including heat exchangers, to convert CBE cycles to CBEX cycles and to convert simple cycles to combined cycles, as shown in chapter 3. When the price of fossil fuels, particularly of natural gas, is low, the operating economics of simple-cycle gas turbines are acceptable, and at the end of the 1990s there is no strong movement toward incorporating recuperators or regenerators. However, the military tank engines on the CBEX cycle and the naval CICBEX engines mark a change in direction. Heat exchangers also bestow major advantages on very small engines that otherwise suffer large losses if the alternative route to higher efficiency, higher compressor pressure ratio, is pursued (McDonald, 1997, and McDonald and Wilson, 1996).

In each of these critical areas, and in many others that are less critical, there is more design freedom than in any other aspect of gas-turbine design. This means that there is more scope for design skill. It also means that there are perhaps more chances of failure. For the costs of deficiencies in design in these areas are high. Inadequate blade cooling of some part of the blade profile in some conditions of operation can lead to early and disastrous failure of the entire machine. Overcooling can lead to high thermal stresses and again failure. Poor heat-exchanger design can lead also to high thermal stresses possibly leading to fatigue failures, and to areas where condensation could lead to corrosion failure. The heat exchanger might also be impractically large, or have penalizing pressure losses, or be uneconomically expensive.

In this chapter we develop some simple design methods and guidelines that can be used to arrive at a close approach to an optimum design of critical components. Insofar as these methods are appropriate to preliminary design, they will satisfy, in particular, the requirements of heat exchangers for analysis. They will provide an understanding of the design and operation of cooled turbine blades. But the state of the art of cooled-blade design is now so sophisticated that, for purposes of preliminary design, gas temperatures would be chosen based on existing cooled-blade performances rather than being designed from first principles. As in most other areas of modern design, final optimization of heat exchangers and of cooled blades is normally made using digital-computer programs. These are, however, no better than the principles on which they are based. Computer programs are all too often built upon poor foundations. Even if they are sound, they will usually arrive at a better result if the input is close to an optimum design, rather than if a somewhat random configuration were used.

We shall deal principally with problems of convection heat transfer: conduction through moving fluids. The starting point is Reynolds' analogy between heat transfer and fluid friction.

10.1 Reynolds' analogy between fluid friction and heat transfer

Many of the fundamental relations correlating convective heat transfer are based on the simple statement made a century ago by Osborne Reynolds that the heat-transfer coefficient in certain classes of fluid flow is a simple multiple of the skin-friction coefficient. This rule can be derived by modeling the velocity and temperature profiles in an "attached" boundary layer (in other words, in an unseparated flow, see figure 10.1).

In continuum flow the molecular layer next to the wall is at rest. Outer layers either slide past in laminar flow, or exchange "packets" of high-velocity fluid with low-velocity fluid displaced from the inner parts of the boundary layer (turbulent flow). In both

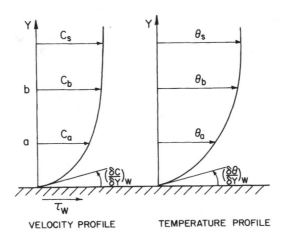

VELOCITY PROFILE   TEMPERATURE PROFILE

Figure 10.1. Boundary-layer velocity and temperature profiles

laminar and turbulent boundary layers, the skin-friction tangential stress at the wall, τ_w, is proportional to the angle of the tangent to the velocity profile at the wall $(\partial C/\partial y)_w$. The viscosity, μ, is defined by this angle

$$\mu \equiv \frac{\tau_w}{(\partial C/\partial y)_w} \tag{10.1}$$

This will be the viscosity at "static" conditions, μ_{st}, but because the Mach number in heat-transfer devices is generally small, we omit the subscripts.

In the same way all (except radiative) heat transfer between the fluid and the wall, $\dot{Q}$, must be conducted through the final layer of stationary molecules. The fluid thermal conductivity, k, is defined in terms of the temperature gradient at the wall $(\partial\theta/\partial y)_w$.

$$k \equiv \frac{\dot{Q}/A_h}{(\partial\theta/\partial y)_w} \qquad \text{where} \qquad \theta \equiv T - T_w \tag{10.2}$$

We use θ as a variable for the temperature difference "driving" heat transfer. It is different from our use here of ΔT, the temperature rise or fall of a fluid flow, the result of heat transfer.

There is clearly an analogy between the transfer of heat and of skin friction through the fluid layer next to the wall in laminar flow. We show in the following that in certain circumstances the analogy also holds in turbulent boundary-layer flow. (The treatment is that of Eckert, 1963.) Consider a single turbulent eddy exchanging "packets" ("austaches") of fluid at a rate $\dot{m}$ between two layers a and b (figure 10.2). The momentum change will cause a tangential stress τ in the fluid:

$$\tau = \frac{\dot{m}}{g_c} \frac{(C_b - C_a)}{A_h} \tag{10.3}$$

where A_h is the area of the wall over which the mass flow rate $\dot{m}$ occurs. The fluid packets will carry with them sensible heat. The effective heat transfer between the two layers is

$$\dot{Q} = \dot{m} C_p(\theta_b - \theta_a) \tag{10.4}$$

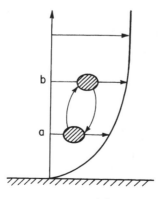

Figure 10.2. Eddy exchange in turbulent flow

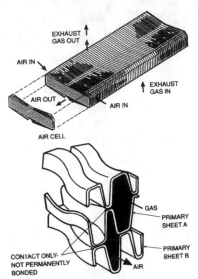

The Solar primary-surface recuperator is designed to reduce greatly the thermal stresses and consequent low-cycle fatigue that are normally problems with gas-turbine recuperators. By "primary surface" is meant the absence of fins. The Solar construction is remarkable in having the minimum of welding and other bonding. The formed sheets are held together in compression as shown here.

Illustration. Primary-surface recuperator. Courtesy Solar Turbines Inc.

The ratio of heat transfer to skin friction within the turbulent boundary layer is then

$$\frac{\dot{Q}/A_h}{\tau} = \frac{g_c C_p (\theta_b - \theta_a)}{(C_b - C_a)} \tag{10.5}$$

Within the laminar layer close to and at the wall this same ratio is

$$\frac{\dot{Q}/A_h}{\tau_w} = \frac{k}{\mu} \left(\frac{\partial \theta}{\partial C} \right)_w \tag{10.6}$$

Obviously, these two ratios are effectively identical when

$$k/\mu = g_c C_p; \qquad \text{that is, when } g_c C_p \mu / k = 1$$

This collection of properties, $g_c C_p \mu / k$, is called the Prandtl number, P_r:

$$P_r \equiv \frac{g_c C_p \mu}{k}$$

Gases have Prandtl numbers sufficiently close to unity for this analogy to be accepted as valid. "Close" can be 0.7 for air, for instance. (Oils can have Prandtl numbers well over 1000, and liquid metals well below 0.005.)

For the case of fluids with Prandtl numbers close to unity, the ratio of heat transfer to fluid tangential stress is the same throughout the laminar and the turbulent parts, if any, of an attached boundary layer. Therefore we are free to locate the planes a and b where we wish. Greatest utility is given by putting plane a at the wall interface, and plane b at the outer limit between the boundary layer and the "free stream". Then $C_a = \theta_a = 0$, and

$$\frac{\dot{Q}/A_h}{\tau_w} = \frac{g_c C_p \theta_{fs}}{C_{fs}} \tag{10.7}$$

where subscript fs denotes the free-stream values. We define the (dimensional) heat-transfer coefficient, h_t, as

$$h_t \equiv \frac{\dot{Q}}{A_h \theta_{fs}} \tag{10.8}$$

and a non-dimensional heat-transfer coefficient, N_u, as

$$N_u \equiv \frac{h_t l}{k} \tag{10.9}$$

where l is a characteristic length (for instance, the hydraulic diameter of a passage or the chord length of a turbine blade). Then

$$
\begin{aligned}
h_t &\equiv \frac{\dot{Q}}{A_h \theta_{fs}} = \frac{g_c C_p \tau_w}{C_{fs}} \\
N_u &\equiv \frac{h_t l}{k} = \frac{g_c C_p}{k} \frac{l \tau_w}{C_{fs}}
\end{aligned}
\tag{10.10}
$$

It is convenient to form a nondimensional skin-friction coefficient, C_f, with the free-stream incompressible dynamic head q_{ip}, where

$$
\begin{aligned}
q_{ip} &\equiv \rho_0 \frac{C_{fs}^2}{2 g_c} \\
C_f &\equiv \frac{\tau_w}{\rho_0 C_{fs}^2 / 2 g_c}
\end{aligned}
\tag{10.11}
$$

$$N_u \equiv \frac{g_c C_p \mu}{k} \frac{l C_{fs} \rho}{\mu} \frac{\tau_w}{(\rho_0 C_{fs}^2 / 2 g_c)} \frac{1}{2 g_c} = P_r \times R_e \frac{C_f}{2} \tag{10.12}$$

Another nondimensional heat-transfer coefficient, the Stanton number, S_t, can be formed from N_u, R_e, and P_r:

$$S_t \equiv \frac{N_u}{R_e \times P_r} \equiv \frac{h_t}{\rho C_p C} \tag{10.13}$$

so that

$$S_t = C_f / 2 \tag{10.14}$$

This is the formal statement of Reynolds' analogy for fluids of Prandtl number near to unity. Von Karman extended the analogy to fluids with Prandtl numbers far from unity by a modification of equation 10.14; more usually, empirical correlations of Prandtl-number effects are given (for example, see Eckert, 1963, and Kays and London, 1964).

10.2   The N_{tu} method of heat-exchanger design

Figure 10.3 is a diagrammatic representation of the changes in temperature of the hot and cold fluids in a two-fluid heat exchanger. The diagram is drawn as if the fluids were in pure counterflow. For the purposes of the following argument, they could equally well be in crossflow or co-current flow, or a mixture of all three types.

The thermal performance of a heat exchanger is usually measured by its effectiveness ϵ_x, defined as

$$\epsilon_x \equiv \frac{\dot{Q}}{\dot{Q}_{mx}} \tag{10.15}$$

where $\dot{Q}$ is the heat actually transferred and $\dot{Q}_{mx}$ is the maximum amount of heat that could be transferred. This maximum occurs when the outlet temperature of the fluid with the smaller heat capacity, $(\dot{m}C_p)_{mn}$, reaches the inlet temperature of the other fluid:

$$\dot{Q}_{mx} = (\dot{m}C_p)_{mn}\theta_0 \tag{10.16}$$

where θ_0 is the overall or maximum temperature difference and the specific-heat capacity is the mean value.

The actual heat transferred can be used to define an overall mean heat-transfer coefficient, $\overline{U}$, a mean temperature difference, θ_m, and a mean heat-transfer area $\overline{A}_h$:

$$\dot{Q} = \overline{U}\,\overline{A}_h\theta_m \tag{10.17}$$

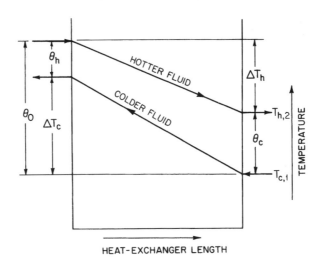

HEAT-EXCHANGER LENGTH

Figure 10.3. Representation of temperature distribution in a counterflow heat exchanger

Therefore,

$$\epsilon_x \equiv \frac{\dot{Q}}{\dot{Q}_{mx}} = \left[\frac{\overline{U A}_h}{(\dot{m} C_p)_{mn}} \right] \left(\frac{\theta_m}{\theta_0} \right) = \frac{(\Delta T)_{mn}}{\theta_0}$$

(where $(\Delta T)_{mn}$ is ΔT_c in figure 10.3)

$$(\Delta T)_{mn} > (\Delta T)_{mx}$$

because the subscript refers to the heat capacity of the fluid.

The dimensionless heat-exchanger size,

$$\overline{U A}_h / (\dot{m} C_p)_{mn}$$

is termed the "number of transfer units" (N_{tu}):

$$N_{tu} \equiv \frac{\overline{U A}_h}{(\dot{m} C_p)_{mn}} \qquad (10.18)$$

Therefore,

$$\epsilon_x \equiv N_{tu} \frac{\theta_m}{\theta_0}$$

The ratio of the mean to the overall temperature difference is a function of the heat-exchanger flow arrangement, the N_{tu} (or alternatively the effectiveness), and the ratio of the smaller heat-capacity rate to the larger, C_{ra},

$$C_{ra} \equiv \frac{(\dot{m} C_p)_{mn}}{(\dot{m} C_p)_{mx}}. \qquad (10.19)$$

Kays and London, who pioneered and developed this method of heat-exchanger design for compact (gas-turbine) heat exchangers, have published charts of effectiveness versus N_{tu}, with C_{ra} as a parameter, for each principal heat-exchanger flow arrangement (Kays and London (1984). Examples are shown in figure 10.4a to 10.4d.

The N_{tu} approach has resulted in an enormous simplification in heat-exchanger design over the old methods where the logarithmic mean temperature difference had to be calculated for all conditions and for variations from these conditions. Heat-exchanger design is still, however, complex. One reason that it is so is the rather strange one that the designer has many degrees of freedom. Most designers are not used to such freedom and hesitate to make what may initially be rather random choices of, for instance, flow configuration, flow surfaces, and flow velocities without reference to analytical methods. The following guidelines for heat-exchanger design give some rules for making these otherwise arbitrary choices.

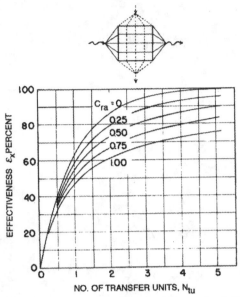

(a) Crossflow heat exchanger (recuperator) with unmixed fluids. From Kays and London (1984)

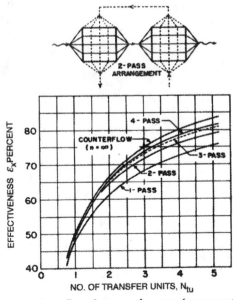

(b) Multipass cross-counter-flow heat exchanger (recuperator), $C_{ra} = 1.0$. From Kays and London (1984)

Figure 10.4. Examples of charts of effectiveness versus N_{tu}

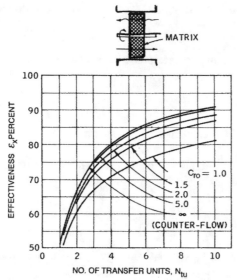

(c) Periodic-flow heat exchanger (regenerator), $C_{ra} = 1.0$. From Kays and London (1984)

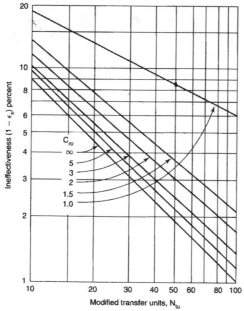

(d) High-effectiveness periodic-flow heat exchanger (regenerator), $C_{ra} = 1.0$. From Kays and London (1984)

Figure 10.4. Concluded

10.3 Guidelines for choice of heat-exchanger passages

Guideline 1

For low pumping-power loss in a single-phase heat exchanger transferring a given flow of heat, the fluid velocities should be low.

Guideline 2

For the overall size of the heat exchanger to be small, the hydraulic diameter of the passages should be small.

The development of these guidelines is given here. They are guidelines rather than rules because frequently other considerations, such as packaging or header (connecting-duct) size, become dominant.

Analysis

The basis for the argument is Reynolds' analogy between heat transfer and skin friction (equation 10.4):

$$S_t = \frac{1}{2}(C_f)$$

where, as before:

$$S_t \equiv \frac{N_u}{R_e \times P_r} \equiv \frac{h_t}{\rho C C_p}$$

the heat-transfer coefficient $h_t \equiv \dot{Q}/A_h\theta$
the heat flow $\equiv \dot{Q}$
the heat-transfer area $\equiv A_h$
the temperature difference driving $\dot{Q} \equiv \theta$
the fluid density $\equiv \rho$
the mean fluid velocity $\equiv C$
the specific heat at constant pressure $\equiv C_p$
the skin-friction coefficient $\equiv C_f \equiv \dfrac{\tau_w}{\rho C^2/2g_c}$
the skin-friction shear stress at the wall $\equiv \tau_w$
constant in Newton's law $\equiv g_c$

Consider heat transfer from an incompressible fluid to the wall of a tube of arbitrary (constant) cross section through which it is flowing (figure 10.5):

$$\dot{Q} = A_h\theta h_t = Zp_e l\theta h_t, \tag{10.20}$$

where

$Z \equiv$ the number of parallel, identical tubes,
$l \equiv$ the length of each tube,
$p_e \equiv$ perimeter of tube (internal) cross section.

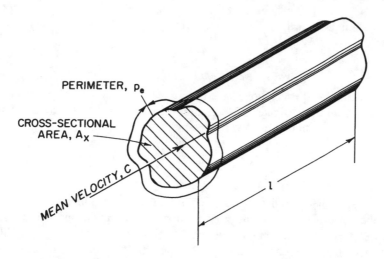

Figure 10.5. Heat-exchanger channel of arbitrary, but constant, cross section

The theoretical pumping power $\dot{W}$ required of an ideal external pump or compressor to overcome the pressure drop is

$$\dot{W} = \dot{V}\Delta p = ZA_x C \Delta p \qquad (10.21)$$

where

$$\dot{V} \equiv \text{volume flow}$$
$$\Delta p \equiv \text{pressure drop across tube length}$$
$$A_x \equiv \text{tube cross-section area}$$

By a force balance $\Delta p A_x = \tau_w p_e l = \tau_w A_h$. Therefore $\dot{W} = ZC\tau_w A_h$ and

$$\frac{\dot{Q}}{\dot{W}} = \frac{Zlp_e\theta h_t}{ZC\tau_w lp_e} = \left[\frac{h_t\theta}{C\tau_w}\right] \qquad (10.22)$$

Introducing Reynolds' analogy

$$\frac{h_t}{\rho C C_p} = \frac{g_c\tau_w}{\rho C^2}$$

Therefore, $\dfrac{h_t}{\tau_w} = \dfrac{g_c C_p}{C}$, and

$$\frac{\dot{Q}}{\dot{W}} = \frac{g_c C_p\theta}{C^2} \qquad (10.23)$$

Therefore, to increase the ratio of heat transferred to pumping power for given C_p and θ for any length, tube size, shape, number of tubes, or for laminar or turbulent flow, the mean fluid velocity should be reduced.

If it is a compressible fluid, $a^2 = g_c C_p T / [(C_p/R) - 1]$ and $M^2 = C^2/a^2$. Therefore,

$$\frac{\dot{Q}}{\dot{W}} = \frac{[(C_p/R) - 1]}{(\theta/T)(1/M^2)} \tag{10.24}$$

Once the boundary specifications have been fixed, the only variable in equation 10.24 under the control of the designer is Mach number, which should be low to increase the ratio of heat transfer to pumping power. This indicates that heat exchangers of large face area (see below) and short length should, where practicable, be used.

Volume for heat transfer

Again consider heat exchange from a fluid to or from a number, Z, of parallel identical tubes of arbitrary constant cross section. How should one choose the length and hydraulic diameter for minimum volume V? This case is one for a given fluid mass flow, $\dot{m}$, temperature differences (driving $\dot{Q}$), θ, and fluid specific heat, C_p:

$$\frac{\dot{Q}}{V} = \left[\frac{h_t \theta Z p_e l}{Z A_x l}\right] = \frac{4 h_t \theta}{d_h} \tag{10.25}$$

using the definition of hydraulic diameter, $d_h \equiv 4 A_x / p_e$

We can also introduce the Stanton number: $S_t \equiv \dfrac{h_t}{\rho C C_p}$

Therefore, $h_t = \rho C C_p S_t$ and

$$\left(\frac{\dot{Q}}{V}\right) = \frac{\rho C C_p S_t \theta Z p_e}{Z A_x} = \frac{4 \rho C C_p \theta S_t}{d_h} \tag{10.26}$$

In turbulent flow the Stanton number changes only slowly with Reynolds number, increasing slowly as the velocity, C, is increased. If the designer has chosen the velocity at a low level to reduce the pumping-power requirements, a small volume will be produced by choosing a small hydraulic diameter for the heat-exchanger passages, regardless of other factors. Another form of this equation is obtained by substituting for the total mass flow:

$$\dot{m} = Z A_x \rho C$$

Therefore,

$$\left(\frac{\dot{Q}}{V}\right) = \frac{\dot{m} S_t C_p \theta}{Z(A_x^2/p_e)}$$

The cross-sectional area, A_x, is proportional to the square of the hydraulic diameter d_h^2 and the perimeter p_e is proportional to the hydraulic diameter, d_h.

Therefore,

$$\left(\frac{\dot{Q}}{V}\right) \propto \frac{\dot{m} S_t C_p \theta}{Z d_h^3} \tag{10.27}$$

The numerator is nearly constant for any case where a given tube shape is being compared at different sizes (d_h), lengths, and so forth, in turbulent flow. The denominator is strongly affected by hydraulic diameter, d_h, and to a lesser extent by the number of tubes, Z.

For a minimum-volume heat exchanger in turbulent flow, therefore, the product Zd_h^3 should be minimized. If a minimum value of mean through-flow velocity, C, has been chosen, ZA_x or d_h^2Z will be constant. Then the volume will be directly proportional to the hydraulic diameter, d_h, other things being equal.

In laminar flow, the Nusselt number is constant:

$$N_u \equiv \frac{h_t d_h}{k}$$

where k is the fluid thermal conductivity. Then,

$$\left(\frac{\dot{Q}}{V} \right) \propto \frac{N_u k \theta}{d_h^2} \qquad (10.28)$$

by the same procedure.

In laminar flow the guideline for reducing heat-exchanger size is unequivocal: the hydraulic diameter, d_h, should be minimized. In laminar flow neither the number of tubes, Z, nor the throughflow velocity, C, influences the heat-exchanger volume[1].

10.4   Guidelines for heat-exchanger design

With the N_{tu} method of approach and data on the performance of various types of heat-exchanger surfaces, it is relatively easy to design a heat-exchanger to fulfill a required thermal performance. The checklist in this section is meant as a guide to the design method. The examples given later will illustrate some of the approaches. There are so many possible configurations of heat exchanger that it would not be practicable to illustrate the design of all of them. The adaptation of the approaches to other configurations is, however, usually obvious. Each configuration has its own set of independent and dependent variables. For instance, a counterflow heat exchanger has the same flow length for the hot flow and the cold flow. The ratio of the flow areas depends solely on the two surfaces chosen. The crossflow rotary regenerator also has intrinsically equal flow lengths, but one can choose any ratio desired for the hot and cold flow areas.

To arrive at an "optimum" design (something which needs very careful definition) is not easy. There are too many nonlinearities for it to be possible to use a good analytical method. There are also many degrees of freedom available in heat-exchanger design, and often some parameter may be varied to advantage.

Optimization therefore becomes essentially a matter of designing a large number of heat exchangers and choosing what seems to be the best. Unless one has some guiding rules as to what is good, one can find oneself tackling thousands of different designs (especially with a computer program) for even a simple case. The guidelines given

---

[1]These guidelines are useful for unseparated flow in tubes and channels, and have somewhat lower validity for situations where there is separated flow, such as in flow across circular tubes.

previously regarding the desirability of choosing low Mach numbers or velocities and small hydraulic diameters should assist in making the initial choices.

The example method given here is for a multipass crossflow configuration. Modifying it to apply to other configurations requires merely care and common sense. We give more-detailed treatment of regenerators later.

Design steps for a multipass crossflow heat exchanger

Assume that the problem is to design a heat exchanger to give required outlet temperatures (obviously these must be compatible with the first law of thermodynamics) of each of two fluids, whose mass flows, inlet temperatures, and other properties are known. Usually there is an additional requirement: that the pressure drops shall not exceed some value. It is in general not possible to specify precisely the pressure drop required in each fluid: to do so introduces an unnatural constraint which could result in the production of sometimes strange designs. For gas-turbine "main" heat exchangers the pressure drop of the hot low-pressure flow is usually limiting. For each design specification, the following process will be necessary.

1. Given the fluid temperatures, find the thermal effectiveness, ϵ_x (using equation 10.17b).

2. Specify the flow rates and properties at inlet of each fluid.

3. Calculate $(\dot{m}C_p)_{mx}$, $(\dot{m}C_p)_{mn}$, and $C_{ra} \equiv [(\dot{m}C_p)_{mn}/(\dot{m}C_p)_{mx}]$.

4. Choose a configuration (e.g., from those in figure 10.6) that seems in advance, from philosophy or experience or plain guesswork, to be favorable (such as counterflow, single-pass crossflow, or, as in this example, five-pass cross-counter flow).

5. Choose compatible surfaces for heat transfer for each fluid with low hydraulic diameters (such as 1/4-inch-bore tubes with 0.025-inch walls, or certain configurations of finned plates); these surfaces should be chosen from among those for which heat-transfer and fluid-friction data exist (Kays and London 1984), because it is almost impossible to generate the data analytically. Choose (low) mean flow velocities and find cross-sectional flow areas.

6. Find the N_{tu} required to give the specified effectiveness using the curves for the appropriate heat-exchanger arrangement and $(\dot{m}C_p)_{mn}/(\dot{m}C_p)_{mx}$ from curves such as those of figure 10.4.

7. From the number of transfer units derive the overall transfer coefficient $\overline{UA}_h = N_{tu}(\dot{m}\overline{C}_p)_{mn}$.

8. Choose the ratio of the conductances, C_r, where

$$C_r \equiv \frac{h_{t,ht}A_{h,ht}}{h_{t,cd}A_{h,cd}} \qquad (10.29)$$

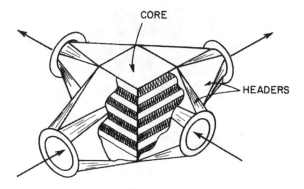

(a) Single-pass crossflow plate fin

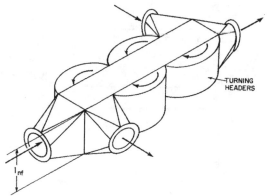

(b) Five-pass counter-crossflow plate fin

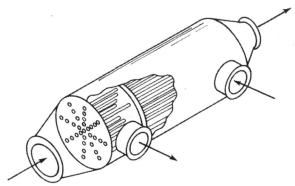

(c) Shell-and-tube multi-pass crossflow

Figure 10.6. Alternative two-fluid heat-exchanger configurations

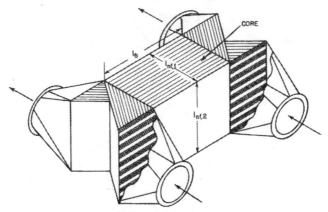

(d) Counterflow plate-fin recuperator

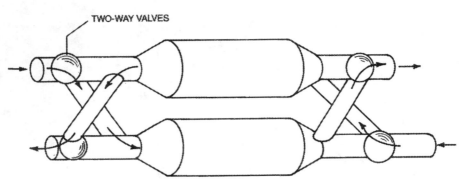

(e) Switching two-chamber periodic-flow regenerator

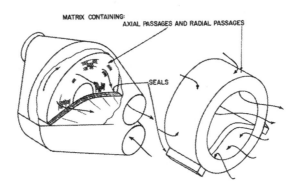

(f) Disk-type rotary periodic-flow regenerator   (g) Drum-type rotary periodic-flow regenerator (ducts omitted)

Figure 10.6. Concluded

In the resistance equation:

$$\frac{1}{\overline{UA_h}} = \frac{1}{h_{t,cd}A_{h,cd}} + \frac{t_w}{k_w A_{h,w}} + \frac{1}{h_{t,ht}A_{h,ht}} \approx \frac{C_r + 1}{h_{t,ht}A_{h,ht}} \qquad (10.30)$$

where t_w is the wall thickness. Good single-phase recuperative heat exchangers usually have negligible wall resistance and comparable film conductances ($1 < C_r < 4$). For small wall resistance,

$$(h_{t,ht}A_{h,ht}) = (C_r + 1)\overline{UA_h} = (C_r + 1)(\dot{m}C_p)_{mn}N_{tu} \qquad (10.31)$$

(For other cases, for instance, surfaces with long fins, refer to Kays and London, 1984.)

9. Choose a middle-of-the-range Reynolds number for both surfaces from the appropriate curves such as in figure 10.7, and find values of the Colburn modulus, $j \equiv S_t \times P_r^{2/3}$, and friction factor, C_f. For a first approximation, use the mean temperature and pressure of each fluid to find the mean velocities.

10. From the estimated Colburn modulus thus found, find the heat-transfer coefficients, h_t, for both surfaces and fluids (hot side and cold side).

11. From the geometry of the two chosen heat-transfer surfaces the heat-transfer area ratio, $(A_{h,cd}/A_{h,ht})$ is known, so that the two values of A_h can be found.

12. Having chosen both Reynolds numbers, both mean flow velocities can be calculated.

13. Using the calculated mean flow velocities and fluid properties, the total cross-sectional "free" passage area, A_{ff}, and the associated "core face" areas, A_f, can be found.

14. These values and $A_{h,cd}$, $A_{h,ht}$ enable the flow lengths $l_{fl,cd}$, $l_{fl,ht}$ and the "nonflow" length, l_{nf}, to be found. The configuration and dimensions of the heat exchanger are known. (In a crossflow heat exchanger the nonflow length is common to both sides[2].)

15. The friction factor for each side can be found from the Reynolds number and the heat-transfer and friction-coefficient curves such as those of figure 10.7. The "core" pressure drops can then be calculated. To these should be added at least the inlet and outlet losses. A simple assumption is that these are 1.5 dynamic heads.

16. The resulting heat exchanger should produce the desired thermal performance. Often, however, it is unsatisfactory for various reasons, such as the shape (it may be very long and thin, for instance). There are obviously many places where new choices can be made. With a little practice one can learn to make changes to move the results in the direction desired, although first attempts are sometimes surprising.

---

[2] A rectangular-solid crossflow two-fluid heat exchanger has two dimensions called "flow lengths", because they are in the directions of flow of the two fluids, and an orthogonal dimension called the "nonflow length". A counterflow or cocurrent heat exchanger has one flow length common to both fluids and, for a rectangular-solid core, two nonflow lengths. See figure 10.6.

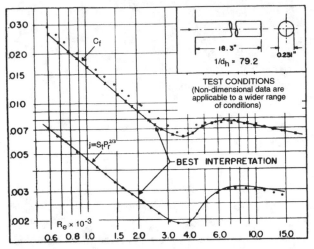

Tube inside diameter = 0.231 in.

Hydraulic diameter = 0.231 in., 0.01925 ft

Flow length/hydraulic diameter, $1/dh$ = 79.2

Free-flow area per tube = 0.0002908 ft^2

(a)

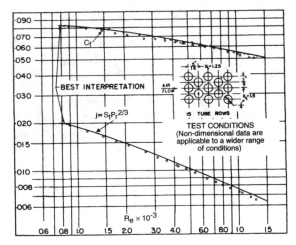

Tube outside diameter = 0.250 in.

Hydraulic diameter, = 0.0166 ft

Free-flow area/frontal area, d_h = 0.333

Heat transfer area/total volume, = 80.3 ft^2/ft^3

Note: Minimum free-flow area is in spaces transverse to flow.

(b)

Figure 10.7. Heat-transfer and skin-friction data for surfaces. (a) Flow inside circular tubes. Heat-transfer and skin-friction data for surfaces. (b) Flow normal to a bank of staggered tubes. From Kays and London (1984)

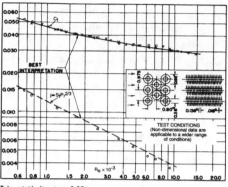

Tube outside diameter = 0.38 in.

Fin pitch = 7.34 per in.

Flow passage hydraulic diameter, $d_h = 0.0154$ ft

Fin thickness (average)* = 0.018 in., aluminum

Free-flow area/frontal area, = 0.538

Heat transfer area/total volume, = 140 ft²/ft³

Fin area/total area = 0.892

Note: Experimental uncertainty for heat-transfer results possibly
somewhat greater than the nominal ±5% quoted for the other
surfaces because of the necessity of estimating a contact
resistance in the bi-metal tubes.

* Fins slightly tapered.

(c)

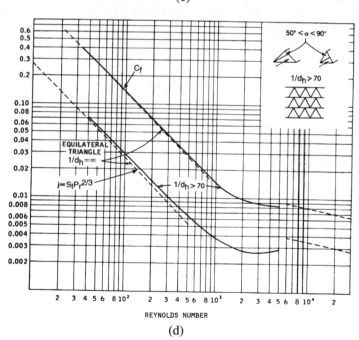

REYNOLDS NUMBER

(d)

Figure 10.7. Concluded. (c) Flow normal to a bank of staggered finned tubes.
(d) Flow along a matrix of triangular passages. From Kays and London (1984)

This simplified approach ignores many complexities that would be accounted for in a full detailed design. The large change of gas properties with temperature in a gas-turbine-engine heat exchanger makes it desirable to divide the heat exchanger into several sections rather than using overall mean properties. Axial conduction and temperature drops in fins might be important. Entrance and exit losses can be quantified, as can flow conditions in headers. All these are treated in more detail in Kays and London (1984). An alternative approach to header design is given by Wilson (1966).

However, despite the simplifications used, comparative trials of this approach and of others using more realistic representations of heat-exchanger conditions have shown rather surprisingly close agreement. The "lumped-parameter" method suggested here seems quite adequate for preliminary design.

10.5 Heat-exchanger design constraints for different configurations

The foregoing guidelines for heat-exchanger design have been written for a multipass-crossflow configuration. They have also emphasized gas-to-gas heat exchangers where equality of conductances $(h_t A_h)$ is a reasonable first choice. The design process is different for other configurations. The design sequence and the constraints for all the principal two-fluid heat-exchanger arrangements, including, for completeness, the cross-flow, are given in this section. In preliminary design the wall resistance, $(t_w/k_w A_{h,w})$ in equation 10.30, is treated as being negligibly small.

Counterflow or cocurrent recuperators

In these the flow length l_{fl} is the same for both fluids, as is the transverse nonflow length $l_{nf,1}$ (figure 10.6). The other nonflow length $l_{nf,2}$ consists of the stacked matrices in the case of plate-fin recuperators and of the combined matrix in the case of, for instance, shell-and-tube recuperators.

Therefore, once the matrices have been selected, the only remaining degree of freedom is the Reynolds number or velocity of one of the two fluids (figure 10.8). The ratio of the two free-flow areas will then determine the other Reynolds number and velocity. How $A_{f,cd}$ and $A_{f,ht}$ are configured is open for choice. They are here defined by the following:

$$(A_{f,cd} + A_{f,ht}) = l_{nf,1} \times l_{nf,2} \qquad (10.32)$$

(These are the two non-flow lengths.) Usually, counterflow or cocurrent heat exchangers incorporate crossflow sections at the end to bring the two fluids out, as shown in figure 10.6.

Crossflow recuperators

Figure 10.9 gives a summary of the earlier notes on heat-exchanger design.

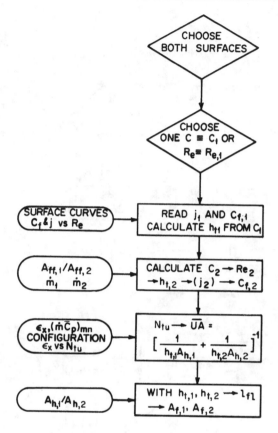

Figure 10.8. Counterflow- or cocurrent-recuperator calculation sequence

10.6 Regenerator design

There are two general regenerator types: the rotating matrix (and its equivalent the rotating duct on a stationary matrix) and the two-chamber switching regenerator (hot stoves for blast furnaces, for example) figure 10.6. In the two-chamber regenerators the chambers are designed to be identical, so that not only the matrix, the hydraulic diameter, and the flow length are the same for both fluids, but also the face area, A_f, the free-flow area, A_{ff}, the heat-transfer area, A_h and the volume, V, are the same for both fluids. Given all these, one's freedom of choice is confined to the Reynolds number or velocity of one side or the other; the other Reynolds number and velocity will be given directly from the now-known free-flow area and mass flow.

While two-chamber switching regenerators have been proposed for special-duty gas-turbine engines, the rotary regenerator has been the type that has been used for virtually all automotive engines. It combines the capability of providing very small hydraulic diameters without the high cost that small passages incur in tubular or plate-fin recuperators, and of allowing large face areas and small throughflow velocities. Thus this type can

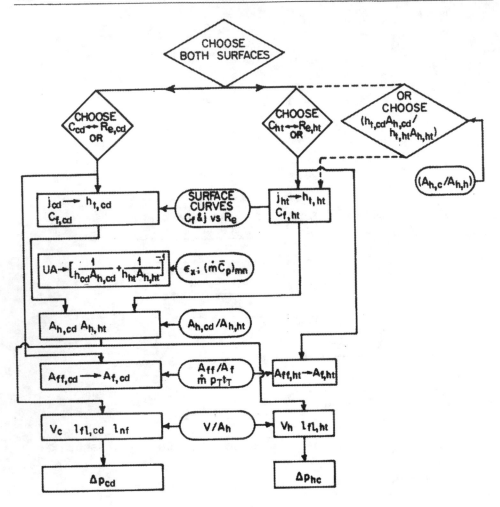

Figure 10.9. Crossflow-recuperator calculation sequence

simultaneously exact low pumping-power requirements and be exceptionally compact. It has the disadvantage of allowing leakage of the compressed air, both through the sliding seals that are required[3], and in the channels that are carried over from the compressed-air duct to the exhaust duct under the seals (called the "carryover leakage").

We shall write here the design progression (figure 10.10) for the more-usual rotary regenerator, in which the area constraints of the switching regenerator do not apply. One therefore has the freedom to choose the Reynolds numbers or mean velocities on both sides. The flow length is common to both sides and is also unknown. However, in practice there are limits that we can use as inputs.

---

[3]Author DGW and MIT have been granted a patent (Wilson, 1993) of considerably reducing seal leakage in regenerators, mentioned at the end of this section.

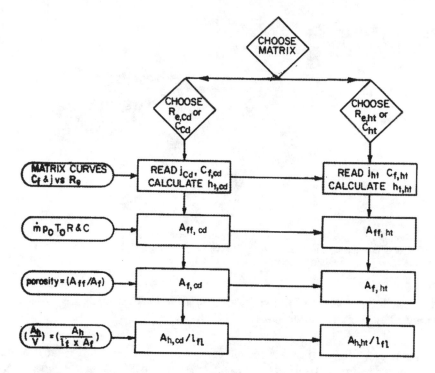

Figure 10.10. Rotating-regenerator calculation sequence

Minimum thickness of regenerator disk

One limit is the average temperature gradient in the matrix. Beyond some level of temperature gradient the axial-conduction loss becomes high, even for ceramic (low-conductivity) materials. The thermal stresses also become high, even for low-expansion materials like ceramics. In the apparent absence of useful data we are using 7.5 K/mm as the maximum allowable temperature gradient of the cold-side fluid, based simply on the dimensions of regenerators that have not experienced cracking.

This limiting value (which is likely to be updated as more information becomes available, and to be different for different materials) enables the designer to specify a minimum cell length, L, or regenerator-disk thickness, directly from the specifications.

Maximum hot-side pressure drop

In chapter 3 it was shown that the pressure-drop parameter that affects the engine specific power and thermal efficiency is the relative pressure drop, the pressure drop divided by the local pressure $(\Delta p_0/p_0)$. The hot-side relative pressure drop is likely to be higher than that of the cold side, partly because the local pressure is the lower, normally in open-cycle machines being near atmospheric pressure, and partly because the lower density of the hot fluid produces a higher absolute pressure drop unless the flow

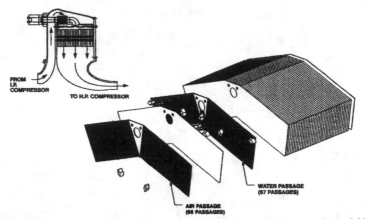

The intercooler for the Northrop-Grumman WR-21 intercooled-recuperated engine (see figure 3.22) is a modular radial-inflow design mounted symmetrically around the engine centerline. Five modules are joined in each half of the engine. The water-glycol intermediate fluid flows radially outwards, except for crossflow headers. Brazed plate-fin construction is used.

Illustration. Intercooler for the Northrop-Grumman WR-21 CICBEX-cycle engine. Courtesy AlliedSignal Aerospace and U.S. Navy

area is made much larger than that for the compressed-air flow. On the other hand, the cold-side pressure drop is liable to be so low, perhaps a small fraction of one percent, that a highly nonuniform flow through the matrix passages must be expected. A nonuniform flow lowers the heat-exchanger effectiveness far below its uniform-flow, designed, value.

We shall show that the hot- and cold-side pressure drops can be brought closer to balance by an appropriate choice of the conductance ratio, and that once that has been selected, the hot-side pressure drop is a function of the mean velocity through the hot side, regardless of the cell cross-section or hydraulic diameter.

A force balance between the pressure drop along a passage and the frictional shear stress at the passage walls (as developed earlier after equation 10.21) yields

$$\Delta p A_x = \tau_w A_h \tag{10.33}$$

where

A_x is the passage cross-section area,
A_h is the cell-wall area (affecting heat transfer and friction), and
τ_w is the frictional shear stress at the wall.

Equation 10.33 is more useful when converted to dimensionless groups:

$$\left[\frac{\Delta p}{p}\right] = \left[\frac{\tau_w}{(\rho C^2/2g_c)}\right]\left[\frac{A_h}{A_x}\right]\left[\frac{\rho C^2}{2g_c p}\right] \tag{10.34}$$

where

C is the mean flow velocity through the passage,

$\left[\dfrac{\tau_w}{(\rho C^2/2g_c)}\right] \equiv C_f$ is the friction coefficient, and

ρ is the mean density of the fluid.

Therefore,

$$\left[\frac{\Delta p}{p}\right] = C_f \left[\frac{A_h}{A_x}\right]\left[\frac{C^2}{2g_c RT}\right] \tag{10.35}$$

Introducing the continuity equation for a passage, $\dot{m} = \rho A_x C$, and Reynolds' analogy (equations 10.13 and 10.14) which will be valid for regenerator passages having wholly unseparated flow *

$$S_t \equiv \frac{h_t}{\rho C C_p} = \frac{C_f}{2} \times \frac{1}{2}$$

produces

$$\left[\frac{\Delta p}{p}\right] = \left[\frac{h_t A_h}{\dot{m}\,\overline{C_p}}\right]\left[\frac{C^2}{g_c RT}\right] \times 2 \tag{10.36}$$

The relative hot-side pressure drop on the left is equal to a quantity that is equivalent to a hot-side N_{tu} and a flow Mach-number function. All the quantities on the right side are fixed by the cycle specifications except the hot-side flow velocity and conductance.

The conductance ratio C_r appears in equation 10.31, and it has a strong effect on the hot-side conductance. We recommend above choosing a conductance ratio of unity as a starting point for the design of recuperators. This choice produces equal temperature drops on the two sides of a recuperator passage wall, which is thermodynamically desirable. The very different densities of the two fluids, one at high pressure and one at low, can be accommodated in a recuperator by using a heat-exchanger "surface" with a large flow cross-section on the low-pressure side, and a surface with a small cross-section on the high-pressure side.

In a rotary regenerator, the same flow "surface" is used alternately by the high- and low-pressure fluids. If the designer chooses to use a conductance ratio of unity, the heat-transfer areas will be almost equal (only the difference of the thermal conductivities of the two fluids would cause a deviation from equality), and if the heat-transfer areas are almost equal the matrix volumes and face areas on the two sides must also be almost equal. Therefore the total flow area (the free face areas, A_{ff}) will be almost equal. Consequently the mean flow velocity in the hot side will be high, and in the cold-side passages will be low, the ratio being approximately equal to the mean density ratio.

Inevitably, either the pressure drop on the hot side will be too high for a high thermal efficiency, or the pressure drop on the cold side will be too low to produce good flow distribution through the "core", or both.

A solution that works for rotary regenerators, at least at low cycle pressure ratio (say 4:1 and below) is to choose a conductance ratio higher than unity. The first choice could be equal to the design cycle pressure ratio up to a maximum conductance ratio of 4.0.

---

* Data in, e.g., fig. 10.7d are closer to $S_t \approx \dfrac{C_f}{4}$.

This choice makes the hot-side face area considerably larger than that of the cold-side, and produces core pressure drops that are more nearly balanced. The ideal seems to be to fix the hot-side design-point pressure drop at some maximum tolerable amount between two and five percent and aim to have the cold-side pressure drop at around one percent, enough to produce, with careful duct and "header" (transition-duct) design, uniform flow in the regenerator core.

The overall size of the regenerator, represented by the number of transfer units, is influenced by another design choice: the rotation rate of the matrix. If the rotation is very slow, the cell walls change greatly in temperature during the rotation, and either the effectiveness is reduced or a larger heat exchanger is required. Kays and London (1984) recommended a nondimensional rotation rate (matrix mass × specific heat × switching rate divided by $(\dot{m}C_p)_{mn}$) of 5.0, but work by Hagler and Wilson (1988) indicated that 3.0 might be a better choice. The effects on the number of transfer units required are shown in figures 10.4c and 10.4d.

The sequence from equation 10.36 is as follows. We start with equation 10.31:

$$[h_t A_h]_{ht} = (C_r + 1)(\dot{m}\overline{C_p})_{mn} N_{tu}$$

The hot-side conductance, $(h_t A_h)_{ht}$, is thus determined from the input specifications and from the designer's choice of C_{ro} (which fixes N_{tu} for the specified effectiveness) and of C_r. The conductance is used in equation 10.36, rearranged to produce the required mean hot-side velocity:

$$C_{ht}^2 = \frac{g_c \overline{R T_{ht}}}{2} \left[\frac{\dot{m}\overline{C_p}}{h_t A_h} \right]_{ht} \left[\frac{\Delta p}{p} \right]_{ht} \tag{10.37}$$

This determination of the hot-side velocity required to produce a specified core pressure drop is completely independent of the shape or size of the passages through the matrix or of the overall porosity (which is related to passage-wall thickness) of the matrix. (In this use of the word, porosity is defined as the passage area divided by the face area, A_{ff}/A_f.)

Calculation of the cold-side pressure drop

The mean cold-side flow velocity can be calculated if the pressure drop is first estimated, so that an approximation to the mean cold-fluid density can be made.

The free-flow area on the hot side can be calculated from the continuity equation and the known mean density:

$$A_{ff,ht} = \left[\frac{\dot{m}}{\overline{\rho_{ht} C_{ht}}} \right]$$

The ratio of the hot-to-cold free-flow areas will be equal to the ratio of the heat-transfer areas. These in turn can be found from the chosen conductance ratio and from the ratio of the heat-transfer coefficients:

$$\left[\frac{A_{ff,ht}}{A_{ff,cd}} \right] = \left[\frac{A_{h,ht}}{A_{h,cd}} \right] = \left[\frac{(h_t A_h)_{ht}}{(h_t A_h)_{cd}} \right] \left[\frac{h_{t,cd}}{h_{c,ht}} \right] = C_r \left[\frac{h_{t,cd}}{h_{t,ht}} \right]$$

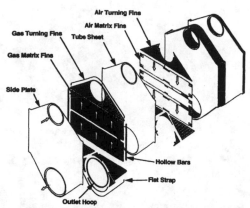

The two-module plate-fin recuperator is mounted in a hollow-wall steel enclosure case. Each module consists of four cores welded together and supported from a common plenum. The main part of each core is counterflow, exhaust gas to air, plus crossflow sections as distributor headers. There are 46 air and 47 gas passages; the core material is 14Cr-4Mo stainless steel.

Illustration. Recuperator for the Northrop-Grumman WR-21 CICREX-cycle engine. Courtesy Allied Signal Aerospace and the US Navy

The Nusselt number is a function of cross-section shape and therefore will be identical for the two sides, as will the hydraulic diameter. Therefore the ratio of heat-transfer coefficients becomes:

$$N_u \equiv \frac{h_t d_h}{k} \Rightarrow \left[\frac{h_{t,cd}}{h_{t,ht}}\right] = \left[\frac{\overline{k_{cd}}}{\overline{k_{ht}}}\right] \Rightarrow \left[\frac{A_{ff,ht}}{A_{ff,cd}}\right] = C_r \left[\frac{\overline{k_{cd}}}{\overline{k_{ht}}}\right]$$

$$C_{cd} = \left[\frac{\dot{m}_{cd}}{\overline{\rho_{cd}} A_{ff,cd}}\right] = \left[\frac{\dot{m}_{cd}}{\overline{\rho_{cd}}} \frac{C_r}{A_{ff,ht}}\right] \left[\frac{\overline{k_{cd}}}{\overline{k_{ht}}}\right]$$

Equation 10.36 gives the pressure drop on the cold side:

$$\left[\frac{\Delta p}{p}\right]_{cd} = \left[\frac{h_t A_h}{\dot{m} \overline{C_p}}\right]_{cd} \left[\frac{C_{cd}^2}{g_c R \overline{T_{cd}}}\right] \times 2$$

This pressure drop can be used to improve on the earlier approximation and to correct the mean density. However, if the value is much larger or smaller than desired, another choice for the conductance ratio should be made.

Calculation of the size of the core, the rotation rate and the carryover loss

With the two free-flow areas and the passage length determined, the designer merely needs to choose an appropriate matrix porosity for the size of the active face area to be fixed. The maximum porosity that has been achieved by matrix manufacturers has been

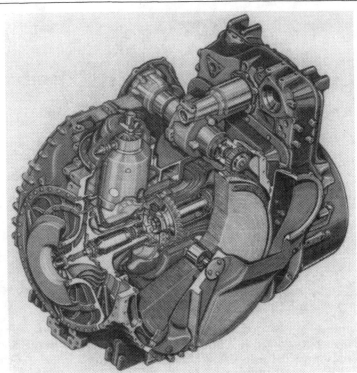

This approximately 250-kW two-shaft engine has been sold with metal-matrix regenerators (and used to power the US Army's Patriot missile ground equipment) and used in experimental DoE/NASA advanced-technology programs (e.g., to power inter-city buses) with ceramic regenerators. These are driven from the output shaft, geared from the low-pressure axial turbine. The centrifugal compressor (of about 4:1 pressure ratio) is driven by the high-pressure axial turbine. This engine, successful in many ways, was penalized by an approximately equal-area split of the regenerator faces. Consequently there was insufficient pressure drop on the cold side (analysis predicted of the order of one-tenth of one percent) to produce uniform flow distribution after the sharp 180-degree bend from the vaned-diffuser channels; and the pressure drop on the hot side was in the region of fifteen percent.

Illustration. The Allison GT 404-4 engine. Courtesy Allison Engine Company

about 70 percent, and this should be used for lightweight engines. For stationary engines with long operating cycles there are benefits in using a reduced porosity, perhaps as low as 40 percent, because the thicker cell walls resist seal wear better, the rotation rate is reduced, and the carryover loss is reduced.

Another choice affecting the regenerator size is the area of the seals. A starting estimate is 20 percent of the active face area.

Of two other choices, one is whether or not to have a central hole in the matrix disk. Doing so increases the outside diameter somewhat, but also reduces the length of the critical radial seals. It might also reduce thermal stresses. A diameter ratio, inner to outer, of between 0.25 and 0.50 seems appropriate. The second choice is whether to use one disk or two or more. Manufacturers have strict limits to the maximum diameter of an extruded disk, so that this limit, rather than overall engine design (sometimes called "packaging"), usually rules.

Total active free-flow area is $= \left[A_{ff,cd} + A_{ff,ht}\right]$
Total active face area, for a matrix of porosity p_{or}, is $= \left[A_{ff,cd} + A_{ff,ht}\right]/p_{or}$
The face area including seal area is

$$\left[(A_{ff,cd} + A_{ff,ht})/p_{or}\right][1 + \text{seal-area ratio}] \equiv \Sigma(A_f) \qquad (10.38)$$

where the seal-area ratio is the ratio of the seal face area to the active face area.

If the number of regenerator disks to be used is Z, and the inner-to-outer diameter ratio is Λ the disk diameters (not including solid rims) are

$$\text{outer diameter} = \sqrt{\frac{4\Sigma(A_f)}{\pi Z(1 - \Lambda^2)}} \qquad (10.39)$$

$$(\text{inner diameter}) = \Lambda \times (\text{outer diameter})$$

The total mass of the matrix (not including solid rims) is

$$[\Sigma(A_f)L][1 - p_{or}]\rho_w \qquad (10.40)$$

where ρ_w is the density of the wall material.

The rotation period is found from the chosen nondimensional rotation rate and the mean specific heat of the wall material:

$$\frac{[\Sigma(A_f)L][1 - p_{or}][\rho \overline{C}_p]_w}{[\dot{m}\overline{C}_p]_{mn} C_{ro}} \qquad (10.41)$$

The mass of compressed air trapped in the matrix cells (under the doubtful assumption that the seals do not leak) can be approximated by calculating the open volume associated with the rotation rate and specifying that each cell is filled with compressed air at mean density. This is often called "carryover leakage". It is quite small, usually less than one percent of the compressor flow, so that the approximations used in its calculation are not of great significance.

$$\text{Carryover leakage} = C_{ro}\left[\frac{p_{or}}{1 - p_{or}}\right]\left[\frac{[\dot{m}\overline{C}_p]_{mn}}{[\rho \overline{C}_p]_w}\right]\overline{\rho_{cd}} \qquad (10.42)$$

Determination of the passage hydraulic diameter

The hot-side free-flow area, $A_{ff,ht}$, has been found. The number of cell passages in the hot face is thus:

$$\left[\frac{A_{ff,ht}}{A_x}\right]$$

The heat-transfer area on the hot side is

$$A_{h,ht} = \left[\frac{A_{ff,ht}}{A_x}\right] p_e L \tag{10.43}$$

where p_e is the perimeter of the passage cross-section, and L is the passage length.

If we combine this with the definitions of the hydraulic diameter and of the Nusselt number, we arrive at an expression for the value of the hydraulic diameter that will satisfy the specifications for the minimum-length channel, L, found from the criterion for the limiting temperature gradient.

$$d_h \equiv \frac{4A_x}{p_e} \quad \text{and} \quad N_u = \frac{h_t d_h}{k}$$

$$[h_t A_h]_{ht} = \left[\frac{N_u \bar{k}}{d_h}\right]_{ht} \left[\frac{A_{ff,ht}}{A_x}\right] p_e L = \frac{4\bar{k} A_{ff,ht} L}{d_{h,ht}^2} N_u$$

$$\Rightarrow d_h = \sqrt{\frac{4\bar{k}_{ht} A_{ff,ht} L}{(h_t A_h)_{ht}}} \sqrt{N_u} \tag{10.44}$$

This value of the hydraulic diameter is a minimum value that will not contravene the minimum-length criterion. It may turn out that the value obtained is smaller than is available in producible cores. The design specifications can then be satisfied by choosing the nearest appropriate core configuration (yielding a value of N_u) and adjusting the cell length L so that (L/d_h^2) is constant.

Choice of regenerator passage shapes

The first gas-turbine regenerators were made from two strips of stainless steel, one flat and one crimped, wound around a mandrel. The same technique was used for the first ceramic regenerators, except that the strips were of lightweight paper soaked in ceramic slurry, and the assembly was baked and fired (figure 10.11). The passage shape was roughly triangular. A table published by Kays and London (1984) gives the nondimensional heat-transfer coefficient, the Nusselt number, for various shapes, some of which are repeated in table 10.1. A triangular passage has the lowest Nusselt number on the list, at 3.00.

Corning developed another method of ceramic-regenerator manufacture, which is to extrude the matrix through a die (figure 10.12). This procedure allows a variety of cell shapes to be produced to considerable accuracy. It might seem better, then, to use a rectangular cell giving a Nusselt number of 6.49, or parallel plates with $N_u = 8.23$. However, doing so does not necessarily lead to better regenerators.

Suppose that in equation 10.44 the specified values gave the hydraulic diameter the following value (which happens to be the value resulting from the design example below; thus table 10.1 serves double duty):

$$d_h = 0.2897 \sqrt{N_u}$$

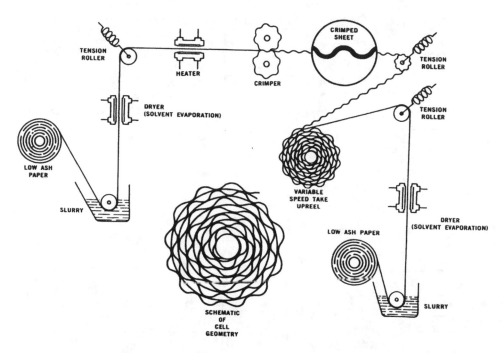

Figure 10.11. Wrapping of ceramic regenerator "core". From Cook et al. (1978)

Figure 10.12. Extrusion of ceramic regenerator "core". Courtesy of J. Paul Day, Corning, Inc.

Table 10.1. Dimensions of alternative passages for a regenerator

Passage cross-section shape	(b/a) for rectangular	N_u	d_h, mn [mm]	(a/d_h)	a [mm]
(triangle, side a)	–	3.00	0.502	1.732	0.869
(square, $b=a$, side a)	1	3.61	0.550	1.000	0.550
(rectangle, width b, height a)	4	5.33	0.669	0.625	0.418
(rectangle, width b, height a)	8	6.49	0.738	0.562	0.415
(flat, height a)	Infinity	8.23	0.831	0.500	0.416

With this relation we find the hydraulic diameter to be used with each passage shape given in table 10.1. We have also calculated the ratio of a significant dimension, a, of the passage shape to the hydraulic diameter, and hence show values of a.

The results are to some extent counterintuitive. While the increased Nusselt number of the 8:1 rectangle, for instance, would give it an advantage over the triangular-passage core if it were used at the same hydraulic diameter, it would produce a reduced-volume regenerator only by reducing the passage length (the disk thickness). The matrix would then be susceptible to thermal cracking, if the temperature-gradient limit proposed here is valid. By imposing this limit on the temperature gradient, and hence fixing the passage length (disk thickness), and by specifying a hot-side pressure drop and hence fixing the mean flow velocity and free-flow area, both the performance and the size of the regenerator have been roughly fixed. The choice of matrix-passage cross-sectional shape could well be left to the core manufacturer.

Example 1   Design of a rotary regenerator

Make a preliminary design of a rotary regenerator for a solar-heated Brayton cycle. This will use pure air on both sides of the heat exchanger and will have a low pressure ratio. The cycle conditions are as follows.

Compressor mass flow	1.0 kg/s
Compressor-inlet stagnation temperature	300 K
Compressor-inlet stagnation pressure	100 kN/m^2
Compressor-outlet static pressure	200 kN/m^2
Compressor stagnation-to-static polytropic efficiency	0.82
Turbine-inlet stagnation temperature	1,400 K
Turbine stagnation-to-static polytropic efficiency	0.90
Heat-exchanger effectiveness	0.95
Heat-exchanger outlet static pressure	100 kN/m^2
Heat-exchanger hot-side relative pressure drop (core)	0.025
Heat-exchanger cold-side relative pressure drop (core)	0.010 − 0.015
Heat-exchanger ceramic solid density	2259 kg/m^2
Heat-exchanger ceramic specific heat	1146 J/(kg · K)
Maximum gradient of cold-side temperature	7.5 K/mm
Sum of cycle relative drops in stagnation pressure (goal)	0.08
Leakage between compressor outlet and regenerator inlet	0.02 kg/s

"Preliminary design" means that we will do all our calculations on a lumped-parameter mean-property basis. We will follow the calculation sequence of figure 10.10 and of the development of the previous section.

Calculate mean properties

Figure 10.13 shows the specified temperature changes through the regenerator with values inserted for temperatures and properties calculated from the following.

Compressor-outlet temperature (regenerator-inlet temperature, cold side)

$$\frac{T_{0,2}}{T_{0,1}} = \left(\frac{p_{0,2}^{\otimes}}{p_{0,1}}\right)^{\left[\left(\frac{R}{C_{p,c}}\right)\frac{1}{\eta_{p,c}}\right]} = 2^{[286.96/(1010 \times 0.82)]}$$

After iterating on mean C_p, $T_{0,2} = 381.7$ K

Turbine-outlet temperature (regenerator-inlet temperature, cold side)

$$\frac{T_{0,4}}{T_{0,5}} = \left(\frac{p_{0,4}}{p_{0,5}^{\otimes}}\right)^{\left[\left(\frac{R}{C_{p,e}}\right)\eta_{p,e}\right]} = [(1 - 0.08)2]^{[(286.96 \times 0.90)/1190]}$$

After iterating on mean C_p, $T_{0,5} = 1226.5$ K

The mass flows are equal on each side of the regenerator (0.98 kg/s), but the lower specific heat at the lower temperature of the cold side will give this the lower heat capacity. Therefore the cold-side regenerator-outlet temperature is given by:

$$T_{0,3} - T_{0,2} = 0.95(T_{0,5} - T_{0,2}) \Rightarrow T_{0,3} = 1184.3 \text{ K}$$

The mean temperature is $\overline{T_{cd}} = (1184.3 + 381.7)/2 = 783.8$ K. The associated mean properties of the cold side are $\overline{C_{p,c}} = 1094.7$ J/(kg · K) and $\overline{k_{cd}} = 0.05692$ W/(m · K).

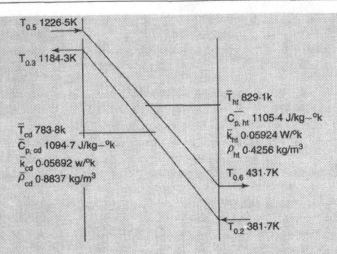

$T_{0.5}$ 1226·5K

$T_{0.3}$ 1184·3K

$\overline{T}_{ht}$ 829·1k

$\overline{C}_{p,\,ht}$ 1105·4 J/kg–°k

$\overline{k}_{ht}$ 0·05924 W/°k

$\overline{\rho}_{ht}$ 0·4256 kg/m³

$\overline{T}_{cd}$ 783·8k

$\overline{C}_{p,\,cd}$ 1094·7 J/kg–°k

$\overline{k}_{cd}$ 0·05692 w/°k

$\overline{\rho}_{cd}$ 0·8837 kg/m³

$T_{0.6}$ 431·7K

$T_{0.2}$ 381·7K

Figure 10.13. Properties along the regenerator length

The mean density is calculated on an estimate that the relative-pressure drop on the cold side will be 0.0125. It can be updated later if this estimate is far off.

$$\overline{\rho_c} = \frac{200,000[1 - 0.0125/2]}{286.96 \times 783.8} = 0.8837 \text{ kg/m}^3$$

For the mean properties on the hot side we first need to find the outlet temperature, iterating on the mean specific heat.

$$\frac{0.98\overline{C_{p,ht}}(1226.5 - T_{0,6})}{0.98 \cdot 1094.7(1184.3 - 381.7)} \Rightarrow$$

$$T_{0,6} = 431.7 \text{ K} \quad \text{and} \quad \overline{T_{ht}} = 829.1 \text{ K}$$

$$\overline{C_{p,ht}} = 1105.4 \text{ J/(kg} \cdot \text{K)} \quad \text{and} \quad \overline{k_{ht}} = 0.05924 \text{ W/(m} \cdot \text{K)}.$$

The mean density on the hot side can be calculated for the specified core pressure drop.

$$\overline{\rho_{ht}} = \frac{100,000[1 + 0.025/2]}{286.96 \times 829.1} = 0.4256 \text{ kg/m}^3$$

Choose rotation rate, N_{tu} and conductance ratio

We choose a nondimensional rotation rate, C_{ro}, of 3.0. The value of N_{tu} required to give an effectiveness of 0.95 is read from figure 10.4d as 23.

We follow the guidelines given earlier of choosing the conductance ratio equal to the compressor pressure ratio for values up to 4.0. Therefore $C_r = 2.0$.

Hot-side conductance and mean velocity

From equation 10.31, the hot-side conductance is

$$[h_t A_h]_{ht} = (C_r + 1)(\dot{m}\overline{C_p})_{mn} N_{tu} = 3 \times 0.98 \times 1094.7 \times 23 = 74 \text{ kW/K}$$

The hot-side mean velocity is given by equation 10.37:

$$C_{ht}^2 = \frac{g_c R T_{ht}}{2}\left[\frac{\dot{m} C_p}{h_t A_h}\right]_{ht}\left[\frac{\Delta p}{p}\right]_{ht}$$

$$= \frac{286.96}{2} \times 829.1 \frac{0.98 \times 1105.4}{3 \times 0.98 \times 1094.7 \times 23} \times 0.025 \Rightarrow C_h = 6.60 \text{ m/s}$$

Free-flow areas and cold-side mean velocity

$$A_{ff,ht} = \left[\frac{\dot{m}_{ht}}{\rho_{ht} C_{ht}}\right] = \frac{0.98}{0.4256 \times 9.33} = 0.247 \text{ m}^2$$

$$\left[\frac{A_{ff,ht}}{A_{ff,cd}}\right] = C_r\left[\frac{h_{t,cd}}{h_{t,ht}}\right] = \frac{2 \times 0.05692}{0.05924} = 1.9217 \Rightarrow A_{ff,cd} = 0.128 \text{ m}^2$$

$$C_{cd} = \left[\frac{\dot{m}_{cd}}{\rho_{cd} A_{ff,cd}}\right] = \frac{0.98}{0.8837 \times 0.128} = 8.635 \text{ m/s}$$

Cold-side pressure drop

This is obtained from equation 10.36; the cold-side conductance is found from the hot-side conductance and the conductance ratio.

$$\left[\frac{\Delta p}{p}\right]_{cd} = \left[\frac{h_t A_h}{\dot{m} C_p}\right]_{cd}\left[\frac{C_{cd}^2}{g_c R T_{cd}}\right] = \left[\frac{3}{2} \times \frac{0.98 \times 1094.7 \times 23}{0.98 \times 1094.7}\right]\left[\frac{8.635^2}{286.96 \times 783.8}\right] = 0.0114$$

This value is within the acceptable range of 0.010 to 0.015. The previous estimate of the cold-side mean density will be good enough.

Core number, size, mass, porosity, rotation rate, and carryover leakage

We choose to have two cores with central holes of 0.25 inner-to-outer diameter ratio. We also choose a seal-area ratio of 0.2 and a porosity of 0.7. The total face area is obtained from equation 10.38:

$$\Sigma(A_f) = \frac{0.247 + 0.128}{0.7} \times 1.2 = 0.6429 \text{ m}^2$$

Diameter of disks:

For two disks, $Z = 2$
outside diameter $= 661$ mm
inside diameter $= 165$ mm
Core length (thickness), $L = \dfrac{1184.3 - 381.7}{7.5} = 107$ mm.
Mass of the active disks plus that under the seals (equation 10.40):
$0.6429 \times 0.107 \times 0.3 \times 2259 = 46.31$ kg (both disks).
Rotation period (equation 10.41):

$$\frac{46.31 \times 1146}{0.98 \times 1094.7 \times 3} = 16.495 \text{ (3.64 rev/min)}.$$

Carryover leakage (equation 10.42):

$$3 \times \frac{0.7}{0.3}\frac{0.98 \times 1094.7}{2259 \times 1146}\frac{0.8837}{0.98} = 0.0026 = 0.26\%.$$

Core-passage hydraulic diameter and passage shape

The hydraulic diameter is calculated from equation 10.44. If we had already chosen a passage shape, the Nusselt number for that shape would be inserted. We prefer to examine several possible shapes, with the results shown in table 10.1. The shape of the passage does not change the size or performance of the regenerator. Therefore the choice can be made on the basis of what core shapes and sizes are available, or what is the easiest and lowest cost to manufacture. The only passage of those examined that would be unattractive for a ceramic rotary regenerator, because of manufacturing and structural difficulties, is the parallel-plate configuration.

Discussion of overall design

This preliminary design has met the design objectives (the effectiveness and the two pressure drops) well. It is likely to be conservative because the number of transfer units was read from a chart for a capacitance ratio of unity, whereas the actual rate is less, for which a smaller number of transfer units is sufficient. Whether the overall pressure-loss goal can be met is in some doubt. The pressure drops of the two flows in the regenerator cores add to under 4 percent. If the combustor has a pressure drop of 3.5 percent, the duct and air-filter losses are likely to exceed the small balance left.

The two disks have an outside diameter of the active matrix of about 660 mm, which are too large for current extrusion dies (about 300 mm, 1995). They could presumably be fabricated from two extruded half-disks, cemented. Or it might be decided to reduce the outside diameter, taking increased pressure drops as the inevitable consequence.

Another arrangement is to extrude a segmented part of the disk or to press-form the segment casings and pack them with ribbed sheets before firing (figure 10.14a and 10.14b). The segments would be bound mechanically into disks. This construction would reduce thermal stresses. It would also favor discontinuous motion of the disks, the seals being pneumatically clamped to the disks while it is stationary, and lifted very slightly for rapid movement through, e.g., 30° or 45° (figure 10.15, Wilson (1993); Wilson and Pfahnl (1997)).

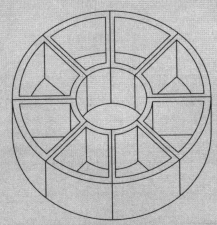

(a) Extruded casing segment for regenerator disks

Figure 10.14. Extruded casing segment and packed ribbed sheets for regenerators

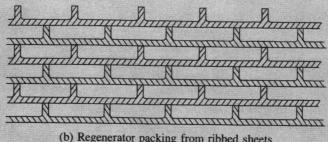

(b) Regenerator packing from ribbed sheets

Figure 10.14. Concluded

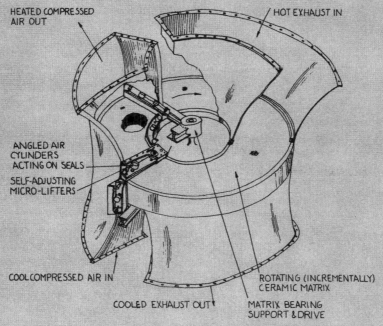

ROTARY REGENERATOR WITH CLAMPABLE SEALS.

Figure 10.15. Rotary regenerator with clampable and releasable seals

When it is decided that the preliminary design is satisfactory, it would be sensible to carry out a full design by computer, treating the regenerator disk as a fairly large number of slices, in each of which average properties, velocities and so forth are calculated. The general experience is that this changes the preliminary-design estimates of pressure drops and diameters by no more than five percent.

10.7   Turbine-blade cooling

The contribution of turbine-blade cooling to the increase in cycle thermodynamic efficiency of gas turbines was described in chapter 1 and illustrated in figure 1.10. In this chapter we shall limit the discussion to the fundamentals of the heat-transfer phenomena, and to the design problems involved in blade cooling.

The heat transfer that occurs as a result of a hot gas stream flowing at high velocity past and through a row of cooler turbine blades is dominated and controlled by the boundary layers. In general, turbines have accelerating flow, which normally produces laminar boundary layers. However, turbulent boundary layers occur even in a generally accelerating flow if, for instance, there is a local area of adverse pressure gradient producing decelerating flow, or an area of flow separation, either of which might be caused by high flow incidence angles, or by poor blade profiles including slope-of-curvature discontinuities, see sections 7.4 and 7.5, or as a result of manufacturing inaccuracies or of corrosion or deposition, or by the wakes from upstream blade rows (boundary-layer/wake interaction).

In addition, the high first cost of cooled blades, coupled with the deleterious effect on the cycle efficiency of supplying (usually) high-energy compressed and cooled air, gives the designer a strong incentive to use high loading coefficients to avoid the need for additional rows of cooled blades. Accordingly, the reaction at the blade hub might be low (chapter 6) which could lead to local adverse pressure gradients and consequent turbulence.

Turbulent boundary layers may also become established downstream of the passage throat. We are showing (figure 10.16) some results of heat-transfer and static-pressure measurements around such blades not only because they are from author DGW's thesis work (Wilson, 1952), but because, as will be shown below, modern blades experience quick transition of the boundary layers to turbulence (though for different reasons). The blade profiles are shown in figure 10.16a. The static-pressure distribution for a range of air-inlet angles (the design inlet-flow angle is 30°), and the heat-transfer distribution at the design incidence at several Reynolds numbers on an orthogonal plot are shown in figures 10.16b and 10.16c. The design-incidence heat-transfer distribution on a polar plot is shown in figure 10.16d to indicate the high heat-transfer coefficients along the thin trailing-edge region, where the provision of cooling from internal flows is difficult[4]. The abscissa for the orthogonal plots is (x/c), where x is the distance from the leading-edge

---

[4]Some results of this pioneering work by author DGW (Wilson, 1952; Wilson and Pope, 1954) are further explained by author TK's later (Korakianitis, 1993a; Korakianitis and Papagiannidis, 1993b) investigations on the effect of blade-surface-curvature distribution on performance. TK's explanation of the transition shown on the suction surface of figure 10.16d is that transition there is the result of the large "spikes" in suface-pressure and surface-velocity distributions near that point, where figure 10.16a shows the circle is joined to the straight line near the throat (see sections 7.4 and 7.5). On the suction surface and at lower Reynolds (and Mach) numbers the Mach-number distribution "spikes" introduced by joining the leading-edge circle to the suction-surface circle are smaller, and therefore transition is delayed to the larger similar "spike" near the throat. At the higher Reynolds number (6.75×10^5) the suction-surface leading-edge Mach-number "spike" is higher, and therefore transition is triggered at the suction-surface leading-edge Mach-number "spike". The change of curvature introduced by joining the leading-edge circle to the pressure-surface circle (inverse of local radii) is larger than the corresponding one on the suction surface. Therefore the leading-edge Mach-number "spike" on the pressure surface triggers transition at that location at all Reynolds numbers.

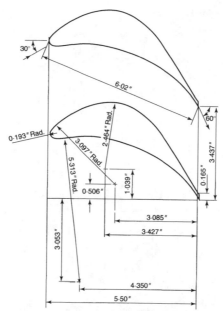

(a) Cascade geometries. From Wilson (1952)

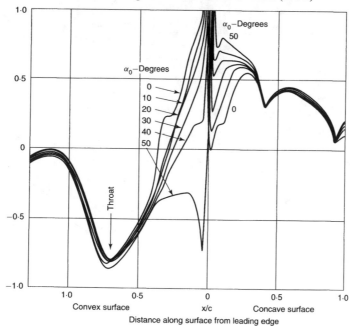

(b) Static pressure distribution at various incidences. From Wilson (1952)

Figure 10.16. Distribution of static pressure and heat transfer around a gas-turbine blade. From Wilson (1952), and Wilson and Pope (1954)

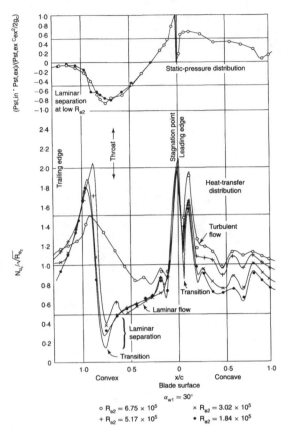

(c) Distribution of static pressure and heat transfer at various Reynolds numbers. From Wilson (1952)

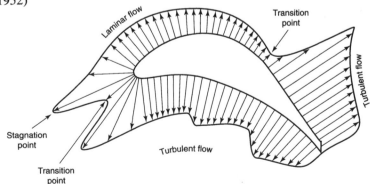

(d) Distribution of heat-transfer coefficient around the turbine blade of figure 10.16a at design incidence and Reynolds number. From Wilson and Pope (1954)

Figure 10.16. Concluded

stagnation point, and c is the blade chord. The ordinate for the heat-transfer plot is the Nusselt number based on the blade chord, divided by the root of the Reynolds number, based on the blade chord and the relative velocity at passage exit. This combination of variables is used because it is constant for laminar boundary layers, but increases with Reynolds number for turbulent boundary layers.

The boundary layer at the leading edge is thus seen to be laminar at all Reynolds numbers, as would be expected, and stays laminar until the blade throat for the three lower Reynolds numbers (up to 5.17×10^5), at which the heat-transfer coefficient is very low. The pressure minimum and subsequent pressure rise at this point is sharp enough to produce a laminar-separation "bubble", clearly seen (at low Reynolds numbers) in the modification it introduces into the pressure-distribution curve above. As the main flow leaves the controlling influence of the wall it snaps into turbulence, producing very high heat-transfer coefficients at reattachment, at a point where the blades are thin. At the highest Reynolds number, almost 7×10^5, the boundary layer becomes unstable without an obvious pressure-minimum trigger, and the heat-transfer coefficient at the throat is consequently much higher than for the laminar condition. However, the boundary layer is thereby made thicker, so that the (turbulent) heat-transfer coefficient just downstream of the throat is considerably lower than for the late-transitioning lower-Reynolds-number flows. Thus this last curve at high Reynolds number and with early transition to turbulence results in a more-even distribution of heat transfer. It may also be more representative of modern practice, as will be shown below.

It had been initially expected that the boundary layers on the concave (pressure) surface would have been almost totally laminar. However, a very sharp pressure minimum again produced transition to turbulence very close to the leading edge. This was so unsuspected (the profile appeared to be smooth) that an early form of finite-element analysis, Southworth's relaxation method, was used (requiring weeks of hand work) and the pressure minimum was quite precisely confirmed. It was, in fact, the product of the early multiple-circular-arc blade profiles that were required on account of then-current manufacturing methods (usually forgings that were subsequently milled). Boundary layers are sensitive to the rate of change of curvature (see section 7.4). Two arcs can meet smoothly, but the curvature has a step change, and this can cause boundary-layer transition. The extent to which the resulting pressure minimum persists over a range of flow inlet angles can be seen in figure 10.16b. The development of greatly improved casting alloys and manufacturing methods (including numerically-controlled milling machines) gave new design freedoms. Profiles with prescribed velocity distributions (PVD profiles, inverse-design methods) were produced with the aid of computer programs. Polytropic turbine efficiencies increased as a result of the introduction of these PVD profiles. Some test results on such profiles are reported next.

Measurements of unsteady rotor-blade heat transfer in a very advanced turbine experiment (a so-called "blow-down" tunnel using a mixture of Argon and Freon-12 that passes for a fraction of a second through a turbine rotor that is run up to speed in a vacuum) giving exact duplication of Reynolds number, Mach number, metal-to-gas temperature ratio, and other parameters, were made by Abhari et al. (1992). They showed that for a modern highly-loaded turbine stage (figure 10.17) the time-averaged rotor

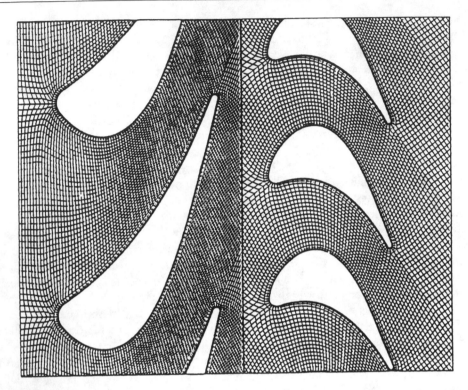

Figure 10.17. Nozzle and rotor profiles studied, and computational grid. From Abhari et al. (1992)

measurements were most closely approached by calculations of a fully turbulent boundary layer, either in steady-state or time-averaged unsteady calculations (figure 10.18). The heat-transfer distribution at $-10°$ incidence is shown in figure 10.19, and the measured and calculated pressure distributions in figure 10.20. Calculations of laminar heat transfer were less than half the measured value except at the leading edge, where measurements agreed very closely with those of Wilson and Pope (1954) for older multiple-arc blades (figure 10.16). (In figure 10.16 the Reynolds number is based on rotor-blade chord, whereas in figure 10.17 the Reynolds number is based on the chord of the nozzle guide vanes.)

The unsteady calculations were based on computer programs developed by Giles (1988) that produced predictions of the shock structure (figure 10.21) and of the nozzle-guide-vane wakes (figure 10.22) interacting with the rotor blades. The accuracy of the resulting predicted heat transfer, while not exact, is a great improvement over anything that has previously been available. Figure 10.21 shows that it is the suction surface of the rotor blades, on which the boundary layer would be expected to be more susceptible to transition to turbulence than the pressure surface, that experiences almost all the shock interactions, Figure 10.22 shows that the wake involves the suction surface fully and

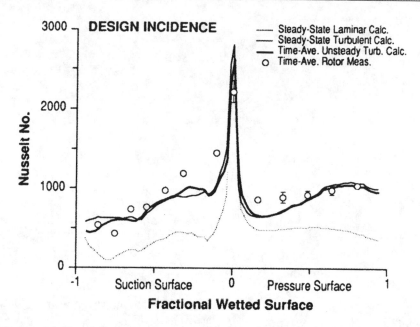

Figure 10.18. Comparison of measured turbine-blade heat transfer with calculations for various boundary layers. From Abhari et al. (1992)

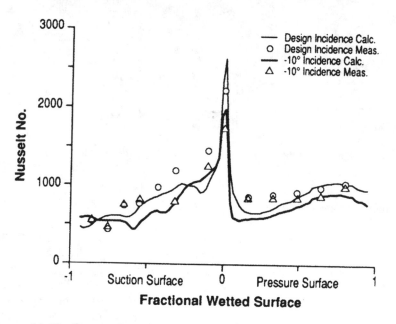

Figure 10.19. Comparison between measurements and time-averaged unsteady calculations at design and $-10°$ incidence. From Abhari et al. (1992)

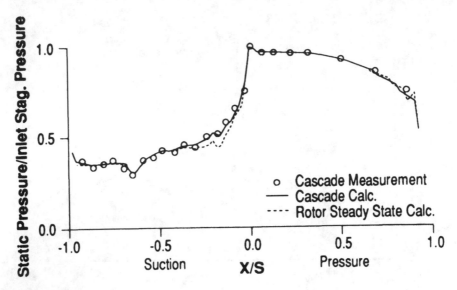

Figure 10.20. Steady-state static-pressure distribution calculations compared with blade measurements. From Abhari et al. (1992)

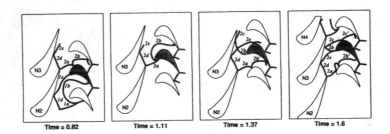

Figure 10.21. Diagram of shock structure in stator-rotor interaction. From Abhari et al. (1992)

produces a travelling "transition trigger", helping to explain that a fully turbulent model is appropriate for heat-transfer prediction.

Korakianitis et al. (1993) used two reduced-frequency (non-dimensional) parameters, one for unsteady flow, one for unsteady heat transfer, to show analytically that in turbine stator-rotor interactions (see section 7.5) the unsteady flow fields must be computed with unsteady (truly time-dependent) calculations, but that the resultant heat transfer is practically quasi-steady. This means that once the time evolution of the velocity profiles has been obtained with an unsteady code, the resultant heat transfer can be computed using the instantaneous velocity profile in steady-flow heat-transfer calculations. They tested the analysis by comparing the results of suitable computations with experiments.

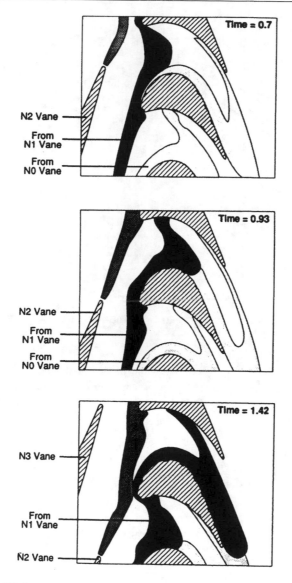

Figure 10.22. Wake propagation in stator-rotor interaction. From Abhari et al. (1992)

In their computational results they used Giles' (1988) unsteady code (a different version of the code used by Abhari et al. (1992)). Korakianitis et al. (1993) used the results of their computations to explain the physical phenomena of the unsteady (quasi-steady) heat-transfer trends. While the accuracy of the calculations is not exact, the results show good agreement with the results of the experiments.

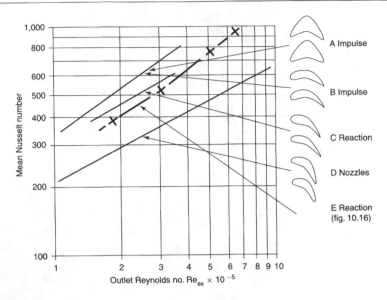

Figure 10.23. Mean heat transfer on turbine blades at design incidence. From Wilson and Pope (1954)

Mean heat-transfer rates for different turbine blades at design incidence are shown as a function of rotor-blade Reynolds number (using the rotor-blade chord as the length) in figure 10.23. This chart dates from 1954, but appears to represent modern blades adequately.

The three-dimensionality of the flow around turbine blades further complicates efforts to predict local heat-transfer rates, necessary if cooling is to be configured to produce as constant a metal temperature as possible. The horseshoe vortex, figure 10.24, produces local variations in heat transfer against which compensatory measures are particularly difficult.

10.8　Heat transfer with mass transfer

The considerations in section 10.7 concern heat transfer in the absence of mass transfer, i.e., blades with nonporous walls. Internal passages alone cannot produce the intensity of cooling needed to maintain metal temperatures at 1350 K when peak gas temperatures may be well over 1800 K. Virtually all air-cooled blades now incorporate labyrinthine passages that discharge into the blade boundary layers through angled laser-drilled holes, spaced more or less closely to match the predicted external heat-transfer coefficients (figure 10.25). The high leading-edge heat-transfer rate is matched by a combination of so-called "impingement cooling" (analogous to internal stagnation-point heat transfer) and film-cooling holes.

Employing mass transfer has two large potential advantages over solid-wall cooling. First, the coolant in its intimate contact with the wall will transfer more heat from the

Secondary Flows in a Turbine Nozzle Cascade

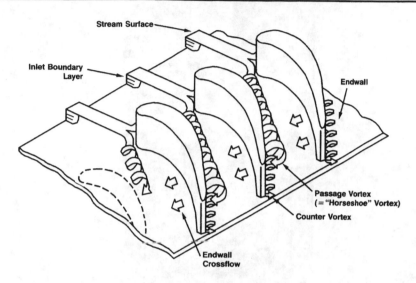

Figure 10.24. Horseshoe vortex in blade passage. Courtesy of GE Aircraft Engines

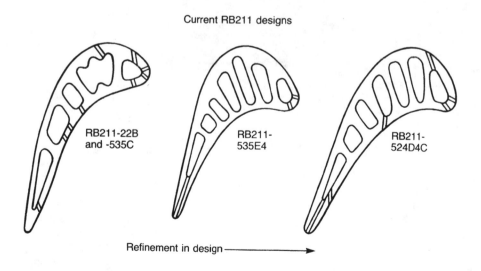

Figure 10.25. Cooling passages in high-temperature turbine blades. Courtesy Rolls-Royce plc.

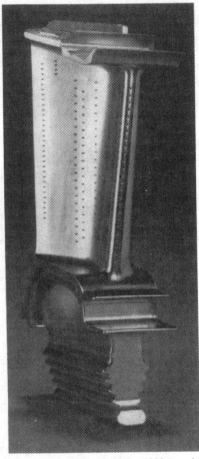

The blade cross-sections in figure 10.25 show blades currently (1990s) used in the RB-211 engines. These are "shrouded" designs, an example of which is shown in the photograph. The film-cooling holes, laser-drilled to be as tangential to the surface as possible (except in the leading-edge region) so as to energize sluggish boundary layers, can be seen on the blade flanks.

Illustration. Air-cooled turbine blades. Courtesy Rolls-Royce plc

blade. Second, the actual heat-transfer coefficients around the blade can be reduced because of the modification of the temperature distribution in the "thermal" boundary layer. Therefore the blade-surface temperature can be greatly reduced for the same coolant flow, or a reduced coolant usage can be realized for the same surface temperatures as for solid-wall cooling. Experimental turbines with blades formed from rolled mesh to give a controlled porosity for cooling air have been run at design conditions with inlet gas temperatures up to 4, 000 °F (2,480 K).

On the other hand, mass transfer into the boundary layer adds low-momentum fluid that can make otherwise almost-laminar boundary layers fully turbulent and thereby increase flow losses and heat transfer. Film cooling applied as in some of the blades of figure 10.25 can be less effective, for these reasons, than solid-wall conduction cooling.

10.9 Internal-surface heat transfer

With air used for blade cooling, there has been a natural design progression from the crude hollow blades used in some versions of the Junkers Jumo 004 jet engine in Germany in the Second World War (figure 10.26a). These were introduced principally to allow the use of non-heat-resistant metals (see the brief history at the beginning of the text). The effectiveness of the heat exchange between the cooling air and the blade was low, and a large quantity of cold air was discharged from the blade tips, disrupting the turbine flow and requiring a large increase in compressor flow. As indicated in equations 10.26 and 10.28, the hydraulic diameter of the heat-transfer passages should be reduced when it is desired to increase the heat transfer to be accomplished in a given volume. The methods used to produce finer and more intricate flow passages have changed as new manufacturing methods have been developed.

An early method used in Britain was to cast longitudinal wires into small billets of high-temperature material, to etch out the wires, and to forge the billets into the final blade forms (figures 10.26b and 10.26c). The once-round holes would be deformed into favorably shaped slits which could be concentrated in the leading and trailing edges, regions of high external-surface heat transfer. Later, improved methods of precision casting were used to produce cooling channels that could make several passes along the blade before being discharged at the tips or along the trailing edges. These blades were often formed of an intricate internal strut and a brazed-on outer shell.

With the introduction of inverse blade design the trailing-edge heat-transfer peak was reduced, but the very high heat transfer at the leading-edge stagnation point remained as a problem. The design requirement is not merely to keep the blade material to a

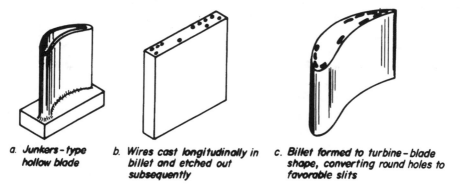

a. *Junkers-type hollow blade*

b. *Wires cast longitudinally in billet and etched out subsequently*

c. *Billet formed to turbine-blade shape, converting round holes to favorable slits*

Figure 10.25. Early types of air-cooled turbine blades

temperature below which the physical properties will be adequate but also to reduce the temperature gradients to a level below which fatigue failure will not result from thermal cycling. (This results in so-called "low-cycle" fatigue, because the stress reversals occur in general only upon start-up and shutdown or upon rapid power changes in high-performance engines, rather than in "high-cycle" fatigue, experienced principally through stator-rotor interactions.) Therefore the high external heat-transfer coefficient at the leading-edge stagnation point must be matched by high heat transfer from the internal cooling passages. At the present time this is accomplished generally by producing stagnation-point heat transfer from impingement jets from the primary coolant supply (figure 10.25).

Water can also be used for rotor-blade cooling, but the very high pressures in the centrifugal field, and the tendency for scale and other deposits to clog passages, allows less design freedom. The first known modern water-cooled rotor was designed by Schmidt in Germany to have a free water surface at the rotor interior (figure 10.27a). The multistage rotor and blades were machined out of a solid forging, the cooling holes drilled, and end caps were brazed on. Very high internal heat-transfer rates were assured because at design speed the water would be above critical pressure over the whole length of the blades and, therefore, no vapor could be present there. It was intended that the steam produced at the water surface would be expanded in a steam turbine, condensed, and recycled for cooling. However, a wave instability of the free water surface produced dangerous rotor vibrations (DGW has an exciting but uncomfortable memory of being left working at a neighboring rig by the scampering test engineers when the turbine vibrated the floor violently as it encountered an instability at about 15,000 rpm). In addition there were failures in the integrity of the coolant channels, and the experiments were terminated. (Unsuccessful attempts were made in Britain and the United States after 1946 to overcome these problems.)

Another form of water cooling that has been studied but not, so far as is known, built and tested is the closed thermosiphon (figure 10.27b). In one form of this system each blade carries its own heat exchanger at an inner radius, and the closed cooling passage has sealed within it a small quantity of water or other coolant. The liquid boils from the free surface near the blade tip and condenses in the hub heat exchanger. The centrifuged condensate and vapors cool the remaining parts of the passage.

The latest advance in cooling for large industrial turbines has been with the use of closed-circuit steam cooling. Steam has several advantages over air for internal cooling. The specific heat is higher. The molecular weight is lower, so that the speed of sound is higher, and Mach numbers are generally low. The required pressures can be generated in a small steam generator by waste heat with little shaft power required. The plumbing and turbine-casing modifications are more complex than for air cooling, but the substantial gains in turbine-rotor inlet temperature that have been achieved (of the order of 250 K) in a very short period after the introduction of this technology has made the change highly cost-effective.

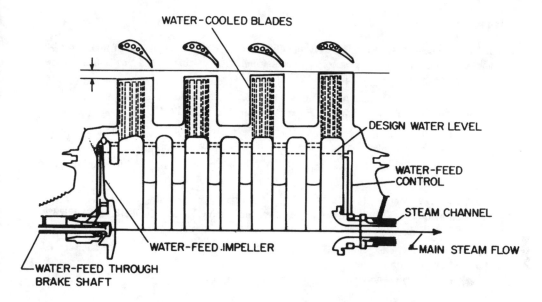

WATER-COOLED BLADES

DESIGN WATER LEVEL

WATER-FEED CONTROL

STEAM CHANNEL

MAIN STEAM FLOW

WATER-FEED IMPELLER

WATER-FEED THROUGH BRAKE SHAFT

(a) Schmidt turbine. From Smith and Pearson (1950)

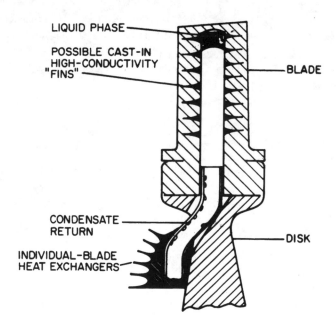

LIQUID PHASE

POSSIBLE CAST-IN HIGH-CONDUCTIVITY "FINS"

BLADE

CONDENSATE RETURN

INDIVIDUAL-BLADE HEAT EXCHANGERS

DISK

(b) Closed thermosiphon

Figure 10.26. Closed-circuit water-cooled blades

References

Abhari, R.S., Guenette, G.R., Epstein, A.H., and Giles, M.B. (1992). Comparison of time-resolved turbine-rotor-blade heat-transfer measurements and numerical calculations, Trans. ASME, Journal of Turbomachinery, vol. 114 p. 818–828, October.

Bayley, F.J., J.W. Cornforth, and A.B. Turner. (1973). Experiments on a transpiration-cooled combustion chamber. Proc. Inst. Mech. Engrs., London, UK 187:158–169.

Bayley, F.J., and G.S.H. Lock. (1964). Heat-transfer characteristics of the closed thermosyphon. Trans. ASME J. Heat Transfer, paper 64-HT-6, pp. 1–10. New York, NY.

Bayley, F.J., and A.B. Turner. (1970). Transpiration-cooled turbines. Proc. Inst. Mech. Engrs., London, UK. 185:943–951.

Brown, A., and B.W. Martin. (1974). A review of the bases of predicting heat transfer to gas turbine rotor blades. Gas Turbine Conference & Products Show, Zurich, Switzerland, March 30-April 4, 1974. ASME, New York, NY.

Caruvana, A., W.H. Day, G.A. Cincotta, and R.S. Rose. (1979). System status of the water-cooled gas turbine for the high-temperature turbine-technology program. Gas Turbine Conference & Exhibit & Solar Energy Conference, San Diego, Calif., March 12-15, 1979. ASME, New York, NY.

Cook, J.A., C.A. Fucinari, J.N. Lingscheit, and C.J. Rahnke (1978). Evaluation of advanced regenerator systems. DOE/NASA/0008/78/4, NASA report CR 159422, Cleveland, OH.

Dakin, J.T., M.W. Horner, A.J. Piekarski, and J. Triandafyllis. (1978). Heat transfer in the rotating blades of a water-cooled gas turbine. In Gas Turbine Heat Transfer. ASME, New York, NY.

Day, J.P. (1995). Extruded ceramic regenerator material/process development. Proc. Annual Automotive Technology Development Contractors' Coordination Meeting, US Dept of Energy, SAE publication P-239, Warrendale, PA.

Eckert, E.R.G. (1963). Introduction to heat and mass transfer. McGraw-Hill, New York, NY.

Giles, M.B. (1988). Calculation of unsteady wake/rotor interactions. AIAA Journal of Propulsion and Power, vol. 4, no. 4, July/August. Washington, DC.

Gladden, Herbert J. 1974. A cascade investigation of a convection- and film-cooled turbine vane made from radially stacked laminates. Technical memorandum TM X-3122. NASA, Washington, DC.

Hagler, Carla D., and Wilson, David Gordon (1988). Considerations for the design of high-effectiveness ceramic rotary regenerators for regenerated low-pressure-ratio gas-turbine engines. 24th Joint Propulsion Conference, AIAA paper 88–3191, Boston, MA.

Horner, M.W., D.P. Smith, W.H. Day, and A. Cohn. (1978). Development of a water-cooled gas turbine. Gas Turbine Conference & Products Show, London, April 9-13, 1978. ASME, New York, NY.

Kays, W.M., and A.L. London. (1984). Compact Heat Exchangers[5]. Third-edition. McGraw-Hill, New York, NY.

Korakianitis, T. (1993a). Prescribed-curvature-distribution airfoils for the preliminary geometric design of axial-turbomachinery cascades. Journal of Turbomachinery, vol. 115, no. 2, April, pp. 325–333, ASME, New York, NY.

---

[5]This book is being reprinted for 1998 publication by Kreiger Publishing Co., Melbourne, FL.

Korakianitis, T. and Papagiannidis, P. (1993b). Surface-curvature distribution effects on turbine-cascade performance. Journal of Turbomachinery, vol. 115, no. 2, April, pp. 334–342, ASME, New York, NY.

Korakianitis, T., Papagiannidis, P. and Vlachopoulos, N.E. (1993c). Unsteady-flow / quasi-steady heat transfer calculations on a turbine rotor and comparison with experiments. ASME Journal of Turbomachinery (in print). ASME paper 1993-GT-145, ASME, New York, NY.

LeBrocq, P.V., B.E. Launder, and C.H. Priddin. (1973). Experiments on transpiration cooling. Proc. Inst. Mech. Engrs., London, UK. 187:149–169.

London, A.L., M.B.O. Young, and J.E. Stang. (1970). Glass-ceramic surfaces, straight triangular passages-heat-transfer and flow-friction characteristics. ASME, J. Eng. Power, vol. A 92. New York, NY.

Martin, B.W., A. Brown, and S.E. Garrett. (1978). Heat-transfer to a PVD rotor blade at high subsonic passage-throat Mach numbers. Proc. Inst. Mech. Engrs., London, UK. vol. 192.

McDonald, C.F. (1997). Ceramic heat exchangers—the key to high efficiency in very small gas turbines. ASME paper 97-GT-463, New York, NY.

McDonald, C.F. and D.G. Wilson (1996). The utilization of recuperated and regenerated engine cycles for high-efficiency gas turbines in the 21st century. Jl. Appl. Thermal Energy vol. 16 nos. 8/9, pp. 635–653, Elsevier, London, UK.

Meitner, P.L. (1978). A computer program for full-coverage film-cooled blading analysis including the effects of a thermal barrier coating. In Gas Turbine Heat Transfer. ASME, New York, NY. pp. 31–38.

Smith, A.G. (1948). Heat flow in the gas turbine. Proc. Inst. Mech. Engrs., London, UK. vol. 159, WEP no. 41:245–254.

Smith, A.G., and R.D. Pearson. (1950). The cooled gas turbine. Proc. Inst. Mech. Engrs. London, UK. 153 (WEP no. 50):221–234.

Wilson, D.G. and Pfahnl, A. (1997). A look at the automotive-turbine regenerator system and proposals to improve performance and reduce cost. SAE paper 970237, Warrendale, PA.

Wilson, D.G. (1993). Heat exchangers containing a component capable of discontinuous movement. US patent no. 5,259,444, assigned to MIT, Cambridge, MA.

Wilson, D.G. (1966). A method of design for heat-exchanger inlet headers. Paper 66-WA/HT-41. ASME, New York, NY.

Wilson, D.G., and J.A. Pope. (1954). Convective heat transfer to gas-turbine blade surfaces. Proc. Inst. Mech. Engrs., London, UK. 168:861–876.

Wilson, D.G. (1952). The chordwise variation of heat-transfer coefficient across a typical gas-turbine blade in cascade. Ph.D. thesis, Mechanical Engineering department, University of Nottingham, UK.

Yeh, F.C., H.J. Gladden, J.W. Gaunter, and D.J. Gaunter. (1974). Comparison of cooling effectiveness of turbine vanes with and without film cooling. Technical memorandum TM X-3022. NASA, Washington, DC.

Problems

1. Would Reynolds analogy be useful to you if you wanted to use the specified pressure drop in a combustion chamber to calculate the heat transfer to the walls of the chamber? Give reasons.

2. Assuming that it is desirable to use low fluid velocities in heat-exchanger passages if pumping-power losses are to be minimized, and that it is desirable to use small hydraulic diameters in the passages if the heat exchanger volume is to be minimized, give some, say four, practical limits to how small you could specify the velocity and the hydraulic diameter in the heat exchanger of a gas-turbine engine.

3. (a) Calculate the entropy increase in each of several categories in the gas-turbine-engine heat exchanger, whose conditions are given in figure P10.3. The effectiveness is 0.9, the mass flow 1 kg/s; the specific heat can be taken as constant at 1,010 J/(kg · K) the gas constant is 286.96 J/(kg · K), and the fluids can be treated as air and as perfect gases. Use Gibbs' equation ($Tds = du + pdv$) to find the increase of entropy at constant pressure:

     from a to b,
     from e to f,
     from b to c,
     from f to g.

   (b) Add the two categories, then find the total, and comment.

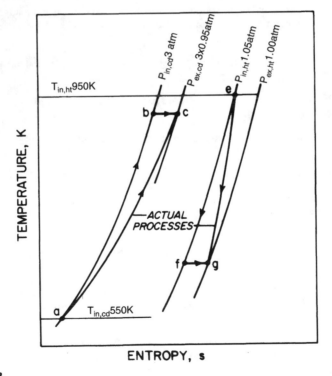

Figure P10.3

4. Using figure P10.4, find the effects on cycle thermal efficiency and specific power of cooling the first-stage turbine nozzle vanes and rotor blades of a heat-exchanger-cycle gas-turbine engine (CBEX) of the following specifications: $T_{0,1} = 540$ °R, $\dot{m} = 150$ lbm/s, $r = 5.0$, $T_{0,2} = 876.6$ °R, $T_{0,4} = 3,011.4$ °R, $\Sigma(\Delta p_0/p_0) = 0.12$, $\eta_{p,e} = 0.90$, and heat addition from fuel 314 Btu/lbm air, $p_{0,1} = 10^5$ N/m^2.

(a) The cooling air is bled from the compressor discharge. After leaving the turbine blades and vanes it neither contributes to the turbine work nor adds any losses.

(b) The vanes and blades must be maintained at a mean temperature of $1,800$ °F. The cooling air is fed to the blades at compressor-delivery temperature and discharges at $1,700$ °F.

(c) The hot gas approaches each blade row with a relative (axial) velocity of 600 ft/s and leaves with a relative velocity of 1,000 ft/s. The effective temperature of the gas is the static temperature at blade-row exit plus 0.88 of the difference between static and stagnation temperature (use the steady-flow energy equation).

(d) Assume a mean Nusselt number ($h_t c/k$) of 900 for both vanes and blades. To save effort, find the blade chord length based just on calculating the arithmetic mean diameter from the inner and outer diameters at nozzle-vane exit (or rotor-blade entrance). Use air properties. Specify the inner-to-outer diameter ratio as 0.85, the number of blades and vanes as 48 and 49, respectively, the space-chord ratio s/c as 0.75 for both, and the blade or vane perimeter as $2.5\,c$.

(e) Ignore the effects of fuel addition.

(f) Use 50-percent reaction.

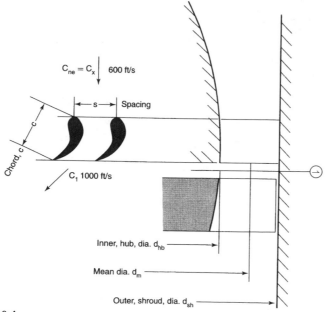

Figure P10.4

(g) Assume that 2/3 of the cycle pressure losses occur before the turbine.

(h) There is a 1-percent drop in stagnation pressure from nozzle entrance to nozzle exit.

5. Find the number of cooling holes one millimeter in diameter, and the quantity of compressor-delivery air used, to cool the first-stage rotor blades of the solar Brayton-cycle engine given in the example in section 10.6.

(a) A four-stage turbine is used of equal enthalpy drop per stage. The first-stage rotor has 13 blades at a mean-diameter solidity of 1.5. The mean hub-tip ratio is 0.85. The mean-diameter velocity diagram is that of the medium-work turbine in figure 5.11 ($R_n = 0.5$, $\psi = 1.0$, $\phi = 0.5$). This diagram may be used to find the blade speed, relative outlet velocity, Reynolds number, heat-transfer coefficient, and so forth.

(b) Use curve C, extrapolate if necessary, in figure 10.23 to obtain the mean Nusselt number of the external flow (based on chord and conductivity at local static conditions).

(c) Specify that the blade metal temperature shall be uniform at 1,200 K, for the purpose of calculating both the external and the internal heat transfer.

(d) Specify that the cooling holes make three passes in the blade. Specify also that the effectiveness of the crossflow heat exchanger, which the flow between the cooling air (bled from the compressor discharge) and the blade can be regarded as, shall be 85 percent. Use figure 10.4b to find the NTU required. Note that as with a constant blade temperature, $C_{ra} = 0$.

(e) The flow in the cooling holes will be laminar. Assume that the mean Nusselt number, based on diameter, is 3.65. (In fact Coriolis acceleration would affect the heat-transfer coefficient on different sides of the passage.) Specify that the flow will be metered into each hole so that the desired velocities are obtained.

(f) For the external heat-transfer area, use a ratio of perimeter to chord of 2.6.

(g) Use an expander (turbine) stagnation-to-static polytropic efficiency of 0.90. Ignore the effect of heat transfer on the turbine expansion for this first approximate calculation.

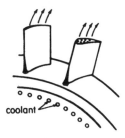

Figure P10.5

6. Find possible core dimensions for three alternative designs of ship-borne intercoolers for a main-propulsion gas turbine. Also find the pressure drops as proportions of the inlet stagnation pressure of each fluid. Common specifications for all designs are these:

      Air mass flow, 30 kg/s
      Air inlet pressure, 3 atm

Air inlet temperature, 411 K
Air outlet temperature, 300 K
Sea-water inlet temperature, 290 K
Sea-water outlet temperature, 310 K

The intercooler is to be designed as a three-pass cross-counter-flow unit. The three designs should use these surfaces:

(a) Compressed air inside straight tubes of 0.231 in bore (figure 10.7a). Water flows in three passes across staggered tube banks of 0.250-in outside diameter (figure 10.7b).

(b) The water flows in the tubes and the air outside the tubes of the first case.

(c) The water flows inside the tubes as in case (b), and the air flows in the same arrangement as in case (b) but the outside of the tubes are finned as in figure 10.7c.

Take all properties at the mean fluid temperatures of each side; neglect compressibility of the air; neglect any tube-wall temperature difference or fin temperature drop; number at the low end of turbulent flow, and a low water velocity of a very few feet per second. Go onto a first iteration, and comment on how you would change your selections if you were to go to a second iteration. Use the results of this first case to guide you in the second and third cases.

7. Find the ratio of the temperature differences $(\theta_{ht}/\theta_{cd})$ from the hot fluid to the wall θ_{ht} to that from the cold fluid to the wall θ_{cd} in a parallel-plate heat exchanger with laminar flow both sides. Use the data in table P10.7, and refer to the construction details in figure P10.7:

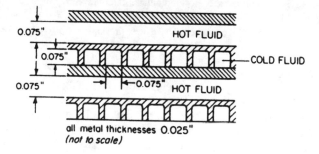

Figure P10.7

Table P10.7

	Hot	Cold
Temperature, °F	1,000	200
Pressure, atm	1.2	4.0
$N_u \equiv (h_t d_h)/k$	6.5	3.63
$R_e \equiv (\rho C d_h)/\mu$	1,500	900
k (btu $-$ ft)/(h ft² °F)	0.035	0.020
$\mu \times 10^5$ lbm/(s ft)	2.44	1.44

8. Make a preliminary design, and one subsequent design with your own chosen conditions of flow, for regenerative heat exchanger for a CBEX cycle with the specifications that follow. The outputs of such preliminary design are the outside diameter of the regenerator disk, the

axial thickness, the pressure losses, and the carry-over leakage (the air and gas trapped in the matrix). Cycle specifications:

$\dot{m} = 1.0$ lbm/s, $r = 5.0$, $T_{0,1} = 540$ °R, $T_{0,4} = 2,700$ °R, $\epsilon_x = 0.9$, $\Sigma(\Delta p_0/p_0) = 0.15$, $\eta_{p,e} = 0.80$

Some other suggested inputs are these. Use a capacity-rate ratio of 3.0 to find the NTU. Use the matrix surface the results for which are given in figure 10.7d. Take equilateral triangles for the passages with the wall thickness determined by a ratio of passage area to total matrix face of 0.75. Use a hydraulic diameter of 0.030 in. Take the specific heat of the ceramic-core material as 0.308 Btu/(lbm °R), and the density as 141 lbm/ft³. Choose the inner (open) diameter of the matrix to be 0.4 of the outer diameter, and the seal (radial) area to increase the face area by 20 percent.

For this preliminary calculation neglect all leakage as far as determining mass flows is concerned, and find the flow conditions for each fluid at the mean of the inlet and outlet temperatures for that fluid. Also use air properties for both fluids. These choices will lead to some inconsistencies with the temperatures previously calculated: you may ignore them or remove them as you wish. The effects will be small.

Take the pressure drops as the calculated core pressure drops plus two dynamic heads, calculated again on mean conditions. All of the following choices can be used for both hot and cold passages to be 0.02. Comment on the hot-cold conductance $(h_t A_h)$ ratio and the relative pressure-drop ratios $(\Delta P_o/P_o)$ as well as the regenerator-disk size which are given by this and the other choices. Then for your second calculation, make another choice of Mach numbers, Reynolds numbers, or conductances; calculate new outputs; and comment on them. (A pressure ratio of five is approximately the upper limit for rotary regenerators to have acceptable seal and carryover leakage. A lower pressure ratio would give unacceptably high temperatures at heat-exchanger inlet.)

9. Find the percentage of compressor air required to give a mean blade temperature of 1,600 °F in a turbine where the local effective gas temperature is 2,200 °F. The coolant enters the blade root at 600 °F and is discharged at 1,500°F. The blade Reynolds number of the external gas flow is 5×10^5, based on blade chord and the outlet velocity, giving a Stanton number of 0.0021. Use the blade and passage configuration shown in figure 10.16a. Simplify the calculations with such assumptions as equalizing the compressor and turbine mass flows. Take the ratio of perimeter to chord as 2.37 and the solidity (c/s) as 1.6. Describe the procedure you would follow, but do not make any calculations, to choose the hydraulic diameter and the configuration of the internal cooling passages.

10. The illustration (fig. P10.10) is of a type of heat exchanger, a circulating-pebble regenerator, that has been proposed for the steel industry, here applied to a gas turbine. Your task is to write down, in brief, the series of steps required to arrive at a viable design.

Ceramic spheres 10-mm diameter, are fed into an upper bed that is held up by a perforated plate. The turbine exhaust flows up through the bed as the spheres pass downwards through the bed (we'll assume that the motion is uniform and steady and controlled to be of approximately the same thermal capacity rate as that of the exhaust flow). The spheres are removed from the bottom of the bed by a motor-driven screw feeder and passed into a rather similar bed (but of differing dimensions) through which the compressed air from the compressor outlet passes. A feeder at the bottom of this bed re-circulates the cool spheres to the top of the upper bed.

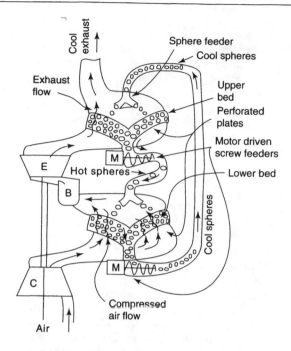

Fig. P10.10. Circulating-pebble regenerator

Assume that you have available the N_{tu} vs. effectiveness chart and the C_f and j (heat-transfer) factors for this type of heat exchanger and matrix. In this type of heat exchanger there are many degrees of freedom because each bed may have a different diameter and depth. Show how you would handle these freedoms—if in fact they are degrees of freedom—in your method of approach. Also comment on some advantages and disadvantages of this type of regenerator for land-based gas turbines compared with rotary ceramic-matrix regenerators.

11. To reduce the size of a counterflow heat exchanger, high fluid velocities may be used. Is energy thereby lost? Identify the location and character of any energy loss.

12. Explain, as if to a fellow engineering student, why a single-pass cross flow arrangement for a heat exchanger can give a high effectiveness when used for a water-cooled intercooler but not if used for the main cycle heat exchanger.

13. Calculate the proportion of the overall temperature drop (from hot fluid to cold fluid) that occurs in the hot-fluid boundary layer of a counterflow gas-turbine heat exchanger for each of two design philosophies. One has equal Reynolds numbers chosen at 3000, for which the cold-side mean velocity is 0.5 m/s. The other has both mean velocities set at 10 m/s.

The heat exchanger consists of parallel unfinned plates with alternate passages for hot and cold fluids. The capacity-rate ratio is approximately unity and the Stanton numbers are equal and constant in the relevant range. The density of the cold fluid is three times that of the hot fluid, but the Prandtl numbers, absolute viscosities, specific heats, and thermal conductivities can be taken as equal for the two fluids. The temperature drop in the walls can be neglected.

14. Calculate the number of tubes required for the following counterflow gas-turbine tubular heat exchanger to give an effectiveness of 0.95, and the hot and cold heat-transfer areas and the tube lengths. The tubes (carrying the compressed air) are 3-mm bore and 0.05-mm in wall thickness. The mean temperature difference between the air and the tube wall is 22.5 K. On the cold side the Nusselt number for the internal flow is 3.2; the air mass flow is 15 kg/s; the mean density is 3.5 kg/m^3; the mean specific heat is 1030 J/(kg · K); the mean viscosity is 0.00002675 kg/(m · s); the thermal conductivity is 0.0404 W/(m · K); and the mean air velocity is 6 m/s. The conductance ratio, hot-to-cold, is 2.0. The exhaust mass flow is 15.65 kg/s. The number of transfer units required is 13.

15. An internally cooled gas-turbine blade can be modeled as a parallel- or co-current-flow heat exchanger, as if the coolant were introduced inside the leading edge of the blade and passed towards the trailing edge. Sketch the temperature-length diagram of this heat-exchanger model with the following specifications. The hot-gas flow approaches the blade at 1800 K and leaves at 1758 K. The blade metal is at a uniform temperature of 1200 K. The coolant enters at 800 K and leaves at 1100 K. Your diagram should, therefore, have three temperature lines, for the hot gas, the blade metal, and for the coolant, plotted against the blade length, with the leading edge on the left and the trailing edge on the right.
(a) What is the value of $[(\dot{m}C_{p,mn}/\dot{m}C_{p,mx})]$? (b) What is the effectiveness of the blade as a heat exchanger? (c) Although figure 10.4a is not strictly appropriate for this type of parallel-flow heat exchanger, it will serve for the low effectiveness required. Using it, what is the approximate value of N_{tu} needed? (d) If you know the mass-flow rate and specific heat of the coolant, what relation would you use to give the overall conductance (overall heat-transfer coefficient times mean area)? (e) Give the equation relating this overall conductance to the external and internal conductances. (f) What could you do to reduce the blade-metal temperature insofar as the terms of this equation give you guidance?

Chapter 11

Gas-turbine starting and control-system principles

Because this text is concerned primarily with the design of turbomachinery, this chapter is merely of a background review nature and does not cover the design of control systems.

Gas-turbine control systems have three purposes (not all of which are pursued in all systems):

- to safeguard the integrity of the machinery;

- to reduce the operator's workload; and

- to choose optimum working points or transients between working points to minimize fuel consumption, or peak stress, or to optimize some other parameter.

The tendency is for engines to have increasingly sophisticated control systems as their design power levels increase, and for engines of any size to have more sophisticated control systems with time. Thus an early small engine made in Britain by Budworth in the 1950s was started by hand-cranking. When the speed was judged to be high enough, a lighted rag was pushed through a port into the combustor casing, and the fuel was turned on and adjusted to bring the engine to design speed. Presumably even this simple machine had an over-ride control to limit shaft speed so that a rotor burst could not take place.

Modern small engines have automatic starting systems that limit the temperature excursions that can occur under manual starting, and that control at least speed and turbine-inlet temperature during load transients and changes in inlet (atmospheric) temperature and pressure.

11.1 Starting

The reason for the importance of control of the starting sequence can be seen from figure 11.1. The discussion of the power relationship between compressor and turbine in a gas-turbine engine in chapter 1 points out that the normal positive net power pro-

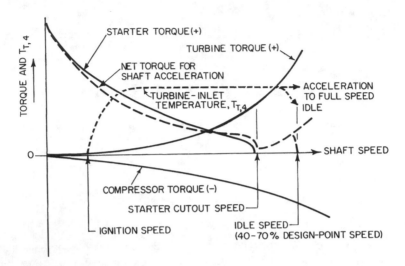

Figure 11.1. Torque input and output of a gas-turbine engine during starting

duced at design point is greatly affected by relatively small changes in the component efficiencies. In the extreme off-design conditions of starting, the compressor and turbine efficiencies are very low, and power must be supplied externally. To a first approximation, the contribution of the turbine power during the first, low-speed, phase of starting can be neglected. The starting device must supply the compressor power, which varies approximately as the cube of the shaft speed, and must also overcome the acceleration and friction torques. The design-point compressor power itself might, especially in older engines, be greater than the net output of a single-spool shaft-power engine. In other words, the compressor might take more than half the gross output of the turbine at full power. The required starter-motor power can therefore be very large, and there is a strong incentive to maximize the contribution from the turbine by using, during the starting transient, a high turbine-inlet temperature, often, at least in older engines, above the design-point value, to enable the maximum starter-motor power to be reduced. (The temperature is shown in figure 11.1 reaching its design-point value and remaining at this temperature until steady running has been achieved.)

The turbine can be run at a higher-than-design temperature for two reasons: the speed is low, so that the critical blade-root stresses will be low; and the duration is short. Nevertheless, the penalty in shortening of engine life from a few degrees further over-temperature is expensive. The starting control sequence should be able to guarantee that the chosen temperature schedule is not exceeded.

Author DGW remembers another penalty of exceeding a (rather undefined) temperature limit when he had the responsibility of starting, under manual control, an experimental 2.5-MW Brush gas turbine in the early 1950s. The starter motor had a nominal steady-state rating of 6 kW, but it was grossly over-driven and at the peak of the starting cycle it was producing over 60 kW. The windings increased in temperature very rapidly.

In his endeavor to prevent burnout of the overloaded starter motor, he frequently let the turbine temperature go too high, which pushed the compressor into nonrecoverable rotating stall, and completely prevented starting until after the shaft had been allowed to come almost to rest. (Nonrecoverable stall is discussed briefly in chapter 8.)

Modern high-temperature gas-turbine engines have less-critical energy balances because (a) the turbine-inlet temperature is so much higher than could be used earlier; (b) turbine output at design point is much larger than (in the region of twice) the compressor input; (c) and the cycle efficiency is much less sensitive to changes in compressor and turbine efficiencies, such as during starting, than is an engine with a low-temperature turbine, most of the output of which goes to drive the compressor.

Starting of multispool and variable-geometry machines

The previous discussion implicitly concerned single-spool engines. A "two-shaft" engine has a high-pressure turbine that drives the compressor, and a separate low-pressure or "power" turbine that drives the load (figure 11.2). (This category obviously applies only to shaft-power engines, including some turboprops and turbofans.) The starter is connected to just the "gasifier", or compressor-and-high-pressure-turbine, shaft. A two-spool engine has two compressors in series, each driven by a turbine, generally also in series. A three-spool engine is usually one in which a two-spool configuration has an added power turbine, possibly driving a fan. The Avco-Lycoming AGT-1500 engine for the US Army M1 tank is an example of a three-spool engine.

In all these multispool or multishaft cases, starting is carried out just on the high-pressure shaft. (Some starting systems are shown in figures 11.3-5.) In other words, the starter motor will spin only the high-pressure compressor and turbine, fuel will be added to the combustor between the compressor and the turbine, and this shaft will "start" rather as if it were a low-pressure-ratio engine. The exhaust gases will then begin to drive the lower-pressure turbine(s), which in turn drive the low-pressure compressor and possibly the fan, increasing the overall pressure ratio gradually to the design value. The control system that governs the scheduling of the fuel flow must account for the same basic parameters as in a single-shaft set, but many other parameters must also be taken into account. These are discussed later.

In a single-shaft machine with a compressor having variable settings for some of the stator rows, the control system has the further task of scheduling the changing settings of those stator blade rows that are adjustable. Normally these will be the low-pressure stators to between one-third and a half of the stages. Also, normally, the stators in an independently driven compressor are connected together by linkages so that one actuator varies all the stators in one action, the angle variation being greatest for the first-stage stators and decreasing towards the high-pressure end of the compressor. (The reasons for this variation distribution should be explained by the discussion in chapter 8 of off-design operation of compressors.) In early variable-stator compressors, the stators often had two settings, one for starting and, when a speed somewhat above idle had been reached, one for running. As pressure ratios have increased, the control system has been called upon to modulate the stator settings throughout the off-design range of operation.

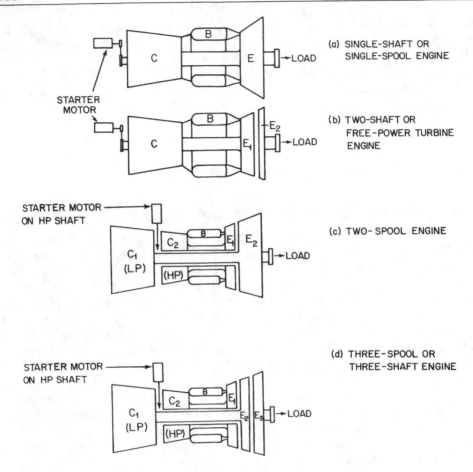

Figure 11.2. Single-spool and multispool engine arrangements

Starting and control of heat-exchanger engines

The discussions of chapter 3 showed that when a heat exchanger is incorporated to transfer turbine-exhaust heat to compressor-discharge air, the pressure ratio for maximum efficiency is considerably decreased. The pressure ratio chosen depends on the extent to which the design aims for low fuel consumption versus high power density, and on the design-point effectiveness of the heat exchanger. Thus we reviewed some ultra-high-efficiency cycles using a heat-exchanger effectiveness of over 95 percent, for which the optimum pressure ratio for maximizing cycle thermal efficiency was between 2.5 and 4. The designer's choice for high power density, on the other hand, is well illustrated by the Avco-Lycoming AGT 1500 engine mentioned above, having a heat exchanger of under 70 percent effectiveness and a design-point pressure ratio of almost 14:1. This engine will need all the control sophistication of an unrecuperated three-spool engine, plus those additional complications brought in by the use of a heat exchanger.

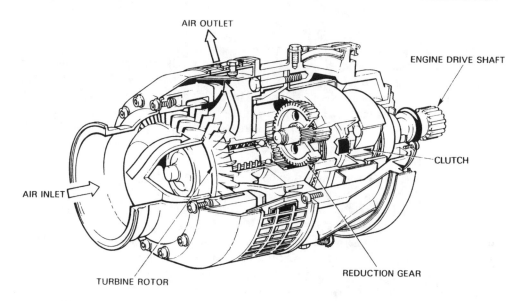

AIR OUTLET

ENGINE DRIVE SHAFT

CLUTCH

AIR INLET

REDUCTION GEAR

TURBINE ROTOR

Figure 11.3. Geared air-turbine starter. Courtesy Rolls-Royce plc

Two principal additional parameters must be monitored when a heat exchanger is used. The first is the heat-exchanger-inlet temperature. This is also the turbine-outlet temperature, and is normally measured as a surrogate for the turbine-inlet temperature. Metallic heat exchangers have temperature limits of under 1400 °F, 760 °C. If the engine pressure ratio and turbine-inlet temperature are chosen appropriately, the design-point turbine-outlet temperature can be designed to have a comfortable margin below this limiting value. However, during starting the turbine-inlet temperature may, as stated above, actually rise above the design-point value; and the temperature drop through the turbine in low-speed conditions will be considerably less than that at design point. Therefore the heat exchanger is at risk during starting, particularly in hot-day conditions, and the control system must be designed to protect it.

The second factor to be accounted for by the control system is the thermal capacity of the heat exchanger. During the initial phases of starting the heat exchanger will perform more as a heat absorber than a heat exchanger. More fuel will need to be added than if no heat exchanger is used, because the heat exchanger will initially actually cool the compressor-outlet air. Then, more or less rapidly depending on the design and thermal mass of the heat exchanger, design-point temperatures will be approached and a greatly reduced fuel flow will be needed to reach the same turbine-inlet temperature. At sudden decreases of load the converse is true: the compressed-air heat-exchanger-outlet temperature will fall only slowly, so that the reduction of fuel flow must be more rapid and go to a lower level than for an engine working on a non-recuperative cycle. Some limit on deceleration rate might be necessary to avoid flame extinction.

Starter types

Many commercial gas-turbine engines, including those on the smaller aircraft, are started by DC electric motors geared to the compressor (or the high-pressure-compressor) shaft. Larger jet and fan-jet engines are more frequently started by a geared air turbine connected to the accessory-drive gearbox. The air supply for the first engine started may come from a ground power unit or a small gas turbine called an APU or auxiliary power unit, to provide starting power in addition to cabin air-conditioning while on the ground. Power for the second and later engine turbine starters may then be bled from the compressor of the first engine. Occasionally the air for the first engine will come from stored compressed air on the aircraft. Sometimes, to minimize the storage requirement, the air is heated by a fuel-burning combustion chamber before admission to the turbine. Air usage will be inversely proportional to the absolute stagnation temperature of the air entering the turbine.

Some marine helicopters and other marine turbines use hydraulic-motor starters. A few large engines are started by a shaft-power small gas turbine (figure 11.4). A more frequent system for military aircraft is a cartridge starter, in which a monopropellant is used in place of an external supply of compressed air for a turbine starter.

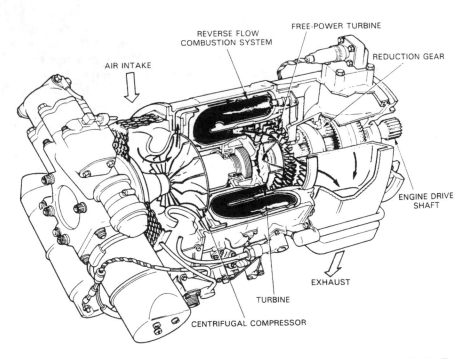

Figure 11.4. Small gas-turbine engine used as a starter. Courtesy Rolls-Royce plc

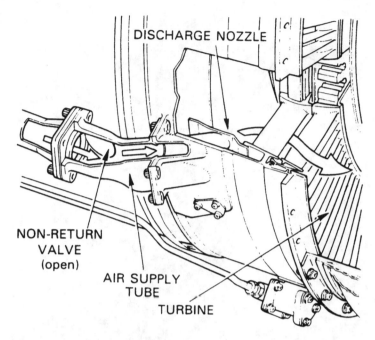

DISCHARGE NOZZLE

NON-RETURN VALVE (open)

AIR SUPPLY TUBE

TURBINE

Figure 11.5. Air-impingement starting. Courtesy Rolls-Royce plc

The principal alternative to gearing a starter motor to the compressor shaft is to direct a jet of air directly on to the blades of the compressor or high-pressure turbine (figure 11.5). One or more tangential nozzles coming through the shroud are used. This system is called "impingement starting" and has been used principally for small engines.

For aircraft in flight, restarting an engine after a "flame-out" or other shut-down can sometimes be accomplished through the "windmilling" of the high-pressure shaft from the ram effect of the inlet air.

11.2 Ignition systems

Modern engines are all ignited through some form of electric spark or discharge. The more common form is a high-energy (4–20 Joules) surface-discharge concentric spark plug placed close to the primary zone of the combustion system, but sufficiently removed so that during normal running it will not overheat (figure 11.6). A type more popular in earlier engines, particularly in Britain, was the "torch igniter", in which a lower-energy electrical discharge was used to ignite an auxiliary fuel jet. The resulting flame can be of such a length that the igniter can be situated at a greater distance from the primary zone. Nevertheless, the presence of stagnant fuel in the torch igniter requires a cooler environment to prevent cracking and coking, especially during "heat soak" after a sudden shut-down.

Normally two igniters are used in an engine, to give some redundancy. When separate flame tubes are used, neighboring tubes are connected by ducts near the primary-zone

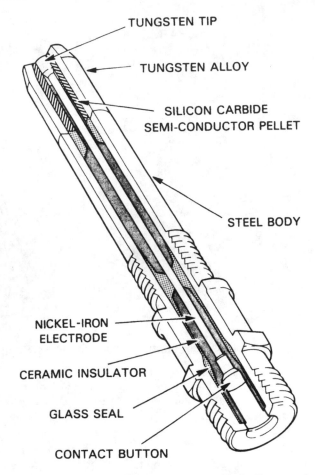

TUNGSTEN TIP

TUNGSTEN ALLOY

SILICON CARBIDE
SEMI-CONDUCTOR PELLET

STEEL BODY

NICKEL-IRON
ELECTRODE

CERAMIC INSULATOR

GLASS SEAL

CONTACT BUTTON

Figure 11.6. Igniter plug. Courtesy Rolls-Royce plc

vortex (see chapter 12) that form a circular ring. When the fuel in one chamber ignites, the high temperature produced leads to a reduced mass flow being able to pass through the associated turbine nozzles at the common pressure ratio, and flaming gases are diverted in both directions to the neighboring flame tubes. In this way ignition propagates very fast to all the separate flame tubes of the combustion system, and a new pressure-ratio balance with the nozzle flow is achieved. There is then again no flow through the interconnecting ducts.

In an annular combustion system cross ignition is direct, and no additional measures are needed.

The igniters are normally energized by the control system just after the starter is switched on, and before the fuel is admitted. The igniters are switched off after the engine has reached a speed at which it can accelerate itself, and before the starter motor or turbine is cut out.

11.3   Safety limits and control of running variables

While the fundamental control is that of fuel flow, the control system has interposed an increasing number of over-ride commands and auxiliary functions, such as the modulation of the linkage of a variable-geometry compressor or of turbine stators or exhaust nozzles, so that the direct connection between the operator's setting and the fuel-flow control has become somewhat remote.

The primary variable being sensed to provide fuel control is power output for industrial and marine shaft-power engines, and thrust, derived from pressures, for aircraft engines. Previously, thrust was derived from high-pressure-shaft speed, according to Harman (1981). The over-rides and auxiliary functions are designed to avoid operating-safety and engine-life limits.

Shaft speed(s)

Turbomachinery rotors of new design are normally spin-tested under vacuum in a heavily shielded "spin pit" at 20-percent overspeed. In some cases spin tests are used for production turbine disks or assembled rotors. The control system in a fully operational engine is designed to limit the overspeed to perhaps five percent. A shaft-mounted eccentric mass, spring loaded to hold to the shaft surface until the shaft speed reaches ten-percent overspeed, is often used as a final safeguard in the case of a control-system failure or some other emergency such as a shaft or coupling failure. The mass flying out to a new stable position at a larger radius (after reaching the "overspeed-trip" speed), puts into action fast-acting devices to cut off the fuel and perhaps to open bypass ducts.

Compressor surge

Compressor surge, particularly at high speed and pressure ratio, must be avoided under all circumstances. Surge is a condition of repeated flow breakdown (chapter 8) so catastrophic that it propagates back through the compressor, releasing the stored high-pressure air downstream of the compressor exit. In addition to causing very high cyclic stresses in the long low-pressure compressor blading, surge will almost certainly extinguish combustion: this is known as "flame-out". Many airplane losses have been ascribed to surge followed by flame-out.

The characteristics of a compressor working in the as-new condition and under nominal operating settings can be established in tests. The surge line is unfortunately greatly affected by many factors that change the operating conditions and the detailed configuration of the compressor. At altitude the Reynolds number will be much lower than in sea-level tests. The casing will change size with changes in temperature and pressure difference, and the rotor tips will change diameter with temperature and rotational speed. These transient changes will occur, in general, at different rates in the rotor and in the casing. Therefore the tip clearances will grow or shrink, considerably affecting the efficiency, particularly of the small high-pressure blade rows. These high-pressure blades will also be those most affected by erosion from ingested particles, making them more susceptible to stall as their profile deteriorates. It was argued in chapter 8 that high-speed surge is initiated by stall in the high-pressure stages. Foreign-object damage in the low-pressure or fan stages could also affect the surge line.

No one has yet been able to devise an instrument that could detect the onset of surge before it actually occurs and take avoiding action, although a group in the MIT Gas Turbine Laboratory is measuring instability waves that just precede a stall. Therefore the control system must limit engine operation, particularly during acceleration, to leave a considerable margin between the maximum-acceleration line and the nominal surge line.

Besides limiting the fuel input, the control system has, on many engines, other means of affecting the compressor characteristics. The setting angles of the compressor inlet guide vanes and sometimes of several rows of the low-pressure stators on either the low-pressure compressor or the high-pressure compressor or both are frequently adjustable. However, such variations are principally aimed at improving the low-speed stall conditions, as is apparent from the discussion in chapter 8. The exhaust nozzle on aircraft engines, particularly those for military aircraft employing afterburners (devices that allow fuel to be injected and ignited in the duct between the turbine exhaust and the propelling nozzle) is usually adjustable, and increasing the nozzle area moves the operating line away from surge. Some compressors incorporate one or two locations where flow can be bled off through rings of holes, or an annular slit, in or adjacent to a stator blade row. However, these are used principally to enable a high-pressure compressor to accomplish the low-speed starting phase without pushing the low-pressure stages into deep stall, and cannot be used to influence the surge line favorably.

Turbine temperature

The most critical temperature on the engine is the gas temperature at entry to the high-pressure nozzle blade row. Increase in this temperature improves engine efficiency and specific power. It also tends to drive the compressor towards surge. If the temperature of the rotor blades is also increased, which of course is highly likely despite the sophisticated cooling systems that are incorporated in modern turbine blades, then at least the life of the turbine will be shortened and, if the temperature increase is excessive, there will be a danger to the turbine integrity.

Because turbine-entry temperatures are as high as can be reached at the current state of the art, which includes application of the best possible cooling schemes, no temperature probe immersed in the flow and designed to measure gas-stream stagnation temperature can be relied on to have a long life. Attempts have been made to measure the speed of sound in the turbine-inlet gas stream, but the integrity of even wall-mounted or shielded transducers cannot be safeguarded in this hostile environment. On some engines optical pyrometers focussing on the high-pressure turbine blades themselves are used, with the lens-cell system well removed from the casing.

However, in general the temperature at outlet from the turbine is measured as a surrogate for the inlet temperature, and the control system calculates the inlet temperature from speed, mass-flow and pressure readings.

Other control functions

Many other requirements can be placed on the control system. For instance, rate of deceleration to idle may be limited if there is any possibility that flame instability and extinction could occur. Water injection, deicing, oil and fuel temperatures, variable inlet geometry, transition to reverse thrust, the switch to continuous ignition in severe precipitation, the removal of unburned fuel after a false start, and the limitation of compressor-outlet pressure to safeguard the casing integrity are all possible functions in advanced aircraft engines.

Types of control systems

The predominant form of control system has been hydromechanical, and has reached a high degree of specialization. Hydromechanical controls survived a short-lived challenge from fluidic systems. Future control systems are likely to be entirely electronic in their logic because of the great flexibility, versatility, reliability, and compactness which can be given.

Greenhalf (1980) noted that even industrial control systems have to withstand shade temperatures from 59 °C in parts of the Middle East to −50 °C in cold regions. Solid-state systems were considered the best able to withstand these extremes. Lewis (1987) stated the case for military aircraft: "Future engine controls in military aircraft will be microprocessor-based systems. Traditional hydromechanical systems do not have the computational and communication capabilities required to control next-generation engines". Davis et al. (1987) foresaw the eventual need for optical systems in certain cases: "Advanced integrated flight and propulsion control systems may require the use of optic technology to provide enhanced electromagnetic immunity and reduced weight".

References

Anon. (1986). The jet engine. Rolls-Royce plc, Derby, UK.

Davies, W.J., R.A. Baumbick, and R.W. Vizzini. (1987). Conceptual design of an optic-based engine-control system. ASME paper 87-GT-168, New York, NY.

Greenhalf, P.D. (1980). Recent developments on gas-turbine control systems. Jl. of Engg. for Power, vol. 102, pp. 249–256; trans. ASME, New York, NY.

Harman, Richard T.C. (1981). Gas-Turbine Engineering. John Wiley & Sons, New York, NY.

Lewis, Timothy J. (1987). Highly reliable, microprocessor-based engine control. AIAA/ SAE/ ASME/ ASEE 23rd Joint Propulsion Conference, San Diego, CA; AIAA, New York, NY.

Chapter 12

Combustion systems and combustion calculations

The fuel that is burned in the combustion systems of the most common form of gas-turbine engine, the aircraft jet engine, is liquid, a close relative to kerosine and Diesel oil. Radical changes in aircraft fuels are unlikely for the near future. For stationary gas turbines there is a continuing move from various grades of oil fuels to gas. New discoveries of natural gas, and the use of methods of recovering gas that was formerly burned in an open-air "flare", have led to an increase in the supply and a decrease in the price of natural gas. It is now being specified as the fuel for most new stationary turbines. This trend is substantially aided by the increasingly stringent regulations on the emissions of nitrogen oxides and other pollutants (in California, new plants must produce no more than 5 ppm NOx at 15% O_2). These limits can be met at present only in plants fueled by natural gas. Gas turbines have also been powered for many decades by gas collected at sewage-treatment plants, and the rather similar gas emitted during the anaerobic decomposition of solid wastes in landfills is being used for the same purpose by "capping" the larger landfills and drilling for gas.

Despite the advantages of gas fuels, most people concerned with planning our future power supplies believe that they must prepare for the time when natural gas again becomes in short supply and prices increase. Many countries, including the US, have huge quantities of coal, normally a very "dirty" fuel. Thus, although the cost of coal is relatively low, the cost of meeting environmental regulations is high, and only a tiny amount of coal is burned in gas turbines anywhere in the world. There are several programs, including those reviewed in chapter 3, aimed at burning coal, refuse-derived fuels, and biomass in general in stationary gas-turbine engines with low emissions and at low overall costs. A brief review of coal-combustion systems for gas turbines is included in this chapter.

Flame speeds

Before combustion systems are described it is helpful to consider the range of conditions under which aircraft combustion systems, these having the widest range of any application, must operate, and the very low flame speeds that must be taken into account for all gas-turbine applications. The following ranges are for all aircraft combustors, not necessarily the range for any one combustor:

> fuel-air mass ratio 0.002 to 0.020;
> inlet pressure 0.2 to 60 bar; and
> inlet temperature 245 to 1000 K.

These requirements seem particularly difficult when the low flame speeds that can be obtained are considered (figure 12.1).

The maximum speed normally reached in turbulent flames is shown in figure 12.1 to be little over 10 m/s. Typical axial velocities in axial-flow compressors are in the range of 125 to 200 m/s. An efficient diffuser at the outlet of an axial-flow compressor cannot be designed to reduce this velocity by much more than a half (chapter 4). Therefore even in the most favorable circumstances there will be a need for the velocity to be reduced to one-sixth of diffuser-outlet velocity if a flame of even maximum speed is not to be blown downstream. We do not yet know how to reduce fluid velocities by this amount efficiently. The designer is forced to reduce the flow velocity inefficiently, by means of various forms of baffle, to create a flame holder. As a result the combustion system is usually responsible for the largest single pressure loss in a gas-turbine engine.

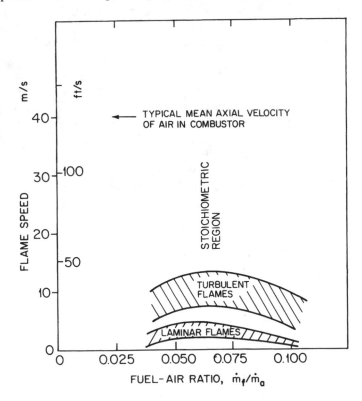

Figure 12.1. Flame speed versus fuel-air ratio

The flow must also be divided, because, as figure 12.1 shows, the highest flame speeds occur at near chemically correct ("stoichiometric") mixture strength. The gas turbine is the only major engine whose overall design-point fuel-air ratio is much weaker than stoichiometric.

12.1 Combustion-system types

There are five distinctly different types of combustion system that are being used in turbine engines or on which experiments are being conducted. We have called the first type (as used on aircraft engines) the "standard" type because of its widespread use and effective performance. It suffers from producing relatively high NOx levels, although these can be greatly reduced in stationary plant through water or steam injection (see below and chapter 3). NOx can also be reduced without water injection by "lean" combustion in the so-called "dry low-NOx" systems, although the lowest levels of NOx requires selective catalytic reduction (SCR) of the exhaust stream. The lowest levels of NOx appear to be given by catalytic combustors that are receiving much attention. Experimental pressure-gain combustion systems are also receiving considerable attention and government funding: one type is mentioned in section 3.10 and will not be further discussed here other than a reference to the work of Kentfield and O'Blenes (1988). Coal and other solid fuels are being successfully burned with low emissions in various types of fluidized bed. Coal-gasification producing a low-sulfur fuel (mentioned briefly in chapter 3) has been successfully developed; any combustion system that can burn natural gas can be easily adapted to this system.

Standard system: stoichiometric combustion plus dilution

This may be termed the "standard" or aircraft type of combustion system because it is also the lightest and most-compact. The incoming air is divided into two streams (figure 12.2a). The "primary" stream has just enough air to provide approximately the chemically correct amount required for combustion (i.e., it is a "stoichiometric" mixture.) It is injected in a direction to form a toroidal vortex. The fuel, liquid or gas, is normally injected at the axis of the vortex at a position and at a direction such that the fuel has maximum contact with the vortex. Combustion is initiated by a high-energy electric spark or by a pilot flame. Circulation within the vortex results in hot combustion products continuously igniting neighboring streams of fuel-air mixture even though they may be moving at well above flame speeds.

Typical forms of the standard combustion chamber are shown in figure 12.2. A combustion system may have a single large combustion chamber, or several smaller chambers in parallel, or an annular chamber with several burners. The cross section shown in figure 12.2a applies quite closely to all these types. Combustion takes place in an enclosure within the main stream, situated in many engines at the outlet of the compressor diffuser. Holes in the upstream part of the enclosure meter a proportion of the total flow to produce approximately stoichiometric conditions in this upstream region, which is called the "primary zone". Usually the enclosure (sometimes called the "flame

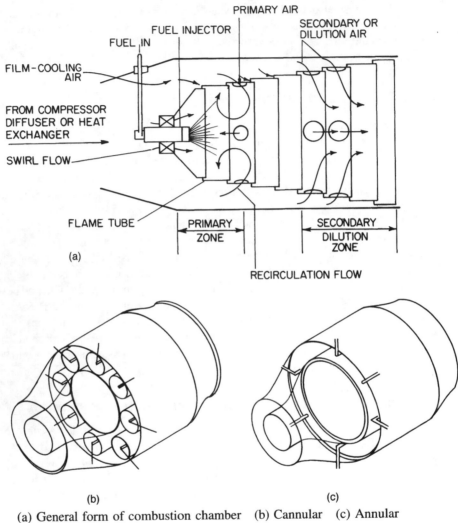

(a) General form of combustion chamber   (b) Cannular   (c) Annular

Figure 12.2. Combustor types

tube") is formed from light-gauge heat-resisting metal, tubular in most early engines and annular in most current aircraft engines.

Where the holes are placed is a matter of experience and experiment. About 15 percent of the total airflow is usually passed partly through a small ring of turning vanes surrounding the fuel nozzle and partly through small holes and slits in the upstream end of the flame tube. The purpose of this wall-cooling flow is to keep the material to well below the gas temperature and sometimes also to deflect the fuel spray from the walls. (The impact of liquid, or solid, fuel on a comparatively cool wall is very likely to cause

This simple-cycle single-shaft engine supplied ground power and compressed air for airplanes. A two-stage radial-flow compressor is followed by an annular combustor and a three-stage axial-flow turbine. Heat can be recovered from the exhaust for cabin heating. The air-conditioning radial-flow compressor is integral with the engine, and the shaft will also drive an electric generator.

Illustration. Garrett radial-axial GTCP331 auxiliary power unit. Courtesy Allied-Signal Aerospace

a rapid buildup of soot, or worse, hard coke, which can result in severe alterations of the air-flow pattern.) Another 15 percent of the air flow, the balance of the so-called "primary" or combustion air, is led through two or more holes downstream of the fuel spray to stabilize the flame and to provide recirculation of the hot combustion products with the incoming cool air and fuel.

This recirculation comes about because the swirling air coming around the nozzle, through which the fuel emerges also in a swirling conical spray, forms a toroidal vortex, which has a low-pressure zone and upstream recirculation flow in the same manner as in all intense vortices. The radial holes feed into this vortex recirculation.

The remainder of the air, called the "secondary" or dilution air, is added downstream, principally through large jets that can penetrate to the center of the hot gas stream. It is necessary that they produce sufficient mixing to give a subsequent temperature profile without high-temperature regions. Other secondary air is used for liner cooling.

Reduction of emissions through control of flame temperature

Lefebvre (1995) reviews the technology of low-emission combustion. The normal characteristics of gas-turbine combustion are shown in figure 12.3. Water or steam injection (mentioned in chapter 3) can be used to reduce the flame temperature. Lefebvre quotes Hung (1974) in predicting the reduction in NOx emissions as:

$$\frac{\text{(NOx with water injection)}}{\text{(NOx without water)}} = e^{-[0.2(m/m_F)^2 + 1.41(m/m_F)]}$$

where m/m_F is the mass ratio of injected water to fuel.

The combustor designer must be concerned about other emissions besides NOx, however. If the flame temperature drops far below 1700 K, carbon-monoxide emissions increase (figure 12.4). The temperature range to produce low emissions is, therefore, narrow: from about 1680 K to 1900 K (vs. 1000 K to 2500 K for conventional combustors, according to Lefebvre). Special measures must be taken to keep within this range over the full extent of operation of gas-turbine combustors (variations of load for sea-level engines, and this coupled with altitude variations and accelerations for aircraft engines). Mechanical changes in the combustor configuration (so-called "variable geometry" combustors) have been tried, such as in the first automotive-turbine (GM) engine to meet full emissions standards in the 1970s, and in some large industrial units in the 1990s. Manufacturers in general have preferred to use an alternative: "staged" combustion, in which multiple fuel burners within a combustion chamber are brought into action successively as the temperature rises above that for rapid NOx production. Each new burner lowers the "equivalence ratio" (the ratio of the fuel/air ratio to that for exact, stoichiometric, combustion) which in turn lowers the flame temperature un-

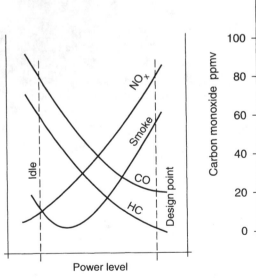

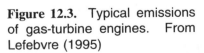

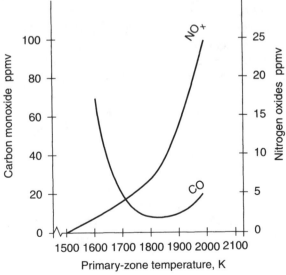

Figure 12.3. Typical emissions of gas-turbine engines. From Lefebvre (1995)

Figure 12.4. Influence of primary-zone temperature on CO and NOx emissions. From Lefebvre (1995)

til the load is increased to the point where another burner is fired. Staged combustion can also produce more-stable overall combustion because each upstream burner acts as an igniter for nearby or downstream burners. The use of staged combustion (for a full treatment of which the reader should refer to Lefebvre, op. cit.) coupled with lean premixed burners (introduced below) is necessary if low emissions are to be attained without water injection. Such systems are called "dry low-NOx" or "DLN" combustors.

Lean pre-mixed systems

The lean-premix systems operate as a consequence of the wide flammability limits that are found as the temperature of the reactants is increased. A sketch of the ABB silo combustor is shown in figure 12.5. The combustion air is brought counterflow along the outside of the flame tube, cooling it and giving maximum preheat to the air. This is mixed with the fuel in intense vortices formed in conical swirl tubes (figure 12.6). Combustion occurs at vortex breakdown downstream, away from the burner walls. Tests of this combustor showed an ability to reach below 10 ppmv NOx in optimal conditions (corrected to 15% oxygen) (Ainger and Mueller, 1992).

Catalytic combustors

Another method of avoiding "temperature overshoot" is to achieve reaction between fuel and air catalytically. The most-favored catalyst is palladium oxide on metal or ceramic (cordierite) supports (Dalla Betta et al., 1994). This catalyst has an operating range that requires the inlet mixture to be at least 725 K. At 1050 K the catalyst begins to decompose to the metal (op. cit.). The combustor described in the Dalla Betta paper is for the GE MS9001E turbine. A diagrammatic version is shown in figure 12.7.

To increase the temperature of the inlet flow to the minimum 725 K required a so-called "pre-burner" to be used. This is where the small quantity of NOx was produced. We assume that this small temperature rise will be produced in future by heat-exchange with downstream parts of the system, as is done with the lean-combustion systems discussed above.

The turbine-inlet temperature is, even for industrial turbines, well above the maximum tolerable catalyst temperature, requiring that continued combustion take place downstream of the catalyst. This will be similar to the combustion in the lean-combustion systems whereby combustion in very weak mixtures can take place if the starting temperature is sufficiently high. The test results showed overall NOx levels of 3-5 ppm, all produced in the pre-burner (op. cit.).

A further potential advantage of catalytic combustors is a pressure drop of 1-2.5%, lower than the 4% usually required in standard combustors (op. cit. and Fujii et al., 1996).

Fluidized-bed combustion systems

In fluidized-bed combustion systems, granular or lumpy solids such as coal, coke, sand etc. are confined within walls, usually circular, over a porous or perforated grate.

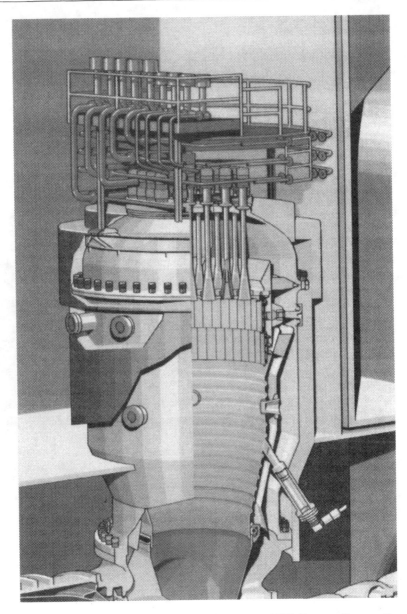

Figure 12.5. ABB "EV" silo combustor. Courtesy ABB Power Generation, Inc.

Combustion air is admitted through this grate at a sufficient velocity for the particles to become airborne, or "fluidized". If the granules cover a small range of sizes there will appear a top surface that will seem like the surface of a boiling liquid. This quasi-liquid can flow so long as the upflow is maintained, and one arrangement is that of a "circu-

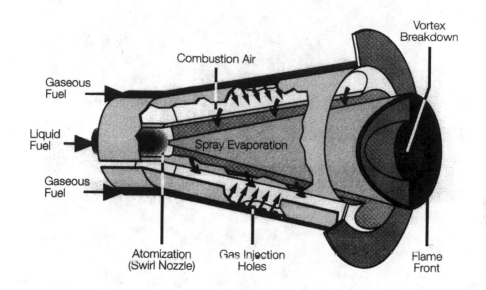

Figure 12.6. ABB low-NOx "EV" burner. Courtesy ABB Power Generation, Inc.

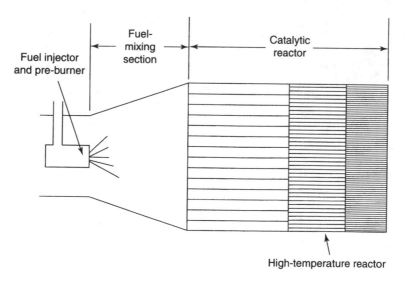

Figure 12.7. Catalytic combustor (diagrammatic)

This annular combustor liner is from a very successful military jet engine. At the top in this photo are the locations for the 18 fuel nozzles, surrounded by swirl stators. Much of the primary air is admitted through the larger "D"-shaped holes that produce jets to stabilize the recirculation vortex. The secondary-air holes are small, as would be expected in a high-temperature engine. We can see four rows of very small holes (five rows on the inside surface) to inject cooling films, and two distributed rings of cooling holes on the outer downstream wall. At the downstream end of the inner wall there is a ring of large oval holes: these may add cooling air to lower the temperature of the highly stressed rotor-blade roots. The pipe coming through the outer wall is one of two for igniters.

Illustration. Annular combustor liner of the GE F404 engine. Courtesy GE Aircraft Engines

lating fluidized bed" in which the motion is made more-or-less coherent to accomplish a circuit.

There are normally three types of particles in a fluidized bed: the fuel particles; reactants such as limestone or dolomite that react with sulfur in the fuel; and inert particles such as ash or sand that might be included (or allowed to remain) to assist in heat transfer to steam-generator tubes sunk in the bed, for instance.

The application of a pressurized circulating fluidized bed (PCFB) combustion system to coal-burning gas turbines is described by Pai (1996) in connection with the cycle shown in figure 12.8. This cycle follows the recommendations of Beer (MIT) in having the coal pass into a "carbonizer" to be split into a carbonaceous char and a (clean) fuel gas, which is burned in a low-NOx "multiannular swirl burner", figure 12.9, patented by

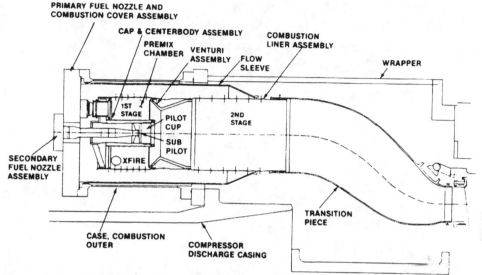

The use of staged combustion is well-illustrated in this diagram of the combustion system for a GE MS7001E turbine. Low NOx emissions can thus be reached without water or steam injection.

Illustration. Dry low-NOx combustion system

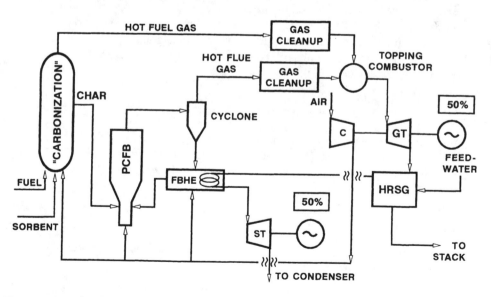

Figure 12.8. Cycle incorporating a pressurized circulating-fluidized-bed combustor burning coal and a "topping" combustor burning gas. From Pai (1996)

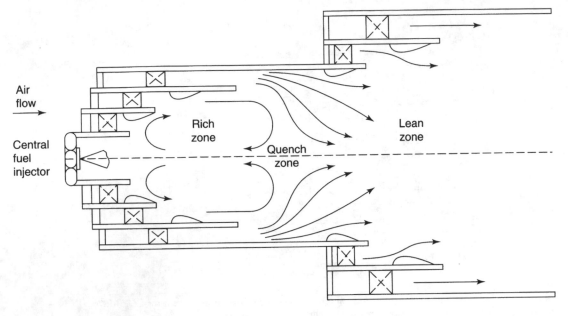

Figure 12.9. Low NOx multi-annular swirl burner. From Beer (1989)

Beer (1989) in a so-called "topping combustor". (See section 3.10 for the supplementary-fired exhaust-heated cycle, in which the use of a clean-fuel-fired topping combustor was also suggested by Beer.) The performance of this cycle is predicted as approaching 50%, figure 12.10, and is considerably better than that of the integrated gasification combined cycle (section 3.10) (op. cit.). Pai also states that the PCFB cycle would be considerably smaller and less expensive than the IGCC.

12.2 Conservation laws for combustion

In addition to the laws of conservation of energy and of mass, a third conservation law must be observed in combustion: the conservation of atomic species.

The determination of the molal (volume) analysis of the combustion products for the burning of a hydrocarbon fuel in air is simple if these rules are followed.

1. Find the number of moles of oxygen required to burn all the hydrogen in the fuel to H_2O and all the carbon to CO_2.

2. Add nitrogen to both sides of the equation, in the amount of 3.76 moles of nitrogen for each mole of oxygen added.

3. Add excess air (oxygen and nitrogen in the above proportions) to both sides of the equation to the amount specified.

For example, to find the molal analysis of the products of combustion of benzene (C_6H_6) with 400 percent theoretical air, first we write the equation with 100 percent

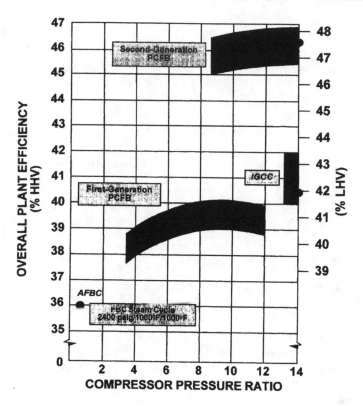

Figure 12.10. Predicted PCFB-cycle efficiencies. From Pai (1996)

theoretical oxygen:

$$C_6 H_6 + \text{oxygen} \rightarrow 6\,CO_2 + 3\,H_2O$$

Therefore,

$$C_6 H_6 + 7.5\,O_2 \rightarrow 6\,CO_2 + 3\,H_2O \qquad (12.1)$$

Then we add 100-percent theoretical nitrogen. We obtain the equation for the stoichiometric (theoretical, that is, no excess air) mixture burning in air. The principal components of air are nitrogen and oxygen, in proportion of 3.76 moles N_2 to one mole O_2:

$$C_6 H_6 + 7.5\,O_2 + 7.5(3.76)N_2 \rightarrow 6\,CO_2 + 3\,H_2O + 7.5(3.76)\,N_2 \qquad (12.2)$$

Therefore 1 kg O_2 is associated with $3.76 \times 28.016/32.0 = 3.292$ kg nitrogen.

Last, we multiply the air components by four to obtain the equation for 400-percent theoretical air:

$$C_6 H_6 + 30\,O_2 + 30(3.76)\,N_2 \rightarrow 6\,CO_2 + 3\,H_2O + 22.5\,O_2 + 112.8\,N_2 \qquad (12.3)$$

Molal analysis

The number of moles of products in the foregoing example is $(6+3+22.5+112.8) =$ 144.3. The molal proportions of the constituents are then as follows.

$$\text{The mole fraction of } CO_2 = \frac{6}{144.3} = 0.0416$$

$$\text{The mole fraction of } H_2O = \frac{3}{144.3} = 0.0208$$

$$\text{The mole fraction of } O_2 = \frac{22.5}{144.3} = 0.1559$$

$$\text{The mole fraction of } N_2 = \frac{112.8}{144.3} = 0.7817 \qquad (12.4)$$

The total of the mole fractions is = 1.0. This, then, is the molal analysis.

Dew point of water in exhaust

If the water is all in the vapor state (and, being in a small proportion, it can be treated as a perfect gas even when about to condense), the molal analysis is also the volume analysis and the partial-pressure analysis.

We can use this partial-pressure analysis to determine the temperature at which the water would start to condense. This temperature is significant, in some circumstances, for the following reason. It is usually desirable to recover as much heat from the exhaust of industrial gas turbines as possible, first in the cycle heat exchanger if the CBEX cycles is used, and/or in a waste-heat "boiler" or process heat exchanger. However, corrosion will result if the temperature is reduced to the point where water or acid could condense out, either in the heat exchanger or in the exhaust stack.

Suppose that the stagnation pressure of the exhaust in the example just given is $103.4 \, \text{kN/m}^2$ ($15 \, \text{lbf/in}^2$) absolute (a little above atmospheric pressure). Then the partial pressure of the water vapor is $0.0208 \times 103.4 = 2.151 \, \text{kN/m}^2$ ($0.312 \, \text{lbf/in}^2$) for the combustion of benzene with 400-percent theoretical air.

Steam tables will show that the condensing temperature for this pressure is about $18.3 \, °C$ ($65 \, °F$). In engines using higher turbine-inlet temperatures the excess air will be less, and the condensation temperature will be correspondingly higher. However, gas-turbine-exhaust condensation temperatures will still be considerably below those for steam plant or for any other combustion engine using approximately stoichiometric mixtures.

This constitutes therefore an advantage of Brayton-cycle engines in that it is easier and less expensive to recover heat from their exhaust than it is for other types of engine. This is true, however, only if there is no sulfur in the fuel. Sulfur oxides in combination with excess air raise the dewpoint temperature very significantly, producing liquid sulfuric acid at well above the dewpoint temperature. Exhaust ducts and heat exchangers can be quickly corroded in this way. In a steam plant it is traditional to limit the stack-gas temperature to a minimum of about $150 \, °C$ ($300 \, °F$), which constitutes a significant heat loss. The same limit applies to the burning of fuels containing sulfur in gas-turbine and other engines.

Lower and higher heating values of fuels

Because in general the latent heat in the water vapor in the exhaust is not recovered, Brayton-cycle efficiencies are calculated using the so-called "lower" or "net" heating value of the fuel, rather than the "higher" or "gross" heating value. The difference is given by calculating, from the combustion equation 12.3), the kg of water per kg of fuel, and multiplying this by the latent heat, or heat of vaporization, of water (2,465 kJ/kg at 288 K).

One mole of benzene has a mass of $(6 \times 12.01 + 3 \times 2.016) = 78.108$ kg. It produces three moles of water $(3 \times 18.016) = 540.48$ kg. Therefore 1 kg fuel yields 0.692 kg water, which has a heat of vaporization of 1,706 kJ/kg.

This, then, is the difference between the higher and the lower heating values for the combustion of benzene. For most hydrocarbon fuels, this difference is about 4 percent.

Adiabatic flame temperature

The fundamental method of finding the flame temperature after complete combustion is to add the enthalpies of the reactants and then to add the enthalpies of the products for a range of temperatures. The actual adiabatic temperature can then be found either by iteration or by graphic interpolation. It is absolutely necessary to use thermodynamic tables with a common base temperature for the various constituents or to make corrections for any tables with deviations. There are in fact several base temperatures in common use, and using mixed-base tables will give erroneous results.

If we use tables whose standard reference or base conditions are 25 °C (77 °F) and 1 atm, and if we further take the example where the reactants are initially at 25 °C, then the enthalpies of oxygen and nitrogen are, by definition, zero (for the reactants only) as shown in figure 12.11. We can then apply the first law for a steady-flow process:

$$\frac{Q_{in} - \dot{W}_{ex}}{\dot{m}} = 0 = \Delta_{rc}^{pd} h_0$$

Therefore,

$$\sum (\hat{n}\hat{h})_{rc} = \sum (\hat{n}\hat{h})_{pd}$$

where

$$\hat{n} \equiv \text{ number of moles} \equiv m/M_w$$
$$\hat{h} \equiv \text{ molal enthalpy}$$

Subscripts rc and pd refer to reactants and products respectively.

Let us use again the case of benzene burning with 400-percent excess air. For the reactants only the benzene has a finite enthalpy.

$$\hat{n} = 1 \text{ mole}$$
$$\hat{h} = 83.76 \text{ MJ/kg mole}$$
$$\sum (\hat{n}\hat{h})_{rc} = 83.76 \text{ MJ/kg mole} \qquad (12.5)$$

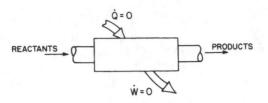

Figure 12.11. Diagramatic representation of combustion process

Table 12.1. Calculation of adiabatic flame temperature

Constituent	$\hat{n}$	[Btu/(lbm °R)] $\hat{h}$ at 1700 R	[Btu/(lbm °R)] $\hat{h}$ at 1800 R
CO_2	6	$-158,111.1$	$-154,821.0$
H_2O	3	$-93,780.9$	$-92,803.3$
O_2	22.5	$8,930.5$	$9,760.7$
N_2	112.8	$8,449.4$	$9,226.8$
	$\sum(\hat{n}\hat{h})_{pd}$	$-75,980.7$	$+53,062.9$

For the products we can find the value of $\sum(\hat{n}\hat{h})$ at various temperatures and interpolate (see table 12.1). By straight-line interpolation of the value of stagnation modal enthalpy of the reactants given in equation 12.5 the adiabatic temperature is estimated to be 1,786.5 °R (992.5 K).

Calculations of this type are tedious. Fortunately, hydrocarbon fuels have rather similar enthalpies of formation for the most part, and various methods have been developed for finding the adiabatic combustion temperature for standard fuels, with average mean specific heats or enthalpies of the products and with the ability to apply corrections for variations from the average or standard conditions. One method is given in the Gas Tables (Keenan and Kaye, 1948). Another particularly suited to the use of computers and calculators is by Chappell and Cockshutt (1974), which is summarized in appendix A1.

Combustion-system size and performance

The scaling laws for hydrocarbon-burning combustion (Clarke, 1955) lead to this relationship for the volume of combustion systems.

Heat release rate per unit volume:

$$\frac{\dot{Q}_b}{V_b} \propto \frac{p_{0,in}\sqrt{T_{0,in}}\sqrt{(\Delta p_0/p_{0,in})}}{(l_b/d_b)d_b} \tag{12.6}$$

where V_b, l_b, and d_b are the volume, length and diameter of a combustor.

Typical combustion-intensity data have been plotted in figure 12.12. Lines tentatively showing constant performance to agree with this scaling law have been drawn in. An important measure of performance is, however, missing: the temperature profile at combustor outlet. These data are for combustors having an approximately stoichiometric primary section, followed by a secondary dilution region in which the gas temperature is brought down to the permissible turbine-inlet temperature. This secondary section involves a large proportion of the combustor volume. Jets of dilution air must penetrate the hot stream from the primary section and mix out sufficiently well to give an acceptable temperature distribution at the turbine nozzles, figure 12.13. As the data are all for developed engine combustors, the outlet temperature profiles will have a somewhat higher temperature at the blade tips than at the blade roots where the centripetal stresses are maximum, so that an approximation to constant creep life along the length of the blade will be given (see also chapter 13 and figure 13.12). The temperature profile may be reversed, with higher temperatures at the inner diameter for blades that are strongly cooled, because the coolant temperature will be much lower at the hub where it enters the blade.

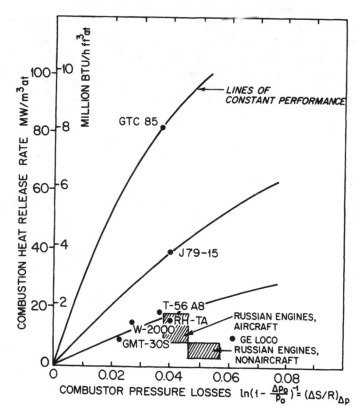

Figure 12.12. Combustion intensity versus losses

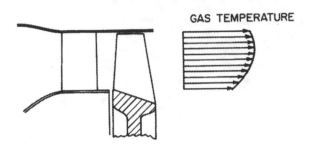

Figure 12.13. Desirable temperature distribution at combustor outlet

12.3 General combustor design

As intimated, combustor design involves scientific general principles, but there is a very large necessary component of experience, judgment, and development. Many aspects of combustor design and performance have been given little or no mention here, such as initial ignition of the fuel, and re-ignition at high altitudes in an aircraft engine; flame holding during compressor surges and other transients; flame-tube cooling, thermal stresses, and temperature transients; and many others. We cannot do justice to these topics in a book concerned with preliminary design, and we refer readers to the references for further guidance. A particularly useful general text is Lefebvre, *Gas-Turbine Combustion* (1983). Some aspects of starting are briefly discussed in chapter 11 of Lefebvre.

References

Aigner, M., and G. Mueller (1992). Second-generation low-emission combustors for ABB gas turbines: field measurements with GT11N-EV. Paper 92-GT-322, ASME, New York, NY.

Beer, J.M. (1989). US patent no. 4845940.

Borghi, Roland, Francis Hirsinger, and Helene Tichtinsky. (1978). Methodes disponibles a l'ONERA pour le calcul des chambres de combustion. In Entropie, 18:3–14. Paris, France.

Chappell, M.S., and E.P. Cockshutt. (1974). Gas-turbine cycle calculations: Thermodyanic data tables for air and combustion products for three systems of units. Report NRC 14300. National Research Council, Ottawa, Canada.

Clarke, J.S. (1956). A review of some combustion problems associated with aero gas turbine. J. Roy. Aero. Soc., London, UK. 60:221–240.

Deacon, W. (1969). A survey of the current state of the art in gas-turbine combustion-chamber design. Paper 3. Symposium on Technical Advances in Gas-Turbine Design. Inst. Mech. Eng., London, UK.

Dalla Betta, Ralph A., et al. (1996). Development of a catalytic combustor for a heavy-duty utility gas turbine. Paper no. 96-GT-485, ASME, New York, NY.

Dooley, Philip G. (1964). Design and development of combustion chambers for turbine engines. Paper 64-WA/GTP-8. ASME, New York, NY.

Graves, Charles C., and Grobman, Jack S. (1957). Theoretical analysis of total-pressure loss and airflow distribution for tubular turbojet combustors with constant annulus and liner cross-sectional areas. Report 1373. NACA, Washington, DC.

Fujii, T., et al. (1996). High-pressure test results of a catalytic combustor for gas turbine. Paper no. 96-GT-382, ASME, New York, NY.

Hazard, Herbert R. (1976). Combustor design. In Sawyer's gas-turbine engineering handbook. vol. 1. 2d ed. Gas-Turbine Publications, Inc., Stamford, CT.

Hung, W.S.Y. (1974). Accurate method of predicting the effect of humidity or injected water on NOx emissions from industrial gas turbines. ASME paper 74-WA/GT-6, New York, NY.

Keenan, Joseph H., and Joseph Kaye (1948). Gas Tables. Wiley, New York, NY.

Kentfield, J.A.C., and M. O'Blenes (1988). Methods for achieving a combustion-driven pressure gain in gas turbines. Jl. of Engg. for Gas Turbines & Power vol. 110, ASME, New York, NY.

Lefebvre, A.H. (1995). The role of fuel preparation in low-emissions combustion. ASME paper 95-GT-465, New York, NY.

Lefebvre, Arthur H. (1983). Gas-Turbine Combustion. Hemisphere, New York, NY.

Lefebvre, A.H., and E.R. Norster (1969). The design of tubular gas-turbine combustion chambers for optimum mixing performance. Paper 15. Symposium on Technical Advances in Gas-Turbine Design. Inst. Mech. Eng., London, UK.

Pai, David H. (1996). Advanced pressurized fluidized-bed combustion. Heat Engineering, vol. LIX, no. 4, Foster-Wheeler Corp., Clinton, NJ.

Pechelkin, Yu. M. (1971). Combustion chambers of gas-turbine Engines. Transl. from Russian. N71-38542, AD 727 960 Foreign-Technology Div. Wright-Patterson AFB, OH.

Sattelmayer, T., et al. (1992). Second-generation low-emission combustors for ABB gas turbines: burner development and tests at atmospheric pressure. Jl. Engg. for Gas Turbines & Power, vol. 114, ASME, New York, NY.

Sullivan, D.A. (1974). Gas-turbine combustor analysis. Paper 74-WA/GT-2. ASME, New York, NY.

Szaniszlo. A.J. (1979). The advanced low-emissions catalytic-combustor program: Phase 1-Description and status. Paper 79-GT-192. ASME, New York, NY.

Tacina, Robert R., and Jack Grobman. (1969). Analysis of total-pressure loss and airflow distribution for annular gas-turbine combustors. Technical note TN D-5385. NASA, Cleveland, OH.

Problems

1. Why is it necessary, at the present state of combustion technology, to produce a near-stoichiometric mixture in a gas-turbine combustion chamber? What new developments could relax this requirement?

2. Since the air velocity leaving a compressor diffuser and entering a combustion chamber is in the range of 125 to 225 m/s, why doesn't the flame blow out?

3. By writing down the equation for the complete combustion of ethyl alcohol (C_2H_5OH) in 250 percent theoretical air, find the percentages by volume of oxygen and nitrogen in the high-temperature combustion products. You may take air as a mixture of one mole oxygen to 3.76 moles nitrogen. Ignore dissociation effects.

4. If the products of combustion found in problem 12.3 were expanded through a turbine and were passed into a waste-heat boiler at 14.8 lbf/in² absolute, what would be the dew-point temperature at which the water in the mixture would start to condense? Figure P12.4 shows a partial chart of the dry-saturated temperature and pressure of water vapor. (If there were sulfur in the fuel, sulfuric-acid condensation could occur at temperatures well above the water dew point.)

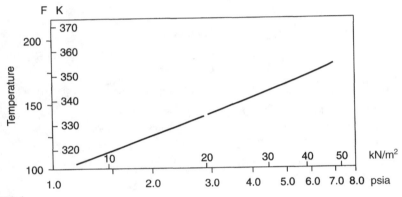

Figure P12.4

5. Describe briefly the steps you would use to find the adiabatic temperature of the combustion products of problem 12.3, given the initial temperature of the reactants and the gross or higher calorific value of the ethyl alcohol.

6. Calculate the final adiabatic combustion temperature, and the higher heating (calorific) value of the fuel, for the combustion of C_7H_{16} with 200 percent excess air (three times the amount of air necessary for combustion). The lower heating value of the fuel is 19,000 Btu/lbm, the initial temperature of the reactants is 100 °F and the mean C_p of the products can be taken as 0.26 Btu/lbm °F over the temperature range concerned.

7. Write the combustion equation for the combustion of normal octane, C_8H_{18}, with 300 percent theoretical air. (b) By adding atomic weights, find the ratio (mass of products/mass of fuel). (c) Find the difference between the so-called higher and lower heating values of octane per pound of fuel if this difference is due only to the latent heat of water, which is 1,050 Btu/lbm. (The higher heating value of a fuel is used if the water in the products is all condensed, and the lower heating value is used if the water is not condensed.) The atomic weights needed are carbon, 12.011; hydrogen, 1.008; oxygen, 16.00; nitrogen, 14.008. In air, each mole of oxygen is accompanied by 3.76 moles of nitrogen.

8. Calculate the percentage of carbon dioxide by volume in the room-temperature dry products of combustion of kerosene, $C_{12}H_{26}$, and 250% theoretical air. (Each molecule of oxygen in air is accompanied by 3.76 molecules, or volumes, of nitrogen.)

Chapter 13

Mechanical-design considerations

In this chapter we briefly review some of the unique aspects of turbomachinery design, in particular those aspects that influence, and are influenced by, the aerothermodynamic design. We first look at some of the choices for the overall machine. We then look at design criteria, material selection and forming techniques, and vibrations. At the end of the chapter we discuss some engine designs. It is not possible to do more than bring some of these aspects to the reader's attention for deeper analysis, guided perhaps by other treatises. Mechanical design of turbomachinery is highly specialized, and in many respects it is at the pinnacle of the engineer's art in any field. Even today, design criteria and methods are still changing, but the changes that have already occurred have greatly improved the tools available to the designer.

13.1 Overall design choices

Our starting point is after the completion of the preliminary thermodynamic and fluid-mechanic design. Let us use a gas-turbine engine as an example, because it is more complex than most single turbomachines. The overall cycle (simple, heat-exchanger, intercooled and so forth) will have been chosen, together with the compressor pressure ratio(s) and temperatures. The number of shafts is an important choice that may have already been made. This choice can be made after the thermodynamics of the cycle have been decided. The compressor and turbine design might, however, be affected by the choice of engine configuration. For instance, in a two- or three-spool machine the turbines must be chosen to drive the associated compressors. Here are some considerations.

Number of shafts

In a single-shaft set, the turbine (or turbines) drive the compressor (or compressors) and also the load (figure 11.2a). The exception is a non-fan jet engine: there is no shaft load (except for auxiliaries like oil and fuel pumps) and the turbine exhausts at a relatively high pressure. This higher-pressure flow exhausts through the propelling nozzle at the aft end of the engine.

A single-shaft set may have an intercooled compressor and possibly a reheated turbine (i.e., a second combustor between turbine stages or groups of stages), although it is rather unlikely for the reasons connected with compressor stability given below.

Single-shaft versus two-shaft designs

The usual initial choice between a one-shaft or two-shaft configurations has nothing to do with compressor stability. In the usual two-shaft configuration, one shaft connects the compressor and the part of the turbine driving the compressor (figure 11.2b). This combination is sometimes called the "gasifier" because it supplies hot gas to the so-called "power turbine" in the same way as aircraft jet engines have been occasionally used as standby units to supply gas to a separate power turbine. It is also called the "high-pressure spool" or the "core engine".

This type of two-shaft arrangement has been used for most experimental automotive gas turbines because of the great flexibility it confers on the power turbine. This can go from a stationary high-torque state to the design-point speed up to a high-speed zero-torque condition with almost no effects on the high-pressure spool. The use of variable-setting nozzle vanes gives further control of the torque-speed characteristic. A two-shaft arrangement is, therefore, equivalent to having an efficient "torque converter" on the engine.

The disadvantages to this arrangement, compared with a single-shaft design, are that one or more additional bearings in a hot environment are required; and the turbine must be split into two turbines, one of which must produce a torque that matches that of the compressor. The machine therefore becomes larger and more expensive. Manufacturers of modern experimental automotive gas turbines have tended towards single-shaft designs, relying on the promise of continuously variable transmissions or of electric drive to produce the coupling of a high-speed shaft to wheels that must start from rest.

Two-spool designs

This term refers to an arrangement in which two compressors are separately driven by two turbines (figure 11.2c). The term was used first for aircraft engines in which the shafts are coaxial, the high-pressure turbine driving the high-pressure compressor and the low-pressure turbine driving the low-pressure compressor and possibly the fan. There is sometimes a third shaft by which a lower-pressure power turbine drives the load (figure 11.2d). In the Rolls-Royce RB-211 fan-jet engine the load is the propulsion fan, whereas in the Lycoming AGT-1500 engine the load is the transmission going to the tracks of an army tank. The terms "two-spool" and "three-spool" are nowadays sometimes applied to arrangements in which the "spools" are not coaxial.

The reasons for going to a multi-spool design come almost entirely from the type of compressor instability that is found as the design-point pressure ratio is raised, as described in chapter 8. Compressors of pressure ratio up to about twelve to fifteen to one can be started and run successfully using "bleed" or "blow-off" valves to keep the early stages out of stall during starting. At this pressure-ratio range designers must contemplate choosing either two disconnected spools each having a compressor with close to the square root of the overall pressure ratio, the starting of which would be no problem; or the use of a single-shaft design with the first few rows of compressor stator vanes having variable setting angles. These alternatives suffice (with blow-off valves) to enable overall pressure ratios of up to around 25:1 to be handled. At higher pressure ratios designers must use multi-spool compressors together with variable-stator designs.

As designers of high-performance aircraft engines contemplate using design-point pressure ratios of 50:1 and 100:1 the compressor-stability problem becomes extremely challenging. We assume that splitting the compressor among three spools would be necessary. The individual spool pressure ratio could then be below 5:1. The blade length in a concentric high-pressure spool would be so small that the polytropic efficiencies of these stages would be low (because of tip-clearance and hub boundary-layer losses). The third spool is therefore likely to be on a small very-high-speed shaft "outside" the low- and medium-pressure spools. The third spool will include the combustion system, which will also be small because of the high-pressure environment. To start the engine only this high-pressure spool need be rotated, requiring much less energy than if the whole rotating system had to be brought up to starting speed. This is an advantage of multi-spool designs that partially offsets the considerable cost in complex components and time of developing a viable unit.

Rotor design

Very small engines (under 25 kW)

In very small engines (e.g., below 25 kW) both compressor and turbine are usually radial-flow machines. The rotor of the centrifugal compressor is normally connected to that of the radial-inflow turbine to form what is sometimes termed a "monorotor" (figure 13.1). This implies that a single rotor forging or casting is used, as is sometimes the case. There are thermal and stability advantages and disadvantages in having a single piece of metal. The compressor rotor is unduly heated by the turbine rotor—but this might permit a higher gas temperature to be used at turbine entrance, increasing both efficiency and power output.

Monorotor designs, whether or not the rotor is in one piece, are usually "overhung": that is, they are supported by a pair of bearings on the "cold" (compressor) side. This degree of shaft overhang coupled with the thermal growth of the rotor and stator as it goes through transients can lead to shaft-instability problems (forms of "whirling") and to the need for high blade-to-casing clearances. These clearances reduce the rotor efficiency considerably. These clearances are increased if journal bearings are used. Hence, most small turbines, and all aircraft-engine bearings, are rolling-element, ball and roller.

Small engines (30—300 kW)

Engines in this size range have centrifugal compressors and radial-inflow turbines, at least at the lower end of the range (figure 13.2). Single-stage axial-flow turbines are an alternative throughout the range and are almost mandatory for the larger power levels. (One reason is that the massive hubs required in radial-inflow turbines produce,

This small simple-cycle power unit has overhung back-to-back impellers, coupled to what is sometimes called a "monorotor". There is a single offset combustion chamber, feeding the turbine-inlet plenum tangentially to give preliminary swirl to the flow into the nozzle blades. As is usual with small engines the reduction gearing, controls, and starter are larger, heavier and probably more costly than the basic gas-turbine engine.

Figure 13.1. Example of a monorotor design, the Garrett GTC36-50/100. Courtesy Garrett Turbine Engine Company, now Allied Signal Aerospace

in turbines of above some limiting size, high transient thermal stresses that can lead to low-cycle-fatigue failures. Also the rotating inertia is much higher for radial than for axial turbines (figure 9.4)). The bearings are normally well separated: one almost inevitably upstream of the compressor inlet; and one either upstream or downstream of the turbine rotor, usually rolling-element. The upstream location is needed for radial-inflow turbines, which do not carry a downstream shaft, and is often used for axial turbines despite the high-temperature environment resulting from the closeness of the hot inlet flow.

Moderate-power engines (250—1000 kW)

This power range includes many helicopter engines, auxiliary power units and "ground" power units for aircraft servicing, and experimental truck engines. At the lower end of

This all-radial-flow single-shaft machine is unusual in several respects. The impellers are not back-to-back. The arrangement used enabled a single central combusion chamber to be used. It also required special measures, for instance the circumferential seal near the periphery of the compressor backface, to reach approximate axial-thrust balance. The shaft and the compressor rotor are metallic; much of the rest of the engine is ceramic. One rotary ceramic regenerator is used (versus two for many prototype automotive engines) and it is annular, surrounding the combustor, instead of the usual solid-disk form. The design-point peripheral speed of the ceramic rotor was over 700 m/s in the initial configuration. This required such a high centripetal acceleration of any particulates in the gas flow that the rotor was frequently eroded. A later design used a mixed-flow rotor with low-reaction blades, allowing a substantial reduction in blade speed and the near-elimination of the erosion problem. The radial diffuser following the radial-inflow turbine recovers much of the swirl kinetic energy that will occur at off-design operation.

Figure 13.2. The AGT101 100-hp automotive gas-turbine engine. Courtesy Garrett Turbine Engine Company, now Allied signal

the range the compressors are centrifugal, while at the upper end of the range there are fully axial compressors. In the past there have been popular engines with two centrifugal stages in this range, the best known probably being the Rolls-Royce "Dart". Modern engines are more likely, however, to be axial-radial; i.e., several axial-flow stages followed by a single radial-flow high-pressure stage (figure 13.3). The turbines of these engines will be universally axial-flow. The pressure ratios will generally be below 10:1 because the turbine blades are too small to be air- or steam-cooled. Accordingly, the turbine-rotor inlet temperature is likely to be too low to make a high pressure ratio viable. Ceramic turbine blades may soon become feasible for this size range, but the compressor pressure ratios are unlikely to increase greatly because of the very small size of the high-pressure blading that would result, with consequent high losses.

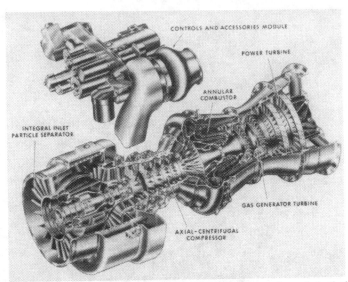

This is a "straight" aero engine when used for helicopter propulsion, and an "aero-derivative" when used for other purposes. The compressor has five axial stages followed by a radial-outflow stage. The setting angles of the inlet guide vanes and of the first two axial stator-vane rows can be varied. Two axial-flow turbine stages drive the compressor, and there is a two-stage power turbine driving the output shaft. The compressor rotor is built up from disks engaged as radial-serration couplings at the outer diameter, giving great rigidity and concentricity (with precise machining). There is a conventional ball-thrust bearing at the low-pressure end of the compressor, but the location of the roller bearing inboard of the gas-generator turbine (and wholly enclosed by the annular combustor) is highly unusual. Obviously, great pains have been taken to cool and drain the bearing cavity.

Figure 13.3. T-700 shaft-power engine. Courtesy GE Aircraft Engines

Therefore the turbines will have one stage for the lower-pressure-ratio single-shaft units, two or possibly three stages for all two-shaft units, and three stages for the higher-power higher-pressure-ratio engines. Aircraft engines will universally have rotors assembled from individual disks, one per stage in both the compressor and the turbine. The compressor blades are thin, and often made of titanium alloy so that the disks are likewise of very thin section. Turbine blades are normally of dense high-nickel-chromium-cobalt alloys of substantial section. At least the lower-pressure blades are likely to be solid, so that the turbine disks must be more substantial. Some air cooling of the first-stage turbine blades is, however, possible at the higher power levels in this range. Compressor disks are sometimes welded together near the periphery using electron-beam or laser welding. Turbine disks are normally separate and carry some form of radial spline on their two faces. These splines engage mating splines on the intermediate disks that space the rotor disks apart and provide a sealing surface for the turbine stators (nozzles). A stub shaft can also carry mating splines so that an assembly of disks and stub shaft (that includes the bearing) can be held together by one or more central through bolts. This form of construction, that can also be used for multistage axial compressors, ensures exact concentricity and alignment (assuming that machining tolerances are adhered to).

Industrial machines usually employ thicker sections for the rotating and stationary parts, The compressor rotor may often be a forged or cast drum carrying circumferential dovetail grooves for the rotor blades. As the power level of interest is increased, so there is a greater likelihood that the industrial machines will use journal ("plain") radial and thrust bearings.

Large machines (over 1000 kW)

Compressors will be all-axial. The turbine blades for machines of perhaps 5 MW upwards can be fully cooled with air or steam so that the maximum state-of-the-art turbine-inlet temperatures may be used. Industrial turbines will almost universally use plain bearings (figure 13.4). It is likely that magnetic bearings will take over both for industrial and aircraft bearings once their reliability can be assured. The absence of an oil-cooling circuit, the greater shaft stability, the small blade-tip clearances that are made possible, and the negligible level of bearing drag all have favorable effects on the overall safety and effectiveness of gas turbines.

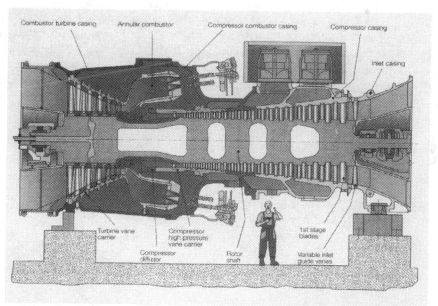

The design of this engine has many noteworthy features. It is in sharp contrast to the "aero-derivative" approach. There are only two plain bearings. The rotor is constructed from welded forgings. It includes a 20-stage axial compressor of 15:1 pressure ratio with no variable stators apart from the variable inlet guide vanes. Starting is accomplished using three "blow-off" manifolds (see chapter 8). Both the compressor and the turbine have constant inner diameters without extreme hub-shroud diameter ratios (chapter 7). An annular combustor with low-NOx "EV" burners is used giving NOx emissions below 25 ppm (dry) burning natural gas (see chapter 12). The simple-cycle efficiency is about 35.5% (in 1997).

Figure 13.4. The ABB GT13E2 164-MW gas turbine. Courtesy ABB Power Generation, Inc.

Stator design

The aspects of mechanical design that have been discussed to this point are connected with the aero-thermodynamic design that is the principal topic of this text. Stator design has almost no influence on the aero-thermodynamic design, and vice versa. The notes in this section are, therefore, intended merely to give the reader the very broadest of reviews of a highly specialized topic.

Engine casings

The great majority of engine casings (and all larger industrial engine casings) are split horizontally, usually requiring a flanged joint. The flanges are massive for high-pressure machines such as large steam turbines and their feed pumps.

Pressures are lower for aircraft engines, even those having pressure ratios approaching 50:1, and flanges are relatively much smaller (figure 13.5). Also, the need to keep weight down leads to the use of a large number of small bolts or studs in aircraft-engine flanges, both those in the horizontal and the longitudinal splits, whereas flanges for steam turbines, feed pumps and industrial high-pressure machines in general use a much smaller number of large bolts and consequently thicker sections in the flanges and in the casing walls to keep deflections small. Despite the relatively small size of flanges in aircraft engines, they nevertheless form a pronounced discontinuity in the casing that will tend to distort the internal cross-section from the ideal circular shape under changes in internal pressure and in transient thermal growth. Accordingly, some aircraft engines have no horizontal splits, the casings being complete rings. Such engines must be assembled longitudinally stage by stage.

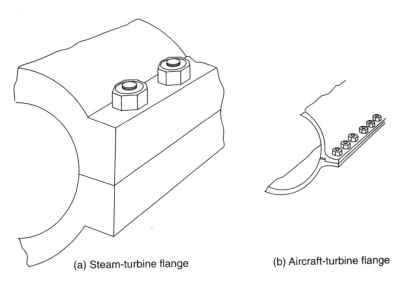

(a) Steam-turbine flange   (b) Aircraft-turbine flange

Figure 13.5. Horizontally split casings of large steam turbines and of aircraft turbines, showing major differences in flanges

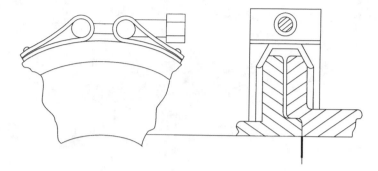

Figure 13.6. Clamp-ring flange closure used in small engines

For the longitudinal splits in small engines and turbochargers, flanges are often held together by a clamp ring of the type shown in figure 13.6, requiring a single bolt instead of a large number of individual bolts, and providing an almost-uniform flange-closing force circumferentially.

Double casings are used for many steam turbines and some industrial gas turbines. The outer casing takes the differential pressure with the atmosphere, but is kept at a relatively low temperature, while the inner casing defines the flow path and is subject to something approaching fluid temperature. Thus the inner casing can be much thinner than the outer casing, and distorts less during thermal transients.

13.2   Material selection

This is another highly specialized topic. There is little interaction between materials and aerothermodynamic design except in the limits. Gas-turbine designers cannot specify turbine-inlet temperatures above that which can be provided by the best available materials and cooling methods. Steam-turbine designers cannot specify stop-valve temperatures above those at which superheater tubes will give an acceptable life.

However, there are other more-subtle issues in material selection. For instance, author DGW is enthusiastic about the possibility of mass-producing small gas-turbine engines (e.g., for automobiles) from low-cost materials such as fiber-reinforced plastics for the compressor and turbocharger-grade ceramics for the turbine and other hot parts. The statement above about the design interaction being at the limits is still true, but the limits may be those of materials considerably removed from the highest-possible grade available.

In this section we will make general comments on materials and quote those chosen by designers for some recent engines. These will be the "core" stress-carrying materials. Almost all surfaces of modern gas turbines are coated, in the hot sections by thermal-barrier coatings and in other places by coatings intended to reduce corrosion and deposition or to provide abradable deposits for tip-clearance maintenance, for instances. There is a short section on coatings later in this section.

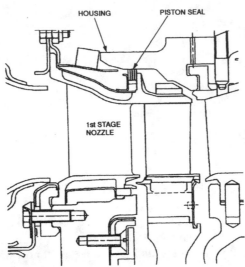

Turbomachinery usually has axial and radial gaps between components to allow for movement resulting from differential thermal growth, centrifugal loading and pressure loading. It is good practice to follow gaps in the flow path with rounded entrances to the next component. This cross-section of an axial-flow turbine also shows the axial restraints on the nozzle movement that also allow radial expansion. Sheet-metal surfaces enclosing coolant passages can be seen. Turbines often experience considerable axial movement because the thrust bearing is usually placed at the compressor inlet. Having a flared fixed shroud around the turbine blades would therefore result in a large tip clearance if differential thermal expansion moved the turbine away from the shroud. In the design above, a cylindrical fixed shroud, giving no increase in tip clearance with axial movement, has been used. Rather unusually, the inner flowpath of the turbine rotor has also been made cylindrical, so that the axial velocity will increase through the blade row. The locking clip that holds the rotor-blade root in the fir-tree slots of the disk is shown on the right of the blade root.

Illustration. Flow "overlaps". Courtesy Solar Turbines, Inc.

Casing materials

Throughout the size range, the compressor casing of an industrial compressor will often be of nodular cast iron, machined on the inner surface and on flanges and "bosses" carrying the bearings for variable-angle stator blades, for instances. Aircraft and other lightweight engines will use a fully machined aluminum or titanium alloy for the casings of aircraft compressors up to the point where the pressure ratio is around 25:1. Where the pressure ratio exceeds this approximate limit the temperatures require the use of steel or stainless steel.

Rotor, blade, vane and other materials: case studies

Brandt (1988) gave the materials used in the GE MS7001F gas turbine rated at 135 MW and having a turbine-rotor inlet temperature of 1530 K. It is therefore representative of end-of-the-century machines, even though higher-temperature more highly rated units, steam-cooled, are on order.

The compressor inlet guide vanes and the vanes and blades of stages 1 through 8 are made of a stainless steel called "custom 450". Tests showed that in a corrosive environment this material, uncoated, survived better than AISI 403 with various coatings. For the higher-pressure compressor stages (where corrosion is less of a problem) a higher-strength version of AISI 403 with columbium addition is used.

Each stage of the three-stage turbine has 92 investment-cast rotor blades ("buckets" in GE parlance) of GTD-111. All are coated, the first two stages with proprietary "Plasmaguard" alloys and the third with a high-chromium coating All nozzle vanes are investment-cast in FSX-414, in pairs for the first two stages and in groups of three for the third stage. The three turbine disks, spacer disks and aft shaft are of M152 alloy (12CrNiMoV steel).

The first sixteen compressor disks and the stub shaft are forged from NiCrMoV steel, while the last two disks use CrMoV steel. The compressor inlet is of grey cast iron, the compressor casings are of ferritic nodular cast iron, and the compressor discharge casing, combustor outer casings, turbine shell, and exhaust frame are fabricated of SA516 carbon steel.

Love (1987) describes the design of an advanced replacement axial-flow compressor using nine stages to give a design-point pressure ratio of 8:1. The first four stages use titanium blading, and the later stages use a nickel alloy.

Akao et al. (1995) tested two ceramic materials for the regenerator matrix, magnesium and lithium aluminum silicate (MAS and LAS). It was found necessary to stress-relieve the matrix by cutting radial slits from the periphery to 80% of the diameter. Better mechanical performance was found with the LAS. For the diametral face seals, copper graphite was found best for the cold side, and flame-sprayed $NiO - BaTiO_3$ for the hot side.

A wide variety of ceramic materials is being investigated for monolithic radial and axial turbines ("blisks", for "blades and disks"), for ceramic blades inserted into metal disks, and for combustion chambers. The overwhelmingly-favored material at the time of writing is silicon nitride. (Yoshimo et al. [1995] wrote "silicon-nitride nozzles should exceed feasible temperature limits [for a 1500 °C-class gas turbine] without cooling".) Many other more-complex high-temperature nonmetallic materials show promise but have not yet been generally adopted. Two of these are whisker-toughened ceramics, and carbon fibers in a graphite matrix (so-called "carbon-carbon"). This latter material actually increases in strength with increase in temperature up to near the stoichiometric level of combustion of turbine fuels in air. No way of protecting carbon-carbon from being changed to carbon dioxide at high temperatures has yet been found. One would think that by operating slightly fuel-rich there would be no oxygen to combine with the carbon, but to ensure that this was always the situation would require a complex control system.

Design with brittle materials

New techniques are required to design with materials as brittle as are most ceramics. A useful early primer on this art was written by McLean and Fisher (1981).

13.3 Design with traditional materials

The state of refinement of a design method can usually be judged by the value of the safety factor (the ratio of permissible material stress to calculated stress) that is used. Present-day safety-factor values, which are often in the range of 1.1 to 1.5, show that design methods are considerably more refined than they were two or three decades ago, when safety factors of 2.0 and above were common. This topic is discussed first.

Design criteria

In the early stage of the development of design methods of any device whose loading is more complex than simple tension, it is common to use so-called safety factors which might have values of five or ten. In other words, the designer allows a calculated stress (for instance) only one-fifth or one-tenth of the value that would produce failure. These are not, of course, true factors of safety. Rather they are factors of ignorance used because the methods available for analyzing loading are highly deficient. A large safety factor might be used if a turbine disk or turbine blade were designed on the basis solely of ultimate tensile strength, for instance. We now know that turbine blades and disks can fail because one of several criteria, besides ultimate tensile strength, is exceeded. Some of the criteria that will be involved in the design of the disk of a high-pressure turbine rotor are indicated in figure 13.7a. Criteria used for the design of high-temperature turbine blades are given in figure 13.7b. Turbine disks and blades are the most critically stressed components of gas turbines, and this section will be confined to a simplified discussion of these criteria.

Low-cycle fatigue

Low-cycle fatigue ("LCF" in figures 13.7a and 13.12b) is a term used for large variations in stress applied a relatively small number of times. Every time a turbomachine is started, the stresses due to rotation go from a low value to the peak value and so constitute one component of low-cycle fatigue, provided that peak value is larger than the so-called "fatigue limit". Usually, in addition, there is some plastic strain in a gas-turbine disk during the first run to the overspeed "proof" point. The bore stretches plastically, and the rim crushes plastically around the blade dovetail roots. Then, when the machine is shut down, there is a locked-in compressive stress in the bore and a tensile stress at the rim. In subsequent starts and stops, while there should be no further plastic strain, the stress excursions may be enough to begin accumulating a fatigue history.

A more important component of low-cycle fatigue is, however, thermal stress, especially in high-temperature turbines. High-temperature materials tend to have lower thermal conductivities and higher thermal-expansion coefficients than metals used for service under 775 K (931 °F), greatly intensifying the thermal stresses. (In 1987 Inco Alloys International announced a series of superalloys based on iron, nickel, and cobalt. For instance, Inco Alloy 907, has about half the thermal-expansion coefficients of conventional superalloys.) Large fluctuations in thermal stress can occur not merely upon start-up and shut-down but upon load changes. As aircraft-jet-engine performance, espe-

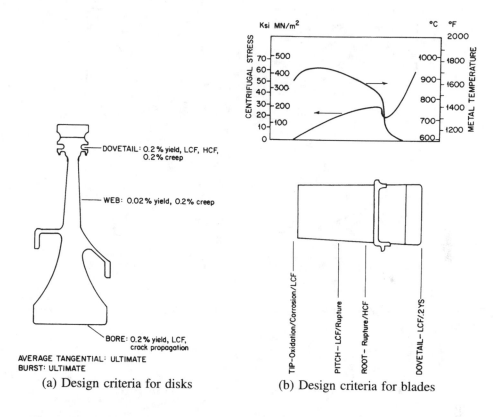

(a) Design criteria for disks   (b) Design criteria for blades

Figure 13.7. Design criteria for disks and blades. From Anderson (1979)

cially thrust-mass ratio, increases, the thermal-stress fluctuations tend to increase. This is because when engines of lower performance were used they tended to be run at near their peak-power condition throughout their missions. Higher-performance engines, especially when used in military aircraft, cruise at a lower proportion of their peak thrust and might be boosted to maximum-thrust level perhaps ten times during an average mission (figure 13.8), adding fatigue-loading cycles. Some fatigue data are given in figure 13.9.

Creep

When loaded continuously at high temperatures, metallic materials continuously deform, figure 13.10. The stress at which long-duration deformation is significant can be far below the ultimate tensile stress. For instance, a popular turbine-blade material has, at room temperature, an ultimate tensile stress of 965 MN/m^2 (140,000 psi), which reduces to 453 MN/m^2 (65,750 psi) at 1175 K (1,650 °F). The stress for 0.1 percent creep strain at 1090 K (1,500 °F) in 1,000 hours is 69 MN/m^2 (10,000 psi). This stress level is close to that producing rupture in 10,000 hours at the same temperature.

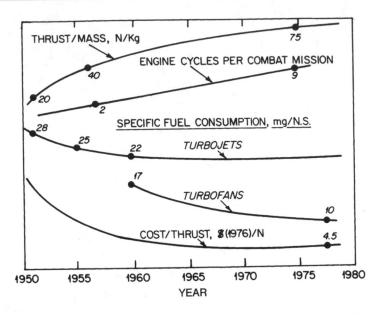

Figure 13.8. Aircraft-engine developments. From Anderson (1979) and Dixon (1979)

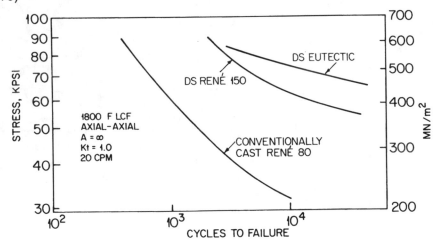

Figure 13.9. Fatigue strength of turbine-blade materials. From Anderson (1979)

A limiting value of creep often used for design is 0.2 percent strain. The stress that can be used at various combinations of temperature and life for various disk alloys is shown in figure 13.11. Large aircraft engines are currently being designed for lives of from 20,000 to 40,000 hours. Large industrial engines are usually designed for 100,000 hours.

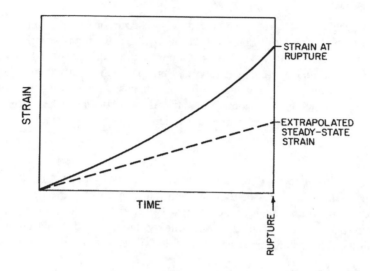

$$\text{CREEP DUCTILITY} \equiv \left[\frac{\text{STRAIN AT RUPTURE}}{\text{EXTRAPOLATED STEADY-STATE STRAIN}}\right]$$

Figure 13.10. Creep-strain phenomena

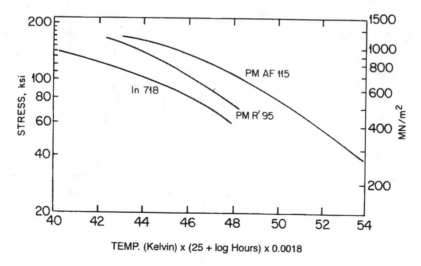

Figure 13.11. 0.2-percent creep strength of disk alloys. From Anderson (1979)

Crack propagation

Crack propagation is a relatively recent design criterion; it is applied at the bore of the disks of high-pressure turbines. It recognizes that for high-performance engines, particularly those for military service, it is impracticable or uneconomic to avoid that combination of thermal low-cycle fatigue and creep which results in the formation of cracks. The materials are not notch-sensitive, and cracks continue to grow slowly. Limiting lifetimes for crack propagation are established for particular disk and blade designs, and no simple general design rules are yet available.

High-frequency fatigue

The natural frequency of vibration of blades and disks depends on the component size and shape, and the mode of the vibration, but can be of the order of 10,000 Hz. The number of reversals is then of the order of 10^8 per hour, a period sufficient to reach the fatigue limits for many materials. Accordingly, fatigue modes that produce stresses greater than the fatigue limit must be avoided, as will be discussed more fully below.

Ultimate tensile stress

The minimum cross-sectional area of a disk is set by the ultimate tensile stress at the appropriate temperature acting to carry the rotational load of the blades and rim at overspeed. The overspeed used for the burst criterion is often 20 or 25 percent over design speed. The ultimate tensile stress is appropriate because, when the disk approaches burst speed, the bore and inner parts of the disk will have yielded.

Oxidation-corrosion

Modern high-temperature turbine blades are coated to protect the base material against corrosion. The clearance between blade tip and shroud is set at the minimum practicable value because aerodynamic losses in this region are proportional to clearance. If a tip rub occurs, the protective coating is removed, and the base alloy is then subjected to a hostile corrosive environment. The coating may also be removed through low-cycle fatigue acting on the unrestrained blade tip. The turbine life is then set by the oxidation-corrosion rate of the base alloy without its protective coating.

Vibration characteristics

In conditions of extreme centripetal loading, it is almost impossible to incorporate mechanical damping into blade and disk construction other than provided by the material itself. If a blade or disk receives an excitation, usually from an aerodynamic source, at a frequency close to one of the fundamental natural frequencies, large amplitudes and high vibratory stresses can result. Failure in high-frequency fatigue can then be rapid. An intense effort must be made at the design stage to avoid such excitations, because detuning is expensive if severe vibratory stresses are found after prototype manufacture. Nevertheless, prototype testing normally incorporates the extensive use of strain gauges, the readings from which can provide real-time and accurate information on stress fluctuations.

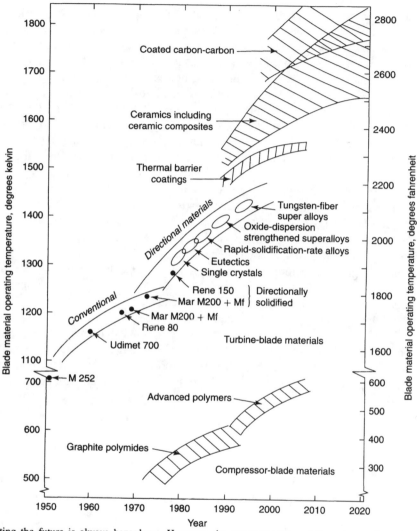

Forecasting the future is always hazardous. However, the gas temperature at the entry to the turbine rotor blades has increased on average 25 K per year since the start of industrial and aircraft gas turbines in the years around 1940, as shown in figure 1.10. Much of this increase has been the result of blade cooling, described in chapter 10. The collection of information in the chart above shows that there are still gains to be made in material operating temperatures. Less-stringent demands are made for compressor-blade materials. As pressure ratios have increased, the intermediate-pressure stages have been made of titanium alloys (often also required for the low-pressure stages for marine use), while the high-pressure stages have used stainless steels. The data for the polymer composites have been included because of author DGW's enthusiasm for highly regenerated low-pressure-ratio engines that could incorporate injection-molded compressor components.

Illustration. Turbine-blade and compressor materials. From Machlin (1979), and NASA sources (1990), including Meader, Michael A. "High-temperature polymer matrix composites" in the NASA conference "Aeropropulsion '87," publication 3049, Cleveland, OH.

Avoidance of excitation of fundamental modes of compressor and turbine blades can be assisted by the construction of a Campbell diagram (figure 7.22), The principal degree of freedom in the mechanical design, once the velocity diagrams and other aspects of the aerodynamic design have been decided upon, is the number of blades in each blade row, rotor and stator. Changing the number of blades changes the principal excitation frequency for neighboring blade rows, and also changes the natural frequency of each blade through the change in the blade chord and, hence, the aspect ratio. (The change in aspect ratio will have a minor effect on stage efficiency, figure 7.32, so long as the change in blade number is not drastic, for instance, doubling or halving the number.)

The natural frequencies of prototype blades in various modes (fundamental bending, or "first flap"; fundamental torsion, second bending, second torsion, and so forth) can be measured in the laboratory. But the stiffening effect of rotational stress cannot be simulated accurately without rotation. Accordingly, more reliance is given nowadays to computer programs that can calculate the various frequencies of, at least, solid blades, over a full range of rotational speeds, as shown in figure 7.22.

The intersection of a fundamental mode with an exciting frequency in the usual running range of a turbomachine is normally cause for a recalculation using a different number of blades in one or more blade rows. Excitations that can be run through rapidly during accelerations up to speed can usually be tolerated, although strain-gauge readings are desirable during prototype testing to find if low-cycle or high-frequency fatigue limits could be approached during a lifetime of frequent starts and shutdowns.

Special measures must often be taken for long jet-engine fan blades, whose vibrational characteristics are often changed through the incorporation of part-span shrouds. Rolls Royce pioneered the development of long-chord hollow blades that, with the elimination of the part-span shrouds, increased the fan (and therefore the engine) efficiency by one to two percent (figure 13.12).

13.4   Engine examples

The mechanical arrangement of some production engines and turbomachines is discussed in this section, principally by comments on cross-sections or illustrations of the machines themselves. There is also an example of an experimental turbine using an advanced material, carbon-carbon.

We also include some notes on perhaps an unusual use of new materials: the production of automotive gas turbines largely by injection molding, (Wilson, (1997) Wilson and Pfahnl, (1997)). The argument is that automotive turbines have) not been adopted mainly because of the high cost of materials. For instance, a highly successful Chrysler turbine was installed in 50 automobiles in 1963 that were loaned to 203 members of the public for up to three months. They suffered no breakdowns. However, they cost $8.00 per horsepower in materials costs alone (1963 dollars) versus a contemporary total production cost of spark-ignition engines of $1.50 per horsepower. Later, ceramics were introduced for regenerators and turbines in the belief that the cost was going to be a fraction of a dollar per kilogram. Instead, they were designed into very-high-stress rotors that suffered

The wide-chord fan blades on this 190-kN-thrust engine can be clearly seen. The diameter of the fan is 1895 mm. The high aerodynamic loading at the hub section is also evident, while the tip profiles are supersonic. Rolls-Royce claims lower noise and greater resistance to foreign-object damage in addition to higher efficiency for this design. The Kevlar wrapping around the fan shroud can contain a fan blade (but not a burst disk). The engine is remarkable in that it has an overall pressure ratio of almost 26, yet there are no variable-angle stators in the compressor. The engine uses, instead, a three-spool design: the fan is driven by the low-pressure turbine; the low-pressure compressor by the intermediate-pressure turbine; and the high-pressure compressor by the high-pressure turbine. All seek their optimum speeds. The compressor blades are designed to give controlled diffusion. The artist has shown the extensive use of honeycomb in the fan-duct walls. The high pressure ratio of modern aircraft engines shrinks the combustor to an almost vestigial size.

Figure 13.12. Rolls-Royce RB211-535 fan-jet engine. Courtesy Rolls-Royce plc.

continual failures. Designers countered by going to highly expensive ceramic materials and methods of preparation.

The alternative approach proposed is to lower rotor speeds and stresses by using three turbine and three compressor stages instead of one; by using a higher regenerator effectiveness than previously designed for; and by taking the lower optimum pressure ratio that follows from this higher effectiveness (figures 13.13a and 13.13b; also see figure 3.20). The resulting blade speeds are so low that the calculated stresses indicate a fiber-reinforced injection-molded compressor rotor would be feasible, as would axial turbines molded from the grade of silicon nitride used to mass-produce turbocharger turbines. The overall turbine and compressor efficiencies would be greatly increased because of the reduction in kinetic-energy leaving losses, as illustrated in figure 5.28.

The performance of two hypothetical 100-kW engines was calculated, one being a low-pressure-ratio engine "LPR" designed as recommended here, and one being a conventional high-pressure-ratio engine (5.5:1) ("HPR") as has become almost standard (table 13.1). The compressor and turbine polytropic efficiencies were taken from the correlations of figures 3.12 and 3.14. The regenerators were designed to have about one-

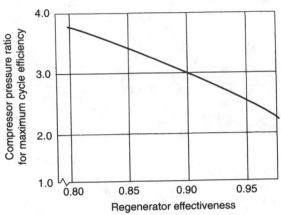

(a) Optimum compressor pressure ratio falls with increase of regenerator effectiveness.

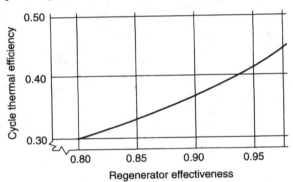

(b) Cycle thermal efficiency increases with regenerator effectiveness.

Conditions for calculations:

turbine-inlet temperature: 1650 K   pressure losses: 8%
compressor-inlet temperature: 288 K   compressor efficiency: $0.821 - r_{op}/94$
leakage: 4%   turbine efficiency: $0.859 - r_{op}/190$
$r_{op} \equiv$ optimum pressure ratio for each effectiveness

Figure 13.13. Strong influence of regenerator effectiveness on engine performance

and two-percent core pressure drops on the air and exhaust sides respectively, according to the methods of chapter 10, and were deemed to have discontinuous rotation and clampable seals, as in figures 10.14 and 10.15. The resulting cross-sections of the two engines are shown against that of a spark-ignition ("S.I.") engine of equal power in figure 13.14, and the results of the design calculations are given in table 13.1. The LPR engine is considerably larger than the HPR, but still smaller than the SI engine. The shaft speed is reduced from over 150,000 RPM to 32,500 RPM. The predicted full-power efficiency

Table 13.1. Results of preliminary design of alternative engines

Specifications, choices and results	single-stage HPR	three-stage LPR
Net output, kW	100	100
Compressor stagnation-pressure ratio	5.5	2.5
Compressor stagnation-static $\eta_{p,c,ts}$	0.83	0.87
Turbine type	radial	axial
Turbine stagnation-static $\eta_{p,e,ts}$	0.84	0.88
Windage, bearing friction, auxiliaries, kW	2	2
Regenerator effectiveness	0.90	0.975
Estimated leakage/compressor flow	0.01	0.01
Predicted full-power thermal efficiency	0.46	0.56
Predicted specific power	1.10	0.74
Required air mass flow, kg/s	0.31	0.47
Compressor-rotor peripheral speed required, m/s	630	240
Shaft speed, rev/min	158,000	32,500
Compressor-rotor diameter, mm	77	142
Turbine-rotor diameter, mm (mean dia. for axial)	97	192
Turbine-rotor peripheral (mean) speed, m/s	798	326
Two-rotor regenerator outside diameter, mm	325	572
Regenerator-rotor inside diameter, mm	81	143
Regenerator-disk thickness (flow length), mm	80	130
Mass of active disk (does not include rims), kg	9	42
Rotation period, sec	9	30

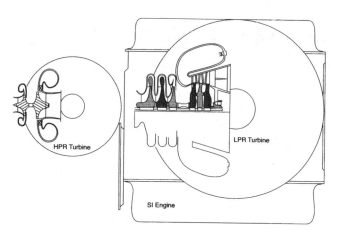

Figure 13.14. Relative sizes of HPR, LPR and S.I. engines

This engine was the first (and we believe the only one) of the AGT series (the AGT program was sponsored by the U.S. Department of Energy and NASA) to attain full speed and temperature (1645 K). It existed previously as an experimental General Motors engine, and was taken over (when Allison was still a division of GM) because the company's earlier AGT-100 engine, having two ceramic radial-inflow turbines in series, suffered repeated turbine-rotor erosion. The AGT-5 has two axial-flow turbines, one driving the compressor and the other driving the output shaft, and two ceramic regenerator disks, an arrangement that can be taken as the basic automotive-turbine layout.

Illustration. Allison AGT-5 engine. Courtesy Allison Engine Corp.

is 0.59 for the LPR engine and 0.49 for the HPR engine (with the low-loss regenerator). The part-power efficiency of the LPR engine would be further improved over that of the HPR engine, because of the better off-design performance of compressors designed for low pressure ratios (chapter 8). The acceleration times for the larger LPR engine would be greatly reduced over those for the HPR engine because of the far lower rotating kinetic energy that has to be supplied.

The summarized results of this study are offered as an example of the use of the methods given in this text, and of the sometimes unusual (and apparently highly advantageous) conclusions that may be reached.

The remainder of this chapter consists of illustrations of different machines and comments on their construction and design.

General Electric is unique among gas-turbine manufacturers in that it makes a wide range of jet engines and "aero-derivative" versions of these engines for, for example, marine uses, and it also makes so-called "heavy-frame" industrial gas turbines, an example of which is shown here. Wide-chord blades are used in the compressor, giving higher efficiency and greater robustness to foreign-object damage and fouling. (Wide-chord blades would make an aero-engine unacceptably long and heavy. The short-chord compressor blades used, for instance, in the GE LM5000 aero-derivative engine sometimes require part-span shrouds or "snubbers" to limit vibration. These add losses to the flow.) The compressor has no variable-angle stators: starting is accomplished using two bypass (blow-off) bands. The simple blade roots required for lightweight compressor blades are contrasted with the complex turbine-blade roots. For these, the attachment to the disk is by "fir-tree" engagements. There is a long shank portion between the disk and the blade-root platforms, used partly to duct cooling air or steam into the blades, and partly to allow a temperature gradient from the hot blades to the cooler roots. The first-stage turbine rotors are cooled and unshrouded; the second and third stages carry integral shrouds and the blade profiles are probably uncooled. The casing thickness and flange sizes are intermediate between the lightweight construction of aero-derivative engines and the very heavy construction required of steam turbines.

Illustration. "Heavy-frame" industrial gas turbine. Courtesy General Electric Co.

This experimental GE turbine was tested in 1992. The carbon-carbon composite is 35% lighter than metal and needs no cooling. The material is theoretically capable of high strength up to about 2400 K. Material degradation (including oxidation) occurs in the presence of oxygen, so that use of this material requires either self-healing coatings that can withstand high turbine-inlet temperatures, or operation in a working fluid devoid of oxygen.

Illustration. Experimental carbon-carbon composite turbine blades in metal disk. Courtesy GE Aircraft Engines

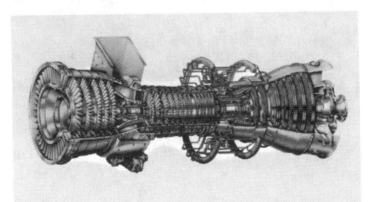

This engine is based on the CF6 series of aircraft engines. It produces about 42 MW at a thermal efficiency between 40% and 43% (depending on whether it is operated "dry" or with water or steam injection). This is claimed, with reason, to be the highest simple-cycle thermal efficiency available from any engine world-wide. It can also be used in a combined cycle. It is a two-spool engine, with power take-off from either end of the low-pressure shaft. The high-pressure compressor, part of the original core engine, has six rows of variable-setting stators, including the inlet guide vanes; only the IGVs are variable in the low-pressure compressor. There is a substantial blow-off band between the two compressors to assist starting. Part-span shrouds or snubbers are used on the first stage of the high-pressure compressor. The contrast between this GE engine and the heavy-frame engine of a previous illustration is striking.

Illustration. GE LM6000 engine. Courtesy GE Marine & Industrial Engines

Steam turbines operate at far higher pressure ratios than do gas turbines, and the nature of steam results in very high density ratios. The low molecular weight also gives a high sonic velocity and consequently low Mach numbers in all but the lowest-pressure blading. Therefore many stages (of the order of thirty) of turbine expansion are required, in contrast to the two-to-six required in gas turbines. Single shafts are used because geared shafts at the high power levels used are economically infeasible. The high-pressure blading is consequently very short (specific speed below the optimum), while the low-pressure blading is as long as can be retained mechanically. Even so, the blading can handle the huge volume flow only through being passed in parallel through two (as here) and sometimes four sets of outward-flow low-pressure blading.

Illustration. Tandem-compound two-flow reheat steam turbine. Courtesy General Electric

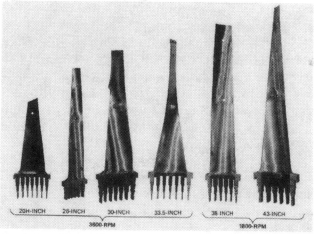

Steam turbines are massively constructed because of the very high pressures (sometimes over 300 bar, 4500 psi) required. Huge single-piece forged rotors are normally used for each of the two, three or more casings (for the high-pressure, intermediate-pressure and low-pressure units) employed, joined together by couplings that allow a very small degree of misalignment. Steam turbines must be warmed up slowly to avoid thermal distortion. The rotor-blade roots have evolved to be circumferential "fir trees" with the blades entered through a slot, whereas gas turbines are almost universally fastened with axi-symmetric axial fir trees. Another form of steam-turbine blade-root fastening is the multi-tang construction shown here. These mate with peripheral grooves in the turbine disk and are fastened by axial rods. The longest blades require two circumferential "lacing wires" to avoid vibrational excitations.

Illustration. Steam-turbine construction details. Courtesy General Electric

This is a typical example of a compressor for the process industry. Centrifugal compressors are used at up to much larger sizes than is the case for gas-turbine engines, principally because of their robustness and reliability. The blading has high sweepback and is shrouded. Intercoolers are used between stages. These help to produce a large drop in specific speed of the stages from inlet to outlet. The blade lengths become so short that tip clearances resulting from thermal growth and pressure loading would produce severe losses in the high-pressure stages if open impellers were used.

Illustration. Intercooled process compressor. Courtesy Sulzer

References

Akao, Y., et al. (1995). Development of a regenerator for a 100 kW automobile gas turbine. Paper 38, Proceedings of the International Gas Turbine Congress, Yokohama, Japan.

Brandt, D. (1988). The design and development of an advanced heavy-duty gas turbine. Jl. of Engg. for Gas Turbines and Power, vol. 110 pp. 243–250, ASME, New York, NY.

Anderson, Richard J. (1979). Material properties and their relationships to critical jet-engine components. In IDA paper P-1421. Proc. Workshop on High-Temperature Materials for Advanced Military Engines. Institute for Defense Analyses, Arlington, VA.

Dixon, James T. (1979). High-temperature combustor and turbine-design problems. In IDA paper P-1421, Proc. Workshop on High-Temperature Materials for Advanced Military Engines. Institute for Defense Analyses. Arlington, VA.

Freche, John C., and G. Mervin Ault. (1979). The promise of more heat-resistant turbine materials. Product Engineering (July).

Love, C.M. (1987). Design and development of an advanced F100 compressor. AGARD current paper CP 421, no. 36, Brussels, Belgium.

Machlin, Irving. (1979). High-temperature materials in turbofan engines for V/STOL aircraft. In IDA paper P-1421. Proc. Workshop on High-Temperature Materials for Advanced Military Engines. Institute for Defense Analyses, Arlington, VA.

McLean, A.F. and Fisher, E.A. (1981). Brittle materials design, high temperature gas turbine. Report AMMRC TR 81–14, AMMRC, Watertown, MA.

Wilson, D.G. (1997). A new approach is low-cost high-efficiency automotive gas turbines. Paper 970234, SAE, Warrendale, PA.

Wilson, D.G. and Pfahnl, A.C. (1997). A look at the automotive-turbine regenerator system and proposals to improve performance and reduce costs. Paper 970237, SAE, Warrendale, PA.

Yoshimo, S. et al. (1995). Design and test of an air-cooled ceramic nozzle for a power-generating gas turbine. Paper 80, Proceedings of the International Gas Turbine Congress, Yokohama, Japan.

Appendix A

Properties of air and combustion products

For the most precise data on air and gas properties, use *Gas Tables* by Keenan et al. (1983), or *Thermodynamic Properties in SI* by W. C. Reynolds (1979). In this appendix we are giving some data more appropriate for preliminary design of gas-turbine engines. First, there is a table (A.1) of thermodynamic properties for dry air, taken from a U.S. N.B.S. circular with inputs from Gas Tables. We have added a column of (C_p/R) to be more suitable for the form of thermodynamic expressions used particularly in chapters 2 and 3.

We are also repeating a correlation from Chappell and Cockshutt (1974), who have given us permission to reproduce a modified form of their data for SI units in table A.2. With the correlation equations below, these provide a very useful method of calculating thermodynamic data, including the specific heat at constant pressure and the enthalpy as functions of temperature for dry air and for combustion products. The fuel/air ratio required for a specified combustion temperature to be reached from a given initial temperature can also be calculated. These give errors for enthalpy and entropy of less than a half percent up to about 1750 K. Because they also convey consistency, the resulting errors in cycle-efficiency predictions, for instance, are much less than the approximations to enthalpy values. (There are also considerable differences in tabulated values. For instance, the specific heat of dry air at 1200 K in J/kg is given by the National Bureau of Standards as 1179 and by *Gas Tables* as 1173.8. We inserted the Gas Tables values because they were considerably later than those of the NBS).

A.1 Chappell and Cockshutt correlations

In this NRC, Canada, work, sets of polynomial coefficients for generating thermodynamic data for different unit systems were produced. Andre By and Robert Bjorge translated these into SI coefficients, given in table A.2. The data and the polynomials are based on the treatment of dry air and combustion products as semi-perfect gases, so that the specific heat, enthalpy and entropy functions are dependent solely on temperature and are independent of pressure. The combustion products are for a standard fuel of 86.08-percent carbon and 13.92-percent hydrogen by mass, which gives a molal mass of combustion products identical to that of dry air (Fielding and Topps, 1959).

Table A.1. Thermodynamic property data for dry air in SI units

Temperature K	Specific heat C_p J/kg-deg K	(C_p/R)	Viscosity (absolute) kg/ms $\times 10^5$	Thermal conductivity mW/m deg K	Prandtl number
100	1001.9	3.4914	0.06924	9.246	0.770
150	1002.0	3.4946	1.02833	13.735	0.753
200	1002.2	3.4925	1.3289	18.09	0.739
250	1002.8	3.4946	1.488	22.27	0.722
300	1004.5	3.5005	1.983	26.24	0.708
350	1007.9	3.5123	2.075	30.03	0.697
400	1013.1	3.5305	2.286	33.65	0.689
450	1020.3	3.5555	2.484	37.07	0.683
500	1029.2	3.5866	2.671	40.38	0.680
550	1039.4	3.6221	2.848	43.60	0.680
600	1050.7	3.6615	3.018	46.59	0.680
650	1062.5	3.7026	3.177	49.53	0.682
700	1074.5	3.7444	3.332	52.30	0.684
750	1086.5	3.7862	3.481	55.09	0.686
800	1098.2	3.8270	3.625	57.79	0.689
850	1109.5	3.8664	3.765	60.28	0.692
900	1120.4	3.9044	3.899	62.79	0.696
950	1131.3	3.9424	4.023	65.25	0.699
1000	1140.4	3.9741	4.152	67.52	0.702
1100	1158.2	4.0361	4.44	73.2	0.704
1200	1173.8	4.0905	4.69	78.2	0.707
1300	1187.5	4.1382	4.93	83.7	0.705
1400	1199.6	4.1804	5.17	89.1	0.705
1500	1210.2	4.2173	5.40	94.6	0.705
1600	1219.7	4.2504	5.63	100	0.705
1700	1228.1	4.2797	5.85	105	0.705
1800	1235.7	4.3062	6.07	111	0.704
1900	1242.6	4.3302	6.29	117	0.704
2000	1248.9	4.3522	6.50	124	0.702
2100	1254.7	4.3724	6.72	131	0.700
2200	1260.0	4.3909	6.93	139	0.707
2300	1264.9	4.4079	7.14	149	0.710
2400	1269.5	4.4240	7.35	161	0.718
2500	1273.8	4.4389	7.57	175	0.730

Sources: U.S. National Bureau of Standards Circular 564 (1955) and *Gas Tables*, Keenan et al. (1983) There is considerable disagreement between these two sources of data. The data for the second and third columns are taken from *Gas Tables* and the remaining data from the NBS.

Table A.2. Polynomial coefficients for generating thermodynamic in SI data units, J/kg

Symbol	Temperature 200–800 K	Temperature 800–2200 K
C_0	$+1.0189134E+03$	$+7.9865509E+02$
C_1	$-1.3783636E-01$	$+5.3392159E-01$
C_2	$+1.9843397E-04$	$-2.2881694E-04$
C_3	$+4.2399242E-07$	$+3.7420857E-08$
C_4	$-3.7632489E-10$	0.0
CH^a	$-1.6984633E+03$	$+4.7384653E+04$
CF	$+3.2050096E+00$	$+7.0344726E+00$
CP_0	$-3.5949415E+02$	$+1.0887572E+03$
CP_1	$+4.5163996E+00$	$-1.4158834E-01$
CP_2	$+2.8116360E-03$	$+1.9160159E-03$
	$-2.1708731E-05$	$-1.2400934E-06$
CP_4	$+2.8688783E-08$	$+3.0669459E-10$
CP_5	$-1.2226336E-11$	$-2.6117109E-14$
H_0	$+6.2637416E+04$	$-1.7683851E+05$
H_1	$-5.2903044E+02$	$+8.3690644E+02$
H_2	$+3.2226232E-00$	$+3.6476206E-01$
H_3	$-2.1670252E-03$	$+2.5155448E-04$
H_4	$+2.4951703E-07$	$-1.2541337E-07$
H_5	$+3.4891819E-10$	$+1.6406268E-11$

[a]For a continuous enthalpy function, use $CH = +4.7378825E+04$.
Source: Chappell and Cockshutt (1974) and Andre By, NREC, and Robert Bjorge, GE.

Below is an abbreviated guide to the use of the polynomials, omitting an entropy function that is not needed if the polytropic-efficiency methods given in chapters 2 and 3 are followed.

Guide to the use of the polynomials

The symbols used here are principally those of Chappell and Cockshutt, and are not used in the remainder of the text, nor are they included in the list of symbols.

Specific heat and enthalpy of dry air

The specific heat of dry air at temperature T (K) is given by:

$$C_{p,a,T} = C_0 + C_1 T + C_2 T^2 + C_3 T^3 \ldots$$

The enthalpy of dry air at T (K), to a base of absolute zero, is:

$$h_{a,T} = C_0 T + \frac{C_1}{2} T^2 + \frac{C_2}{3} T^3 + \frac{C_3}{4} T^4 + \cdots + CH$$

Effective calorific value (ECV) of standard fuel at T (K)

$$ECV_T = \Delta H_R - (h_{a,T} - h_{a,288K}) - (\theta_{h,T} - \theta_{h,288K})$$

where $\Delta H_R \equiv$ reference enthalpy of reaction at 288 K = 43,124 kJ/kg

$$\theta_{h,T} = H_0 + H_1 T + H_2 T^2 + H_3 T^3 + \cdots$$

Fuel/air ratio by mass, f

For a process starting with dry air at T_1 K and reaching T_2 K after combustion,

$$f \equiv \frac{\dot{m}_{fuel}}{\dot{m}_{air}} = \frac{h_{air,T,2} - h_{air,T,1}}{ECV_{T,2} + (\text{sensible heat of injected fuel})}$$

The fuel sensible heat is small and is normally neglected. If there should be a second combustion process in a reheat combustor or afterburner, the products of the first combustion include partly vitiated air. The second fuel/air ratio, f_2, is given by

$$f_2 - f_1 = \frac{(1 + f_1)(h_{p1,T,2} - h_{p1,T,1})}{ECV_{T,2}}$$

where $h_{p1} \equiv$ enthalpy of the products from the first combustion. The sensible heat of the fuel has been omitted.

Specific heat and enthalpy of products of combustion at T (K)

$$C_{p,prod,T} = C_{p,air,T} + \frac{f}{1+f} \theta_{C_p,T}$$

$$\text{where } \theta_{C_p,T} = CP_0 + CP_1 T + CP_2 T^2 + CP_3 T^3 + \cdots$$

$$h_{prod,T} = h_{air,T} + \frac{f}{1+f} \theta_{h,T}$$

References

Chappell, M.S., and E.P. Cockshutt (1974). Gas-turbine cycle calculations: thermodynamic data tables for air and combustion products for three systems of units. National Research Council, Canada, report no. 14,300, Ottawa.

Fielding, D., and J.E.C. Topps (1959). Thermodynamic data for the calculation of gas-turbine performance. Aeronautical Research Committee, UK, R&M 3099, HMSO, London, UK.

Keenan, Joseph H., Jing Chao, and Joseph Kaye (1983). Gas Tables, international version, second edition (SI units). John Wiley and Sons, New York, NY.

Reynolds, William C. (1979). Thermodynamic properties in SI. Department of mechanical engineering, Stanford University, Stanford, CA.

Appendix B

Collected formulae

Steady flow energy equation (the first law for a flow process):

$$\frac{\dot{Q}_{in} + \dot{W}_{in} - \dot{Q}_{ex} - \dot{W}_{ex}}{\dot{m}} = \Delta_1^2 \left(h_{st} + \frac{C^2}{2g_c} + \frac{gz}{g_c} \right)$$

Gibbs' equation for a simple substance:

$$T\,ds = du + p\,dv$$

Equation of state for a perfect gas:

$$pv = RT$$

Other relations for a perfect gas are:

$$
\begin{aligned}
C_p &= C_v + R \\
dh &= C_p\,dT \\
du &= C_v\,dT \\
a^2 &= -g_c v^2 \left(\frac{\delta p}{\delta v} \right)_s \\
p\,dv + v\,dp &= R\,dT \\
&= R\,dT + Cv\,dT - T\,ds \\
v\left(\frac{\delta P}{\delta v} \right)_s &= -\frac{pC_p}{C_v} = -\frac{pC_p}{C_p - R}; \qquad a_{st}^2 = \frac{g_c T_{st}}{\left(\dfrac{1}{R} - \dfrac{1}{C_p} \right)} = \frac{g_c C_p T_{st}}{\left(\dfrac{C_p}{R} - 1 \right)}
\end{aligned}
$$

Equations for one-dimensional flow in a perfect gas, or between total and static conditions at any point in any flow:

$$\frac{T_0}{T_{st}} = 1 + \frac{M^2}{2\left(\dfrac{C_p}{R} - 1 \right)}$$

$$\frac{p_0}{p_{st}} = \left(\frac{T_0}{T_{st}}\right)^{C_p/R}$$

$$\frac{v_0}{v_{st}} \equiv \frac{\rho_{st}}{\rho_0} = \left(1 + \frac{M^2}{2\left(\frac{C_p}{R} - 1\right)}\right)^{-\left(\frac{C_p}{R} - 1\right)}$$

Continuity equations for a perfect gas:

$$\frac{\dot{m}\sqrt{RT_0/g_c}}{Ap_0} = M\left(1 - \frac{R}{C_p}\right)^{-\frac{1}{2}} \left(1 + \frac{M^2}{2\left(\frac{C_p}{R} - 1\right)}\right)^{-\left(\frac{C_p}{R} - \frac{1}{2}\right)}$$

$$\frac{C}{\sqrt{g_c RT_0}} = \sqrt{\left\{2\frac{C_p}{R}\left(1 - \left(1 + \frac{M^2}{2\left(\frac{C_p}{R} - 1\right)}\right)^{-1}\right)\right\}}$$

$$a_{st} = \sqrt{\frac{g_c C_p T_{st}}{\left(\frac{C_p}{R} - 1\right)}}$$

$$= \sqrt{\frac{g_c C_p T_0}{\left(\frac{C_p}{R} - 1\right)\left(1 + \frac{M^2}{2\left(\frac{C_p}{R} - 1\right)}\right)}}$$

$$= \sqrt{\frac{g_c C_p T_0}{\left(\frac{M^2}{2} + \frac{C_p}{R} - 1\right)}}$$

Stagnation-temperature ratio in polytropic processes:

$$\frac{T_{0,2}}{T_{0,1}} = r^{\left[\left(\frac{R}{C_p}\right)\frac{1}{\eta_{p,c}}\right]}$$

for compressors, and

$$\frac{T_{0,2}}{T_{0,1}} = r^{\left[\left(\frac{R}{C_p}\right)\eta_{p,e}\right]}$$

for expanders (turbines).

Relations between polytropic and isentropic efficiencies for perfect gases:
  For compression,

$$\eta_{s,c} = \frac{-C_p T_{0,1}(r^{R/C_p} - 1)}{-C_p(T_{0,2} - T_{0,1})} = \frac{r^{R/C_p} - 1}{\left\{\left(\frac{T_{0,2}}{T_{0,1}}\right) - 1\right\}}$$

$$\eta_{s,c} = \frac{r^{R/C_{p,c}} - 1}{r^{[(R/C_{p,c})/\eta_{p,c}]} - 1}$$

$$\eta_{p,c} = \frac{\ln(r)^{R/C_{p,c}}}{\ln\left(\frac{r^{R/C_{p,c}} - 1}{\eta_{s,c}} + 1\right)}$$

For expansion,

$$\eta_{s,e} = \frac{1 - r^{[(R/C_{p,e})\eta_{p,e}]}}{1 - r^{R/C_{p,e}}}$$

$$\eta_{p,e} = \frac{\ln(1 - \eta_{s,e}(1 - r^{R/C_{p,e}}))}{\ln r^{R/C_{p,e}}}$$

Specific power of a CBE or CBEX cycle:

$$\dot{W}' \equiv \frac{\dot{W}}{\dot{m}_c \overline{C_{p,c}} T_{0,1}} = \left[\frac{\dot{m}_e \overline{C_{p,e}}}{\dot{m}_c \overline{C_{p,c}}}\right] E_1 T' - C.$$

CBEX cycle thermal efficiency:

$$\eta_{th} = \frac{[\dot{m}_e \overline{C_{p,e}}/\dot{m}_c \overline{C_{p,c}}]E_1 T' - C}{[\dot{m}_b \overline{C_{p,b}}/\dot{m}_c \overline{C_{p,c}}][T'(1 - \varepsilon_x(1 - E_1)) - (1 + C)(1 - \varepsilon_x)]}$$

$$E_1 \equiv (T_{0,4} - T_{0,5})/T_{0,4}$$

$$= 1 - \left\{\left[\left(1 - \sum(\Delta p_T/p_T)\right)r\right]^{(R/\overline{C_{p,e}})\eta_{p,e}}\right\}$$

$$T' \equiv (T_{0,4}/T_{0,1})$$

$$C \equiv (T_{0,2} - T_{0,1})/T_{0,1} = \left[r^{[(R/\overline{C_{p,e}})/\eta_{p,c}]} - 1\right]$$

The actual pressure-rise coefficient of a diffuser is:

$$C_{pr} \equiv \frac{p_{st,2} - p_{st,1}}{p_{0,1} - p_{st,1}}$$

The theoretical pressure-rise coefficient is:

$$C_{pr,tl} = 1 - \left(\frac{W_{ex}}{W_{in}}\right)^2$$

Euler's equation:

$$\frac{\dot{W}}{\dot{m}} = \frac{1}{g_c}(u_1 C_{u,1} - u_2 C_{u,2})$$

$$g_c \Delta_1^2 h_0 = \Delta_1^2 (u C_u)$$

The work coefficient for an adiabatic stage is defined as

$$\psi \equiv \frac{-g_c \Delta_1^2 h_0}{u^2} = -\left[\frac{\Delta_1^2 (u C_u)}{u^2}\right]$$

For "simple" diagrams,

$$\psi = -\left[\frac{\Delta_1^2 C_u}{u}\right]$$

The flow coefficient, ϕ.

$$\phi \equiv \frac{C_x}{u}$$

The strict definition of reaction is the ratio of the change in static enthalpy to the change in total enthalpy of the flow passing through the rotor:

$$R_n \equiv \left(\frac{\Delta h_{st}}{\Delta h_0}\right)_{rr} = \frac{\Delta h_{st,rr}}{\Delta h_{0,se}}$$

For a simple velocity diagram:

$$\begin{aligned}
R_n &= 1 - \left[\frac{1}{2}\right]\frac{[C_x^2 + C_{u,2}^2 - C_x^2 - C_{u,1}^2]}{u(C_{u,2} - C_{u,1})} \\
&= 1 - \left[\frac{1}{2}\right]\frac{(C_{u,2} - C_{u,1})(C_{u,2} + C_{u,1})}{u(C_{u,2} - C_{u,1})} \\
&= 1 - \frac{C_{u,1} + C_{u,2}}{2u}
\end{aligned}$$

The following geometrical equations can be derived from a simple velocity diagram of a compressor or turbine:

$$\begin{aligned}
\tan \alpha_{C,1} &= [\psi/2 + (1 - R_n)]/\phi \\
\tan \alpha_{C,2} &= -[\psi/2 - (1 - R_n)]/\phi \\
\tan \alpha_{W,1} &= -[\psi/2 - R_n]/\phi \\
\tan \alpha_{W,2} &= [\psi/2 + R_n]/\phi \\
(C_1/u)^2 &= [(1 - R_n) + \psi/2]^2 + \phi^2 \\
(C_2/u)^2 &= [(1 - R_n) - \psi/2]^2 + \phi^2 \\
(W_1/u)^2 &= [\psi/2 - R_n]^2 + \phi^2 \\
(W_2/u)^2 &= [\psi/2 + R_n]^2 + \phi^2
\end{aligned}$$

$$1 - \left(\frac{d_{hb}}{d_{sh}}\right)^2 = \frac{d_{sh}^2 - d_{hb}^2}{d_{sh}^2} = \frac{A_a}{\pi d_{sh}^2/4}$$

$$= \frac{\dot{V}}{C_x \pi d_{sh}^2/4}, \qquad \text{because } \dot{V} = A_a C_x$$

$$= \frac{\dot{V} N^2}{\phi u_2 (\pi d_{sh}^2/4)(\omega 60/2\pi)^2}, \qquad \text{because } \phi \equiv C_x/u_2,$$

$$N \text{ rpm} = 60\omega \text{ rads/s}/2\pi,$$
$$\omega/2 = u_2/d_{sh}$$
$$\psi \equiv g_c \Delta h_0/u_2^2$$

$$= \frac{\pi}{900} \frac{N^2 \dot{V}}{\phi u_2^3}$$

$$= \frac{\pi}{900} \frac{\psi^{3/2} N^2 \dot{V}}{\phi (g_c \Delta h_0)^{3/2}}$$

$$\left[1 - \left(\frac{d_{hb}}{d_{sh}}\right)^2\right]^{1/2} = \frac{\sqrt{\pi}}{30} \frac{|\psi|^{3/4}}{\phi^{1/2}} \left[\frac{N\sqrt{\dot{V}}}{(g_c \Delta h_0)^{3/4}}\right] - \frac{|\psi|^{3/4}}{\pi^{1/2}\phi^{1/2}} N_{s1}$$

where the nondimensional specific speed is defined as:

$$N_{s1} \equiv \frac{2\pi N \sqrt{\dot{V}}}{60|g_c \Delta h_0|^{3/4}}$$

The equation of simple radial equilibrium (SRE):

$$\frac{1}{r^2}\frac{d}{dr}(r^2 C_u^2) + \frac{d}{dr}C_x^2 = 0$$

Slip-factor correlation (by Wiesner) for radial-flow compressors and turbines:

$$\left[\frac{C_{u,ac}}{C_{u,tl}}\right] = 1 - \frac{\sqrt{\cos\beta}}{Z^{0.7}}$$

where Z is the number of rotor blades and β is the blade angle at the rotor periphery (figure 9.13).

Reynolds' analogy between heat transfer and skin friction is:

$$S_t = C_f/2$$
$$S_t \equiv \frac{N_u}{Re \times P_r} \equiv \frac{h_t}{\rho C_p C}$$
$$C_f \equiv \frac{\tau_w}{\rho_T C_{fs}^2/2g_c}$$

Polytropic relations:

$$\left[\frac{\rho_{0,in}}{\rho_{0,ex}}\right] = \left[\frac{T_{0,in}}{T_{0,ex}}\right]^{\left(\frac{C_p}{R}\frac{1}{\eta_{p,e}} - 1\right)}$$

$$\frac{\rho_{0,ex}}{\rho_{0,in}} = \left[\frac{T_{0,ex}}{T_{0,in}}\right]^{\left(\frac{C_p}{R}\eta_{p,c} - 1\right)}$$

$$\frac{\rho_{0,ex}}{\rho_{0,in}} = \left[\frac{p^{\otimes}_{0,ex}}{p_{0,in}}\right]^{\left(1 - \frac{R}{C_p}\eta_{p,e}\right)}$$

Some constants

Molal mass,	Air		28.970
(kg/kgmole)	Carbon	C	12.01
	Carbon dioxide	CO_2	44.01
	Carbon monoxide	CO	28.01
	Helium	He	4.0026
	Hydrogen	H_2	2.0159
	Mercury	Hg	200.6
	Nitrogen	N_2	28.016
	Oxygen	O_2	32.000
	Potassium	K	39.10
	Sodium	Na	22.99
	Sulfur	S	32.06
	Water	H_2O	18.016

Universal gas constant

8313.219 J/(kmole · K)
1.98592 Btu/(lbmole °R)
1545.37 ft · lbf/(lbmole °R)

Gas constant for air

286.96 J/(kg · K)
0.068549 Btu/(lbm °R)
53.344 ft · lbf/(lbm °R)
1716.0 (ft/s)^2 °R

Air constituents
(approximate)

3.76 moles nitrogen/mole oxygen

Constant in
Newton's law, g_c

32.1739 lbm · ft/(lbf s^2)
9.80665 kg · m/(kgf s^2)

Appendix D

Conversion factors

Energy	1 kWh	3412.76 Btu
	1 Btu	778.161 ft·lbf
	"	1054.9 J
	"	0.29302 Wh
	1 J (1 W·s)	0.7376 ft·lbf
	1 kW	0.947989 Btu/s
	1 hp	0.7456 kW
	1 calorie	4.1868 J
	1 kg cal	3.969 Btu
	1 quad	10^{15} Btu
Enthalpy, specific energy	1 Btu/lbm	2.3260 kJ/kg
	"	25,037 (ft/s)^2
Power	1 hp/lbm/s	1.6447 kW/kg/s or kJ/kg
Refrigeration power	1 ton	12,000 Btu/h = 3.516 kW
Specific heat	1 Btu/lbm·deg R	4.187 kJ/(kg · K)
Thermal conductivity	1 Btu/ft·h·deg R	1.73073 W/(m · K)
Viscosity	1 lbm/s·ft	1.48817 N·s/m^2
	"	1488.17 centipoise
Mass	1 lbm	453.59 g
Density	1 lbm/ft^3	16.0166 kg/m^3

Length	1 in	25.40 mm
	1 ft	304.8 mm
	1 mile	1.609 km
	1 nautical mile	1.852 km
Area	1 ft^2	0.09290 m^2
Volume	1 ft^3	0.02832 m^3
	1 liter	0.001 m^3
	1 US gallon	3.785 liter; 0.003785 m^3
	1 Br. gallon	4.546 liter
Flow rate	1 US gpm	63.0902×10^{-6} m^3/s
Force	1 lbf	4.44822 N
	1 kgf	9.80665 N
Torque	1 lbf-ft	1.3558 N-m
Pressure	1 lbf/in^2	6.89476 kPa or kN/m^2
	1 bar	100 kPa
	1 atm	101.325 kPa

Index

Printed in the United States
by Baker & Taylor Publisher Services